PHYSICS FOR A MODERN WORLD

PHYSICS A PRACTICAL APPROACH
MECHANICS: COMPUTER SOFTWARE
PHYSICS FOR A MODERN WORLD
John Wiley & Sons Canada Limited High School Physics Series with accompanying Teacher's Guides
SI METRIC

Cover Photo courtesy of Ken Davies/Masterfile

The cover shows a computer enhanced picture of a laser beam reflecting off a compact storage disk. Information, be it music or computer codes, is stored on the disk as a series of colourless or black microscopic dots. The disk has an infinite life span since no needle or magnetic head need ever touch it. Instead, reflected light is used and decoded by computer. Resultant "perfect" music or mass-stored information is transmitted quickly and free of error.

PHYSICS
FOR A MODERN WORLD

ALAN J. HIRSCH

John Wiley & Sons
Toronto   New York   Chichester   Brisbane   Singapore

Copyright © 1986 by John Wiley & Sons Canada Limited.
All rights reserved.

No part of this publication may be reproduced by any means, stored in a retrieval system, or transmitted in any form or by any means, electronic, mechanical, photocopying, recording or otherwise, without the prior written permission of the publisher.

Care has been taken to trace ownership of copyright material contained in this text. The publisher will gladly receive any information that will enable it to rectify any reference or credit line in subsequent editions.

Communications Branch, Consumer and Corporate Affairs Canada has granted permission for the use of the National Symbol for Metric Conversion.

Reviewers:
Mr. Doug Abe – Sir Wilfrid Laurier Collegiate Institute, Scarborough
Mr. Doug Bannister – Woodlands School, Mississauga
Prof. Ernie McFarland – University of Guelph
Mr. Barry McGuire – Western Canada Senior High, Calgary
Mr. Bob Powell – Mayfield Secondary School, Brampton
Dr. Bill Prior – Malvern Collegiate Institute, Toronto
Mr. Bruce Robb – Northern Secondary School, Toronto
Prof. J. Vanderkooy – University of Waterloo

Canadian Cataloguing in Publication Data
Hirsch, Alan J.
　Physics for a modern world

For use in secondary schools.
Includes index.
ISBN 0-471-79747-2

1. Physics. I. Title.

QC23.H58 1985　　　530　　　C85-099719-4

Text and cover design by Michael van Elsen Design Inc.

Illustrations by James Loates *Illustrating* and
Margaret Kaufhold

Typesetting by Compeer Typographic Services Limited

Printed and bound in Canada by The Bryant Press Limited

1 2 3 4 5　BP　98 97 96 95 94

TABLE OF CONTENTS

UNIT 1 – INTRODUCING PHYSICS	1
CHAPTER 1: THE STORY OF PHYSICS	2
1.1 Our Changing World	3
1.2 Science in Early Civilizations	4
1.3 The Greek Civilization	7
1.4 The Roman Influence and the Middle Ages	9
1.5 The Emergence of the Scientific Method	10
1.6 From Galileo's Time to Ours	11
Review Assignment	14
CHAPTER 2: THE LANGUAGE AND SKILLS OF PHYSICS	16
2.1 The Language of Measurement	17
2.2 Scientific Notation and the Laws of Exponents	21
2.3 Using Metric Prefixes	23
2.4 Accuracy of Measurement	26
2.5 Calculations Based on Measurements	28
2.6 Solving Equations and Analysing Units	29
2.7 Graphing and Analysing Scientific Data	31
Experiment 2A: The Bending of a Metre Stick	34
2.8 Estimating Quantities	35
Review Assignment	37
Relationships between Sets of Variables (Supplementary)	38
UNIT II – MECHANICS	41
CHAPTER 3: SPEED AND VELOCITY	42
3.1 Uniform and Non-Uniform Motion	43
3.2 Scalar and Vector Quantities	43
3.3 Speed and Velocity	46
3.4 Graphing Uniform Motion	48
3.5 Measuring Time	51
Experiment 3A: Uniform Motion at Varying Speeds	54
Review Assignment	57
Relative Velocity (Supplementary)	58
CHAPTER 4: UNIFORM ACCELERATION	64
4.1 Comparing Uniform Motion and Uniform Acceleration	65
4.2 Calculating Acceleration	66
4.3 Using Velocity-Time Graphs to Find Acceleration	68
4.4 Investigating Acceleration in the Laboratory	70
Experiment 4A: Uniform Acceleration	73
4.5 Acceleration Near the Earth's Surface	76
Experiment 4B: Acceleration due to Gravity	77
4.6 Solving Uniform Acceleration Problems	78
4.7 Applications of Acceleration	84
Review Assignment	87

CHAPTER 5: FORCE AND NEWTON'S LAWS OF MOTION — 91
5.1 Forces in Nature — 92
5.2 Measuring Force — 93
5.3 Balanced and Unbalanced Forces Acting on a Single Object — 94
5.4 Newton's First Law of Motion — 95
Experiment 5A: Force, Mass, and Acceleration — 97
5.5 Newton's Second Law of Motion — 101
5.6 The Force of Gravity — 103
5.7 Newton's Third Law of Motion — 105
5.8 Applications of Forces — 108
Review Assignment — 116
Newton's Law of Universal Gravitation (Supplementary) — 119
Impulse and Momentum (Supplementary) — 121
Experiment 5B: Momentum in Collisions (Supplementary) — 123
The Law of Conservation of Momentum (Supplementary) — 124

CHAPTER 6: MECHANICAL ENERGY AND POWER — 127
6.1 The Importance of Energy — 128
6.2 Using Force to Transfer Energy — 129
Experiment 6A: Work — 132
6.3 Gravitational Potential Energy — 134
6.4 Kinetic Energy — 136
6.5 The Law of Conservation of Energy — 138
6.6 Power — 143
Review Assignment — 145

CHAPTER 7: ENERGY IN A MODERN WORLD — 148
7.1 The Importance of Energy — 149
7.2 The Consumption of Energy — 151
7.3 Non-renewable Resources — 156
7.4 Renewable Energy Resources — 159
7.5 Advantages of Using Energy — 165
7.6 Disadvantages of Using Energy — 166
7.7 Society's Responsibilities to Itself — 168
7.8 Our Personal Responsibilities in Solving Energy Problems — 170
Review Assignment — 171
The Physics of Heat (Supplementary) — 173

UNIT III – FLUIDS — 183

CHAPTER 8: FLUIDS AT REST — 184
8.1 General Properties of Fluids — 185
8.2 Pressure — 187
8.3 Atmospheric Pressure — 189
8.4 Measurement of Pressure — 191
8.5 Pascal's Law — 195
8.6 Buoyancy in Fluids — 197
Experiment 8A: Measuring Buoyant Force — 198
8.7 Archimedes' Principle — 200
8.8 Applications of Buoyancy — 201
8.9 Interaction of Particles of Liquids — 204
8.10 Applications of the Interaction of Liquid Particles — 206
Review Assignment — 207

CHAPTER 9: FLUIDS IN MOTION	211
9.1 Viscosity and Turbulence	212
9.2 Streamlining	215
9.3 Bernoulli's Principle	217
9.4 Applications of Bernoulli's Principle	219
Review Assignment	221

UNIT IV – WAVES, SOUND, AND MUSIC — 223

CHAPTER 10: VIBRATIONS AND WAVES	224
10.1 Vibrations	225
10.2 Frequency and Period of Vibration	227
Experiment 10A: The Pendulum	229
10.3 Transfer of Energy	230
Experiment 10B: Pulses on a Coiled Spring	231
10.4 Periodic Waves	234
10.5 The Universal Wave Equation	236
10.6 Interference of Pulses	237
10.7 Viewing Waves on Water	240
Experiment 10C: The Diffraction of Water Waves	241
10.8 Mechanical Resonance	242
10.9 Standing Waves – A Special Case of Both Interference and Resonance	243
Review Assignment	246

CHAPTER 11: SOUND ENERGY AND HEARING	250
11.1 Production and Transmission of Sound Energy	251
11.2 The Speed of Sound in Air	253
11.3 The Speed of Sound in Various Materials	254
Experiment 11A: Interference of Sound Waves	254
11.4 Beat Frequency	257
11.5 Resonance in Sound	258
11.6 Hearing and the Human Ear	259
11.7 Infrasonics, Ultrasonics, and Echo Finding	263
11.8 The Doppler Effect and Supersonic Speeds	264
Review Assignment	267

CHAPTER 12: MUSIC, MUSICAL INSTRUMENTS, AND ACOUSTICS	270
12.1 "Seeing" Sound	271
12.2 Pitch and Musical Scales	272
12.3 Intensity and Loudness of Sounds	276
12.4 Quality of Musical Sounds	278
Experiment 12A: The Frequency of Vibrating Strings	279
12.5 Stringed Instruments	281
12.6 Vibrating Columns of Air	285
Experiment 12B: Sound in a Column of Vibrating Air	288
12.7 Wind Instruments	289
12.8 Percussion Instruments	292
12.9 The Human Voice	293
12.10 Electrical Instruments	295
12.11 Electronic Instruments	296
12.12 Acoustics	298
Review Assignment	301

UNIT V – LIGHT AND COLOUR	305

CHAPTER 13: THE NATURE AND REFLECTION OF LIGHT — 306
13.1 Sources of Light Energy — 307
13.2 The Transmission of Light — 309
13.3 The Interaction of Light with Matter — 313
Experiment 13A: Plane Mirrors — 315
13.4 Ray Diagrams for Plane Mirrors — 317
13.5 Applications of Plane Mirrors — 319
13.6 Curved Mirrors — 321
Experiment 13B: Converging Mirrors — 322
Experiment 13C: Diverging Mirrors — 324
13.7 Ray Diagrams for Curved Mirrors — 325
13.8 Applications of Curved Mirrors — 326
Review Assignment — 328

CHAPTER 14: REFRACTION AND LENSES — 332
14.1 Refraction and the Speed of Light — 333
Experiment 14A: Refraction and the Index of Refraction — 336
14.2 Snell's Law of Refraction — 338
Experiment 14B: Total Internal Reflection — 340
14.3 Applications of Total Internal Reflection — 342
14.4 Lenses — 345
Experiment 14C: Converging Lenses — 346
Experiment 14D: Diverging Lenses — 349
14.5 Ray Diagrams for Lenses — 350
Review Assignment — 352

CHAPTER 15: OPTICAL INSTRUMENTS — 356
15.1 The Functions of Optical Instruments — 357
Experiment 15A: The Pinhole Camera — 358
15.2 Lens Cameras and Photography — 361
15.3 The Human Eye — 367
15.4 Vision Defects and Their Corrections — 369
15.5 The Microscope and the Refracting Telescope — 372
Experiment 15B: The Microscope and the Refracting Telescope — 374
15.6 Ray Diagrams for Other Optical Instruments — 376
Review Assignment — 378

CHAPTER 16: COLOUR AND LIGHT THEORY — 382
16.1 The Dispersion and Recombination of White Light — 383
Experiment 16A: Adding Light Colours and Viewing Colour Shadows — 385
16.2 Additive Colour Mixing — 387
Experiment 16B: Subtracting Light Colours — 388
16.3 Subtractive Colour Mixing — 389
16.4 Applications of Colour — 392
16.5 Light Theories and the Electromagnetic Spectrum — 395
16.6 The Laser: An Incredible Application of Light — 398
Review Assignment — 405

UNIT VI – ELECTRICITY AND ELECTROMAGNETISM — 409

CHAPTER 17: STATIC ELECTRICITY — 410
17.1 The Force of Electricity — 411
Experiment 17A: The Laws of Electric Charges — 413
17.2 The Atomic Theory of Matter — 414
17.3 Transferring Electric Charge by Friction — 417
17.4 Electric Conductors and Insulators — 418
17.5 Mapping Electric Fields — 420
17.6 Distribution of Charges on Insulators and Conductors — 422
17.7 Charging Electroscopes by Conduction and Induction — 425
Experiment 17B: Induction of Electric Charges — 427
17.8 Static Electricity Generators — 428
17.9 More Applications of Static Electricity — 431
Review Assignment — 434
Coulomb's Law and the Elementary Charge (Supplementary) — 436

CHAPTER 18: CURRENT ELECTRICITY — 440
18.1 Electric Charges in Motion — 441
18.2 Sources of Electrical Energy — 443
18.3 Electric Current — 448
18.4 Electric Potential Difference — 450
18.5 Electric Resistance — 452
18.6 Electric Circuits — 454
Experiment 18A: Electric Current, Potential Difference, and Resistance — 456
18.7 Ohm's Law — 457
Experiment 18B: Resistors in Series — 458
Experiment 18C: Resistors in Parallel — 460
18.8 Analysing Electric Circuits — 461
Review Assignment — 466

CHAPTER 19: USING ELECTRICAL ENERGY — 470
19.1 Direct and Alternating Currents — 471
19.2 Current Electricity in the Home — 472
19.3 Electric Power — 474
19.4 The Cost of Electrical Energy — 476
19.5 Electrical Safety — 478
19.6 Electrical Energy for Optimists — 480
Review Assignment — 484

CHAPTER 20: MAGNETISM AND ELECTROMAGNETISM — 487
20.1 The Force of Magnetism — 488
20.2 The Domain Theory of Magnetism — 489
20.3 Magnetic Fields — 491
20.4 The Magnetic Effects of Electricity — 496
Experiment 20A: The Magnetic Field around a Straight Conductor — 497
Experiment 20B: The Magnetic Field in a Coiled Conductor — 499
20.5 Electromagnets — 502
20.6 The Magnetic Force on Moving Charges — 504
20.7 The Design of Electric Motors — 507
20.8 Constructing and Using Electric Motors — 509

| Review Assignment | 510 |

CHAPTER 21: ELECTROMAGNETIC INDUCTION — 514
21.1 Using Magnetism to Produce an Electric Current — 515
Experiment 21A: Inducing Current in a Coiled Conductor — 516
Experiment 21B: The Direction of Induced Current — 517
21.2 Lenz's Law — 518
21.3 Electric Generators — 520
21.4 Using Electricity to Generate Electricity — 526
21.5 Using Transformers to Distribute Electrical Energy — 529
21.6 Electromagnetic Waves — 532
Review Assignment — 532

UNIT VII – ATOMIC PHYSICS — 535

CHAPTER 22: ATOMS AND RADIOACTIVITY — 536
22.1 From Classical to Modern Physics — 537
22.2 The Discovery and Use of X Rays — 538
22.3 The Discovery of Radioactivity — 542
22.4 Detecting Radioactive Emissions — 544
22.5 Models of Atoms — 548
22.6 The Structure of the Nucleus — 552
22.7 Transmutations — 556
22.8 Half-Life — 559
Experiment 22A: Simulation of Half-Life — 560
22.9 Uses of Radioactivity — 561
22.10 Hazards of Radiation — 567
Review Assignment — 567

CHAPTER 23: USING NUCLEAR ENERGY — 570
23.1 Nuclear Reactions — 571
23.2 Mass and Energy in Nuclear Reactions — 574
23.3 Nuclear Fission — 576
23.4 The First Nuclear Fission Reactor — 578
23.5 Using Nuclear Fission to Generate Electricity — 580
23.6 The Nuclear Reactor Debate — 589
23.7 Nuclear Fusion — 594
23.8 Nuclear Weapons — 598
23.9 The Ever-Changing Theories of Physics — 602
Review Assignment — 605

EPILOGUE AN OVERVIEW OF PHYSICS — 608

APPENDICES — 612
Appendix A The Metric System of Measurement — 612
Appendix B Periodic Table of the Elements — 614
Appendix C Physical Constants — 616
Appendix D Terrestrial Data — 616
Appendix E Trigonometry — 616
Appendix F Table of Trigonometric Values — 617
Appendix G The Greek Alphabet — 618

Answers to Numerical Problems — 619

Photo Credits — 629

Index — 633

Preface

Introducing Physics

If the science of physics had never been developed, our world would be very different than it is. We would have to live without automobiles, trains, boats, and airplanes. We would not be able to watch colour television or listen to stereo sound systems. We would have no computers, telephones, thrill rides at fairs, watches, or air conditioning systems. Nor would we have the pollution and energy supply problems that appear to accompany the development of technology. Indeed, without physics, almost nothing we have in our modern world would have advanced to its present stage.

Physics is the study of **matter** and **energy**, and the interaction between matter and energy. Everything – from the smallest particle you can imagine to the galaxies of stars that make up our universe – is composed of matter. Physics helps us measure matter, describe its motion, and understand the forces causing motion in matter. It also helps us apply these forces in the use of tools, transportation vehicles, kitchen appliances, and thousands of other devices.

The great Andromeda galaxy consists of many millions of stars. It is similar in shape to our own Milky Way galaxy. There are thousands of galaxies in the universe.

Wherever there is matter there is energy. The human senses of hearing and sight would not function without the energies of sound and light. Machines develop mechanical energy to help us do much of our work, and our food is prepared by the transfer of heat. Electrical energy has been put to use almost everywhere in the world, and nuclear energy is being applied to increasing numbers of uses each year.

This book is designed to help you study both the principles and the applications of matter and energy. Meeting the challenge of studying physics will give you a better understanding of what occurs in the world around you, and will enable you to learn to adapt to an ever-changing world.

Features of the Text

The 23 chapters in the text are grouped into 7 units, each of which opens with a list of careers that apply the knowledge and skills gained through a study of that unit. Careers are also emphasized in the biographical sketches of contemporary Canadian scientists whose work relates to certain aspects of physics. Each chapter begins with a brief introduction intended to help you relate your experiences to the topics to be presented. This introduction includes a photograph, as well as questions to stimulate interest and curiosity.

Historical aspects of physics form a major part of Chapter 1 and are woven into the rest of the text to help you appreciate the process of science and understand the place of physics in our world.

The principles of physics are developed in a concise and logical manner, with numerous illustrations and photographs to assist understanding. New words or phrases are emphasized in bold type and are defined when first used. The reading level of the text is carefully controlled. Sample Problems, which are clearly outlined, will help you develop problem-solving skills. Experiments are inserted within the text at locations where they will be of maximum benefit. In the Experiments, detailed instructions are followed by analysis questions that direct your thinking toward the conclusions. In some cases, application questions are also added. Less formal experiments and projects, called Activities, are inserted among the Practice and Review Questions. Activities have fewer detailed instructions given, so they require you to exhibit greater thinking, research, and organizational skills than the formal experiments do.

Practice Questions appear at the end of each theory section, and a Review Assignment is placed near the end of each chapter. Questions and problems are a very important part of the text, and appear in a variety of types, including recall, substitution, manipulation of equations, thought, application, estimation, synthesis, and projection.

Special attempts have been made to produce a book which is as interesting as possible. The book has an international flavour, and incorporates numerous applications of physics principles from throughout North America, as well as photographs from approximately 20 countries. Applications are described directly in the textual material, and are also included in the Practice and Review Questions. These applications cover the span of human history – from the past to the present and into the future – and are of interest to students of both sexes and all national origins. In both the theory and applications, various disciplines are included – language, art, music, sports, technical skills, cooking, sewing, biology, chemistry, history, geography, geology, archaeology, medicine, astronomy, cosmology, and space science among them.

Several other features add to the book's usefulness. At the end of each chapter is a list of Key Objectives which can be used to review and summarize the most important concepts in the chapter. Answers to all the numerical problems are found at the end of the book. The SI metric system of measurement is used throughout the text. A set of colour plates enhances the topic of light. Supplementary Topics near the end of some chapters offer challenges beyond the main stream of textual material. The Epilogue placed after the last chapter offers predictions for the future of physics. A set of Appendices is placed conveniently at the end of the text, and a comprehensive Index makes the book easy to use.

Studying physics can be challenging as well as rewarding. May you enjoy both the challenges and rewards as you study *Physics for a Modern World*.

Alan Hirsch

Acknowledgements

I sincerely thank the reviewers of my manuscript whose names are listed on the copyright page of this textbook. They provided me with valuable suggestions for improvement throughout the book's development. However, I alone am responsible for the final content of the book. I also wish to thank the editorial staff at John Wiley & Sons Canada Limited for their assistance throughout the project. I especially thank Judy Evans for her assistance, understanding, and support as the project was developed.

Lastly, I wish to thank the scientists who, at our request, have provided excellent autobiographies for inclusion. I feel that the material they have written about themselves will excite the imaginations of many students. Thus, it is with a great deal of pleasure that I thank Dr. Gretchen Harris, Ms. France Legault, Dr. Hans Kunov, Ms. Claudette McKay-Lassonde, Dr. Carla Miner, Dr. Bruce Pennycook, Dr. Alexander Szabo, and Dr. Henke Wevers for permitting me to present profiles of them in this textbook.

A.J.H.

Dedication
*This book is dedicated to you,
the students, who with searching minds
love to learn of nature's wonders
and of human intervention in applying
these wonders to improve our lives.*

UNIT I
Introducing Physics

The information in this Unit will be especially useful if you plan a career in astronomy, history, archaeology, surveying, science education, or library science.

INTRODUCING PHYSICS

CHAPTER 1
The Story of Physics

This close-up view of the moon is made possible by advances in the field of physics.

Thousands of years ago people treated the moon as a god. It was loved and yet feared, honoured and yet unexplained. Today, with the exception of small numbers of people, we no longer fear the moon. We can now explain its make-up and motion. Sophisticated equipment is available to take photographs of the moon's surface; astronauts have gone to the moon to explore it; and we can reflect laser beams off a small mirror on the moon to keep an accurate record of its distance from the earth. These changes could not have occurred without the gradual accumulation of scientific knowledge over a period of thousands of years. How has physics helped us progress from those early stages of moon worship to our present advanced stage of understanding the very nature of the universe in which we live?

1.1 Our Changing World

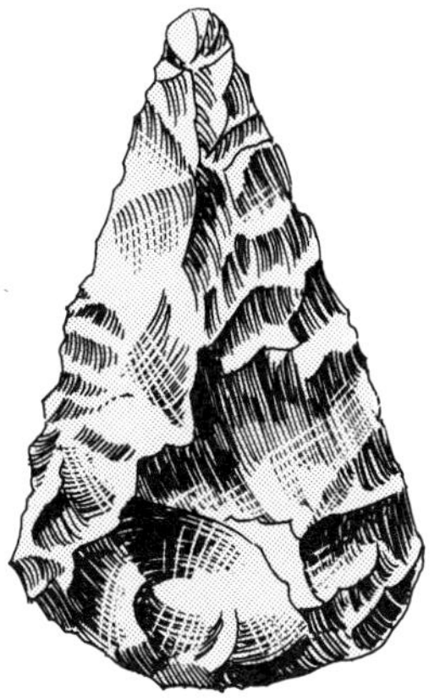

If you were asked to name an electrical appliance in your home having features that 20 years from now would be completely outdated, which would you choose? You could probably name any of several appliances, and you would likely be right. In our exciting and rapidly changing world of science and its applications, we can expect great improvements over even a single generation. It would take both imagination and luck to predict accurately the technical changes that will occur over the next 100 years.

Such has not always been the case. Consider, for example, someone living 30 000 years ago being asked the question, "What tool that you are using today will be outdated in 20 summers?" The answer almost certainly would be, "Not one." The club for hunting and the stone for pounding and cutting (Figure 1-1) were two important tools which remained in use in their basic forms for tens of thousands of years. This era in history is called the **Paleolithic Era** or Old Stone Age. During this period, people were nomadic hunters and food gatherers who lived in caves and perhaps developed a knowledge of fire.

The beginnings of what we call "civilization" occurred when people learned how to farm and raise domesticated animals. Slowly the stone tools became sharpened and polished, and weaving, pottery making, and carpentry emerged (Figure 1-2). We call this period the **Neolithic Era** or New Stone Age. During this time, knowledge still grew very slowly as information was passed from one generation to the next and from one group of people to another.

Toward the end of the Neolithic Era, approximately 5000 to 7000 years ago, the rate at which knowledge increased gradually began to pick up speed. In the remainder of this chapter you will read about how the growth of knowledge in general, and of physics in particular, continued to accelerate, until today we are faced with more rapid changes than at any previous period in human history. By studying the development, applications, and future implications of physics, you will be better able to understand and adapt to these changes.

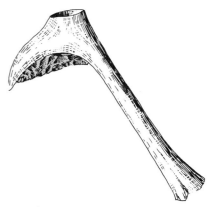

Figure 1-1
The diagram shows a type of hand axe, a Paleolithic tool used tens of thousands of years ago.

Figure 1-2
Some Tools of the Neolithic Era

(a) *This type of knife was used in Egypt around 4500 B.C. The blade was made of a type of stone called flint, set into a bone.*

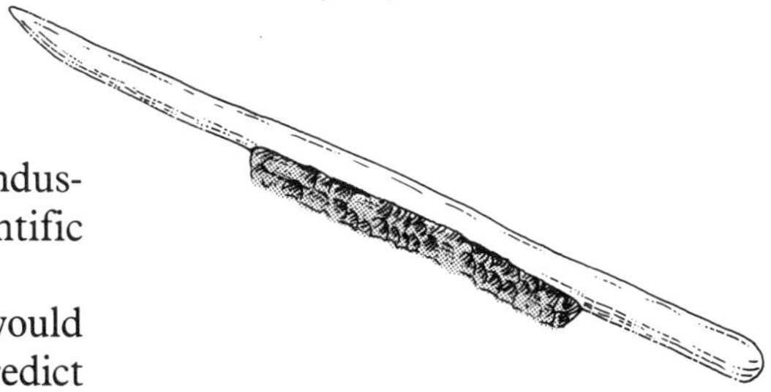

(b) *This type of tool, used to harvest grain, was common in northern Europe around 2500 B.C. The blade was made of flint.*

PRACTICE

1. Discuss what you know about changes in the entertainment industry from the time of silent films to the present. What scientific advances influenced these developments?
2. Suggest some convenient appliances and gadgets which you would like to have in the home that are not yet manufactured. Predict whether they will be possible within the next ten years, given current scientific knowledge and technological advancements.
3. Compare the acceleration of new ideas during the Neolithic Era, the past century, and the near future.

1.2 Science in Early Civilizations

In order to study the development of science, it is useful to know about the time-lines of various civilizations and eras. Figure 1-3 shows the major time-lines of both early civilizations and later periods, so you can see where we fit into the historical picture.

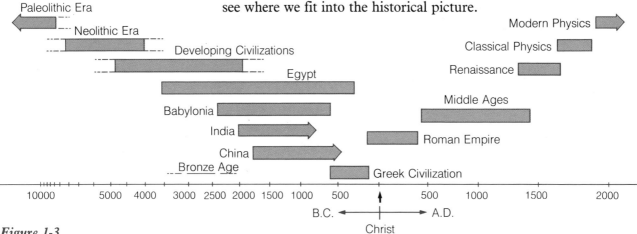

Figure 1-3
Historical Time-lines

We can divide scientific developments into two categories: science resulting in applications, or **technological advances**; and science involving the search for an explanation of the universe, or **philosophical advances**. The early civilizations had many overwhelming technological achievements, as you will soon see. However, they did not arrive at the stage where they formulated any sophisticated scientific explanations.

From 5000 B.C. to about 600 B.C., the most advanced civilizations in the world were found in Egypt, Babylonia, India, and China. The majority of people lived along important rivers, such as the Nile, Tigris, Euphrates, and Indus. See Figure 1-4. During this time, people dis-

Figure 1-4
Historical Regions

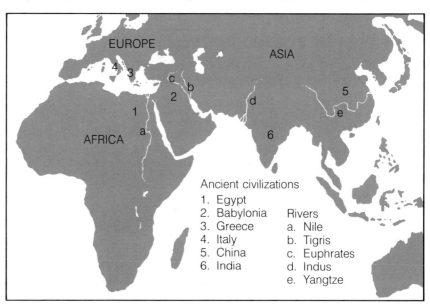

covered that metals such as copper, tin, and iron could be extracted from ore through the use of fire. They also discovered that a mixture or alloy of copper and tin, called bronze, was a useful metal. This scientific discovery was so important that we call the period of history in which bronze was commonly used the **Bronze Age** after this important alloy. See Figure 1-5.

Even today, the ancient Egyptians are famous for their scientific achievements. One of the greatest was a form of writing which was originally done on stone and clay tablets. This invention led to the recording of important events and the passing on of scientific information. The Egyptians also developed a calendar that had 365 days, but no leap years. Furthermore, it took much scientific ability to construct the pyramids (Figure 1-6). These tombs for Egyptian Pharaohs or kings, built between 2600 B.C. and 1800 B.C., were aligned in precise directions using quite sophisticated knowledge of astronomy. As well, the ground had to be levelled and huge stone blocks moved into position. The Egyptians transported these stones over great distances by water. They also shipped

Figure 1-5
"The Charioteer" is a bronze statue fashioned in Delphi, Greece, about 500 B.C. The fact that such a piece of art can last for 2500 years shows the durability of bronze.

Figure 1-6
Although the smooth surface covering this Egyptian pyramid was removed long ago, the stone blocks that remain provide evidence of a civilization with advanced building techniques.

enormous stone obelisks some 200 km down the Nile River from the stone quarries to their beautiful temples at a city called Luxor. See Figure 1-7. In arithmetic, the Egyptians used a decimal system and some fractions, but they had no concept of zero. By the time they were crushed by their enemies in about 400 B.C., the Egyptians had greatly influenced other nearby lands by passing on their scientific knowledge.

Babylonia, another powerful civilization, was a contemporary of the Egyptian civilization. Babylonia was located between the Tigris and Euphrates Rivers, in the area we now call the Middle East, as Figure 1-4 shows. The Babylonians had a calendar whose days and weeks were

Figure 1-7
This obelisk is one of several that were erected by the ancient Egyptians. It is found in the city of Luxor, near the Nile River.

similar to ours. They developed a standard system of measurement about 2500 B.C. They also engineered a canal system to aid in agriculture and in the growing of their fabulous hanging gardens, which were one of the wonders of the ancient world.

India and China, too, had relatively advanced civilizations. In India the main settlements flourished along the Indus River from 2200 B.C. to 1700 B.C. The Indians used writing, and their cities had drainage systems and a standardized construction code. Although China was isolated from the other great civilizations for many centuries, it developed steadily after about 2000 B.C. The Chinese had writing, urban development, silk production, irrigation, and an advanced knowledge of astronomy.

As you can see, the early civilizations achieved many technological advances. However, their philosophical explanations were based on their religious beliefs, not on scientific observations. Scientific philosophies, made possible by the foundations laid by these early civilizations, emerged later.

PRACTICE

4. Discuss how and why our calendar differs from the 365 day Egyptian calendar. (*Hint:* Research the need for leap years.)

5. The largest Egyptian pyramid has a base about 250 m on each side. Obviously, before the base could be built, the ground had to be level. Describe a technique which the Egyptians may have used to level out such a huge area.

6. The Egyptians transported stone blocks and obelisks along the Nile River during the annual flood season. However, they also had to move the stones along the ground, in order to get them from the quarry to the river, and from the river to the building site. Suggest some possible means of moving such huge objects on both water and land.

7. **Activity** This activity may be performed at home using five flat wooden toothpicks. It has some interesting applications, one of which relates to the Egyptians.
   (a) Snap but don't break apart each toothpick at its middle. Arrange the five toothpicks on a hard, smooth surface in a pattern like that shown in Figure 1-8. Add two or three drops of water to the middle of the toothpick pattern. Describe and explain the results.
   (b) The Egyptians made use of the absorption of water by wood fibres and the subsequent swelling when they made uniform stone blocks from a huge rock. Assuming that they were able to drill or chip part-way into the rock, describe how they would have made a stone cube one metre on a side.
   (c) Relate what you learned in this activity to problems that are likely to arise if a homeowner fails to paint or varnish wooden doors and door frames.

Figure 1-8
Toothpick Arrangement

1.3 The Greek Civilization

The ancient Greek civilization, which gained prominence about 600 B.C., is one of the greatest direct influences on our technology and philosophy. The Greeks borrowed some technical achievements from earlier civilizations, in addition to developing their own. Furthermore, although many Greeks used their religion to account for natural occurrences, others provided the beginnings of a natural philosophy which used scientific reasoning to explain the universe.

Many Greek architects and philosophers contributed to the advancement of science, especially between 600 B.C. and about 200 B.C. The scientific philosophers developed their ideas not by experimentation, but by reasoning and discussion. Their method of reasoning was to state some important principle or law, then draw conclusions based on that principle. Of course, their conclusions could be true only if the original principle was true. For example, one of the reasoned principles was that the earth was the centre of the universe. This led to the conclusion that the sun travelled around the earth, a conclusion we now know is false. As you read about the ideas and achievements of some of the scientific philosophers, keep in mind this method of reasoning.

Thales of Miletus (about 636 B.C.–546 B.C.), who is sometimes called the first Greek philosopher, was a merchant, mathematician, and astronomer. After visiting Egypt and seeing their system of surveying land, he is said to have originated the science known as *deductive geometry*. This science was developed further by later Greek and Egyptian philosophers, and is studied and used to this day.

Pythagoras (about 582 B.C.–497 B.C.) founded a mystical order that believed in the importance of numbers. He related numbers to musical sounds and discovered that the ratio of the lengths of strings that sounded pleasant together was 6:4:3, as illustrated in Figure 1-9(a). The Pythagorean theorem, which formed one of the fundamental statements of deductive geometry, was named after him, although he likely borrowed the concept from Egyptian methods of measurement. See Figure 1-9(b). Pythagoras and his followers recognized that the earth was a sphere, but they did not think it revolved around the sun.

Did You Know?
Many scientific words used today have their origin in the Greek language. The word "physics", for example, stems from the Greek word *physis*, which means "nature".

Did You Know?
There are two different processes of reasoning or logical thinking. One process, called **deductive reasoning**, starts with a general principle or law, applies it to a particular situation, and draws a conclusion. The other process, called **inductive reasoning**, starts with a series of observations about several cases which, through experimentation, leads to a conclusion or general principle. The Greeks generally used deductive reasoning, whereas modern scientists generally use inductive reasoning, although the two processes are often difficult to distinguish clearly.

Did You Know?
The prefix "geo" means "land" or "earth"; thus, "geometry" means "land measuring".

Figure 1-9
The Pythagorean Influence

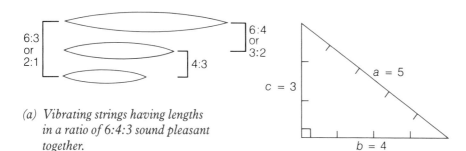

(a) Vibrating strings having lengths in a ratio of 6:4:3 sound pleasant together.

(b) Pythagoras proved that, for a right-angled triangle, $a^2 = b^2 + c^2$, where a is the hypotenuse, and b and c are the other two sides. In the diagram,

$$5^2 = 4^2 + 3^2$$
$$25 = 16 + 9$$
$$25 = 25.$$

INTRODUCING PHYSICS

Figure 1-10
Aristotle (about 384 B.C.–322 B.C.)

Democritus (about 470 B.C.–380 B.C.) was a leader among the group called the *atomists*. They believed that matter was made up of small particles called *atoms*, although they were unable to prove it. All atoms, Democritus said, were made of the same substance; they differed only in their size, shape, and movement.

Aristotle (about 384 B.C.–322 B.C.) became the best-known and most important scientific philosopher in Greece (Figure 1-10). He believed that the earth was the centre of the universe and that all matter on earth consisted of four pure substances or elements — earth, air, fire, and water. Objects beyond the earth, namely stars, were made of a fifth pure substance, quintessence. See Figure 1-11. Aristotle rejected Democritus' atomic theory. He also thought it logical to assume that heavy objects fall more rapidly toward the earth than light objects. These are only a few of Aristotle's many views that influenced scientific thought until about 1500 A.D. Some were later shown by experiment to be correct; many others were disproved.

Figure 1-11
Aristotle's View of the Earth and Its Surroundings

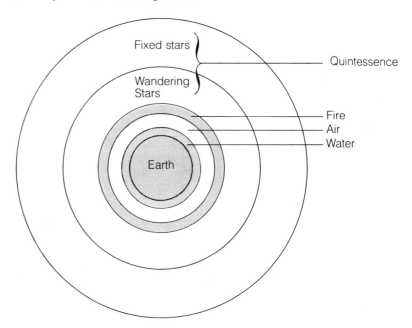

Did You Know?
Aristotle studied under the famous philosopher Plato. Later, he became the tutor of Alexander the Great, one of the greatest military leaders of all time.

Another Greek scientist, **Archimedes** (about 287 B.C.–212 B.C.), was a great inventor, mathematician, and physicist. He laid the foundations of mechanics, through his use of levers and pulleys, for example, as well as of hydrostatics, through his understanding of density, fluids, and hydraulic systems (Figure 1-12). Although he tended to accept Aristotle's views, Archimedes achieved many great technical feats and is now considered to have been the Greek with the most modern scientific outlook.

Thus, scientific knowledge continued to increase, even though the methods used were not, by our standards, scientific. Because they conducted few experiments to test their ideas, many of the conclusions the Greek philosophers reached have been proven incorrect. Nevertheless, they laid the foundations of both technical and philosophical science — the study of physics had begun.

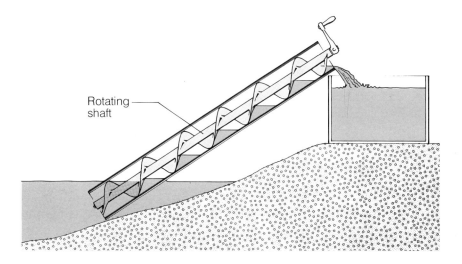

Figure 1-12
One Application of Archimedes' Inventions. Water can be raised from the lower level to the higher level by the continuous turning of the spiral shaft. Drills and grain augers operate on this principle.

PRACTICE
8. One of the proofs used by the Greek philosophers that water and air were primary substances was that "water will not enter a jug until air is removed". Is this statement true? How could you prove whether or not it is true?

1.4 The Roman Influence and the Middle Ages

The great civilization after that of the Greeks was the Roman Empire. For over 600 years, from about 200 B.C. to shortly after 400 A.D., the Romans copied and improved ideas from the Greek and other nations. The wide use of the Latin language was probably their most important contribution to science because it helped spread existing knowledge throughout the Empire. The Romans were famous for their military organization, administrative ability, medical service, and technological achievements in the form of architecture and the great aqueduct systems carrying fresh water. However, they did not create much that was new in science, especially philosophical science.

During the Middle Ages, between the fall of the Roman Empire and approximately 1500 A.D., little inventive work in science occurred in most of Europe. Europe was dominated by the Christian Church and the most powerful philosophers tended to be theologians. They were more concerned with the ideas of religion than with those of science. Their beliefs fit in well with Aristotle's claim that the earth was the centre of the universe.

During the same period, followers of another religion, Islam, played an important role in unifying scientific knowledge. During the seventh century A.D., the Moslems began to translate scientific information from many nations, such as China, India, Egypt, and Greece, into the Arabic language. In the ninth century, they improved the system of numbers, called Hindu or Arabic numbers, by adding a symbol for zero.

INTRODUCING PHYSICS

This system replaced the clumsy Roman numeral system, and is the same number system we use today. The Moslems also developed schools of chemistry that added to their already advanced knowledge of medicine.

In other areas of the world, too, scientific developments were taking place at this time. For example, the Chinese discovered and refined the use of gunpowder, paper, and the medical practice of acupuncture. They also invented the magnetic compass, for use in navigation, and in the twelfth century passed on this knowledge to the Europeans during trade.

By the thirteenth century, some European thinkers began to realize that experimentation added greatly to thought and discussion. Centres of learning called universities were set up. Then, in the last half of the fifteenth century, the printing press was invented, allowing information to be more readily available. These developments set the stage for a new way of solving problems that began to emerge toward the end of the Middle Ages.

PRACTICE
9. One important step in Europe's emergence from the Middle Ages was the discovery of what we now call America. The Chinese contributed to this discovery in two ways. What were they?

1.5 The Emergence of the Scientific Method

By the early 1500s, scientific philosophers no longer accepted old ideas without question. Even the teachings of the great Aristotle were criticized. Scientists realized the importance of experimenting before stating a general law of nature, rather than first guessing at the law, as the Greek philosophers had done. This period in European history is called the *Renaissance*, a French word meaning "rebirth". The term "rebirth" is appropriate for the art and architecture that developed from the classic Greek and Roman styles. Science, however, was not reborn. On the contrary, scientists started to question ancient explanations. Thus, when considering science during this period, the expression "new thinking" applies more aptly than "rebirth".

One great scientist who influenced the new thinking was the Italian **Galileo Galilei** (1564-1642). See Figure 1-13. He combined careful observation and experimentation with mathematical reasoning to form the first true method of physical research. Some of his greatest contributions to science were in the study of motion, a topic about which he disagreed with Aristotle. In a simple experiment he proved that objects, whether heavy or light, fall to the ground with the same acceleration, assuming no air resistance. You can perform a similar experiment by dropping two coins of different masses from the same height. Galileo constructed the first telescope in 1609 and used it to provide evidence that the earth revolves around the sun. His discovery angered many

Figure 1-13
Galileo Galilei (1564-1642)

leading theologians and philosophers. In a Church trial, Galileo was forced to deny this fact in order to avoid death. Other achievements of Galileo are mentioned in later chapters.

Although there were many setbacks — people were imprisoned or even executed for their scientific beliefs — scientific knowledge developed rapidly during and after Galileo's time. Experimentation led to what we know as the **scientific method**. A scientist applying this method would follow this procedure:

1. State a problem.
2. Make a **hypothesis** about the problem (a scientific "guess" based on research of related facts).
3. Perform a controlled experiment to test the hypothesis, making careful observations.
4. Come to a logical conclusion, often in the form of a mathematical or physical model.
5. Test the model and make revisions to it as required.
6. Offer predictions for other related situations.

Did You Know?
Many scientific breakthroughs are discovered by accident, by trial and error, or by experimenting without prior research or hypothesis. Thus, if you were to ask ten different scientists to define the "scientific method", you would get ten different answers.

This method, although often not followed step by step, helped provide a framework for the study of science between the Renaissance and about 1900. Even today, this method is used by some scientists, though probably not consciously, in trying to solve problems. It can also be used in certain everyday situations. For example, imagine you turn the television on and nothing happens. How would you use the scientific method to discover what the problem is?

PRACTICE

10. Describe how you would solve the following problem. Somehow a bird has fallen down the chimney of the fireplace in your home; you can hear its wings flapping just above the closed vent. You want to get the bird out of the house without harm to yourself, the bird, or your house.
11. **Activity** For this activity, you will need a device called a radiometer, shown in Figure 1-14, and a bright light source such as an incandescent light or the sun. Use the steps outlined below to solve this problem: "What happens to the radiometer when a bright light is shone on it, and why?"

    (a) State a hypothesis about the problem.
    (b) Experiment to verify your hypothesis.
    (c) Explain your observations.

Figure 1-14
A Radiometer

1.6 From Galileo's Time to Ours

Thousands of scientists have contributed to the advancement of physics during the last 350 years. Two of the most outstanding among them were Sir Isaac Newton and Albert Einstein.

Sir Isaac Newton (1642–1727), an Englishman, was born in the year Galileo died. He was a physicist and mathematician who carried on the

PROFILE

Gretchen L.H. Harris, Ph.D.
Research Associate Professor
Department of Physics
University of Waterloo

People have often asked me why I chose astronomy as my career priority, but my interest and determination in this direction go so far back that I cannot pinpoint any specific time or event. My "autobiography", written as a 7th grade English assignment, concluded with plans for the future — among which was my desire to be a scientist, preferably an astronomer. The motivation was probably an outgrowth of my interests in science, religion, and philosophy, which were themselves reflected in my favourite recreational reading – science fiction.

During high school I enjoyed mathematics and physics, but despite my junior high school ambitions, I did not study pure science immediately upon going to university. In fact, it was not until studying for my Ph.D. at the University of Toronto that I devoted myself to astronomy full time. The stars have traditionally been associated with romance, and I soon discovered this to be true, for it was while I was still a graduate student that I met my husband. We found we shared not only a love of astronomy but also music (we presently sing in not one but two choirs), sports (we both jog and play tennis), and children (we are now the proud parents of twin girls).

More than a route to romance, astronomy has proved to be an exciting chance to discover clues about how the universe was formed. My own specific research has involved the observation of star clusters – groups of stars with a common origin in time and space. One common type is the globular cluster, a name which comes from the fact that its member stars are generally distributed in a spherical halo. These globular clusters are formed early in the life of a galaxy and, as some of the oldest objects observable, give archaeological clues to the early universe.

In particular, I study the globular clusters in our own galaxy (the Milky Way) and a galaxy that is close by (Centaurus A). I do this by observing the spectra of individual stars through powerful telescopes and by analysing the data collected from electronic instruments that are attached to the telescope. From these observations, I can discover what substances the stars are made of and how these substances are distributed.

In contrast to previous ideas, my colleagues and I are finding that the work of Galileo and other physicists. When he was 23 years old, an outbreak of the plague forced him to move from a crowded city to a small town where he spent a year studying and experimenting. Among the many subjects he wrote about were motion, force, gravity, light, and astronomy. Most of his ideas are still accepted today. In fact, much of the information in this text is based on Newtonian physics. His laws of motion are discussed in Chapter 5.

Albert Einstein (1879–1955) was a German physicist and mathematician who made many contributions to physics. Einstein, who moved to the United States before World War II, is particularly renowned for his theories of relativity. These theories carry Newton's ideas into the realm of high-speed motion, and account for events that cannot be explained by Newton's discoveries. A pacifist, Einstein very much regretted that his ideas were put to use in the development of the atomic bomb.

clouds from which globular star clusters were formed were not composed of a uniform, homogeneous substance. Instead, there seem to be detectable variations in the amounts of iron, calcium, carbon, nitrogen, and so on contained in them. In fact, we are finding that the haloes of different galaxies can be very different from one another. Hopefully, this information will help us learn more about the early formation of galaxies themselves.

I have also spent many years studying the other type of star cluster, the open cluster. In this case my colleagues and I are concentrating on obtaining observations of individual stars in these clusters in order to understand their detailed properties. As I mentioned, globular clusters are extremely old, in fact, almost 15–20 thousand million years old. In contrast, open clusters are mere "babes" with ages from a few million to a few thousand million years, and are located in the plane or disk of our Galaxy. Thus, these objects tell us about the more recent history of the Galaxy, in particular, its disk and most massive stars.

To collect and analyse the data for these projects, we have travelled to observatories in places as far apart as Chile, Hawaii, Australia, Scotland, and the United States.

Of course, an astronomer can't just hop on a plane, fly to one of these observatories, and begin peering through a telescope. The whole process is much more complicated. First you must apply to the observatories for permission to use their telescopes. This usually involves providing detailed documents explaining what you wish to study and what you hope to accomplish. If your application is accepted, you are then allotted a certain amount of observing time on specific dates.

The actual observing is normally a quiet, low-key process, but there can be moments of excitement. One night a colleague and I were trying a last minute project on the 4 m telescope at the Cerro Tololo Interamerican Observatory in Chile when we observed, for the first time, four or five globular clusters around a galaxy previously thought to have virtually none. As the data appeared on the oscilloscope, instead of behaving like sedate scientists, we found ourselves jumping around the control room like maniacs, much to the amusement of our observatory assistant.

Happily, as a child my home environment was one in which reading, argument, expression of opinion, and the ability to listen to others were all stressed. As a result, I grew up with a high regard for rational thought and the testing of ideas. This has proved invaluable to me, for in the final stage of research an astronomer must spend a rather large amount of time analysing the data he or she has collected in order to try to fit it in with the patterns currently known.

People sometimes find it unusual when I tell them that I am a scientist, although women in astronomy are more easily accepted in comparison with women in other scientific disciplines such as physics and engineering. The history of astronomy and the smallness of the population of astronomers mean that women are accepted as equals. However, on the whole, there are relatively few women scientists, an unfortunate situation that will not be overcome until more young women are encouraged to study mathematics and physics in high school.

In the course of the twentieth century, the information provided by physics has grown enormously. There is so much knowledge that scientists who specialize in physics tend to concentrate mainly on one portion of the discipline. For example, there are physicists who spend their entire careers studying the ways in which extremely cold temperatures affect the ability of certain substances to conduct electricity. Compare these specialists to some of the ancient scientists, who were experts in many fields because comparatively little was known in each.

We still categorize physics into two main types — pure and applied — which correspond loosely to philosophical and technological physics. **Pure physics** involves research that increases scientific understanding of matter and energy — nature at the most fundamental level. It leads to theories about many aspects of the universe — from the smallest known particles and the forces that bind those particles together, through the life-span of the earth and the sun, to the origin of the universe itself. These theories are based on thousands of years of accumulated knowledge. We are "advanced" because we have a large foundation on which to build. But there is much room for more advancement as new experimentation exposes new observations, not to mention flaws in current theories, forcing physicists to come up with revised hypotheses.

Applied physics uses the knowledge of physics to develop many things helping to improve our lives. Modern communication systems let us view activities on the opposite side of the earth almost as they happen. Three-dimensional movies, laser disks, and retractable roof tops for stadiums add variety to the exciting world of entertainment. Pacemakers, artificial organs, and laser surgery assist those suffering ill health. Automobiles are now safer and more energy-efficient than at any time since their introduction. The computer, with its countless applications, affects nearly all aspects of our lives. As applied physics continues to progress, the possible improvements are difficult to predict. Also hard to foresee are the negative effects of the misuse or overuse of our technological developments. Because this book is written for a modern world, not only pure physics, but also numerous examples of applied physics accompanied by insights into related problems, will be mentioned in the progression through the chapters.

PRACTICE
12. Research and report on an advancement in physics described in a current magazine or newspaper article. Indicate whether the advancement involved applied physics, pure physics, or both. Be sure to include a description of your source of information.

Review Assignment

1. Place the eight historical ages listed below in the order in which they occurred.
   Greek Civilization   Middle Ages   Paleolithic Era
   Neolithic Era   Bronze Age   Renaissance
   Twentieth Century   Roman Empire
2. List two or three major contributions to the development of science made during each of these eras:
   (a) Greek Civilization
   (b) Roman Empire
   (c) Middle Ages
   (d) Renaissance
3. Name five important scientists who have influenced the growth of physics in the past, and state the period in which each lived.
4. Contrast and compare the so-called "scientific method" with the process of reasoning used by the Greek philosophers.
5. (a) Offer a hypothesis about how the mechanism inside a bathroom scale works. If possible, verify your hypothesis.
   (b) How might the mass of the load carried by a huge transport truck be measured at highway checkpoints?

Key Objectives

Having completed this chapter, you should now be able to do the following:
1. Arrange in chronological order a given set of periods during which the discipline of physics developed.
2. List the main scientific achievements and weaknesses of the civilizations that influenced our present civilization.
3. Recognize the difference between pure physics and applied physics.
4. Name some outstanding scientists and their contributions to scientific knowledge.
5. List the steps of the scientific method and contrast them with the process of reasoning used in early civilizations.
6. Follow the steps of the scientific method in solving science-related problems.
7. Explain unfamiliar phenomena by using a combination of reasoning, discussion, research, and experimentation.

CHAPTER 2
The Language and Skills of Physics

The skills from the study of physics can be applied in numerous situations, including the interpretation of gauges in this control room of an electricity distribution centre.

In our daily lives, the skills of measuring quantities and analysing numbers are important. Temperature forecasts help us decide what clothes to wear; the volume of a bottle of perfume we buy determines the price we pay; highway speed limits help control traffic flow; dials indicate the amount of energy we consume in our homes and schools; and charts and gauges like those shown in the photograph help engineers control the distribution of electricity to a large population. This chapter will address the question, "What skills should physics students develop in order to study the subject successfully?"

THE LANGUAGE AND SKILLS OF PHYSICS

2.1 The Language of Measurement

In order to design, build, operate, and advertise the things we use, we must be able to describe them in detail. Descriptions involving our senses are called **qualitative** descriptions. A baby's skin is smooth; a flower garden is full of aroma; your favourite singer sounds wonderful; and so on. These descriptions are subjective, and may vary depending on the observer.

Descriptions involving measured quantities are called **quantitative** descriptions. Each measured quantity consists of a number and a unit. The three most common fundamental quantities are distance, mass, and time. The east-west distance across Canada is approximately 6000 km (kilometres); the average mass of a newborn baby is about 3 kg (kilograms); and the world record for holding one's breath under water is 389 s (seconds).

For many centuries, humans have been concerned with measuring distance, mass, and time. Historians tell us that one method used to measure time was to refer to objects in the sky, including the sun, moon, and stars. For example, the stars forming the constellation Ursa Major ("The Great Bear", also called "The Big Dipper") change their angular position relative to the North Star about 45° every 3 h. This is illustrated in Figure 2-2. Sundials were also utilized thousands of years ago, but their accuracy was poor until improvements were made in the sixteenth century. See Figure 2-1.

Figure 2-1
The shadow produced by this sundial indicates the approximate time.

Figure 2-2
The photograph shows a time-exposure view of the night sky taken through a telescope. The telescope was aimed at a point about which the stars appear to revolve because of the earth's rotation on its own axis. The straight trail across the photograph is the path of an airplane.

INTRODUCING PHYSICS

Figure 2-3
This exaggerated example indicates the need for a standard measurement.

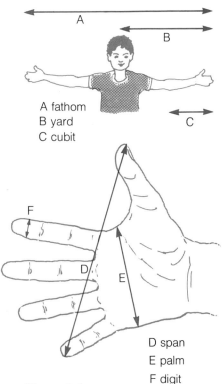

Figure 2-4
Non-uniform Units of Length

Using objects in the sky as measuring devices has obvious limitations, especially during rainy seasons. One way in which past civilizations overcame this problem was to use a *clepsydra* or water clock. A very narrow opening allowed water to drip out of a transparent container that was graduated to the approximate times.

Mass, too, was measured in an inexact fashion. Balances were used to compare unknown masses with known masses; but the known masses were such objects as stones and were obviously not uniform throughout the known world.

Standard units for measurement of distance were also difficult to achieve. The problem illustrated in Figure 2-3 is an exaggeration, but it does show that "this tall" is not an acceptable measurement. Similar problems existed when the human body was used to define standard lengths. Centuries ago, units of length included the digit, hand, span, foot, cubit, and fathom, some of which are illustrated in Figure 2-4. Often the leader of a nation defined these units by taking his own measurements. Of course, with a new leader or in another country, the standard units would be different. By the start of the Renaissance in the fifteenth century, it became obvious that measurement systems had to be standardized.

The **British** or **Imperial system** is a relatively recent system of measurement which has uniform standard units: inches, feet, miles, ounces, pounds, horsepower, and so on (Figure 2-5). However, it is an awkward system for conversion purposes because it is not based on the number 10. For example, there are 12 inches per foot; 3 feet per yard; 1760 yards per mile; and 4840 square yards per acre. Furthermore, some "standard" sizes vary from one country to another—a Canadian gallon is larger than an American gallon. Thus, the Imperial system is obsolete for science and almost obsolete for world trade.

Figure 2-5
These standard British units are displayed at Greenwich, England, at the location of zero longitude.

18

Toward the end of the eighteenth century, after the French Revolution, scientists in France began to develop the **metric system** of measurement. This system has uniform standard units, and is easier to use than the Imperial system because it is based on multiples of 10. Today, the metric system, known internationally as the **Système International d'Unités** or simply the **SI**, is used throughout the world for science and in almost all countries of the world for everyday measurements.

The base unit of length in the metric system is the metre (m). A **base unit** is a unit from which other units may be derived. The metre was originally defined as one ten-millionth of the distance from the equator to the geographic North Pole (Figure 2-6). Then, in 1889, the metre was redefined as the distance between two fine marks on a metal bar now kept in Paris. Over 70 years later, in 1960, the metre was defined once again as: 1 650 763.73 wavelengths of a certain type of light emitted by krypton-86 atoms in a vacuum. In an effort to improve precision, the metre was again redefined in 1983, this time in terms of the speed of light in a vacuum (299 792 458 m/s). Thus, the metre is now defined as the distance light travels in a vacuum in 1/299 792 458 of a second. This quantity, scientists assume, does not change and is reproducible anywhere in the world, so it is an excellent standard.

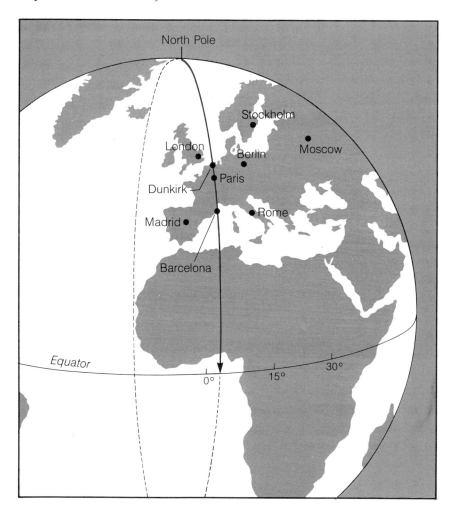

Figure 2-6
The original metre was defined in terms of the assumed distance from the equator to the geographic North Pole. The distance between two European cities, Dunkirk and Barcelona, was measured by surveyors. Then calculations were used to determine the distance from the equator to the North Pole. Finally, that distance was divided by 10^7 to obtain one metre.

INTRODUCING PHYSICS

Figure 2-7
The standard kilogram kept in France was used to make duplicate standards for other countries. Each standard is well protected from the atmosphere. The one shown here is the Canadian standard kept in Ottawa, Ontario.

The base unit of time is the second (s), previously defined as 1/86 400 of the time it takes the earth to rotate once on its own axis. Now it is defined as the time for 9 192 631 770 vibrations of a cesium-133 atom, another unchanging quantity.

The kilogram (kg) is the base unit of mass. It has not yet been defined in terms of a naturally occurring quantity. Currently, the one kilogram standard is a block of iridium alloy kept in France. Copies of the standard kilogram are kept in major centres around the world. See Figure 2-7.

In addition to the metre, kilogram, and second, there are four other base units in the metric system. All units besides these seven are called **derived units** because they can be stated in terms of the seven base units. One example of a derived unit is the unit for force, called the newton (N). Expressed in terms of base units, $1.0 \text{ N} = 1.0 \text{ kg} \cdot \text{m/s}^2$. Refer to Appendix A at the back of the text for a list of the metric system's base units and a partial list of derived units used in physics.

PRACTICE

1. List two qualitative and two quantitative descriptions of the pen you are using to answer this question.

2. Certain techniques once used to measure time have not been discussed in this section, but are described here. Discuss why none of these techniques is a very accurate standard of time.
   (a) An "hour-glass" consists of two transparent cones joined by a narrow opening through which a certain amount of sand or salt falls.
   (b) The ancient Chinese took a rope with knots tied at equal intervals and burned the rope from one end.
   (c) Galileo Galilei was the first scientist who used a simple pendulum to measure time. He took advantage of the fact that the time for each swing of a pendulum remains constant.

3. Estimate the perimeter of your classroom in units of (a) cubits and (b) digits. Compare your measurements with those of other members of the class.

4. (a) Use the data given in this section to convert one acre to square inches.
   (b) Describe the disadvantage of converting units of measurement within the Imperial system.

5. List possible reasons why the original definitions of the metre and the second described in this section were not precise standards.

6. Refer to the metric table in Appendix A and express the derived units listed below in terms of base units. (*Hint*: In this section, the newton was defined in terms of base units.)
   (a) joule
   (b) watt
   (c) pascal

2.2 Scientific Notation and the Laws of Exponents

Extremely large and small numbers are awkward to write in long form. To alleviate this problem, measurements are often recorded using **scientific notation**, also called "standard form". Using this notation, the first digit other than zero is placed before the decimal point and the other digits after it. Then the number is multiplied by the appropriate power of 10. The best way to see the usefulness of scientific notation is to study examples of large and small numbers, like those shown in Table 2-1.

Table 2-1 Examples of Scientific Notation

Quantity Measured	Measurement (most values approx.)	Scientific Notation
Distance to Andromeda Galaxy	19 000 000 000 000 000 000 000 m	1.9×10^{22} m
Distance to nearest star (Proxima Centauri)	40 000 000 000 000 000 m	4×10^{16} m
Distance from earth to sun	150 000 000 000 m	1.5×10^{11} m
Diameter of earth	13 000 000 m	1.3×10^{7} m
Length of a football field	100 m	1×10^{2} m
Width of a fingernail	0.01 m	1×10^{-2} m
Thickness of a credit card	0.0005 m	5×10^{-4} m
Thickness of a human hair	0.000 05 m	5×10^{-5} m
Thickness of a spider web strand	0.000 005 m	5×10^{-6} m
Diameter of a tungsten atom	0.000 000 000 1 m	1×10^{-10} m

Likewise, calculations involving very large and small numbers are much simpler using scientific notation. For example, the product 2 500 000 000 × 3 800 000 000 000 is far more easily written as $2.5 \times 10^9 \times 3.8 \times 10^{12} = 9.5 \times 10^{21}$. The rules described below must be applied when performing mathematical operations using scientific notation.

Addition and Subtraction in Scientific Notation

First, you change all numbers to a common factor, that is, the same power of 10. Then add or subtract them. In general, $ax + bx = (a + b)x$. Remember that $10^0 = 1$.

SAMPLE PROBLEM 1
Add 3.28×10^3 and 2.61×10^5.

Solution
Using a common factor of 10^5,
$$3.28 \times 10^3 + 2.61 \times 10^5 = 0.0328 \times 10^5 + 2.61 \times 10^5$$
$$= 10^5(0.0328 + 2.61)$$
$$= 2.6428 \times 10^5$$
$\therefore$ The sum is 2.64×10^5 (rounded off).

Multiplication and Division in Scientific Notation

First, multiply or divide the non-exponents. Then multiply or divide the powers of 10 by adding or subtracting the exponents. In general, $10^a \times 10^b = 10^{a+b}$, and $10^a \div 10^b = 10^{a-b}$.

SAMPLE PROBLEM 2
Multiply 2.0×10^8 m/s^2 by 4.6×10^{-5} s.

Solution
$$2.0 \times 10^8 \text{ m/s}^2 \times 4.6 \times 10^{-5} \text{ s} = (2.0 \times 4.6)(10^8 \times 10^{-5}) \text{ m/s}$$
$$= 9.2 \times 10^3 \text{ m/s}$$
$\therefore$ The product is 9.2×10^3 m/s.

SAMPLE PROBLEM 3
Divide 8.0×10^6 m by 4.0×10^{10} s.

Solution
$$\frac{8.0 \times 10^6 \text{ m}}{4.0 \times 10^{10} \text{ s}} = \frac{8.0}{4.0} \times \frac{10^6 \text{ m}}{10^{10} \text{ s}}$$
$$= 2.0 \times 10^{6-10} \text{ m/s}$$
$$= 2.0 \times 10^{-4} \text{ m/s}$$
$\therefore$ The value is 2.0×10^{-4} m/s.

PRACTICE

7. Change these measurements to scientific notation:
   (a) 32 000 000 s in one year
   (b) 6 250 000 000 000 000 000 electrons/s
   (c) 9 192 000 000 vibrations/s
   (d) 0.000 000 000 38 kg

8. Write these measurements using ordinary notation:
   (a) 4.2×10^5 kg   (c) 8.2×10^1 s
   (b) 3.1×10^{-7} L   (d) 9.564×10^8 m

9. Express each number using a common factor of 10^6:
   (a) 4.28×10^7 (c) 9.3×10^{10}
   (b) 6.4×10^5 (d) 1.2×10^3

10. Add these measurements:
    (a) 3.78×10^{10} g $+ 4.72 \times 10^8$ g
    (b) 7.43×10^2 m $+ 6.38 \times 10^3$ m $+ 4.38 \times 10^5$ m

11. Subtract these measurements:
    (a) 4.36×10^6 cm $- 2.13 \times 10^4$ cm
    (b) 3.5×10^{-7} s $- 4.0 \times 10^{-8}$ s

12. Multiply:
    (a) $(3 \times 10^5)(4 \times 10^7)(5 \times 10^{12})$
    (b) $(6 \times 10^{-5})(8 \times 10^{11})(10^4)$

13. Divide these measurements:
    (a) 6.0×10^{13} kg $\div 3.0 \times 10^7$ m^3
    (b) 4.4×10^6 cycles $\div 2.2 \times 10^6$ s

2.3 Using Metric Prefixes

Because the metric system is based on multiples of 10, prefixes can be used to denote the multiples. The prefixes, among them milli, centi, and kilo, are placed before the units to give others such as millimetre (mm), centigram (cg), and kilowatt (kW). Refer to Table 2-2 as well as to Appendix A at the back of the text for a list of all the metric prefixes.

Often it is useful to convert one metric measurement to another, especially when calculating quantities in derived units. For example, you may be asked to express the distance from the earth to the sun, which is 150 Gm, in metres. There are several techniques for performing such conversions. If you have successfully learned a method that works for you, go ahead and do the practice questions at the end of this section. If, however, you have experienced difficulties with metric conversions, try learning the technique described here. It is a simple method, based on the use of the "metric memory aid" shown in Table 2-3 and repeated for your convenience in Appendix A. (The appendix extends the use of the memory aid to area and volume conversions.) Notice that, in the memory aid, there are both single and triple sets of curved lines joining the prefixes. The single lines represent a single decimal-place move either to the right or to the left; the triple lines represent a triple decimal-place move. Sample Problems 4, 5, and 6 that follow indicate how to use the metric memory aid.

INTRODUCING PHYSICS

Table 2-2 Metric Prefixes and Their Meanings

Prefix	Abbreviation	Meaning	Example of Use (in most cases, number approx.)
exa	E	10^{18}	6350 Em = diameter of our galaxy
peta	P	10^{15}	9.64 Pm = distance light travels in one year
tera	T	10^{12}	0.3 Tm = diameter of earth's orbit around the sun
giga	G	10^{9}	5 G people = earth's approximate population
mega	M	10^{6}	32 Ms = number of seconds in a year
kilo	k	10^{3}	40 km/h = speed limit in a school zone in a city
hecto	h	10^{2}	2.5 hm = length of someone's fence
deca	da	10^{1}	3.5 dam = length of a certain hallway
Standard Unit	—	10^{0}	3.2 g = mass of a penny
deci	d	10^{-1}	3.0 dm = length of some rulers
centi	c	10^{-2}	1.5 cm = width of a thumbnail
milli	m	10^{-3}	35 mm = diameter of a camera lens
micro	μ	10^{-6}	1 μm = diameter of a bacterium
nano	n	10^{-9}	1 ns = time between operations done in some computers
pico	p	10^{-12}	3 pm = wavelength of some gamma rays
femto	f	10^{-15}	1 fm = diameter of a proton
atto	a	10^{-18}	1 as = time for an electron to spin 100 times on its own axis

Table 2-3 Metric Memory Aid

E P T G M k h da — d c m μ n p f a

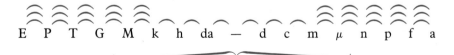

These prefixes are used more frequently than the others.

THE LANGUAGE AND SKILLS OF PHYSICS

SAMPLE PROBLEM 4
Convert 387 mm to decametres.

Solution
To convert millimetres to decametres involves a decimal-place move of 4 places to the *left* (dam m dm cm mm).
Thus, 387 mm = 0.0387 dam
              = 3.87×10^{-2} dam

SAMPLE PROBLEM 5
Convert 9.2×10^3 ML to hectolitres.

Solution
To convert from megalitres to hectolitres involves a decimal-place move of 4 places to the *right*; in other words, we multiply by 10^4 (ML kL hL).
Thus, 9.2×10^3 ML = $9.2 \times 10^3 \times 10^4$ hL
                        = 9.2×10^7 hL

SAMPLE PROBLEM 6
Convert 10^{15} fW to microwatts.

Solution
To convert femtowatts to microwatts involves a decimal-place move of 9 places to the *left*; in other words, we multiply by 10^{-9} (μW nW pW fW).
Thus, 10^{15} fW = $10^{15} \times 10^{-9}$ μW
                  = 10^6 μW

Throughout this text, the most common prefixes used are those in the mega to micro range. However, uses do exist for the extremely large or small prefixes. You can see examples in the right-hand column of Table 2-2.

PRACTICE
14. Convert these measurements:
    (a) 3.8 km = ? hm
    (b) 0.49 hg = ? g
    (c) 4 daW = ?dW
    (d) 57 mL = ? dL
    (e) 305 m = ? hm
    (f) 125 cg = ? dag
15. Convert, expressing your answer in scientific notation:
    (a) 9.4×10^4 Mm = ? m
    (b) 8.5×10^{12} fg = ? g
    (c) 7.1×10^{11} Gm = ? mm
    (d) 2.9×10^{-7} Ps = ? Ms
    (e) 6.5×10^{-12} μm = ? mm

16. **Activity** Measure and memorize the following quantities, in metric units:
    (a) your height (b) your armspan (c) your pace
17. Calculate the number of your paces that would fit into 1.0 km.
18. A certain popular hamburger chain frequently updates the information on the number of burgers it has sold. Do the research necessary to determine an estimate for the average number of McDonald's burgers consumed by each person on earth, assuming each consumes the same number. Is this assumption valid? Explain.
19. The mass of the earth is about 6×10^{24} kg. Calculate the factor by which the earth's mass exceeds your own mass.

2.4 Accuracy of Measurement

If you were to count the number of desks in your classroom, you would obtain an exact number. If you were to measure the length of the room, however, your measurement would be an *approximation*. Measurements can never be exact; they are all subject to error or uncertainty. Just how accurate or close to the true value they are depends on both the instrument used to do the measuring and the person operating the instrument.

Consider, for example, a student using a stopwatch to measure the time for a pendulum to complete 10 swings. Assuming the student has a good reaction time, the measurement may be slightly high in some trials and slightly low in others. This type of error, which results from variation in results about an average value, is called a **random error**. It can be minimized by taking the average of several readings.

Next consider that the stopwatch being used by the student always runs slightly slow. This type of error, which results from a consistent problem with the measuring device or the person using it, is called a **systematic error**. Other examples of systematic errors are a metre stick with worn ends, a dial instrument with a needle that is not properly zeroed, or a human reaction time that is always either too early or too late. These errors can be minimized by calibrating the instrument, by adding or subtracting the known error, or by performing a more complex investigation.

A common source of error in reading scales is **parallax**, the apparent shift in an object's position when the observer's position changes. You can see an example of parallax if you cover your left eye, hold the index finger of your outstretched right arm so it appears to hide a distant object, then switch eyes. The distant object appears to have changed position. To overcome parallax error when reading instruments, you should view the dial and scale at a direct angle, as Figure 2-8 shows.

Every measurement is written with a certain number of **significant digits**, digits that are reliably known. For example, 22 cm has 2 significant digits, and 22.3 cm has 3 significant digits. When taking a measurement, you should record the digits that are known for certain and one estimated digit. If, for instance, you quickly measure your handspan and find it is somewhere between 21 and 23 cm, you would record the

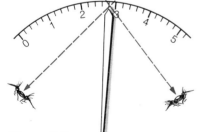

Figure 2-8
The error due to parallax should be minimized when reading instruments. In the example shown, the observer looking from the left reads 2.9, and the observer looking from the right reads 2.7. The true reading, obtained by looking straight at the dial, is 2.8.

measurement as 22 cm ($\pm$ 1 cm). If, however, you perform the measurement more carefully, and find it is between 22.2 and 22.4 cm, you would state the measurement as 22.3 cm ($\pm$ 0.1 cm). In the first reading, the estimated digit is to the closest 1 cm; in the second, to the closest 0.1 cm.

When zeroes are included in a measurement, the number of significant digits can be determined by applying these rules:

(a) Zeroes placed before other digits are not significant; 0.0032 m has 2 significant digits.
(b) Zeroes placed between other digits are always significant; 3007 g has 4 significant digits.
(c) Zeroes placed after other digits behind a decimal are significant; 7.60 has 3 significant digits.
(d) Zeroes at the end of a number are significant only if they are indicated to be so through the use of scientific notation. For example, 8200 km might have 2, 3, or 4 significant digits. However, by using scientific notation, we can tell which digit is the estimated one, so we can determine the number of significant digits.
8.2×10^3 km has 2 significant digits.
8.20×10^3 km has 3 significant digits.
8.200×10^3 km has 4 significant digits.

If the true or accepted value of a measurement is known, the **percent error** of an experimental value can be calculated. In this book, we are concerned only with the absolute value of the error; thus

$$\text{percent error} = \left| \frac{\text{measured value} - \text{accepted value}}{\text{accepted value}} \right| \times 100\%.$$

SAMPLE PROBLEM 7
A student determines the density of pure water at 4°C to be 980 kg/m³. The accepted value is 1000 kg/m³. What is the percent error of the experimental measurement?

Solution

$$\text{percent error} = \left| \frac{\text{experimental value} - \text{accepted value}}{\text{accepted value}} \right| \times 100\%$$

$$= \left| \frac{980 \text{ kg/m}^3 - 1000 \text{ kg/m}^3}{1000 \text{ kg/m}^3} \right| \times 100\%$$

$$= \frac{20}{1000} \times 100\%$$

$$= 2\%$$

$\therefore$ The percent error is 2%.

PRACTICE
20. Some instruments with dials have a small mirror behind the dial, or a dial with a mirrored surface. How does this help avoid parallax error? (*Hint*: The pointer is located between the mirror and the observer.)

INTRODUCING PHYSICS

21. State the number of significant digits in each of the measurements listed:
    (a) 95.2 km
    (b) 3.080×10^5 g
    (c) 0.0067 L
    (d) 0.00670 L
22. A student experimentally determines the acceleration due to gravity to be 9.5 m/s², where the accepted value is 9.8 m/s². Calculate the percent error of the measurement.

2.5 Calculations Based on Measurements

The measurements made in scientific experiments are often used to perform calculations. For example, the surface area of a page can be calculated by multiplying the page length by its width. In this calculation, the number of significant digits should be taken into consideration when writing the final answer.

When you are adding or subtracting measured quantities, your final answer should have no more than one estimated digit.

SAMPLE PROBLEM 8
Add 84.3 mm + 62.4 mm + 31.82 mm.

Solution

       84.3 mm (The "3" is estimated.)
       62.4 mm (The "4" is estimated.)
    + 31.82 mm (The "2" is estimated.)
      178.52 mm (Both the "5" and "2" are estimated.)

Thus, the answer should be rounded off to one estimated digit; the answer is 178.5 mm.

When you are multiplying or dividing measured quantities, the calculated answer should have as many significant digits as the original measurement with the *least* number of significant digits.

SAMPLE PROBLEM 9
Calculate the surface area of a sheet of paper that measures 27.9 cm by 21.7 cm.

Solution
area = length × width
     = 27.9 cm × 21.7 cm
     = 605.43 cm²

This calculated value should be rounded off to three significant digits. Thus, the area is 605 cm².

In each of the sample problems above, notice that the calculations were done first, then the answer was rounded off.

When calculated answers must be rounded off to the correct number of significant digits, the following rules should apply. Each example given is rounded off to two significant digits.

(a) If the first digit to be dropped is 4 or less, the preceding digit is not changed.
e.g., 8.74 becomes 8.7
(b) If the first digit to be dropped is 6 or more, the preceding digit is raised by 1.
e.g., 6.36 becomes 6.4
(c) If the digits to be dropped are a 5 followed by digits other than zeroes, the preceding digit is raised by 1.
e.g., 3.45123 becomes 3.5
(d) If the digit to be dropped is a 5 alone, or a 5 followed by zeroes, the preceding digit is not changed if it is even, but it is raised by 1 if it is odd.
e.g., 2.65 becomes 2.6
2.75 becomes 2.8

PRACTICE
23. Round off each number to two significant digits:
    (a) 6.43
    (b) 8.49
    (c) 2.7538
    (d) 8.25
    (e) 4.55
    (f) 3.54
    (g) 5.07
    (h) 7.15223
    (i) 2.85
    (j) 3.15
24. Add or subtract, following the rules of significant digits:
    (a) 5.1 cm + 6.24 cm
    (b) 8.3 km + 19.8 km + 32.79 km
    (c) 28.7 mL − 25 mL
25. Multiply or divide, taking into consideration significant digits:
    (a) 81.2 m/s × 6.5 s
    (b) 64 cm ÷ 25.5 s
26. **Activity** Calculate the surface area of the front cover of this book. Remember to take into consideration significant digits.

2.6 Solving Equations and Analysing Units

Equations are often derived from the relationships discovered when measurements are taken and calculations done. An equation is usually written so that one variable is expressed in terms of other variables; for example, density = mass/volume or $D = m/V$. It is necessary for you to be able to rearrange such equations to express any one variable in terms of the others. The Sample Problems below illustrate the procedure used to solve for variables in different types of equations.

SAMPLE PROBLEM 10
Given $D = m/V$, solve for (a) m and; (b) V.

Solution
(a) Multiply both sides of the original equation by V.
$$DV = \frac{mV}{V}$$
$$m = DV$$
(b) Divide the equation $m = DV$ on both sides by D.
$$\frac{m}{D} = \frac{DV}{D}$$
$$V = \frac{m}{D}$$
Thus, the required equations are $m = DV$ and $V = m/D$.

SAMPLE PROBLEM 11
Given $a = bc + d$, solve for b.

Solution
First, subtract d: $a - d = bc + d - d$
or $a - d = bc$
Then, divide by c: $\frac{a-d}{c} = \frac{bc}{c}$
$$b = \frac{a-d}{c}$$
Thus, the required equation is $b = (a - d)/c$.

Being able to solve equations is an important skill in the study of physics. Equations usually involve quantities with units. Analysing those units, a process called *dimensional analysis*, provides an excellent means of checking the equations. If the units given in a problem are not consistent, they must be made so before the calculation can be performed.

SAMPLE PROBLEM 12
Mercury, a metal which is liquid at room temperature, has a density of 1.36×10^4 kg/m³. Calculate the mass of 8.0×10^2 cm³ of mercury.

Solution
$m = DV$
$= 1.36 \times 10^4 \frac{\text{kg}}{\text{m}^3} \times 8.0 \times 10^2 \text{ cm}^3$
$= 1.36 \times 10^4 \frac{\text{kg}}{\text{m}^3} \times 8.0 \times 10^{-4} \text{ m}^3$
$= 10.88$ kg
$= 11$ kg

∴ The mass of 8.0×10^2 cm³ of mercury is about 11 kg.

PRACTICE
27. Starting with the equation in Sample Problem 11, solve for d, then c.
28. Solve for x in each equation:
    (a) $x + y = 3$
    (b) $3x - 2y = 0$
    (c) $4xy = m - n$
    (d) $\frac{x}{y} - ab = 0$
29. Given $d = v_i t + \frac{at^2}{2}$, where d is measured in metres, evaluate d in the following instances:
    (a) $v_i = 10$ m/s, $a = 5.0$ m/s^2, and $t = 2.0$ s
    (b) $v_i = 15$ m/s, $a = 3.0$ m/s^2, and $t = 4.0$ s

2.7 Graphing and Analysing Scientific Data

An important operation in physics is finding the mathematical relationship between two variables. One way of discovering such a relationship is to graph measurements involving those variables.

In graphing experimental results, the **independent variable**, the one controlled by the experimenter, is placed along the horizontal axis. The **dependent variable**, which depends on the independent one, is then placed along the vertical axis. The graph should be given a title. Scales should be chosen to occupy as large a portion of the graph paper as possible. The axes should be labelled with the names of the quantities plotted and their units.

Suppose, for example, that an experiment is performed to determine how much a certain rubber band stretches when masses are hung vertically from it. Data recorded in the experiment are shown below:

mass (g)	0	20	40	60	80
stretch (mm)	0	9	21	30	42

In looking at the data, you will notice that, as the mass suspended from the rubber band increases by 20 g, the stretch increases by approximately 10 mm. This relationship becomes more obvious when the data are plotted on a graph. Figure 2-9 shows the graph with stretch plotted as the dependent variable. Notice in the graph that each ordered pair is plotted as a dot surrounded by a circle. The circle roughly takes into consideration experimental error. The data appear to line up fairly closely, so a straight line, called a *line of best fit*, is drawn. Note that the line extends from the origin (0,0), because that point has no experimental error. Scientists and engineers use a complex mathematical analysis to determine where the line of best fit should be.

When one variable increases proportionally as the other variable increases, the relationship is called a **direct variation**. The graph in Figure 2-9 illustrates a direct variation. In mathematical terms, we say that stretch $\propto$ mass, where the symbol "$\propto$" means "is proportional

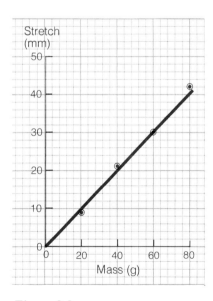

Figure 2-9
Graph of Rubber Band Stretch as a Function of Suspended Mass

INTRODUCING PHYSICS

to". The statement "stretch ∝ mass" can be made more specific by replacing the ∝ with an equals sign and a proportionality constant. Thus,

$$\text{stretch} = (\text{constant})(\text{mass})$$
$$= k \, (\text{mass}) \text{ where k is a constant} \neq 0$$

The advantage of obtaining a straight line on a graph is that the *slope* of the line is the proportionality constant. The equation for slope (m) is written

$$\text{slope} = \frac{\text{rise}}{\text{run}} \quad \text{or} \quad m = \frac{\Delta y}{\Delta x}$$

where Δy is the change in the value plotted on the vertical axis and Δx is the corresponding change in the value on the horizontal axis.

When calculating the slope of a line, always include units, because they indicate the meaning of the slope.

SAMPLE PROBLEM 13

(a) Calculate the slope of the line on the graph in Figure 2-9.

(b) State what the slope represents.

Solution

(a) $m = \dfrac{\Delta y}{\Delta x} = \dfrac{40 \text{ mm} - 0.0 \text{ mm}}{80 \text{ g} - 0.0 \text{ g}}$

$ = 0.50 \text{ mm/g}$

Thus, the slope of the line is 0.50 mm/g.

(b) The slope represents the average stretch for each gram of mass added to the end of the rubber band. The units indicate this fact.

A graph of data can also be used to illustrate examples of interpolation and extrapolation. **Interpolation** is the process of determining values of the dependent variable *between* the plotted points. Sample Problem 14(a) shows an example of interpolation.

Extrapolation is the process of estimating values *beyond* measurements made in an experiment. Extrapolation is possible only if we assume that the line continues in a predictable fashion on the graph. In a rubber band experiment, for example, extrapolation would not be possible if the band stretched to the point of breaking. Sample Problem 14(b) gives an example of extrapolation.

Many physics experiments involve direct variations similar to those shown in this section. Certain experiments, however, involve other types of variations. Analyses of some of them are included as a supplement at the end of this chapter.

THE LANGUAGE AND SKILLS OF PHYSICS

SAMPLE PROBLEM 14
From the distance-time graph shown, give an example of (a) interpolation and (b) extrapolation.

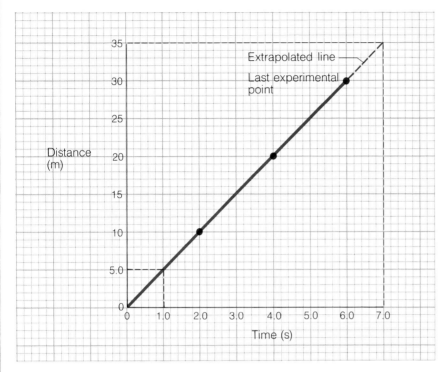

Solution
(a) When $t = 1.0$ s, the distance moved is 5.0 m.
(b) When $t = 7.0$ s, the distance moved would be 35 m.

PRACTICE
30. Plot a graph of each of the sets of data given below. In each case draw the line of best fit. (*Note*: The independent variable is listed first.)

    (a)
diameter of circle (cm)	0.0	4.0	8.0	12
circumference of circle (cm)	0.0	12.5	25.4	37.3

    (b)
time (years)	0.0	1.0	2.0	3.0	4.0
height of tree (m)	0.0	0.32	0.66	1.00	1.30

    (c)
time (s)	0.0	2.0	4.0	6.0
distance (m)	0.0	12	23	37

31. For each of the lines plotted in Question 30,
    (a) calculate the slope, and
    (b) state what the slope represents.
32. For one graph plotted in Question 30, give an example of
    (a) interpolation and (b) extrapolation.

Experiment 2A: The Bending of a Metre Stick

INTRODUCTION
This experiment applies many of the concepts discussed in this chapter. These concepts include quantitative descriptions (measurements), metric units, significant digits, and graphing and analysing data.

The instructions for this experiment are written under the assumption that the metre stick used is relatively stiff. If such a stick is not available, the projection length suggested should be lessened.

PURPOSE
To determine the relationship between the vertical distance through which one end of a metre stick bends and the mass of objects suspended from that end.

APPARATUS
2 metre sticks, one of which should be wooden; various masses; clamp; string or hook

PROCEDURE
1. Predict the relationship described in the Purpose.
2. Clamp a wooden metre stick to the lab bench as shown in Figure 2-10. Use a projection of 80 cm unless your teacher indicates otherwise. Measure and record the vertical distance from the floor to the top of the free end of the metre stick.
3. Use a string or a hook to suspend a 200 g mass from the end of the metre stick. Measure the vertical distance from the floor to the top of the stick, then calculate the vertical distance through which the end of the metre stick is bent.
4. Repeat Procedure Step 3 for masses of 400 g, 600 g, and 800 g. Summarize your measurements and calculations in a table of data.
5. Plot a graph of the data, with the vertical distance bent on the vertical axis. Draw the line of best fit on the graph.

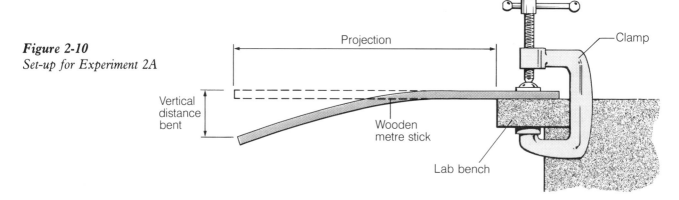

Figure 2-10
Set-up for Experiment 2A

ANALYSIS
1. Explain why the distance bent is plotted on the vertical axis of the graph.
2. According to the graph, what type of relationship exists between the vertical distance bent and the mass suspended from the metre stick?
3. Calculate the slope of the line on the graph. How many significant digits should the slope have? What does the slope represent?
4. Give examples of interpolation and extrapolation on your graph. What assumption is made for extrapolation? Is extrapolation feasible for a 10 kg mass? Explain.
5. Describe sources of error or uncertainty in this experiment.

CONCLUSIONS . . .

2.8 Estimating Quantities

Estimating is an important skill for the weight and age guessers at fairs. But it is also important when you save some money for a vacation or predict how much time you need to complete a certain task. You can likely think of numerous other situations in which estimating skills would be beneficial.

In answering physics or mathematics problems it is always wise to estimate the answers so you can check whether the calculated answer makes sense. For example, two students, A and B, are calculating an answer to the problem, "Determine the approximate volume of air (in litres) in your science classroom." First, both students calculate the approximate volume of the room in cubic metres.

$$\begin{aligned} \text{volume} &= \text{length} \times \text{width} \times \text{height} \\ V &= lwh \\ &= 10 \text{ m} \times 8.0 \text{ m} \times 3.0 \text{ m} \\ &= 240 \text{ m}^3 \end{aligned}$$

Then they recall that there are 1000 L in one cubic metre, so they move the decimal three places and write the final answer as $V = 0.24$ L. Student A rushes on to the next question. Student B, however, looks at the answer, recalls the approximate size of a litre of milk, and realizes that a volume of 0.24 L makes no sense. Then Student B decides to try again and writes

$$\begin{aligned} V &= 240 \text{ m}^3 \times 1000 \frac{\text{L}}{\text{m}^3} \\ &= 2.4 \times 10^5 \text{ L}. \end{aligned}$$

This answer makes more sense.

This example illustrates the advantage of estimating an answer to a numerical problem. It also demonstrates the advantage of including units with the calculations of physical quantities.

Throughout this text you will be asked many estimation questions,

INTRODUCING PHYSICS

to help you develop skill in estimating logically. You can then apply that skill to solving problems both in this course and in your everyday life.

It takes much practice to become good at estimating quantities. When you answer estimation questions, remember to show your reasoning, calculations, and analysis of units.

SAMPLE PROBLEM 15

Estimate the number of breaths you take in one day. This type of estimation problem is called a **Fermi question** after the famous physicist Enrico Fermi (1901–1954). As a professor, Fermi sometimes asked his students to estimate quantities impossible to measure directly. Of course he expected, as your teacher does, an educated, scientific guess with an explanation.

Solution

Assume that you have performed a simple test and discovered that you breathe 12 times per minute. Therefore, in one day you breathe

$$12 \frac{\text{breaths}}{\text{min}} \times 60 \frac{\text{min}}{\text{h}} \times 24 \frac{\text{h}}{\text{d}}$$

$= 1.728 \times 10^4$ breaths/d or about 2×10^4 breaths/d.

A more detailed calculation may take into consideration breathing rates during sleeping, running, and so on.

PRACTICE

33. Determine a logical value for each of the quantities listed below. In each case, state your reasoning.
    (a) Estimate the number of times your eyes have blinked in the past year.
    (b) Estimate the number of people in the world who are awake at this moment.
    (c) Estimate the number of English words you know. (If possible, compare this value to the number of words stored on a commercial computer program that checks spelling.)
34. (a) Without doing any calculations, guess how many words there are in this book.
    (b) Now calculate an estimate of the number of words. Compare your answer to your guess in (a).
35. (a) Estimate the mass of beef (in kilograms) eaten *per annum* by the average Canadian.
    (b) State an estimate of the mass of edible beef available from one cow.
    (c) Calculate the number of cattle killed for food each year in Canada.
    (d) Do research to find information for comparison with your calculations in (a), (b), and (c).

Review Assignment

1. Match each of the following phrases to the appropriate description below: qualitative description; quantitative description; base unit; derived unit.
   (a) a distance of 4.5 m
   (b) a smooth texture
   (c) a temperature of 280 K
   (d) a power of 250 W
2. Change each set of measurements to consistent units, then add or subtract.
   (a) $(2.4 \times 10^5 \text{ m}) + (3.8 \times 10^7 \text{ cm})$
   (b) $(8.7 \times 10^7 \text{ kg}) + (4.1 \times 10^{10} \text{ g}) - (3.9 \times 10^{13} \text{ mg})$
   (c) $(7.6438 \times 10^4 \text{ mm}) + (2.2 \times 10^3 \text{ cm})$
3. Convert each measurement, stating your answer in scientific notation:
   (a) $3.7 \times 10^7 \text{ km} = ? \text{ cm}$
   (b) $4.1 \times 10^{-11} \text{ m} = ? \text{ mm}$
   (c) $6.75 \times 10^3 \text{ mW} = ? \text{ kW}$
   (d) $5.6 \times 10^8 \text{ m/s} = ? \text{ km/s}$
4. A student measures the dimensions of a 5.0×10^2 g block to be 12.3 cm by 6.2 cm by 5.9 cm.
   (a) Calculate the density of the block in grams per cubic centimetre.
   (b) Express your answer in the preferred SI unit.
   (c) If the true density of the block is 1.2 g/cm³, what is the percent error of your answer in (a)?
5. Solve each equation for t:
   (a) $P = \dfrac{E}{t}$ (b) $v_f = v_i + at$ (c) $d = \dfrac{at^2}{2}$
6. Calculate the number of seconds in the month of July. In your calculations, show how the proper cancellation of units results in "seconds".
7. Several students set up an experiment to obtain data about a cart moving at a constant speed in a straight line. The students measure the distance travelled from the starting position after each time interval of 0.10 s. The data obtained are shown below:

time (s)	0.0	0.10	0.20	0.30	0.40
distance (m)	0.0	2.1	4.2	6.1	8.3

   (a) Which variable is dependent? independent?
   (b) Plot a graph of the data and determine the slope of the line of best fit on the graph.
   (c) What does the slope represent? (*Hint*: Consider the units of the slope.)
   (d) Give an example of extrapolation and interpolation, using the graph.
8. A student follows a logical procedure to determine an answer to the problem, "Estimate the number of grains of sand in one cubic

metre." First she counts out 10 average grains and finds that their combined mass is 20 mg. Then she adds 100 cm³ of sand to a 40 g graduated cylinder and finds that the mass of the sand plus cylinder is 290 g.
   (a) Calculate the mass of an average grain of sand.
   (b) How many such grains are needed to fill each of the following volumes?
      (i)   100 cm³
      (ii) 1.0 dm³
      (iii) 1.0 m³
   (c) List sources of error or uncertainty in this student's experiment.
9. Estimate the number of times a tire on an average-sized car rotates during a 100 km trip.
10. Estimate the number of raindrops per minute that strike the roof of your school during a heavy rainfall.

Supplementary Topic: Relationships between Sets of Variables

Section 2.7 dealt with relationships expressed in the form $y \propto x$. In general, several relationships between variables can be expressed in the form $y \propto x^n$, where n can be any positive or negative number. (When $n = 1$, $y \propto x^1$ or $y \propto x$.) For any positive value of n, the relationship is a **direct variation**, which you encountered earlier. Here, an increase in one variable results in an increase in the other variable. For any negative value of n, however, the relationship is called an **inverse variation**. In an inverse variation, an *increase* in one variable results in a *decrease* in the other variable.

Consider, for example, a situation in which measurements are taken to find how long a cyclist takes to travel once around a 1.0 km track at various speeds. In each trial, the cyclist tries to maintain a constant speed. The data obtained are shown below:

speed (m/s)	6.0	9.0	12	15
time (s)	167	111	83	67

It is obvious from the data that, as the independent variable (speed) increases, the dependent variable (time) decreases. Using v as the symbol for speed (from the word "velocity") and t as the symbol for time, this variation can be written $t \propto v^{-1}$ or $t \propto 1/v$, which is read "the time is inversely proportional to the speed". A graph of the data yields a curved line, as shown in Figure 2-11.

In a manner similar to that in Section 2.7, let us replace the "$\propto$" symbol with an equals sign and a constant. Thus, $t = k/v$, an equation that implies that the product vt is a constant. For example,

when $v = 6.0$ m/s, $vt = 6.0$ m/s $\times$ 167 s $= 1002$ m, and

when $v = 9.0$ m/s, $vt = 9.0$ m/s $\times$ 111 s $= 999$ m.

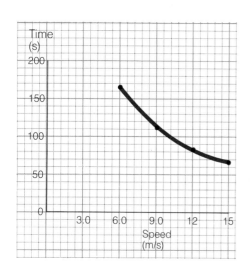

Figure 2-11
An Inverse Variation

THE LANGUAGE AND SKILLS OF PHYSICS

Of course, these results are within the possible errors of an experiment. Can you see that the constant in the equation $t = k/v$ is equal to the distance of 1.0×10^3 m or 1.0 km around the cycle track in the experiment? Thus, since vt = constant, the variation is inverse.

Now recall the fact that if $y \propto x$, a graph of y as a function of x yields a straight line. Similarly, if $y \propto 1/x$, a graph of y as a function of $1/x$ yields a straight line. It is left as an exercise (Practice Question 36) to prove that a graph of t as a function of $1/v$ produces a straight line.

As an example of a direct variation with $n = 2$, consider the relationship between the surface area (A) and radius (r) of a circle. The table below shows data for these variables, and Figure 2-12 shows a graph of the area plotted as a function of the radius. In Practice Question 37, you will verify that $A \propto r^2$, which is read "the surface area is proportional to the radius squared".

radius (cm)	0.0	2.0	4.0	6.0
area (cm²)	0.0	12.6	50	113

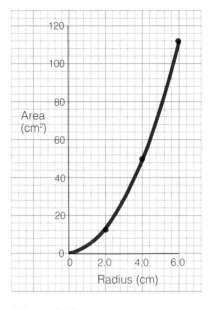

Figure 2-12
Graph of the Area of a Circle

Other relationships exist between sets of variables studied in physics. However, the two general types studied in this chapter — direct and inverse variations — are the types most often encountered in this text.

PRACTICE

36. Refer to the speed-time data in the table for the first example in this section.
    (a) Copy the table into your notebook and add a third row called 1/speed (s/m). Calculate the values to complete the table, expressing your answer in decimals. For instance, when the speed $v = 6.0$ m/s, $1/v = 0.17$ s/m.
    (b) Plot a graph of time as a function of 1/speed and describe the result.
    (c) Calculate the slope of the line of best fit on the graph in (b). What does that slope represent?

37. Refer to the radius-area data in the table for the last example in this section.
    (a) Copy the table into your notebook and add a row called radius² (cm²). Calculate values to complete the table.
    (b) Plot a graph of area as a function of radius² and describe the result.
    (c) Calculate the slope of the line of best fit on the graph in (b). What does that slope represent?

38. The table below gives data for the length of the side of a cube and the volume of a cube.
    (a) Predict the relationship between the volume and the length.
    (b) Plot a graph to obtain a straight line, thereby verifying your prediction in (a).

length (m)	5.0	7.0	9.0
volume (m³)	125	343	729

INTRODUCING PHYSICS

Key Objectives

Having completed this chapter, you should now be able to do the following:
1. Distinguish between qualitative and quantitative descriptions.
2. Name the three most common fundamental quantities of measurement.
3. Recognize the advantages of the metric system.
4. Distinguish between base and derived units of the metric system.
5. Express measurements using scientific notation.
6. Add, subtract, multiply, and divide measurements expressed in scientific notation.
7. Convert one metric measurement to another.
8. Distinguish between random and systematic errors in measuring.
9. Eliminate parallax as a source of error in measuring.
10. Write measured quantities using the correct number of significant digits.
11. Express calculated quantities using the correct number of significant digits.
12. Calculate the percent error of a measurement as compared to an accepted value.
13. Round off calculated answers to an appropriate number of significant digits.
14. Solve simple equations in three or four unknowns for any one of the unknowns, given the others.
15. Recognize which variable is independent and which is dependent in an experiment.
16. Plot a graph of data obtained in an experiment involving two variables.
17. Calculate the slope of a straight line on a graph and recognize what that slope represents.
18. Define and use interpolation and extrapolation in graphing.
19. Recognize a direct variation between two variables which results in a straight line on a graph.
20. Recognize the importance of developing skill in estimating quantities by logical guessing as well as calculating answers.

Supplementary

21. Distinguish between direct and inverse variations and replot graphs of curved lines to obtain straight lines.

UNIT II
Mechanics

The information in this Unit will be especially useful if you plan a career in architecture, urban planning, sports, astronomy, atmospheric physics, geology, geophysics, space physics, environmental studies, natural resources, energy conservation, transportation, aviation traffic control, medical engineering, clinical engineering, bioengineering, surgery, kinesiology, mechanics, machine design, construction, ballistics, aeronautical engineering, civil engineering, industrial engineering, mechanical engineering, mining engineering, or physics education.

MECHANICS

CHAPTER 3
Speed and Velocity

The study of motion in physics leads to important applications such as transportation by aircraft.

The expression "our world is getting smaller" does not, of course, mean that the earth is shrinking. Rather, it means we are able to travel from one place in the world to another more quickly than ever. The pilot of the jet aircraft shown in the photograph is able to tell the passengers the time of arrival at their destination. What makes it possible for the pilot to do so are calculations involving the quantities of displacement, velocity, and time. How do these quantities affect your daily life?

3.1 Uniform and Non-Uniform Motion

Everything in our universe is in a state of motion. Our solar system moves through space in the Milky Way Galaxy. The earth orbits the sun while spinning on its own axis. People, animals, air, and countless other objects move about on the earth's surface. The minute invisible particles that make up all matter, too, are constantly in motion.

These motions, and many others, may seem too complex and too difficult to describe in simple terms with only a few equations. But physicists have developed basic equations and descriptions of simple motion which can be applied to complex motion. It is the understanding of these equations and descriptions that makes up the study of motion.

Scientists call the study of motion **kinematics**, which stems from the Greek word for motion, *kinema*. (A "cinema" is a place in which we watch motion pictures.) Motion may be classified as being either of two types—non-uniform and uniform. **Non-uniform motion** is movement involving changes in speed or direction or both. A roller coaster is an obvious example of such motion—it speeds up, slows down, rises, falls, and travels around corners. Some aspects of non-uniform motion are presented in the next chapter.

Uniform motion is movement in a straight line at a constant speed. A cyclist travelling along a straight path at a speed of 10 m/s is an example of such motion. Because uniform motion is the easiest type of motion to analyse, it will constitute the major portion of this chapter.

PRACTICE
1. Describe each motion listed below as either uniform or non-uniform. If you state that the motion is non-uniform, explain why.
   (a) A steel ball is dropped from your raised hand to the floor.
   (b) A car is travelling a steady 80 km/h due west.
   (c) A communication satellite aboard a rocket is launched into orbit around the earth.
   (d) A motorcycle rider applies the brakes to come to a stop.
   (e) A ball on the end of a rope is whirled around a person's head at a constant speed of 8.0 m/s.

3.2 Scalar and Vector Quantities

Figure 3-1 illustrates the safe way for a pedestrian to cross a busy highway. From the base of one side of the overpass, let us assume that the pedestrian walks 7.0 m up the stairs, then 20 m across the top, and finally 7.0 m down the other side. The total distance travelled is 34 m. But the pedestrian is *displaced* only 30 m straight across from the original position. This example illustrates an important difference between distance and displacement.

MECHANICS

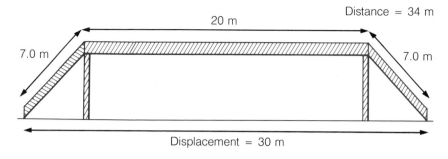

Figure 3-1
The Difference between Distance Moved and Displacement

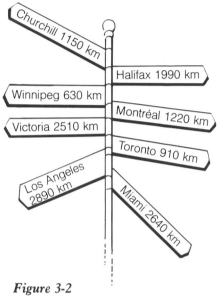

Figure 3-2
Does this signpost represent scalar or vector quantities?

Distance (d) is an example of a **scalar quantity**, a quantity which has magnitude (or size), but no direction. The magnitude is made up of a number and often an appropriate unit; in our example, $d = 34$ m. Other examples of scalar quantities are a mass of 2.0 kg, a time of 7.5 s, and a grade of a mountain highway of 0.11 or 11%.

Displacement ($\vec{d}$) is an example of a **vector quantity**, a quantity with both magnitude and direction. Vector quantities are represented in this book by symbols with arrows over them; their direction is indicated in square brackets. **Displacement** is defined as the distance, in a given direction, that an object is located from a defined position. In our example above, $\vec{d} = 30$ m [across] from the original position. Other vector quantities will be studied later in the book. See Figure 3.2.

Often a vector quantity is represented in a scale diagram by a line in a specific direction, as you will see in Sample Problem 1. The symbols for the directions east, west, north, and south are [E], [W], [N], [S]. Other directions are indicated with angles, as shown in Figure 3-3.

Figure 3-3
Determining the Direction of a Vector Quantity

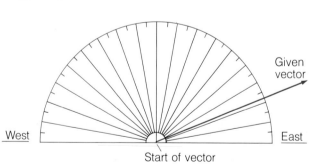

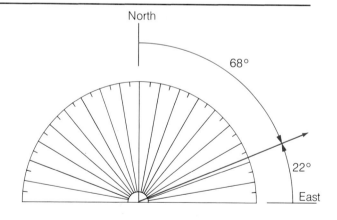

(a) To find the direction of a given vector (shown in colour), place the base of the protractor along the east-west (or north-south) direction and place the origin of the protractor at the starting position of the vector.

(b) Measure the angle between one reference direction (east, west, north, or south) and the one at right angles to it, and write the direction using that angle. In this case, the direction can be written either E 22° N or N 68° E.

SAMPLE PROBLEM 1

A chair, starting at position A, is moved 4.0 m [E], then 3.0 m [S], as in the scale diagram. Determine the following for the chair:
(a) the total distance moved
(b) the displacement from its original position, A

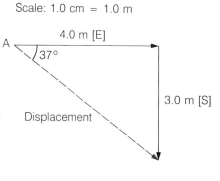

Solution
(a) The total distance travelled is the sum of 4.0 m and 3.0 m. Thus, $d = 7.0$ m.
(b) The displacement is represented by the coloured hypotenuse of the triangle in the diagram. The hypotenuse is 5.0 cm long, so the displacement is 5.0 m in a direction 37° south of east.
Thus, $\vec{d} = 5.0$ m [E37°S].

Notice in Sample Problem 1 that the displacement vector faces from the original position toward the final position. Notice also that the direction [E37°S] is equivalent to [S53°E]. Every direction can be labelled in at least two possible ways.

PRACTICE

2. State which measurements are scalar and which are vector quantities:
   (a) 12.3 s
   (b) 100 km/h [N]
   (c) 3.8 cm [E]
   (d) 87 mg
   (e) 1.8×10^3 km [W]
   (f) 3.2 m [up]

3. Does a car's odometer indicate a scalar quantity, a vector quantity, or both?

4. (a) Can the displacement of an object from its original position ever exceed the total distance moved? Explain.
   (b) Can the total distance moved ever exceed an object's displacement from its original position? Explain.

5. A jogger runs 500 m [N], then turns and runs 220 m [S] before stopping for a short rest. What is the jogger's
   (a) total distance travelled?
   (b) displacement from his original position?

6. A boy starts at position A and walks to B, then C, then D, and back to A, as shown in the scale diagram in Figure 3-4.
   Scale 1.0 cm = 2.0 m
   (a) State the distance travelled when the boy reaches B, C, D, and then A.
   (b) Determine his displacement from his original position when the boy reaches B, C, D, and then A.

7. A girl runs twice around a circular track 100 m in diameter. At the end of the run, what is her
   (a) total distance travelled?
   (b) displacement from her original position?

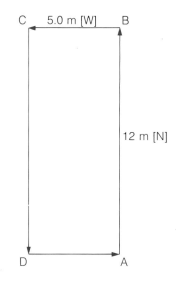

Figure 3-4

3.3 Speed and Velocity

Of course, you are familiar with road signs that read "speed limit 100 km/h". **Speed** (v), a scalar quantity, is defined as the rate of change of distance. (The symbol for speed, v, is taken from the word "velocity".) In this chapter we are concerned with the average speed, v_{ave}, which can be calculated using the equation

$$v_{ave} = \frac{d}{t}$$

where d is the total distance travelled in a total time, t.

For example, if a car travels a total distance of 200 km in 4.0 h, its average speed is 200 km/4.0 h or 50 km/h.

Typical units of speed are metres per second (m/s) and kilometres per hour (km/h). Sample Problems 2 and 3 illustrate how to change from one of these units to the other.

SAMPLE PROBLEM 2
Express 60 km/h in metres per second.

Solution
$$60 \frac{km}{h} = \frac{60\ 000\ m}{3600\ s}$$
$$= \frac{60\ m}{3.6\ s}$$
$$= 17\ m/s \text{ (rounded off)}$$
∴ 60 km/h is equivalent to 17 m/s.

Thus, to change kilometres per hour to metres per second, we divide by 3.6. Conversely, to change metres per second to kilometres per hour, we multiply by 3.6.

SAMPLE PROBLEM 3
Express 30 m/s in kilometres per hour.

Solution
We simply multiply 30 by 3.6 and get 108 km/h, or approximately 110 km/h.

Most people consider that speed, a scalar quantity, and velocity are the same thing, but physicists make an important distinction. **Velocity**, $\vec{v}$, a vector quantity, is the rate of change of displacement. The equation for average velocity is similar to the equation for average speed.

$$\vec{v}_{ave} = \frac{\vec{d}}{t}$$

where $\vec{d}$ is the displacement at time t.

SAMPLE PROBLEM 4

A girl cycles 300 m [E], then turns around and cycles 100 m [W], as illustrated in the diagram. If the trip took 80 s, find her (a) average speed and (b) average velocity.

Solution

(a) $v_{ave} = \dfrac{d}{t}$ (b) $\vec{v}_{ave} = \dfrac{\vec{d}}{t}$

$\phantom{(a) v_{ave}} = \dfrac{400 \text{ m}}{80 \text{ s}}$ $\phantom{(b) \vec{v}_{ave}} = \dfrac{200 \text{ m [E]}}{80 \text{ s}}$

$\phantom{(a) v_{ave}} = 5.0 \text{ m/s}$ $\phantom{(b) \vec{v}_{ave}} = 2.5 \text{ m/s [E]}$

∴ Her average speed is 5.0 m/s, and her average velocity is 2.5 m/s [E].

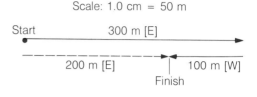

Notice in Sample Problem 4 that the girl could have reached her final destination by either cycling slowly (2.5 m/s) in one direction or by cycling quickly, as described in the problem.

Often the speed or velocity of an object is given, and a different variable (distance, displacement, or time) is unknown. It is left as an exercise to write equations for distance and time in terms of speed, and for displacement and time in terms of velocity.

PRACTICE

8. Does a car's speedometer indicate speed, velocity, or both?
9. For objects moving with uniform motion, compare the average speed with the average velocity.
10. Convert these speeds to metres per second:
    (a) 40 km/h   (b) 250 cm/s   (c) 100 km/h
11. Convert these speeds to kilometres per hour:
    (a) 640 m/h   (b) 10 m/s   (c) 24 m/s
12. At a target-shooting range, a bullet leaves a rifle and in 0.45 s strikes a target 270 m away. Assuming that the bullet's path is straight, what is the bullet's average speed?
13. Electrons in a television tube travel 40 cm from their source to the screen in 4.0×10^{-4} s. Calculate the average speed of the electrons in metres per second.
14. A delivery truck driver travels 140 km straight north, makes a delivery, travels 30 km due south, then stops at a restaurant. If the trip takes 3.5 h, calculate the driver's (a) average speed and (b) average velocity.
15. A canoeist paddles 450 m [W] in a calm lake to get beyond an island, then turns 90° and paddles 360 m [N]. The trip takes 30 min.
    (a) Find the canoeist's average speed in metres per minute.
    (b) Draw a scale diagram of the motion and find the canoeist's average velocity in metres per minute.

MECHANICS

16. Write an equation for each of the following:
    (a) distance in terms of average speed and time
    (b) time in terms of average speed and distance
    (c) displacement in terms of average velocity and time
    (d) time in terms of average velocity and displacement
17. Copy this table into your notebook and calculate the unknown values.

Distance (m)	Time (s)	Average Speed (m/s)
3.8×10^5	9.5×10^{-3}	?
?	2.5	480
1800	?	24

18. In the human body, blood travels faster in the largest blood vessel, the aorta, than in any other blood vessel. Given an average speed of 30 cm/s, how far does blood travel in the aorta in 0.20 s?
19. A supersonic jet travels once around the earth at an average speed of 1.6×10^3 km/h. Its orbital radius is 6.5×10^3 km. How many hours does the trip take?
20. Estimate, in days, how long it would take you to walk non-stop at your average walking speed from one mainland coast of Canada to the other. Show your reasoning.

3.4 Graphing Uniform Motion

In experiments involving motion, the variables measured directly are usually displacement and time. The third variable, velocity, is often found by calculation.

In uniform motion, the velocity is constant, so the change in displacement is the same during equal time intervals. For instance, assume that a boy runs 3.0 m straight west each second for 8.0 s. His velocity is steady at 3.0 m/s [W]. We can make a table describing his motion, starting at 0.0 s.

time (s)	0.0	2.0	4.0	6.0	8.0
displacement (m [W])	0.0	6.0	12	18	24

Figure 3-5 shows a graph of this motion, with displacement from the original position plotted as the dependent variable. Notice that, for uniform motion, a displacement-time graph yields a straight line. This represents a direct variation, as we discussed in Section 2.7. Refer to Sample Problem 5.

We can now use the slope calculation in Sample Problem 5 to plot a velocity-time graph of the motion. Because the slope of the line is constant, the velocity is constant from $t = 0.0$ s to $t = 8.0$ s. Figure 3-6 gives the resulting velocity-time graph.

A velocity-time graph can be used to find the change in displacement during various time intervals. This is accomplished by finding the area under the line on the velocity-time graph (Sample Problem 6).

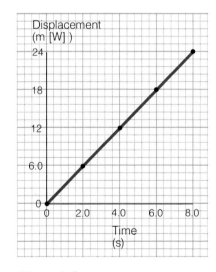

Figure 3-5
A Graph of Uniform Motion

SAMPLE PROBLEM 5
Calculate the slope of the line on the graph in Figure 3-5 and state what that slope represents.

Solution

$$\text{slope} = m = \frac{\Delta y}{\Delta x}$$

$$= \frac{24 \text{ m [W]} - 0.0 \text{ m [W]}}{8.0 \text{ s} - 0.0 \text{ s}}$$

$$= 3.0 \text{ m/s [W]}$$

Thus, the slope of the line is 3.0 m/s [W]. Judging from the slope's units, we know that the slope represents the boy's average velocity. In other words, $\vec{v}_{\text{ave}} = \Delta \vec{d}/\Delta t$.

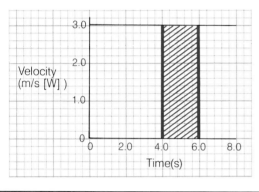

Figure 3-6
A Velocity-Time Graph of Uniform Motion (The shaded region is for Sample Problem 6.)

SAMPLE PROBLEM 6
Find the area of the shaded region on the graph in Figure 3-6. State what that area represents.

Solution
For a rectangular shape,

$$A = lw = \vec{v}(\Delta t)$$

$$= 3.0 \frac{\text{m}}{\text{s}} \text{ [W]} \times 2.0 \text{ s}$$

$$= 6.0 \text{ m [W]}$$

Thus, the area of the shaded region is 6.0 m [W]. This quantity represents the change in the boy's displacement from $t = 4.0$ s to $t = 6.0$ s. In other words, $\Delta \vec{d} = \vec{v}\Delta t$.

Drawing graphs to describe motion has definite advantages. Even a glance at such a graph yields general information about the motion. Also, calculations such as slope and area give specific details. These facts are shown in Sample Problems 7, 8, and 9 that follow.

MECHANICS

SAMPLE PROBLEM 7
Calculate the slopes of lines A and B on the graph shown.

Solution

$m_A = \dfrac{\Delta y}{\Delta x} = \dfrac{\Delta \vec{d}}{\Delta t}$

$= \dfrac{15 \text{ m [E]} - 20 \text{ m [E]}}{2.0 \text{ s} - 0.0 \text{ s}}$

$= -\dfrac{5 \text{ m [E]}}{2.0 \text{ s}}$

$= -2.5 \text{ m/s [E]}$

∴ The slope of line A is -2.5 m/s [E], which is equal to 2.5 m/s [W].

$m_B = \dfrac{\Delta y}{\Delta x} = \dfrac{\Delta \vec{d}}{\Delta t}$

$= \dfrac{15 \text{ m [E]} - 5.0 \text{ m [E]}}{6.0 \text{ s} - 2.0 \text{ s}}$

$= \dfrac{10 \text{ m [E]}}{4.0 \text{ s}}$

$= 2.5 \text{ m/s [E]}$

∴ The slope of line B is 2.5 m/s [E].

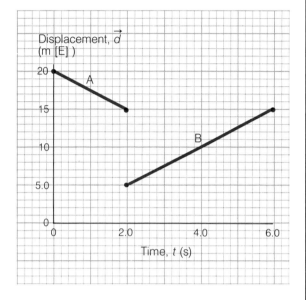

SAMPLE PROBLEM 8
Describe the motion shown in the displacement-time graph.

Solution
The slope of each line segment indicates the velocity.
From 0 to 2 s, the velocity is uniform at a low value.
From 2 s to 3 s, the velocity is uniform at a higher value.
From 3 s to 4 s, the velocity is zero.

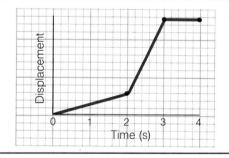

SAMPLE PROBLEM 9
Use the velocity-time graph to find the displacement from the original position when $t = 10$ s.

Solution
The area under the line indicates the displacement.
$A = lw$
$= 5.0 \dfrac{\text{cm}}{\text{s}} \text{ [S]} \times 10 \text{ s}$
$= 50 \text{ cm [S]}$
Thus, the displacement is 50 cm [S] at 10 s.

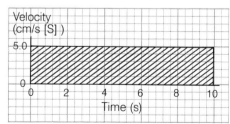

SPEED AND VELOCITY

PRACTICE

21. Refer to the graph in Figure 3-5. Prove that the slope of the line from $t = 4.0$ s to $t = 6.0$ s is the same as the slope of the entire line found in Sample Problem 5.
22. A plane is flying with uniform motion at 60 m/s [S].
    (a) Construct a table showing the plane's displacement from the starting point at the end of each second for a 12 s period.
    (b) Use the data from the table to plot a displacement-time graph.
    (c) Find the slope of the line on the displacement-time graph between the two different sets of points. Is the slope constant? What does it represent?
    (d) Plot a velocity-time graph of the plane's motion.
    (e) Calculate the total area under the line on the velocity-time graph. What does that area represent?
23. By reading the information directly on the displacement-time graph in Figure 3-7, state the object's
    (a) displacement from the original position at $t = 2.0$ s
    (b) displacement from the original position at $t = 4.0$ s
    (c) change in displacement from $t = 2.0$ s to $t = 6.0$ s
    (d) change in displacement from $t = 6.0$ s to $t = 12$ s
    (e) final displacement at $t = 12$ s.

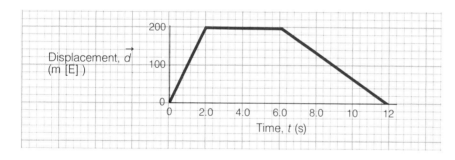

Figure 3-7

24. Use Figure 3-7 to find the object's velocity from
    (a) $t = 0$ s to $t = 2.0$ s
    (b) $t = 2.0$ s to $t = 6.0$ s
    (c) $t = 6.0$ s to $t = 12$ s
25. Plot a velocity-time graph for the motion illustrated in Figure 3-7. (*Hint:* From $t = 6.0$ s to $t = 12$ s, the velocity is negative, so the line is drawn below the horizontal axis.)
26. Find the total area between the lines and the time axis on the graph you drew for Question 25. What does that area represent?

3.5 Measuring Time

Time is another important quantity in the study of motion. You will be required to measure time in various experiments involving motion. However, your technique will be much more advanced than those used by early experimenters such as Galileo Galilei. Galileo tried to use his

MECHANICS

own pulse as a time-measuring device. Later, he discovered that a swinging pendulum had a predictable time period for each vibration. He used this fact to design the first pendulum clock, with which he made measurements of time.

In physics laboratories of today, various methods are used to measure time. A simple **stopwatch** gives acceptable values of time intervals whose duration is over about 2 s. However, if accurate results are required, especially for short intervals, a more elaborate method must be used.

Most physics laboratories have instruments used for demonstration purposes that measure time accurately. A **digital timer** is an electronic device which measures time intervals to a fraction of a second. An **electronic stroboscope** has a light, controlled by adjusting a dial, that flashes on and off at regular intervals. The stroboscope illuminates a moving object in a dark room, while a camera records the object's motion on film. The motion is analysed using the known time between flashes of the strobe (Figure 3-8). A **computer** with appropriate add-on hardware can also be used to measure time intervals.

Another device, called a **recording timer** or **bell timer**, is excellent for student experimentation. Such a timer, shown in Figure 3-9, has a metal arm that vibrates at constant intervals. A needle on the arm strikes carbon paper and records dots on paper tape pulled through the timer. The dots give a record of the motion. The faster the motion, the greater the space between the dots.

Figure 3-8
This photograph of a golf swing was taken with a stroboscopic light. At which part of the swing is the club moving fastest?

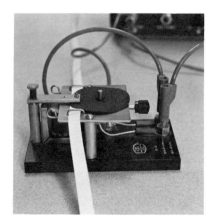

Figure 3-9
A Recording Timer

SPEED AND VELOCITY

Some recording timers make 60 dots each second. We say they have a frequency of 60 Hz or 60 vibrations/s. (The unit hertz, symbol Hz, is named after the German physicist Heinrich Hertz, 1857–1894.) Figure 3-10 illustrates why an interval of six spaces on a recording tape represents a time of 0.10 s.

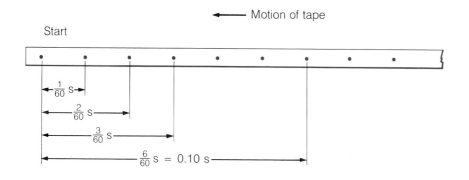

Figure 3-10
Measuring Time with a Recording Timer

SAMPLE PROBLEM 10
A student pulls a tape through a 60 Hz recording timer and obtains the tape shown. Plot a graph of displacement (from the starting position) versus time.

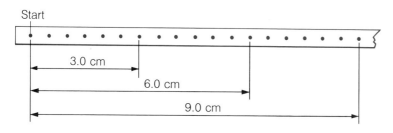

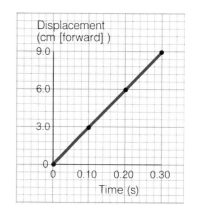

Solution
The experimental data are:

time (s)	0	0.10	0.20	0.30
displacement (cm [forward])	0	3.0	6.0	9.0

The required graph is shown at right.

PRACTICE
27. **Activity** This activity will introduce you to the use of recording timers. You will need a stopwatch as well as the recording timer and its related accessories.

    Obtain a paper tape about 60 cm to 80 cm long and position it in the recording timer. Turn on the timer and use the stopwatch to find how long it takes to pull the tape through at a fairly slow speed. Calibrate the timer by dividing the number of dots by the number of seconds. Your teacher will let you know whether your results are acceptable.

28. (a) The motion of a tape pulled through a 60 Hz recording timer is shown in Figure 3-11. Use the results to plot a graph of displacement from the original position versus time.
    (b) Repeat (a), assuming that the recording timer has a frequency of 30 Hz.

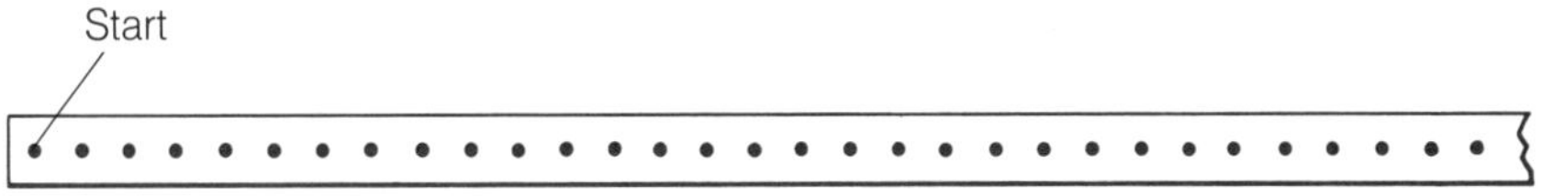

Figure 3-11

Experiment 3A: Uniform Motion at Varying Speeds

INTRODUCTION
Motion experiments can be performed using many different techniques. In this experiment, three different techniques are described, one in each of Procedures I, II, and III. Your choice of procedure will depend on the equipment for measuring time available in your laboratory.

For a graph of displacement *vs.* time for uniform motion, the data should form a straight line. However, because of experimental limitations, you might not obtain such a line. It is acceptable to draw a line that best fits the data.

PURPOSE
To draw and analyse graphs of attempted uniform motion at various speeds.

APPARATUS
General: metre stick; graph paper
- For Procedure I: several stopwatches; bicycle (if desired)
- For Procedure II: recording timer and related apparatus
- For Procedure III: stroboscope; instant camera; linear air track and accessories; pin; overhead projector

PROCEDURE I: STOPWATCH TECHNIQUE
1. Choose an appropriate place to perform the experiment — in the classroom, in the halls, or outside on the track. Mark a starting position, then mark positions at equal intervals from the starting position. For example:
   In the classroom use 0, 1.0, 2.0, 3.0, . . . m
   In the halls use 0, 3.0, 6.0, 9.0, . . . m
   On the track use 0, 10, 20, 30, . . . m.
2. Locate students with stopwatches at the positions named in Procedure Step 1.
3. Choose a member of the group to try to exhibit uniform motion. That student will either walk, run, or ride a bicycle, depending on the location chosen.

4. Start all the stopwatches at the instant that the student, already moving at a relatively low but constant speed, crosses the starting position. Stop the watches, one at a time, at the instant the moving student passes each timer.
5. Repeat Steps 3 and 4 using a medium speed, then a relatively high speed.
6. Set up a table showing the displacement from the starting position and the time for each of the three trials.
7. Plot a displacement-time graph with all three motions on one graph. For each trial, draw the line of best fit.

ANALYSIS
1. Calculate and compare the slopes of the three lines on the displacement-time graph.
2. On one velocity-time graph, plot the average velocities of the three motions. Be sure each line ends at the appropriate time. Explain how this graph relates to the experiment.
3. Calculate the area under each line on the velocity-time graph. What do the areas represent?
4. Describe sources of error in this procedure.

CONCLUSIONS . . .

PROCEDURE II: RECORDING TIMER TECHNIQUE
1. Check whether the recording timer is working properly by pulling a short piece of paper tape through it. If the timer is skipping, double-dotting, or making unclear dots, inform your teacher.
2. Obtain a paper tape about 50 cm long and pull it through the timer at a slow but constant speed.
3. Find a point, A, near the beginning of the motion where the spaces become fairly uniform. From A count 6 spaces to B. Measure AB, which is the displacement [forward], to the end of the first time interval.
4. From B count another 6 spaces to C. Measure AC to obtain the displacement from the original position to the end of the second time interval. Repeat this procedure to the end of the tenth time interval, always measuring the displacement from the original position.
5. Repeat the above steps, using a tape about 80 cm long at a medium speed, then a tape about 100 cm long at a fast speed. For these trials it is acceptable to have fewer than ten intervals.
6. Set up a table showing the displacement from the starting position and the time in seconds for each of the three trials.
7. Plot a displacement-time graph with all three motions on one graph. For each trial, draw the line of best fit.

ANALYSIS
Answer Analysis Questions 1 to 4 in Procedure I.

CONCLUSIONS . . .

MECHANICS

PROCEDURE III: PHOTOGRAPHIC TECHNIQUE

1. Set up and level the air track and hold a metre stick for reference just behind it. (It is best to use a reflecting type of metre stick, marked off in 5 cm or 10 cm segments.) Place the camera about 1.5 m in front of the track and at the same height. Focus the camera on the plastic straw attached to the cart. Position the stroboscope so its light will reflect from the straw to the camera. Adjust the stroboscope to a time interval of 0.20 s between flashes. (This is a frequency of 5.0 Hz.)
2. With the room lights out, set the cart in motion at a slow but constant speed. Cover the light-exposure meter of the camera. Then open the shutter and leave it open for the duration of the motion. Develop the photograph.
3. Poke two small holes along the metre stick in the photograph so that a reference distance can be projected onto the blackboard. (See Figure 3.12 for an example, here, 50 cm.) Poke holes in the photograph wherever the straw appears. With the result projected onto the blackboard, measure the displacement [forward] from the original position to each point.

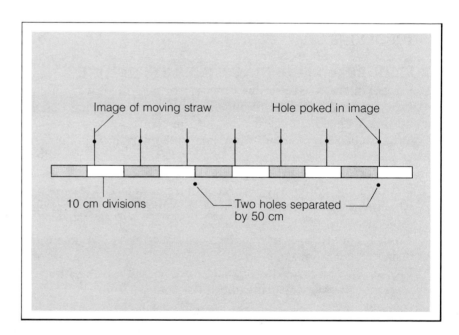

Figure 3-12
Preparing the Air Track Photograph for Measurements

4. Repeat Steps 2 and 3 using a medium and then a fast speed.
5. Set up a table showing the displacement from the original position and the time for each of the three trials.
6. Plot a displacement-time graph with all three motions on one graph. For each trial, draw the line of best fit.

ANALYSIS
Answer Analysis Questions 1 to 4 in Procedure I.

CONCLUSIONS . . .

SPEED AND VELOCITY

Review Assignment

1. Explain the differences between uniform and non-uniform motion. Give a specific example of each type of motion.
2. List three scalar and two vector quantities studied in this chapter.
3. The distance between bases on a baseball diamond is 27.4 m. Use a scale diagram, the law of Pythagoras, or trigonometry to find a runner's displacement from home plate to second base.
4. A fishing boat leaves port at 04:30 h in search of the day's catch. The boat travels 4.5 km [E], then 2.5 km [S], and finally 1.5 km [W] before discovering a large school of fish on the sonar screen at 06:30 h.
   (a) Draw an appropriate scale diagram of the boat's motion.
   (b) Calculate the boat's average speed.
   (c) Determine the boat's average velocity.
5. What is the average speed needed by a bionic track star in order to run 100 m in 8.0 s? Express the speed in both metres per second and kilometres per hour.
6. Figure 3-13 shows one of the first series of high-speed photographs ever produced. The vertical lines in these 1878 photographs are 68 cm apart. Calculate the horse's average speed in metres per second. The time between frames is 0.040 s.

Figure 3-13

7. Light travels from the sun to the earth, a distance of 1.5×10^8 km, in only 500 s. Calculate the average speed of light in metres per second.
8. The highest average lap speed on a closed circuit in motorcycling is about 72 m/s or 258 km/h. If a cyclist takes 56 s to complete one lap of the circular track, what is the track's circumference?
9. Laser light, which travels in a vacuum at 3.00×10^8 m/s, is used to measure the distance from the earth to the moon with great accuracy. On a clear day an experimenter sends a laser signal toward

MECHANICS

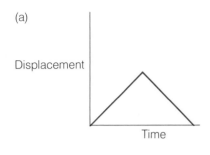

(a) Displacement vs Time

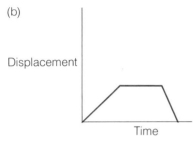

(b) Displacement vs Time

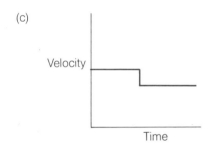

(c) Velocity vs Time

Figure 3-14

a small reflector on the moon. Then, 2.56 s after the signal is sent, the reflected signal is received back on earth. What is the distance between earth and moon at the time of the experiment?

10. The record lap speed for car racing is about 112 m/s (or 402 km/h). The record was set on a track 12.5 km in circumference. How long did the driver take to complete one lap?

11. State what is represented by each of the following calculations:
    (a) slope of a distance-time graph
    (b) slope of a displacement-time graph
    (c) area on a speed-time graph
    (d) area on a velocity-time graph

12. Describe the motion in each graph in Figure 3-14.

13. The table below shows data recorded in a motion experiment.

time (s)	0.00	0.10	0.20	0.30	0.40	0.50	0.60
displacement (cm [W])	0.00	25	50	75	75	75	0.00

   (a) Use the data to plot a displacement-time graph of the motion. Assume uniform motion between points.
   (b) Use the graph from (a) to find the velocity at times 0.10 s, 0.40 s, and 0.55 s.
   (c) Plot a velocity-time graph of the motion.
   (d) Find the area under the line on the velocity-time graph between 0.00 s and 0.50 s. What does that area represent?

14. (a) Pretend that light could somehow travel in a circle around the earth. How many times do you think it could travel around the earth in 1.00 s?
    (b) Now calculate the answer to (a), using the fact that the earth's radius is approximately 6.38×10^3 km.

15. About 140 million years ago, South America and Africa were joined together. Now, because of continental drift, they are about 4000 km apart.
    (a) Guess the average speed of separation of the continents in millimetres per year.
    (b) Calculate the average speed of separation in the same unit.

Supplementary Topic: Relative Velocity

Assume that you are the driver of a car travelling along a highway. A person by the side of the highway observes that you are in motion. However, a passenger in your car observes that, relative to himself or herself, you are not in motion. From one point of view you are moving; from another point of view you are not moving. Thus, we can conclude that the motion observed depends on the observer. In other words, *motion is relative*.

To illustrate relative motion further, consider two trains, A and B, moving toward each other on adjacent sets of tracks. Let E represent an observer standing on the earth near the tracks. If the velocity of train A relative to E is 25 m/s [S] and the velocity of train B relative to E is 45 m/s [N], then the velocity of A relative to B is 70 m/s [S]. In other words, to a passenger in train B, train A appears to be travelling south at a speed of 70 m/s. Similarly, the velocity of B relative to A is 70 m/s [N]. Figure 3-15 shows this motion.

In general, if two objects are travelling in opposite directions, their relative velocities are greater in magnitude than their individual velocities.

Now let us consider two objects moving in the same direction. Car X is travelling 100 km/h [W] on a highway and is overtaking car Y, which is travelling 85 km/h [W] on the same highway. The velocity of X relative to Y is 100 km/h [W] − 85 km/h [W] or 15 km/h [W]. The velocity of Y relative to that of X is thus equal in magnitude but opposite in direction, or 15 km/h [E]. See Figure 3-16. Therefore, if two objects are travelling in the same direction, their relative velocities are less in magnitude than their individual velocities.

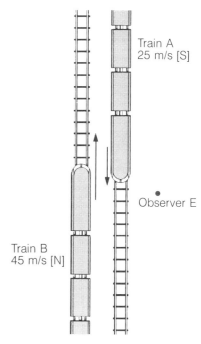

Figure 3-15
Trains A and B are moving in opposite directions. To an observer in B, train A appears to be travelling at 70 m/s [S]. To an observer in A, train B appears to be travelling at 70 m/s [N].

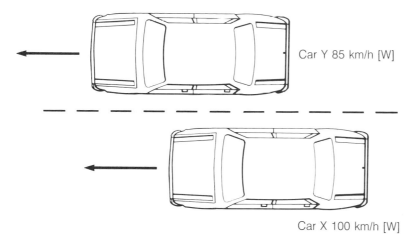

Figure 3-16
The two cars are travelling in the same direction. To an observer in Y, car X appears to be travelling at 15 km/h [W]. To an observer in X, car Y appears to be travelling at 15 km/h [E].

In motion problems, it is often necessary to take into consideration factors such as wind velocity and river currents that affect the motion, as shown in Sample Problems 11 and 12.

SAMPLE PROBLEM 11
A plane is travelling with an air velocity of 370 km/h [S] straight into a wind that has a velocity relative to the ground of 30 km/h [N]. How long will it take the plane to travel 170 km [S]?

Solution
The velocity of the plane relative to the ground is 370 km/h − 30 km/h = 340 km/h [S].

$$t = \frac{\vec{d}}{\vec{v}} = \frac{170 \text{ km [S]}}{340 \text{ km/h [S]}}$$

$$= 0.50 \text{ h}$$

∴ It will take the plane 0.50 h to travel 170 km [S].

Did You Know?
If car A, travelling at 30 m/s, is approaching car B, travelling at 40 m/s, the velocity of B relative to an observer in A is 70 m/s. If, however, car A is approaching a beam of light which is travelling at 3.0×10^8 m/s, the velocity of the beam relative to an observer in A is still 3.0×10^8 m/s. The idea that the speed of light is constant, independent of the motion of the observer, was introduced by Albert Einstein in his Theory of Relativity in 1905.

MECHANICS

SAMPLE PROBLEM 12
A boat that is capable of travelling 4.5 km/h in calm water is travelling on a river, as shown. The river's current is 1.5 km/h [E].
Find the following:
(a) the time the boat takes to travel 3.0 km [downstream] relative to the shore
(b) the time the boat takes to travel 3.0 km [upstream] to return to the starting position
(c) the average speed of motion, relative to the shore, for the entire motion in (a) and (b) (Note: This is a scalar quantity.)

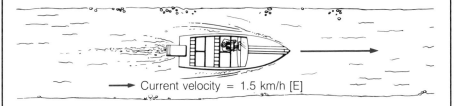
Current velocity = 1.5 km/h [E]

Solution
(a) When the boat is moving downstream (with the current), its velocity relative to the shore is 6.0 km/h [E].

$$t = \frac{\vec{d}}{\vec{v}} = \frac{3.0 \text{ km [E]}}{6.0 \text{ km/h [E]}}$$

$$= 0.50 \text{ h}$$

∴ The time taken to travel downstream is 0.50 h.

(b) When the boat is moving upstream, its velocity relative to the shore is 3.0 km/h [W].

$$t = \frac{\vec{d}}{\vec{v}} = \frac{3.0 \text{ km [W]}}{3.0 \text{ km/h [W]}}$$

$$= 1.0 \text{ h}$$

∴ The time taken to travel upstream is 1.0 h.

(c) $v_{\text{ave}} = \frac{d}{t} = \frac{6.0 \text{ km}}{1.5 \text{ h}}$

$$= 4.0 \text{ km/h}$$

∴ The boat's average speed is 4.0 km/h.

Notice in Sample Problem 12 that the average speed (4.0 km/h) cannot be calculated by finding the average of the boat's downstream and upstream speeds of 6.0 km/h and 3.0 km/h, because the boat travelled for a longer time at 3.0 km/h.

The relative velocity examples given thus far have dealt with motion in a single dimension. Of course, more complex motions occur. The next Sample Problem illustrates motion in two dimensions.

SAMPLE PROBLEM 13

A river 600 m wide has a current with an average velocity of $\vec{v}_2 = 10$ m/min [W], as shown in the diagram. A boat that is able to travel at a speed $v_1 = 40$ m/min relative to the water starts from the south shore and aims due north.
Find the following:
(a) the time needed to cross the river
(b) the boat's landing position on the north shore

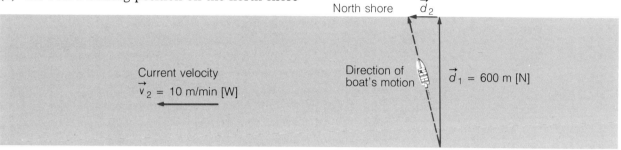

Solution

(a) $t = \dfrac{\vec{d}_1}{\vec{v}_1} = \dfrac{600 \text{ m [N]}}{40 \text{ m/min [N]}}$

   $= 15$ min

∴ The time taken to cross the river is 15 min.

(b) As the boat is aiming due north, the current is pulling it west at 10 m/min. In 15 min the boat will move

$\vec{d}_2 = \vec{v}_2 t = 10 \dfrac{\text{m}}{\text{min}} \text{ [W]} \times 15 \text{ min}$

$= 150$ m [W].

Thus, the boat will land 150 m [W], or downstream, of the original position at which it was aimed.

These examples of relative motion provide a brief exposure to an important topic with numerous applications. Navigation in airplanes, boats, and spacecraft is more complex than what is described here, but the method of analysing such motion is based on the concepts in this section.

PRACTICE

29. Is a passenger in an airplane more concerned about the airplane's air speed (speed relative to the air) or its ground speed (speed relative to the ground)? Explain.
30. A girl is jogging at 4.0 m/s [E] and passes a boy who is jogging at 4.0 m/s [W]. What is the velocity of
    (a) the boy relative to the girl?
    (b) the girl relative to the boy?
    (c) the ground relative to the girl?

MECHANICS

31. A filming crew for a nature film is in a helicopter skimming just above the snow in the Arctic at a velocity relative to the ground of 33 m/s [S]. Suddenly the crew spots a reindeer herd 2.0 km away running south at an average speed of 13 m/s relative to the ground.
    (a) What is the velocity of the helicopter relative to the reindeer herd?
    (b) What is the velocity of the herd relative to the helicopter?
    (c) How long will it take the crew to reach the spot where the reindeer were located when they were first seen?
    (d) How long after the first view of the reindeer will the helicopter take to reach the herd?
32. Why is it advantageous for airplanes to land into the wind?
33. Assume that a jet aircraft has a speed relative to the air of 720 km/h and that on a certain day the average wind velocity in Canada is 40 km/h from the west. If the flight distance between Vancouver, British Columbia and Gander, Newfoundland is 4800 km [E], how long will it take a jet to fly from
    (a) Vancouver to Gander?
    (b) Gander to Vancouver?
34. A tugboat is pulling a huge log pile downstream on a British Columbia river at a speed relative to the water of 0.80 m/s. The river is moving at an average speed of 1.4 m/s. A logger is walking on the log pile at a speed relative to the logs of 1.0 m/s. Calculate the logger's speed relative to the shore if he is walking
    (a) with the current and (b) against the current.
35. Use the data given in Sample Problem 13 to determine the velocity of the boat relative to the land. (*Hint*: This is a vector problem. One way to solve it is by vector addition, similar to the method used to find displacement in two dimensions.)
36. A small airplane is travelling at a velocity relative to the air of 100 km/h [E]. A wind from the north is blowing at 20 km/h.
    (a) Determine the position of the airplane after 2.5 h of this motion.
    (b) Find the average velocity of the airplane during the 2.5 h interval.

Key Objectives

Having completed this chapter, you should now be able to do the following:
1. Define kinematics.
2. Distinguish between uniform motion and non-uniform motion and cite examples of each.
3. Distinguish between scalar and vector quantities and give examples of each.
4. Define distance, displacement, speed, and velocity.
5. Recognize common metric units of distance, displacement, speed, and velocity.
6. Use a scale diagram of two-dimensional motion to find the final displacement of a moving object.

7. Use the equations for average speed and average velocity to solve any one variable, given the others.
8. Given data on the displacement of a moving object at various time intervals, plot a displacement-time graph of the motion.
9. Calculate the slope of a line on a distance-time graph (or a displacement-time graph) and state what that slope represents.
10. Use information obtained from a displacement-time graph to plot a velocity-time graph of the motion.
11. Determine the area between the time axis and the line on a velocity-time graph for certain intervals and state what that area represents.
12. Given a displacement-time or velocity-time graph of a moving object, describe in words the object's motion.
13. Describe various techniques used in a physics laboratory to measure time.
14. Calibrate the frequency of a recording timer.
15. Set up, perform, and analyse an experiment involving uniform motion at various speeds.
16. Describe limitations of motion experiments in the laboratory.

Supplementary

17. Understand the meaning of the statement, "All motion is relative."
18. Given the velocities of two objects moving parallel to each other, determine their velocities relative to each other.
19. Analyse the motion of an object that is influenced by air or water currents for both one-dimensional and two-dimensional motion.

MECHANICS

CHAPTER 4
Uniform Acceleration

While riding through the mountains, this cyclist accelerates frequently.

Kinematics, the study of motion, involves four main quantities. Three of them—displacement, velocity, and time—were discussed in the previous chapter. The fourth quantity—acceleration—is the topic of this chapter. Vehicles like the bicycle in the photograph accelerate when they increase or decrease their velocity, and also when they change direction. Objects that fall freely accelerate toward the earth. Scientists learn about the structure of matter in the universe by using machines that accelerate particles to extremely high velocities. How does acceleration relate to the other quantities of motion, and how can we apply knowledge about acceleration to our lives?

4.1 Comparing Uniform Motion and Uniform Acceleration

You have learned that uniform motion occurs when an object moves at a steady speed in a straight line. For uniform motion the velocity is constant; a velocity-time graph yields a horizontal, straight line.

Most moving objects, however, do not display uniform motion. Any change in an object's speed or direction or both means that its motion is not uniform. This non-uniform motion, or changing velocity, is called **accelerated motion**. Because the direction of the motion is involved, acceleration is a vector quantity. A car ride at rush hour in a city during which the car must speed up, slow down, and turn corners is an obvious example of accelerated motion.

One type of accelerated motion, called **uniform acceleration**, occurs when an object travelling in a straight line changes its speed uniformly with time. Figure 4-1 shows a velocity-time graph for an object whose motion is given in the table. It starts from rest and increases its speed by 5.0 m/s every second in a westerly direction.

time (s)	0.0	1.0	2.0	3.0	4.0
velocity (m/s [W])	0.0	5.0	10	15	20

Uniform acceleration also occurs when an object travelling in a straight line slows down uniformly. In this case, the object is actually *decelerating*, but mathematically it is said to have **negative acceleration**. Refer to the table below and Figure 4-2, which give an example of uniform negative acceleration in which an object slows down uniformly from 20 m/s [E] to 0.0 m/s in 4.0 s.

time (s)	0.0	1.0	2.0	3.0	4.0
velocity (m/s [E])	20	15	10	5.0	0.0

If an object is changing its speed in a non-uniform fashion, its acceleration is non-uniform. Such motion is more difficult to analyse than uniform acceleration, but an example is given in the table below and Figure 4-3 for comparison purposes.

time (s)	0.0	1.0	2.0	3.0	4.0
velocity (m/s [S])	0	10	16	20	22

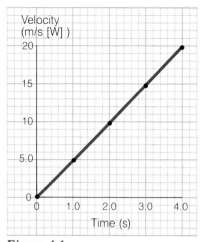

Figure 4-1
Uniform Acceleration

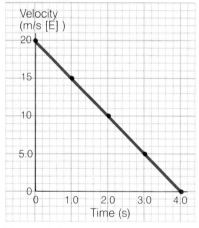

Figure 4-2
Uniform Negative Acceleration

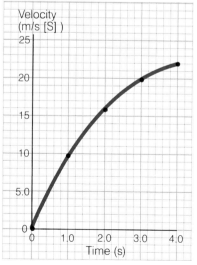

Figure 4-3
Non-uniform Acceleration

PRACTICE

1. The table shows five different sets of velocities at times of 0.0 s, 1.0 s, 2.0 s, and 3.0 s. Describe the motion of each set.

time (s)		0.0	1.0	2.0	3.0
(a) velocity (m/s [E])		0.0	8.0	16	24
(b) velocity (cm/s [W])		0.0	4.0	8.0	8.0
(c) velocity (km/h [N])		58	58	58	58
(d) velocity (m/s [W])		15	16	17	18
(e) velocity (km/h [S])		99	66	33	0.0

MECHANICS

2. Describe the motion illustrated in each velocity-time graph shown in Figure 4-4.

Figure 4-4

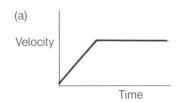
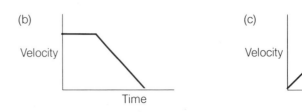

4.2 Calculating Acceleration

Acceleration is defined as the rate of change of velocity. The average acceleration of an object is found using the equation

$$\text{acceleration} = \frac{\text{change of velocity}}{\text{time}} \text{ or } \vec{a} = \frac{\Delta \vec{v}}{t}.$$

Since the change of velocity ($\Delta \vec{v}$) of a moving object is the final velocity ($\vec{v}_f$) minus the initial velocity ($\vec{v}_i$), the equation for acceleration can be written

$$\vec{a} = \frac{\vec{v}_f - \vec{v}_i}{t}.$$

In the metric system, velocity is often measured in metres per second; therefore, acceleration is stated in metres per second per second, or (m/s)/s. This is mathematically equivalent to metres per second squared (m/s²), because

$$\frac{m}{s} \div \frac{s}{1} = \frac{m}{s} \times \frac{1}{s} = m/s^2.$$

It is also common to express acceleration in centimetres per second squared, cm/s², and kilometres per hour per second, (km/h)/s.

SAMPLE PROBLEM 1
A motorbike starting from rest and undergoing uniform acceleration reaches a velocity of 20 m/s [N] in 8.0 s. Find its average acceleration.

Solution
$$\vec{a} = \frac{\vec{v}_f - \vec{v}_i}{t} = \frac{20 \text{ m/s [N]} - 0}{8.0 \text{ s}}$$
$$= 2.5 \text{ m/s}^2 \text{ [N]}$$

Thus, the bike's average acceleration is 2.5 m/s² [N].

In Sample Problem 1, the acceleration of 2.5 m/s² [N] means that the velocity of the motorbike increases by 2.5 m/s [N] every second. Thus, the bike's velocity is 2.5 m/s [N] after 1.0 s, 5.0 m/s [N] after 2.0 s, and so on.

SAMPLE PROBLEM 2
An airline flight is behind schedule, so the pilot increases the air velocity from 135 m/s [W] to 165 m/s [W] in 2.0 min. What is the aircraft's average acceleration in metres per second squared?

Solution
To express the answer in the units required, we must change 2.0 min to 120 s.

$$\vec{a} = \frac{\vec{v}_f - \vec{v}_i}{t} = \frac{165 \text{ m/s [W]} - 135 \text{ m/s [W]}}{120 \text{ s}}$$

$$= 0.25 \text{ m/s}^2 \text{ [W]}$$

∴ The aircraft's average acceleration is 0.25 m/s² [W].

If an object is slowing down, its acceleration is negative. In other words, the acceleration is opposite in direction to the velocity. This is illustrated in the next Sample Problem.

SAMPLE PROBLEM 3
A cyclist, travelling initially at 14 m/s [S], brakes smoothly and stops in 4.0 s. What is the cyclist's average acceleration?

Solution

$$\vec{a} = \frac{\vec{v}_f - \vec{v}_i}{t} = \frac{0.0 - 14 \text{ m/s [S]}}{4.0 \text{ s}}$$

$$= -3.5 \text{ m/s}^2 \text{ [S]}$$

$$= 3.5 \text{ m/s}^2 \text{ [N]}$$

Thus, the cyclist's average acceleration is 3.5 m/s² [N]. Notice in this solution that the direction "negative south" is the same as the direction "positive north".

The equation for average acceleration can be altered to find final velocity, initial velocity, or time, given the other variables. The derivations of the altered equations are part of the Practice questions that follow.

PRACTICE
3. Calculate the average acceleration in each instance, given the following:
   (a) $\vec{v}_f = 72$ m/s [W], $\vec{v}_i = 0.0$, and $t = 6.0$ s
   (b) $\vec{v}_f = 32.1$ m/s [N], $\vec{v}_i = 23.7$ m/s [N], and $t = 0.0500$ s
   (c) $\vec{v}_f = 24$ km/h [S], $\vec{v}_i = 37$ km/h [S], and $t = 20$ s

MECHANICS

4. The world record for motorcycle acceleration occurred when a cycle took only 6.0 s to go from rest to 281 km/h [forward]. Calculate the average acceleration in
   (a) kilometres per hour per second
   (b) metres per second squared
   (If you cannot recall how to change kilometres per hour to metres per second, refer to Section 3.3.)

5. A 1500 kg car travelling at 57 km/h [N] increases its velocity to 99 km/h [N] in 7.0 s. What is its average acceleration?

6. Calculate the average acceleration needed by a train travelling at 13 m/s [S] to stop in 4 min and 20 s. Express your answer in metres per second squared.

7. Rewrite the equation $\vec{a} = \dfrac{\vec{v}_f - \vec{v}_i}{t}$ to solve the following:
   (a) final velocity
   (b) initial velocity
   (c) time

8. Calculate the unknown quantities in the table.

	Acceleration (m/s² [E])	Initial Velocity (m/s [E])	Final Velocity (m/s [E])	Time (s)
(a)	8.5	?	93	4.0
(b)	0.50	15	?	120
(c)	−0.20	21	12	?

9. In the second stage of a rocket launch, the rocket's upward velocity increased from 1.0×10^3 m/s to 1.0×10^4 m/s, with an average acceleration of 30 m/s². How long did the acceleration last?

10. A truck driver travelling at 90 km/h [W] applies the brakes to prevent hitting a stalled car. In order to avoid a collision, the truck would have to be stopped in 20 s. At an average acceleration of −4.0 (km/h)/s [W], will a collision occur? Try to solve this problem using two or three different techniques.

11. Assume that when a ball is thrown upwards it accelerates at a rate of −9.8 m/s² [up]. With what velocity must a ball leave a thrower's hand in order to climb for 2.2 s before stopping?

12. Estimate your own maximum velocity when running, then estimate the average acceleration you undergo from rest to reach that velocity.

13. **Activity** Design and perform an experiment to check your estimates in Practice Question 12.

4.3 Using Velocity-Time Graphs to Find Acceleration

In Chapter 3 you learned that the slope of a line on a displacement-time graph indicates the velocity. Let us now use the equation $m = \Delta y / \Delta x$ to determine the slope of a line on a velocity-time graph.

UNIFORM ACCELERATION

Consider the graph in Figure 4-5. The slope of the line is

$$m = \frac{\Delta y}{\Delta x} = \frac{\Delta \vec{v}}{\Delta t}$$

$$= \frac{30 \text{ m/s [E]} - 0.0 \text{ m/s [E]}}{10 \text{ s} - 0.0 \text{ s}}$$

$$= 3.0 \text{ m/s}^2 \text{ [E]}.$$

The units for this slope represent acceleration. Thus, we can conclude that the slope of a line on a velocity-time graph equals the acceleration. In equation form,

acceleration = slope on a velocity-time graph

$$a = \frac{\Delta v}{\Delta t}.$$

This equation is equivalent to the equation used previously for acceleration:

$$\vec{a} = \frac{\vec{v}_f - \vec{v}_i}{t}$$

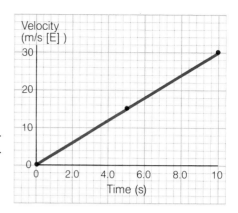

Figure 4-5
Velocity-Time Graph

SAMPLE PROBLEM 4

For the motion shown in the graph, determine the acceleration in segments A, B, and C.

Solution

(a) Segment A: $\vec{a} = \frac{\Delta y}{\Delta x} = \frac{30 \text{ m/s [S]} - 10 \text{ m/s [S]}}{5.0 \text{ s} - 0.0 \text{ s}}$

$= 4.0 \text{ m/s}^2 \text{ [S]}$

∴ The acceleration is 4.0 m/s^2 [S].

(b) Segment B: $\vec{a} = \frac{\Delta y}{\Delta x} = \frac{30 \text{ m/s [S]} - 30 \text{ m/s [S]}}{15 \text{ s} - 5.0 \text{ s}}$

$= 0.0$

∴ There is zero acceleration from 5.0 s to 15 s.

(c) Segment C: $\vec{a} = \frac{\Delta y}{\Delta x} = \frac{0.0 \text{ m/s [S]} - 30 \text{ m/s [S]}}{20 \text{ s} - 15 \text{ s}}$

$= -6.0 \text{ m/s}^2 \text{ [S] or } 6.0 \text{ m/s}^2 \text{ [N]}$

∴ The acceleration is -6.0 m/s^2 [S], which is equivalent to 6.0 m/s^2 [N].

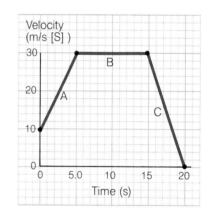

PRACTICE

14. Use the velocity-time graph in Figure 4-6 to determine the following:

    (a) velocity at 0.40 s and 0.80 s
    (b) acceleration between 0.0 s and 0.60 s
    (c) acceleration between 0.60 s and 1.4 s
    (d) acceleration between 0.80 s and 1.2 s

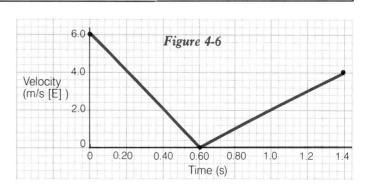

Figure 4-6

15. Sketch a velocity-time graph for the motion of a car travelling south along a straight road having a posted speed limit of 60 km/h, except in a school zone where the speed limit is 40 km/h. The only traffic lights are found at either end of the school zone, and the car must stop at both sets of lights. (Assume that when $t = 0.0$ s, the velocity is 60 km/h [S].)

16. The data in the table below represent test results on a recently built, standard transmission automobile. Use the data to plot a fully labelled, accurate velocity-time graph of the motion from $t = 0.0$ s to $t = 35.2$ s. Assume that the acceleration is constant in each time segment. Then use the graph to determine the average acceleration in each gear and during braking.

Acceleration Mode	Change in Velocity (km/h [forward])	Time Taken for the Change (s)
first gear	0.0 to 48	4.0
second gear	48 to 96	8.9
third gear	96 to 128	17.6
braking	128 to 0.0	4.7

4.4 Investigating Acceleration in the Laboratory

Assume that you are asked to calculate the acceleration of a car as it goes from 0.0 km/h to 100 km/h [W]. The car has a speedometer which can be read directly, so the only instrument you need is a watch. The average acceleration can be calculated from knowing the time it takes to reach maximum velocity. For instance, if the time taken is 12 s, the average acceleration is

$$\vec{a} = \frac{\Delta \vec{v}}{\Delta t}$$

$$= \frac{100 \text{ km/h [W]}}{12 \text{ s}}$$

$$= 8.3 \text{ (km/h)/s [W]}.$$

In a science laboratory, however, an acceleration experiment is not so simple. Objects that move (for example, a cart, ball, or metal mass) do not come equipped with speedometers, so their speed cannot be found directly.

One way to overcome this problem when performing acceleration experiments is to measure the displacement of the object from its starting position at specific times. Then a displacement-time graph can be plotted and a mathematical technique used to calculate the average acceleration.

An example of how to calculate acceleration starting with a displacement-time graph is described below. Similar calculations will be performed in the next two experiments.

UNIFORM ACCELERATION

To begin, consider Figure 4-7, which shows a typical displacement-time graph for a skier starting from rest and accelerating downhill [D]. Notice that the line is curved, not straight as for uniform motion. The reason is that the skier's change of displacement in each time interval increases as time increases.

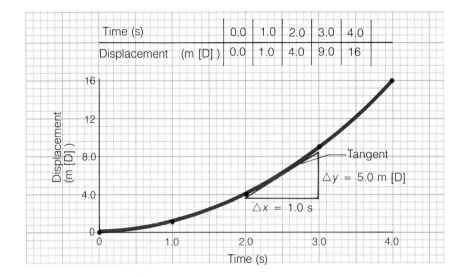

Figure 4-7
Displacement-Time Graph of Uniform Acceleration

Since the slope of a line on a displacement-time graph indicates the velocity, we perform slope calculations first. Because the line is curved, however, its slope keeps changing. Thus, we must find the slope of the curved line at various times. The technique we will use is called the **tangent technique**. A **tangent** is a straight line that touches a curve at a point and has the same slope as the curve at that point. To find the velocity at 2.5 s, for example, we draw a tangent to the curve at that time. For convenience, the tangent in our example is drawn so that its Δx value is 1.0 s. Then the slope of the tangent is

$$m = \frac{\Delta y}{\Delta x}$$
$$= \frac{5.0 \text{ m [D]}}{1.0 \text{ s}}$$
$$= 5.0 \text{ m/s [D]}.$$

That is, the skier's velocity at 2.5 s is 5.0 m/s [D].

This velocity is known as an **instantaneous velocity**, one which occurs at a particular instant. Refer to Figure 4-8(a), which shows the same displacement graph with all the tangents drawn and velocities shown. Verify to your own satisfaction that the velocities (the slopes of the tangents) are calculated correctly.

In Figure 4-8(b), the instantaneous velocities calculated from the displacement graph are plotted on a velocity-time graph. Notice that the line is extended to 4.0 s, the same final time as that found on the displacement graph. The slope of the line on the velocity graph is then calculated and used to plot the acceleration graph, shown in Figure 4-8(c).

71

MECHANICS

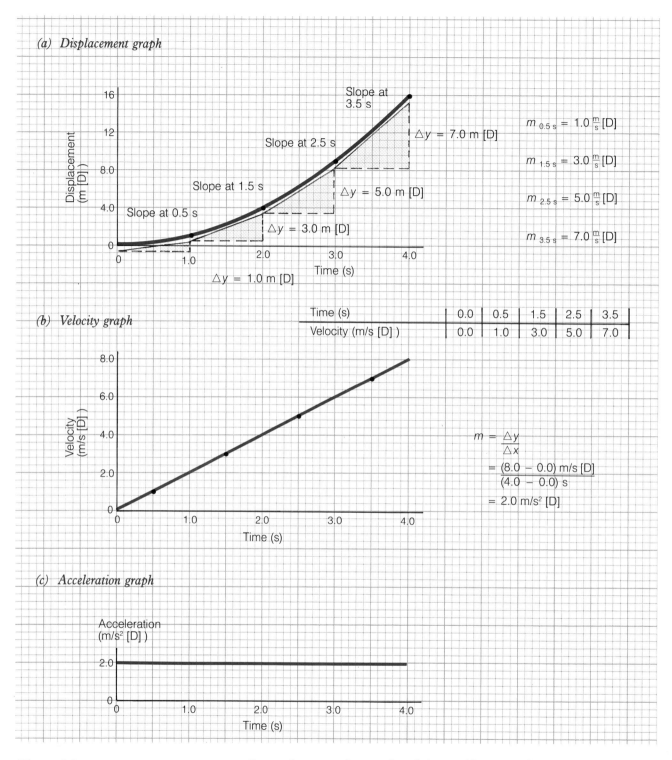

Figure 4-8
Graphing Uniform Acceleration

In motion experiments involving uniform acceleration, the velocity-time graph should yield a straight line. However, because of experimental error, this might not occur. If the points on a velocity-time graph of uniform acceleration do not lie on a straight line, draw a straight line of best fit, then calculate its slope to find the acceleration.

UNIFORM ACCELERATION

PRACTICE

17. In the tables below are two sets of displacement-time data for uniform acceleration. In each case, do the following:
    (a) Plot a displacement-time graph.
    (b) Find the slopes of tangents at appropriate times.
    (c) Plot a velocity-time graph.
    (d) Plot an acceleration-time graph.

    (i)
time (s)	0.0	2.0	4.0	6.0	8.0
displacement (m [N])	0.0	8.0	32	72	128

    (ii)
time (s)	0.0	0.10	0.20	0.30	0.40
displacement (mm [W])	0.0	3.0	12	27	48

Experiment 4A: Uniform Acceleration

INTRODUCTION
This experiment, like Experiment 3A in Chapter 3, may be performed using one of three alternatives—stopwatches (Procedure I), a recording timer (Procedure II), or a flashing light (Procedure III). In each case, acceleration is achieved by allowing an object starting from rest to travel down a ramp elevated at one end.

PURPOSE
To observe and analyse uniform acceleration.

APPARATUS
General: metre stick, graph paper
- For Procedure I: 5 or 6 stopwatches; steel ball; 2.5 m board; 2.5 m drapery track
- For Procedure II: recording timer and related apparatus; cart; masking tape; 2.5 m board
- For Procedure III: stroboscope; instant camera; linear air track and related apparatus; pin; overhead projector

PROCEDURE I: STOPWATCH TECHNIQUE
1. Elevate one end of the board about 4.0 cm. Attach the drapery track to it to act as a track for the steel ball. See Figure 4-9. Mark a starting position near the top of the track and mark positions at 50 cm intervals along the track.
2. Each student with a stopwatch will be assigned to a particular position. Place the ball at the starting position. At a signal, allow the ball to begin rolling and start all the stopwatches. Measure the time it takes the ball to reach each position. Take the average of several trials to improve results.
3. Record the observed results in a table of displacement [forward] from the starting position and time, such as the one below:

displacement (cm [f])	0.0	50	100	...
time (s)	0.0			

Figure 4-9
Apparatus for Procedure I, Experiment 4A

73

ANALYSIS

1. Plot the data on a displacement-time graph, with time along the horizontal axis. Draw a smooth curve to fit the data closely.
2. Use the tangent technique to determine the instantaneous velocity of the object at a minimum of three specific times. Use the results to plot a velocity-time graph of the motion.
3. Use the velocity-time graph to find the acceleration; then plot an acceleration-time graph of the motion.
4. Calculate the area under the line on the velocity-time graph and state what that area represents.
5. Repeat Question 4 for the acceleration-time graph.
6. Plot a graph of displacement *versus* time squared. Draw the line of best fit and find its slope. Explain the shape and meaning of the graph. (You should do this question only if you have studied the Supplementary Topic in Chapter 2.)
7. Describe any sources of error in this experiment.

CONCLUSIONS . . .

PROCEDURE II: RECORDING TIMER TECHNIQUE

1. Check whether the recording timer is working properly by pulling a short piece of paper tape through it. Inform your teacher of any difficulties.
2. Elevate one end of the board about 30 cm; be sure it does not sag in the middle. Use masking tape to attach a 2.0 m length of paper tape to the cart. Feed the tape through the recording timer, as shown in Figure 4-10, so the cart is close to the timer. Hold the cart still, turn on the timer, then release the cart.

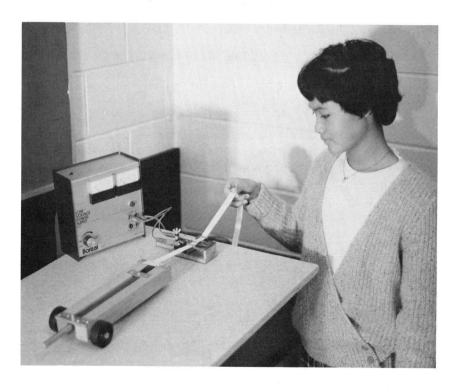

Figure 4-10
Apparatus for Procedure II, Experiment 4A

3. Choose as a starting position the first distinct dot on the tape, point A. Count 6 spaces to the next position, B, and measure AB. See Figure 4-11. From B count another 6 spaces to C and measure AC, also shown in Figure 4-11. Continue this procedure for as many sets of 6 spaces as possible, always measuring displacement [forward] from the starting position, A. Tabulate the observations in a table like the one below:

time (s)	0.0	0.10	0.20	. . .
displacement (cm [f])	0.0			

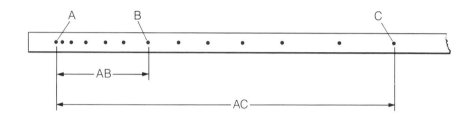

Figure 4-11
Measuring Displacements from the Starting Position

ANALYSIS
Answer Analysis Questions 1 to 7 in Procedure I.

CONCLUSIONS . . .

PROCEDURE III: PHOTOGRAPHIC TECHNIQUE
1. Elevate one end of the linear air track 10 cm to 15 cm. Hold the cart with the straw near the elevated end of the track. Position the camera in such a way that the cart is at one edge of the field of view. Hold a metre stick along the track for reference and focus the camera. Adjust the stroboscope to a flash interval of 0.2 s, which corresponds to a frequency of 5.0 Hz.
2. With the camera's light-exposure meter covered and the room lights out, take a time-exposure photograph of the cart as it accelerates down the track. Develop the photograph.
3. Poke two small holes along the metre stick in the photograph so that a reference distance (*e.g.*, 50 cm) can be projected onto the blackboard. Poke holes in the photograph wherever the straw appears. With the result projected onto the blackboard, measure the displacement [forward] from the start to each point. Tabulate the results in a table like the one below:

time (s)	0.0	0.20	0.40 . . .
displacement (cm [f])	0.0		

ANALYSIS
Answer Analysis Questions 1 to 7 in Procedure I.

CONCLUSIONS . . .

4.5 Acceleration Near the Earth's Surface

If two solid metal objects of different mass (*e.g.*, 50 g and 100 g) are dropped from the same height above the floor, they land at the same time. This fact proves that the acceleration of falling objects near the surface of the earth does not depend on mass.

The acceleration of falling objects does, however, depend on air resistance. Try the following demonstration. Fold a piece of notebook paper once. Hold the paper and a textbook horizontally about 50 cm above a table top. Release them at the same instant. Which lands first? Why? Now place the paper on the top of the book, hold the book horizontally, and drop it. The book eliminates the effect of air resistance.

It was Galileo Galilei who first proved that, if we ignore the effect of air resistance, the acceleration of falling objects is constant. He proved this experimentally by measuring the acceleration of metal balls rolling down a ramp. Galileo found that, for a constant slope of the ramp, the acceleration was constant — it did not depend on the mass of the ball. The reason he could not measure vertical acceleration was that he had no way of measuring short periods of time accurately. You will appreciate the difficulty of measuring time when you perform the next experiment.

Had Galileo been able to evaluate the acceleration of freely falling objects near the earth's surface, he would have found it to be approximately 9.8 m/s² [down]. This value does not apply to objects influenced by air resistance. It is an average value that changes slightly from one location to the other. It is the acceleration caused by the force of gravity.

The quantity 9.8 m/s² [down] occurs so commonly that from now on we will give it the symbol $\vec{g}$, the **acceleration due to gravity**. (Do not confuse this $\vec{g}$ with the g used as the symbol for "gram".) Modern values of $\vec{g}$ are determined by scientists throughout the world. For example, at the International Bureau of Weights and Measures in France, experiments are performed in a vacuum chamber in which an object is launched upwards using an elastic. The object has a system of mirrors at its top and bottom which reflect laser beams used to measure time of flight. The magnitude of $\vec{g}$ obtained at the Bureau using this technique is 9.809 260 m/s². Galileo would have been pleased with the precision!

In solving problems involving the acceleration due to gravity, the value of 9.8 m/s² [down] can be used if the effect of air resistance is assumed to be negligible. When air resistance on an object is negligible, we say the object is "falling freely".

PRACTICE

18. A stone is dropped from a bridge and falls freely under the influence of gravity. Calculate its velocity after
    (a) 0.60 s; (b) 1.2 s; (c) 2.4 s.
19. Repeat Practice Question 18 for a stone that is thrown vertically from the bridge with an initial velocity of 4.0 m/s [down].

Experiment 4B: Acceleration Due to Gravity

INTRODUCTION
Here you will be expected to use your experience from previous motion experiments to design and perform an experiment for determining the acceleration due to gravity at your location. The procedure listed below gives only minimum detail, so you will have to write your own.

PURPOSE
To determine the acceleration due to gravity and to compare the result to the accepted average value of 9.8 m/s^2 [down].

APPARATUS . . .

PROCEDURE
1. Predict how your result for this experiment will compare to 9.8 m/s^2 [down], and explain why.
2. Choose a technique for finding the acceleration of a falling object, but take into consideration these hints:
   (a) If you use a stopwatch, the object should be allowed to fall at least 5.0 m.
   (b) If you use a recording timer, the friction should be reduced as much as possible. See Figure 4-12.
   (c) Have the falling mass land on pieces of cardboard.
   (d) Refer to Experiment 4A for specific details.

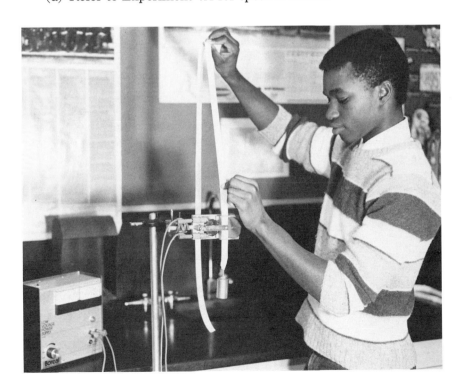

Figure 4-12
Using a Recording Timer to Determine the Acceleration due to Gravity

3. Perform and report on your experiment. A computer program might be available to let you check your results before analysing them. The Analysis from Experiment 4A can be used as a basis for analysis in this experiment. You should also find the percent error of your result, then write your conclusions.

4.6 Solving Uniform Acceleration Problems

Now that you have learned the definitions and basic equations associated with uniform acceleration, it is possible to extend your knowledge so that you can solve more complex problems. In this section you will learn how to derive and use some important equations involving the following five variables: initial velocity, final velocity, displacement, time, and acceleration. Each equation derived will involve four of these five variables, and thus will serve a different purpose.

The process of deriving equations involves three main stages:

1. State the given facts and equations.
2. Substitute for the variable which is to be eliminated.
3. Simplify the equation to a convenient form.

The derivations here involve two given equations. The first is the equation which defines acceleration, $a = (v_f - v_i)/t$. (Notice that the vector notation has been omitted to let you concentrate on the mathematics.) A second equation can be found by applying the fact that the area under the line on a velocity-time graph indicates the change in displacement. Figure 4-13 shows a typical velocity-time graph for an object that undergoes uniform acceleration from an initial velocity (v_i) to a final velocity (v_f) during a time (t). The shape of the area under the line is a trapezoid, so the area is $d = \frac{1}{2}(v_i + v_f)t$. (The area of a trapezoid is the product of the average length of the two parallel sides and the perpendicular distance between them. In Figure 4-13, the area could also be found by adding the area of a triangle to the area of the rectangle beneath it.)

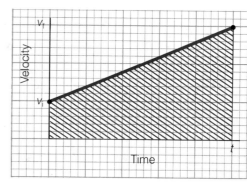

Figure 4-13
A Velocity-Time Graph of Uniform Acceleration

UNIFORM ACCELERATION

These given equations can now be used in the last two stages of deriving equations — substitution and simplification. Figure 4-14 gives the step-by-step process of deriving four other equations using the two given ones. Study the steps carefully. It is a good idea to have a pen and some scrap paper handy so you can verify the steps in these kinds of derivations.

Figure 4-14
Derivations of Equations Involving Uniform Acceleration

Equation ② already has omitted the acceleration. Thus, since the average velocity during uniform acceleration is

$$v_{ave} = \frac{v_i + v_f}{2},$$

$$\boxed{\Delta d = v_{ave}\, t} \quad \text{----} \quad ③$$

↑ Eliminate the acceleration.

From ①, $v_f = v_i + at$. Substituting this into ② will eliminate v_f. Thus,

$$\Delta d = \tfrac{1}{2}(v_i + v_f)\, t$$
$$\Delta d = \tfrac{1}{2}(v_i + v_i + at)\, t$$
$$\Delta d = \tfrac{1}{2}(2v_i + at)\, t$$
$$\boxed{\Delta d = v_i t + \frac{at^2}{2}} \quad \text{----} \quad ④$$

↑ Eliminate the final velocity.

GIVEN

The defining equation for acceleration is

$$a = \frac{v_f - v_i}{t} \quad \text{----} \quad ①$$

The area under the line on a velocity-time graph indicates the change of displacement:

$$\Delta d = \tfrac{1}{2}(v_i + v_f)\, t \quad \text{----} \quad ②$$

↓ Eliminate the time.

From ①, $t = \dfrac{v_f - v_i}{a}$, which can be substituted into ② to eliminate t.

$$\Delta d = \tfrac{1}{2}(v_f + v_i)\, t$$
$$\Delta d = \tfrac{1}{2}(v_f + v_i)\left(\frac{v_f - v_i}{a}\right)$$
$$\Delta d = \frac{(v_f + v_i)(v_f - v_i)}{2a} \quad \text{Factors of the difference of squares}$$
$$\Delta d = \frac{(v_f^2 - v_i^2)}{2a} \quad \text{or} \quad 2a\Delta d = v_f^2 - v_i^2$$
$$\therefore \quad \boxed{v_f^2 = v_i^2 + 2a\Delta d} \quad \text{----} \quad ⑤$$

↓ Eliminate the initial velocity.

You are asked in Practice Question 26 to derive the equation in which the initial velocity has been eliminated.

$$\boxed{\Delta d = v_f t - \frac{at^2}{2}} \quad \text{----} \quad ⑥$$

MECHANICS

The equations for uniform acceleration are summarized for your convenience in Table 4-1.

Table 4-1 Equations for Uniformly Accelerated Motion

Variables Involved	General Equation
a, v_i, v_f, t	$a = \dfrac{v_f - v_i}{t}$
$\Delta d, v_i, a, t$	$\Delta d = v_i t + \dfrac{at^2}{2}$
$\Delta d, v_i, v_f, t$	$\Delta d = v_{ave} t$
$\Delta d, v_f, v_i, a$	$v_f^2 = v_i^2 + 2a\Delta d$
$\Delta d, v_f, t, a$	$\Delta d = v_f t - \dfrac{at^2}{2}$

SAMPLE PROBLEM 5
A car accelerates uniformly from rest at $t = 0.0$ s, at 4.0 m/s². How far does the car travel between 5.0 s and 8.0 s?

Solution
The four variables involved are v_i, a, t, and Δd.

Thus, the equation to be used is $\Delta d = v_i t + \dfrac{at^2}{2}$.

Since $v_i = 0$, this equation reduces to $\Delta d = \dfrac{at^2}{2}$. At 5.0 s the car's displacement from the original position is

$$\Delta d = \dfrac{at^2}{2}$$
$$= \dfrac{(4.0 \text{ m/s}^2)(5.0 \text{ s})^2}{2}$$
$$= 50 \text{ m}.$$

At 8.0 s the car's displacement from the original position is

$$\Delta d = \dfrac{at^2}{2}$$
$$= \dfrac{(4.0 \text{ m/s}^2)(8.0 \text{ s})^2}{2}$$
$$= 128 \text{ m}.$$

∴ The car travels 128 m − 50 m = 78 m between 5.0 s and 8.0 s.

In Sample Problem 6, as in many motion problems, there is likely more than one method for finding the solution. Practice is necessary to help you develop skill in solving these types of problems efficiently.

> **SAMPLE PROBLEM 6**
> A ball is held at rest on a balcony 15 m above the ground. Then it is dropped and accelerates freely. With what velocity does it strike the ground?
>
> **Solution**
> The four variables involved are v_i, v_f, a, and Δd. Thus, the equation to be used is $v_f^2 = v_i^2 + 2a\Delta d$. Since the ball starts from rest, $v_i = 0$, so the equation reduces to $v_f^2 = 2a\Delta d$.
>
> $$\begin{aligned} v_f^2 &= 2a\Delta d \\ &= 2(9.8 \text{ m/s}^2)(15 \text{ m}) \\ &= 294 \text{ m}^2/\text{s}^2 \\ v_f &= \sqrt{294 \text{ m}^2/\text{s}^2} \\ &= 17.1 \text{ m/s} \end{aligned}$$
>
> ∴ The ball strikes the ground at a velocity of 17 m/s [downward].

PRACTICE

20. A girl accelerates uniformly from rest along a track at 0.90 m/s² for 10 s.
    (a) What is her final velocity?
    (b) How far did she travel in the 10 s?
21. In an acceleration test of a sports car, two markers 300 m apart were set up along a road. The car passed the first marker with a speed of 5.0 m/s and the second marker with a speed of 33 m/s. Calculate the car's average acceleration between the markers.
22. A baseball travelling at 30 m/s strikes a catcher's mitt and comes to a stop while moving 10 cm with the mitt. Calculate the average acceleration of the ball as it is stopping.
23. A plane travelling at 50 m/s [E] down a runway begins accelerating uniformly at 2.8 m/s² [E].
    (a) What is the plane's velocity after 5.0 s?
    (b) How far has it travelled during the 5.0 s?
24. A skier starting from rest accelerates uniformly downhill at a rate of 1.8 m/s². How long will it take the skier to reach a point 90 m from the starting position?
25. For a certain motorcycle, the magnitude of the braking acceleration is $|4\ \vec{g}\text{'s}\ |$. If the bike is travelling 40 m/s,
    (a) how long does it take to stop?
    (b) how far does it travel during the stopping time?
26. Derive the uniform acceleration equation in which the initial velocity has been omitted. See Figure 4-14.
27. A car travelling along a highway must uniformly reduce its speed to 12 m/s in 3.0 s. If the distance travelled during that time is 60 m, what is the car's average acceleration? What is its initial velocity?

PROFILE

France A. Legault, P.Eng.
Automotive Safety Engineer
Transport Canada

I was born during the year in which cars sported the largest tail fins ever seen. The oil crisis had not yet been heard of, and a new field in automotive technology, traffic safety, was in its infancy.

Traffic safety is the science of preventing accidents and of protecting car occupants in the event that an accident does occur. These goals are mainly achieved through the study, testing, and installation of such familiar automotive items as seat belts and padded dashes, laminated windshields, and reinforced side doors.

As a child, I was always fascinated by cars. I remember that I was about four years old when my parents took me to my first car show. Later, when I learned to read, I would spend endless hours looking over brochures of the latest models. Then, when I started to get pocket money, I would buy all the automotive magazines available and practically learn them by heart.

I always knew that I wanted to work in the automotive industry but did not quite know what I could do. Then finally, toward the end of high school, the solution presented itself. I had always enjoyed mathematics and physics, especially mechanics. So I decided to study engineering at the University of Ottawa. Becoming a mechanical engineer, I thought, would surely lead to a career working with cars.

While I was still a university student, I managed to arrange to work for two summer terms as an accident investigator for Transport Canada. My job was to determine how fast the vehicles had been travelling when they collided and to analyse the damage that was caused by the collision. I also had to record any pointed objects that the people inside the vehicles had been thrown against so that I could later correlate this with their injuries.

Needless to say, this was quite a sobering experience. During that first summer, I became very aware of the hazards involved in driving and, for a while, was quite paranoid about getting into a car with another driver at the wheel. For this reason, after just a few weeks, I had already decided that I wanted a career in traffic safety. Consequently, upon graduating with my degree in mechanical engineering in 1981, I went to work for a contracting firm that specialized in reconstructing accidents to help settle insurance claims. I later rejoined Transport Canada as an automotive standards engineer, a position I still hold to this day.

At Transport Canada, I am responsible for identifying areas where improvements can be made in the way we protect car occupants. This involves testing both new safety devices and new models of cars, as well as writing and amending government standards that require manufacturers to build safe vehicles. My main responsibilities are child seats, vehicles for handicapped drivers, seat belts and seat anchorages, and side door and roof strength.

Child seats provide an excellent example of how I go about my work. To test child seats, I use specially designed dummies with the physical characteristics of children six months, three years, and six years old. These dummies are strapped one at a time in the seat to be tested and then placed on an impact sled. This sled is designed

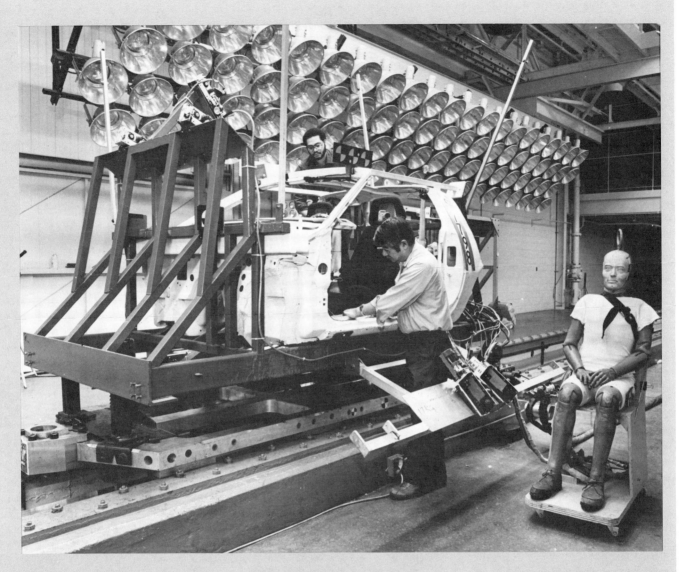

to imitate the effect of a car hitting a concrete wall head-on at 48 km/h.

Once the dummy has been put through the simulated crash, I measure the effects on its head and chest to determine if a child in a similar crash situation would have suffered any injuries from the seat. I also record the forward head movement of the dummy to ensure that a child inside a car during an accident would not hit his or her head violently on an object in front of them.

In addition, I also test child seats to check such features as the strength of the webbing and buckles, the effectiveness of the padding, and the flame resistance of all the materials used in making the seat.

Many things can be done to make a vehicle safe for its occupants, but in the end, the driver must assume the responsibility for ensuring the safety of his or her passengers. For this reason, I strongly support the campaign against people who drink and drive.

Safety engineers perform well-designed experiments to test safety designs. In this case, the dummy at the right will be placed in the impact sled shown being prepared for a test crash. The sled will slam into a solid barrier at a speed of 48 km/h, which is equivalent to hitting a parked car at 96 km/h. The engineers use modern instrumentation, sophisticated dummies, and high-speed photography to analyse the test results.

4.7 Applications of Acceleration

Galileo Galilei began the mathematical analysis of acceleration, and the topic has been studied by physicists ever since. However, only during the twentieth century has acceleration become a topic that relates closely to our everyday lives.

The study of acceleration is important in the field of transportation. Humans undergo acceleration in automobiles, airplanes, rockets, and other vehicles. The positive acceleration in cars and passenger airplanes is usually small, but in a military airplane or a rocket it can be great enough to cause damage to the human body. In 1941, a Canadian pilot and inventor named W.R. Franks designed an "anti-gravity" suit to prevent pilot blackouts in military planes undergoing high-speed turns and dives. Blackouts occur when blood drains from the head and goes to the lower part of the body. To prevent this, Franks designed a suit with water encased in the inner lining to prevent the blood vessels from expanding outwards (Figure 4-15).

Modern experiments have shown that the maximum acceleration a human being can withstand for more than about 0.5 s is approximately $|30\,\vec{g}|$ (294 m/s²). Astronauts experience up to $|10\,\vec{g}|$ (98 m/s²) for several seconds during a rocket launch. At that acceleration, if the astronauts were standing they would faint from loss of blood to the head. To prevent this problem, astronauts must sit horizontally during blast-off (Figure 4-16).

In our day-to-day lives we are more concerned with negative acceleration in cars and other vehicles than with positive acceleration in rockets.

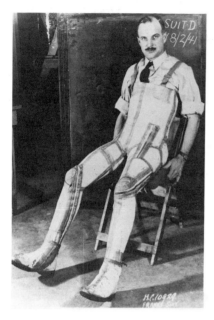

Figure 4-15
This 1941 photograph shows W.R. Franks in the "anti-gravity" suit he designed.

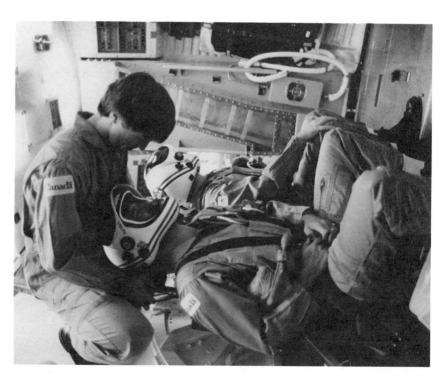

Figure 4-16
Two astronauts participate in a launch simulation exercise as a backup crew member assists.

Studies are continually being made to determine the effect on the human body when a car has a collision or must stop quickly. Seatbelts, headrests, and air bags help prevent many injuries caused by a large negative acceleration (Figure 4-17).

Another application is in the study of the effect of acceleration rates on the gasoline consumption of cars. It is logical that a driver with a "heavy foot" wastes valuable resources by accelerating at excessive rates. Some cars are now equipped with a light that indicates when the acceleration exceeds a certain energy-saving value.

Did You Know?
One area of research into the effects of acceleration on the human body deals with the design of emergency escape systems from high-performance aircraft. In an emergency, the pilot would be shot upward from a sitting position through an escape hatch. The escape system would have to be designed to produce an acceleration for an extremely short period of time.

Figure 4-17
As the test vehicle shown crashes into a barrier, the airbag being researched expands rapidly and prevents the dummy's head from striking the windshield or steering wheel. After the crash, the air bag begins to deflate so that, in a real situation, the driver can breathe.

In the exciting sport of skydiving, the diver jumps from an airplane and accelerates toward the ground, experiencing *free fall*. See Figure 4-18. The parachute remains closed until the diver is a predetermined distance from the ground. During free fall, the skydiver's speed will increase to a maximum amount called **terminal velocity**. Air resistance prevents a higher velocity. At terminal velocity the diver's acceleration becomes zero; in other words, the velocity remains constant. For humans, terminal velocity in air is about 53 m/s or 190 km/h. Of course, after the parachute opens, the terminal velocity is reduced to between 5 and 10 m/s.

Terminal velocity is also important in other situations. Certain plant seeds, such as dandelions, act like parachutes and have a terminal velocity of about 0.5 m/s. Some industries take advantage of the different terminal velocities of various particles in water when they use sedimentation to separate particles of rock, clay, or sand from each other. Volcanic

Figure 4-18
This skydiver experiences "free fall" immediately upon leaving the aircraft.

eruptions produce dust particles of different sizes. The larger dust particles settle more rapidly than the smaller ones. Thus, very tiny particles with low terminal velocities travel great horizontal distances around the world before they settle. This phenomenon can have a serious effect on the earth's climate.

One further application is the acceleration due to gravity on heavenly bodies other than the earth. If an astronaut standing on the moon dropped a ball, the ball would accelerate at about one-sixth of its rate on earth, or about 1.6 m/s². Table 4-2 lists mass, radius, and acceleration due to gravity for the nine planets in our solar system. The planets are listed in the order of their average distance from the sun.

Table 4-2 Statistics for the Planets of the Solar System

Planet	Mass (kg)	Radius of Planet (m)	Acceleration Due to Gravity (m/s²)
Mercury	3.4×10^{23}	2.4×10^{6}	3.8
Venus	4.8×10^{24}	6.1×10^{6}	8.9
Earth	6.0×10^{24}	6.4×10^{6}	9.8
Mars	6.5×10^{23}	3.4×10^{6}	3.7
Jupiter	1.9×10^{27}	7.2×10^{7}	25.8
Saturn	5.7×10^{26}	6.1×10^{7}	11.1
Uranus	8.8×10^{25}	2.4×10^{7}	10.5
Neptune	1.0×10^{26}	2.3×10^{7}	13.8
Pluto (approx.)	3.6×10^{23}	3×10^{6}	unknown

Figure 4-19
Determining Reaction Time

PRACTICE

28. Sketch the general shape of a velocity-time graph for a skydiver who accelerates, then reaches terminal velocity, then opens the parachute. Assume that downwards is positive.

29. Use the data in Table 4-2 to state what factors determine the acceleration due to gravity on a planet.

30. **Activity** The earth's acceleration due to gravity can be used to determine human reaction time (the time taken to react to an event one sees). Determine your own reaction time by performing the following activity. Your partner will hold a 30 cm wooden ruler at a certain position, say the 25 cm mark, in such a way that it lines up with your thumb and index finger. (Refer to Figure 4-19.) Now, as you look at the ruler, your partner will drop the ruler without warning. Grasp it as quickly as possible. Repeat this several times for accuracy and find an average of the distances the ruler falls before you catch it. Since the ruler accelerates at 9.8 m/s² and you know the distance it falls, you can calculate your reaction time using $t = \sqrt{2\Delta d/a}$. Compare your reaction time to that of other students.

31. Use your reaction time, calculated in Practice Question 30, to determine how far a car you are driving at 100 km/h would travel between the time you see an emergency and the time you slam on the brakes. Express the answer in metres.

Review Assignment

1. (a) If the speed of an object remains constant, can its velocity change? Explain.
   (b) If the velocity of an object remains constant, can its speed change? Explain.
   (c) Can an object have a westward velocity while experiencing an eastward acceleration? Explain.
2. Prove that (cm/s)/s is mathematically equivalent to cm/s^2.
3. Describe the motion represented by each of the graphs in Figure 4-20.

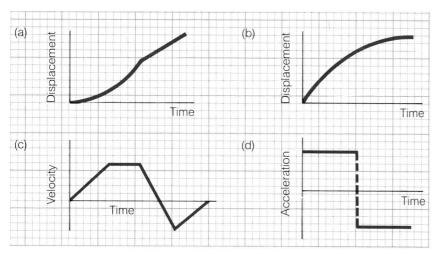

Figure 4-20

4. A boy on a ten-speed bicycle accelerates from rest to 2.2 m/s in 5.0 s in third gear, then changes into fifth gear. After 10 s in fifth gear, he reaches 5.2 m/s. Calculate the average acceleration in the third and fifth gears.
5. For the graph shown in Figure 4-21, state the following:
   (a) velocity at 1.0 s, 3.0 s, and 5.5 s
   (b) acceleration at 1.0 s, 3.0 s, and 5.5 s

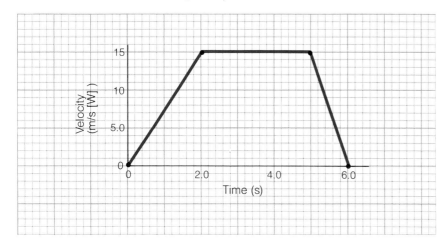

Figure 4-21

MECHANICS

6. How long does it take a freely falling object, starting from rest, to reach a velocity of 63.7 m/s [down]?
7. The results of a motion experiment are summarized in the table. Plot a displacement-time graph of the motion and use it to find the data for plotting the corresponding velocity-time graph. Determine the acceleration and plot an acceleration-time graph.

time (s)	0.0	1.0	2.0	3.0	4.0
displacement (m [S])	0.0	20	80	180	320

8. In a certain acceleration experiment, the initial velocity is zero and the initial displacement is zero. The acceleration is shown in Figure 4-22. From the graph, determine the information needed to plot a velocity-time graph. Then, from the velocity-time graph, find the information needed to plot a displacement-time graph. (*Hint*: You should make at least three calculations on the velocity-time graph to be sure you obtain a smooth curve on the displacement-time graph.)
9. At a certain location the acceleration due to gravity is 9.82 m/s². Calculate the percent error of the following experimental values of $\vec{g}$ at that location:
   (a) 9.74 m/s²   (b) 9.95 m/s²
10. Change the following accelerations to metres per second squared:
    (a) 970 cm/s²   (c) 42 mm/s²
    (b) 8.3 × 10² mm/s²   (d) 9.0 (km/h)/s
11. (a) Is it possible to have zero velocity and still have acceleration? Explain.
    (b) Is it possible to have zero acceleration and still have velocity? Explain.
12. A ball is thrown vertically upwards. What is its acceleration
    (a) after it has left the thrower's hand and is travelling upwards?
    (b) at the instant it reaches the top of its flight?
    (c) on its way down?
13. A student throws a stone vertically upwards, and 2.8 s later it returns to the height from which she threw it. Ignoring air resistance, calculate the following:
    (a) The velocity with which the stone left her hand. (*Hint*: Assume that, when air resistance is ignored, the time taken to rise equals the time to fall for an object thrown upward.)
    (b) The height to which the stone climbed above the student's hand.
14. **Activity** Design and perform an experiment to determine the maximum height you can throw a ball vertically upward. This is an outdoor activity, requiring the use of an appropriate ball, such as a baseball, and a stopwatch.
15. An arrow is accelerated for a distance of 80 cm while it is on the bow. If the arrow leaves the bow with a speed of 80 m/s, what is its average acceleration while on the bow?
16. A carpenter wants to throw a 1.2 kg hammer to a height of 6.0 m so that it will land with zero velocity on a roof. What initial vertical speed must be given to the hammer?

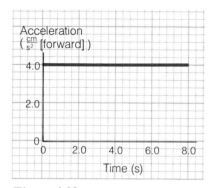

Figure 4-22

17. An astronaut on Mars accidentally drops a camera from a height of 1.6 m.
    (a) How fast will the camera be travelling after 0.50 s?
    (b) How long will the camera take to strike the surface of Mars?
18. A uniformly dripping faucet can be used as a crude timing device. Assume that one drip strikes the sink below the faucet just as the next drip starts falling. What is the vertical distance from the sink to the faucet which results in a dripping period of 0.25 s?
19. An athlete in good physical condition can land on the ground at a speed of up to 12 m/s without injury. Calculate the maximum height from which the athlete can jump without injury.
20. Two cars at the same stoplight accelerate from rest when the light turns green. Their motions are shown in the velocity-time graph in Figure 4-23.
    (a) After the motion has begun, at what time do the cars have the same velocity?
    (b) When does the car with the higher final velocity overtake the other car?
    (c) How far from the starting position are they when one car overtakes the other?

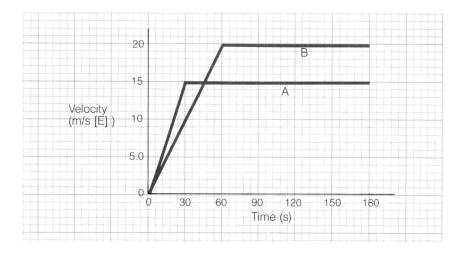

Figure 4-23

21. Estimate how long it would take an imaginary particle starting from rest and accelerating at a rate equivalent to the acceleration due to gravity on the earth to reach the speed of light in a vacuum (3.0×10^8 m/s). Then calculate the time required. (The answer is theoretical, because no particle can reach the speed of light.)
22. Describe why the topic of acceleration has more applications now than in previous centuries.
23. (a) Discuss the factors which are likely to affect the terminal velocity of an object falling in the earth's atmosphere.
    (b) Is there a terminal velocity for objects falling on the moon? Explain.
24. Governments are spending much money and physicists are spending much time on research into the fundamental particles that make up our universe. The properties of these tiny particles are

studied by means of huge particle accelerators. In one experiment, for instance, a particle called a muon, travelling at a speed of 5.00×10^6 m/s, enters a space in which electric forces produce an acceleration of -1.25×10^{14} m/s². How far does the muon travel before it comes to rest?

25. **Activity** Design and build an *accelerometer*, a device which can be used to measure acceleration directly.

Key Objectives

Having completed this chapter, you should now be able to do the following:
1. Recognize the difference between uniform motion and uniform acceleration.
2. Define either positive or negative uniform acceleration.
3. Solve a fourth variable, given any three of the variables: acceleration, time, initial velocity, and final velocity.
4. Recognize and convert common metric units of acceleration.
5. Determine the slope of a straight line on a velocity-time graph and use it to plot an acceleration-time graph.
6. Recognize that the shapes of displacement-time graphs of uniform motion and uniform acceleration differ.
7. Given data on displacement and time for uniform acceleration, plot a displacement-time graph and from it determine velocities at specific times, then plot velocity-time and acceleration-time graphs.
8. Determine experimentally the acceleration of an object which starts from rest and accelerates down a ramp.
9. Calculate the areas under the lines on velocity-time and acceleration-time graphs of uniform acceleration and state what the areas represent.
10. Know the average acceleration due to gravity at the earth's surface.
11. Design and perform an experiment in the classroom to measure acceleration due to gravity.
12. Calculate the percent error of an experimental determination of the acceleration due to gravity.
13. Use a velocity-time graph of uniform acceleration to derive equations involving displacement, velocity, acceleration, and time.
14. Use the motion equations involving acceleration, initial velocity, final velocity, displacement, and time to solve for any one variable, given the others.
15. Describe applications of acceleration.
16. Define terminal velocity and state what factors affect it.
17. Define and measure human reaction time.

CHAPTER 5
Force and Newton's Laws of Motion

Forces are everywhere. If there were no forces in our universe, our earth would not be trapped in its orbit around the sun; the uses of electricity would never have been discovered; we would not be able to operate automobiles or even walk—in fact, we would not exist, because objects need forces to keep their shape. The photograph shows the launch of the space shuttle *Discovery*. What forces are acting on the shuttle at the instant shown? Which of those forces influences your life? What other forces exist?

Large forces are involved in the launching of a space shuttle

MECHANICS

5.1 Forces in Nature

Figure 5-1
Part of an Ancient Egyptian Temple at Karnak

Did You Know?
Although many types of forces exist, physicists consider that there are only four fundamental forces which determine all others. They are gravitational force, electromagnetic force, strong nuclear force, and weak nuclear force.

Figure 5-1 shows a small portion of an Egyptian temple built near the Nile River at Karnak, Egypt, more than 3000 years ago. How did the Egyptians get the huge boulders to the top of the columns? Evidently they must have had a technique of exerting very large forces to overcome nature's downward force on the boulders.

In simple terms, a **force** is a push or a pull. Forces surround us at all times and act on everything, from the smallest particles imaginable to the largest objects in the universe. Some forces can act on objects without having to be in contact with them. Such forces include gravitational, electric, magnetic, and nuclear forces. *Gravitational forces* influence the motion of the planets around the sun and of the galaxies in the universe; they also help prevent the escape of particles from earth to outer space. *Electric forces* between charged particles help objects keep their shape and have a major influence on mechanical forces, which are discussed in the next paragraph. *Magnetic forces*, which are directly related to electric forces, are needed to operate telephones, computers, and numerous other devices. *Nuclear forces* act between the tiny particles that make up the centres of atoms.

FORCE AND NEWTON'S LAWS OF MOTION

Many other forces, called *mechanical forces*, are caused mainly by electric forces. They act when particles or objects contact each other. A basketball undergoes *compression forces* when it strikes a gymnasium floor. Then it bounces up, because *elastic forces* exert an outward force on the ball to help it regain its shape. *Frictional forces* help a car stay on the road. *Tension forces* in the cable of an elevator are needed to help the elevator to ascend and descend, and *shearing forces* cause materials to separate. Can you think of other examples of mechanical forces?

Forces play a major role in the study of physics because they determine how matter interacts with matter. In this chapter you will learn how to measure and add forces; then you will study how forces influence the motion of objects.

PRACTICE
1. List several physical forces that have influenced you today. Describe briefly how each has affected you.

Figure 5-2
Sir Isaac Newton (1642–1727)

5.2 Measuring Force

The title of this chapter refers to one of the greatest scientists in history, Sir Isaac Newton (Figure 5-2). Many of Newton's brilliant ideas resulted from Galileo Galilei's discoveries. Since Newton developed important ideas about force, it is fitting that the unit of force is called the newton (symbol N).

The newton is a derived unit in the Système International (SI), which means that it can be expressed in the metric base units of metres, kilograms, and seconds. For the remainder of your study of the topic of mechanics, it is important when you are performing calculations to express distance in metres, mass in kilograms, and time in seconds. In other words, all units must conform to the **preferred SI units** of metres, kilograms, and seconds.

A common device used to measure force in the physics laboratory is a **spring scale** (Figure 5-3). This device has a spring that extends when a pulling force is applied. The spring is attached to a needle which indicates the force on a graduated scale.

Force is a vector quantity, so its direction is important, as you will see in the next section.

Sir Isaac Newton was both an eccentric and a genius. He had a difficult childhood and in his adult life cared little for his personal appearance and social life. By the age of 26 he had made profound discoveries in mathematics, mechanics, and optics. His classic book, *Principia Mathematica*, written in Latin, appeared in 1687. It laid the foundations of physics that still apply today. In 1705 he became the first scientist to be knighted. In his later years he turned his attention to politics and theology and became the Official Master of England's mint.

PRACTICE
2. **Activity** In order to become accustomed to measuring force in SI units, obtain a spring scale calibrated in newtons. Without overstretching the spring, pull on it so you can "feel" forces of 1 N, 2 N, and so on. Next, hang various known masses on the end of the scale and determine the force required to hold up masses of 100 g, 200 g, and more.
3. Assuming you are not allowed to jump, estimate the maximum force you think you could exert on a strong rope attached firmly to (a) the ceiling and (b) the floor.

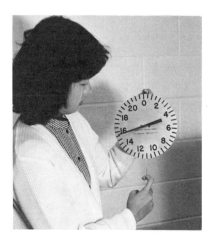

Figure 5-3
Force is measured in newtons, using a spring scale.

MECHANICS

5.3 Balanced and Unbalanced Forces Acting on a Single Object

To study the effects of forces acting on any object, we must distinguish between balanced and unbalanced forces. Imagine performing a simple experiment using a small cart resting on a table, as illustrated in Figure 5-4. Let us assume that friction is so small that we can ignore it. Two horizontal forces act on the cart — one is 3.0 N [E] and the other is 3.0 N [W]. Will the cart move? No — the forces are equal in magnitude and act in opposite directions, cancelling each other out. When such forces act on the same object, they are called **balanced forces**. This definition of balanced forces applies whether the object is stationary or moving.

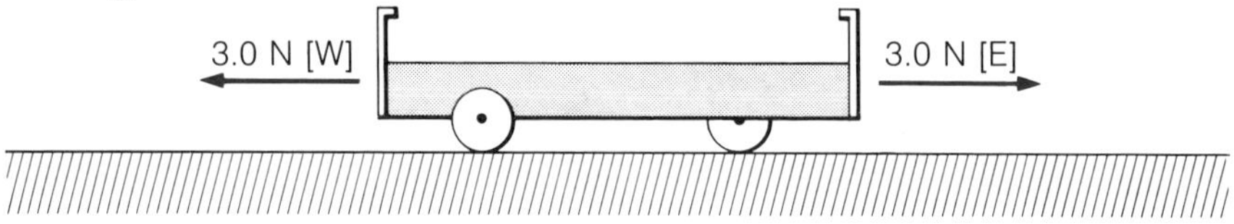

Figure 5-4
The forces on the cart are balanced: they are equal and opposite in direction.

Now consider forces that are not balanced. Figure 5-5 shows the cart from the previous example with two forces acting on it — this time, one is 5.0 N [E] and the other is 3.0 N [W]. Thus, the cart has on it an **unbalanced force**, a force which is greater in one direction than in any other. In this case, the cart will move.

Because forces are vector quantities, they can be drawn to scale in a diagram. Such is the case in Figures 5-4 and 5-5, where the scale is 1.0 cm = 2.0 N. To calculate the unbalanced force, $\vec{F}_{un}$, acting on an object, the force vectors are added in the scale diagram. If the force vectors act in one dimension, as is the case with examples in this text, one direction is called positive and the opposite direction is called negative. The forces are then added directly.

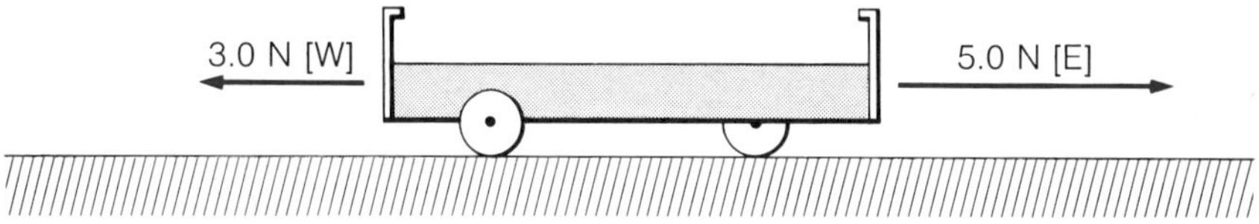

Figure 5-5
The force on the cart is unbalanced: it is greater in one direction than in the other.

SAMPLE PROBLEM 1
Find the unbalanced force on the cart in Figure 5-5.

Solution
$\vec{F}_{un} = \vec{F}_1 + \vec{F}_2$
$\phantom{\vec{F}_{un}} = 5.0 \text{ N [E]} + 3.0 \text{ N [W]}$
$\phantom{\vec{F}_{un}} = 5.0 \text{ N [E]} - 3.0 \text{ N [E]}$ (Assume east is positive.)
$\phantom{\vec{F}_{un}} = 2.0 \text{ N [E]}$
∴ The unbalanced force on the cart is 2.0 N [E].

PRACTICE

4. Calculate the unbalanced force when the following sets of forces act on the same object:
   (a) 2.4 N [N], 1.8 N [N], and 8.6 N [S]
   (b) 65 N [down], 92 N [up], and 74 N [up]
5. Figure 5-6 shows a book held at rest in a person's hand. Two forces are shown in the situation. One is the weight of the book pushing *down* on the hand, and the other is the force of the hand pushing *up* on the book.
   (a) Are the forces balanced? Explain your answer.
   (b) Assume the hand is suddenly removed. Are the forces now balanced? Explain.
6. Four identical force springs, A, B, C, and D, are used to demonstrate balanced forces in Figure 5-7. If the downward force of the mass is 10 N, predict the upward force exerted by each spring. Explain your reasoning.
7. **Activity** Set up and perform an experiment to verify your predictions in Practice Question 6.
8. **Activity** See Figure 5-8. Predict the direction the spool will move in each of the two cases when the thread is pulled gently in the direction shown. Explain your predictions, then verify them experimentally.

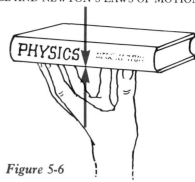

Figure 5-6

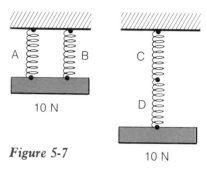

Figure 5-7

5.4 Newton's First Law of Motion

You learned about kinematics, the study of motion, in Chapters 3 and 4. So far in this chapter you have studied forces, in particular, balanced and unbalanced forces. Now you are ready to combine the two topics and study **dynamics**, the study of the causes of motion.

Relative to their surroundings, all objects are either at rest or in motion. An object at rest which has either balanced forces or no forces applied to it will remain at rest. It will begin to move only when the forces become unbalanced. The situation for an object at rest seems logical enough, but now consider a moving object.

Before Galileo's time, the common view of a moving object was that it required an unbalanced force to keep it moving. This view may seem correct when you think of pushing a box along a floor. When the forward force is removed, the box soon stops moving. But Galileo reasoned differently. He believed that a moving object has the tendency to keep moving uniformly. The reason a moving object slows down when the forward force is removed is that an unbalanced force acts on it. What is that unbalanced force? It is what we call **friction**, a force that resists the motion between two surfaces in contact. Galileo concluded that in the absence of friction a moving object maintains constant velocity. His ideas were very bold for his time, because he was not able to verify them experimentally. See Figure 5-9 on page 96.

Nowadays, we can reduce friction on the surface of the earth to try to illustrate Galileo's reasoning. In the physics laboratory, for instance,

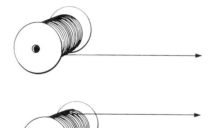

Figure 5-8

Did You Know?
The word "dynamics" stems from the Greek word *dynamis*, which means "power" or "strength". Other words, such as dynamite and dyne, have the same origin. (A dyne is a non-preferred metric unit of force equal to 10^{-5} N.)

MECHANICS

Figure 5-9
Galileo's Views of Motion

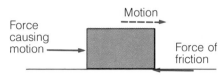

(a) *For a moving object, when the force causing the motion balances the force of friction, the object moves at a constant velocity (constant speed and direction).*

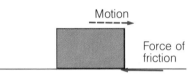

(b) *When the force causing the motion is removed, the force of friction becomes an unbalanced force causing the object to decelerate.*

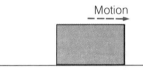

(c) *When no outside or "external" forces, including friction, act on the moving object, the object continues to move at a constant velocity.*

we can use a linear air track or an air puck for motion over short distances. However, we cannot eliminate friction completely, so all moving objects eventually come to rest. In outer space, though, there is virtually no friction. Therefore, once a space probe is moving toward another planet, it glides through space with uniform motion relative to a stationary reference point.

Sir Isaac Newton was the first scientist to put Galileo's ideas into the form of a universal physical law, one obeyed throughout the universe. Newton's **first law of motion** states:

> **An object maintains its state of rest or uniform motion unless it is acted upon by an external unbalanced force.**

Examples of Newton's first law of motion will help clarify its meaning. Consider these examples of objects at rest, which tend to remain at rest unless acted upon by an external unbalanced force:

(a) Some "magicians" can jerk a smooth tablecloth from beneath a table setting of glasses and silverware. If the tablecloth is moved quickly enough, it exerts a very small force on the glasses and silverware. Thus, these objects tend to remain at rest.

(b) A passenger standing on a bus, subway train, or other transportation vehicle appears to fall backward as the vehicle accelerates forward from rest. Actually, the passenger tends to remain at his or her initial position with respect to the stationary surroundings.

Now consider an object in uniform motion. It will continue with constant speed and direction unless an external unbalanced force acts on it. A car in uniform motion is subject to balanced forces, as illustrated in Figure 5-10. The downward force of the car on the road is balanced by the upward force of the road on the car. The backward force of moving friction is balanced by the forward force produced by the engine. As long as the forces remain balanced, the car does not accelerate.

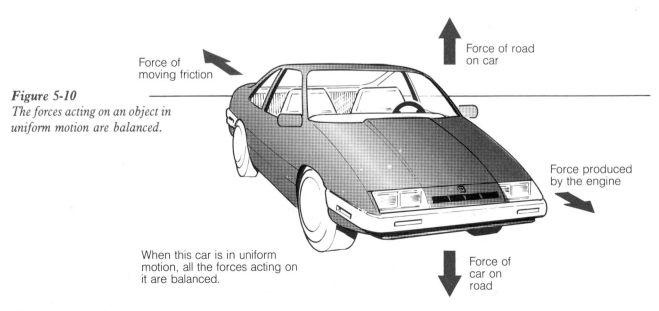

Figure 5-10
The forces acting on an object in uniform motion are balanced.

96

Here are two examples of objects in motion tending to maintain uniform motion:

(a) A speeding car approaching a curve on an icy highway has the tendency to continue in a straight line, thus failing to follow the curve. This explains why drivers must take care when driving on slippery roads.

(b) People in cars and other vehicles have the tendency to maintain uniform motion. This can be dangerous for them if the vehicle comes to a sudden stop and the passengers keep on going. The danger can be reduced through use of at least one restraint system, such as a seat belt.

Let us summarize Newton's first law of motion by stating four important results of it:

(a) Objects at rest tend to remain at rest.
(b) Objects in motion tend to remain in motion.
(c) If the velocity of an object is constant, the forces acting on it are balanced.
(d) If the velocity of an object is changing in either magnitude or direction or both, the change must be caused by an external unbalanced force acting on the object. This fact sets the stage for experimentation in dynamics.

Did You Know?
Safety experts indicate that human injuries in automobile accidents can be further reduced if more than one restraint system is used. The fastened seat belt is an important part of any restraint system, but an air bag in the steering wheel greatly lessens the chance of injury to the driver's head. When a frontal collision occurs, an electronic sensor triggers the inflation of the bag. The bag inflates within 1/30 of a second, and in less than a second deflates again.

PRACTICE

9. Astronauts are placed horizontally in their space capsule during blast-off from the launching pad. Explain why this is a good example of Newton's first law of motion.
10. A thread supports a mass hung from the ceiling; another thread of equal dimension is suspended from the mass. See Figure 5-11. Which thread is likely to break if the bottom thread is pulled slowly? quickly? Explain.
11. Explain how it would be possible to apply the first law of motion when trying to get a heap of snow off a shovel.
12. Some thrill rides at exhibitions or fairs create sensations that may be explained using the first law of motion. Describe two such rides and how the law applies to them.
13. Some people claim to have seen unidentified flying objects (UFOs) stop and start instantaneously or make right-angle turns at high speeds. What arguments can be given for or against such motion?

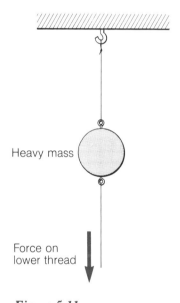

Figure 5-11

Experiment 5A: Force, Mass, and Acceleration

INTRODUCTION
The instructions for this experiment are based on the assumption that you have performed motion experiments using a recording timer or similar device. This experiment, however, is different from previous experiments in this text because it has three, rather than just two, variables.

You must therefore be "scientific" in your approach to controlling these variables.

The three variables in this experiment are unbalanced force ($\vec{F}_{un}$), mass (m), and acceleration ($\vec{a}$). You will determine how the acceleration depends on the other two variables. To accomplish this, you will keep the mass constant as you control or vary the unbalanced force, then you will keep the unbalanced force constant as you control or vary the mass. Thus, you should be able to name the dependent and independent variables.

After you have verified the relationship between acceleration and each of the other variables, you will want to express the results mathematically. To do so, first review Section 2.7, which describes direct variations. (If you have studied the Supplementary Topic in Chapter 2, you should review it as well.)

In previous acceleration experiments you measured displacements and plotted displacement, velocity, and acceleration graphs. Such a procedure would be far too tedious in this experiment, so all you must do here is calculate the acceleration using an equation. Recall that, for uniformly accelerated motion, the displacement moved during acceleration is given by $\vec{d} = \vec{v}_i t + \vec{a} t^2/2$. However, $\vec{v}_i = 0$ in this experiment, so $\vec{d} = at^2/2$. Rearranging this equation yields the following important equation:

$$\vec{a} = 2\vec{d}/t^2$$

Thus, you must measure the time for the acceleration and the displacement during that time.

You are going to use gravity to help you produce a constant force in this experiment. A mass of 100 g, for instance, has a force of about 1.0 N pulling down on it. When connected by a string to a cart and suspended over a pulley, as in Figure 5-12, the mass exerts a force of 1.0 N on the entire system—itself *and* the cart. Thus, the mass of the system being accelerated must include the suspended mass.

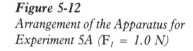

Figure 5-12
Arrangement of the Apparatus for Experiment 5A ($F_1 = 1.0$ N)

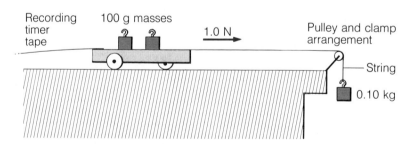

Finally, because force is expressed in the SI unit of newtons, you must express mass in kilograms and acceleration in metres per second squared.

PURPOSE

To determine the relationship between the acceleration of an object and (a) the unbalanced force applied to cause that acceleration and (b) the mass of the object, and to express the relationships in a single equation.

FORCE AND NEWTON'S LAWS OF MOTION

APPARATUS
one dynamics cart; three 100 g masses; two 1.0 kg masses; string; pulley; clamp; recording timer and related apparatus; beam balance

PROCEDURE
Part A: Getting Ready
1. Predict the relationships named in (a) and (b) of the Purpose.
2. Set up a table of observations based on the sample shown. There will be five sets of observations. The direction of the vectors depends on your experiment.
3. Measure the mass of the cart.
4. Verify that the recording timer is functioning properly.

Table of Observations

Trial	Total Mass of System (kg)	Unbalanced Force (N [ ])	Time (s)	Displacement (m [ ])	Acceleration (m/s² [ ])	Ratio of Unbalanced Force to Total Mass (N/kg)

Part B: Acceleration and Unbalanced Force
1. Set up the apparatus as shown in Figure 5-12. In this case you will vary the unbalanced force while keeping the mass of the system constant. Attach a recording timer tape about 80 cm long to the cart and feed it through the timer. Attach the string to the other end of the cart, hang it over the pulley, and suspend a 100 g mass from it.
2. Hold the cart stationary by grasping the tape in the recording timer. Turn on the timer, then release the tape. Catch the cart *before* it crashes into the pulley. Label the tape $\vec{F}_1 = 1.0$ N.
3. Repeat the procedure with an unbalanced force of 2.0 N by transferring one of the 100 g masses from the cart to the string hanging over the pulley. This allows the mass of the system to remain constant, as Figure 5-13(a) shows. Label the tape $\vec{F}_2 = 2.0$ N.
4. Repeat the procedure with an unbalanced force of 3.0 N, as in Figure 5-13(b). This tape will be used in the observations for Part C of the experiment, so label it both $\vec{F}_3 = 3.0$ N and m_1.

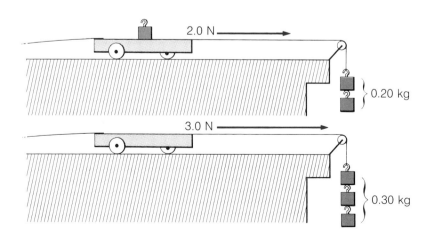

Figure 5-13
Changing the Unbalanced Forces
(The total mass must remain constant.)

(a) *An unbalanced force of 2.0 N* (F_2)

(b) *An unbalanced force of 3.0 N* (F_3, m_1)

MECHANICS

Part C: Acceleration and Mass
1. Keep the unbalanced force constant at 3.0 N, but add a 1.0 kg mass to the cart, as illustrated in Figure 5-14(a). Perform the experiment and label the tape m_2.
2. Add another 1.0 kg mass, as in Figure 5-14(b), repeat the experiment, and label the tape m_3.

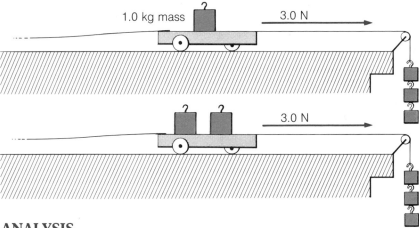

Figure 5-14
Changing the Mass (The unbalanced force must be kept constant.)

(a) Total mass is m_2.

(b) Total mass is m_3.

ANALYSIS
1. For each tape, choose the best starting point and an end point made while the unbalanced force was still being applied. In each case, use a suitable total time such as 0.5 s, 0.6 s, or whatever. Measure and calculate the data required to complete the table of observations. (Recall from the Introduction of this experiment that the acceleration can be calculated using an equation.) In the final column, express the ratio as a decimal number.
2. Plot a graph of the acceleration (vertical axis) as a function of the unbalanced force for the trials in which the mass remained constant. Draw a straight line of best fit and calculate its slope. Relate the graph to the experiment.
3. Plot a graph of the acceleration (vertical axis) as a function of the mass of the system for the trials in which the unbalanced force remained constant. (Theoretically the line on this graph should be a smooth curve.) Relate the graph to the experiment.
4. Plot a graph of acceleration (vertical axis) as a function of the reciprocal of the mass ($1/m$) for a constant unbalanced force. Draw a straight line of best fit and calculate its slope. Relate the graph to the experiment.
5. From the observation table, compare the calculated acceleration in each case to the ratio $\vec{F}_{un}/m$.
6. Use the calculations in the last two columns of your observation table to write an equation relating all three variables in this experiment.
7. How could you determine whether or not friction had an effect on the results of this experiment? Were the "unbalanced forces" (1.0 N, 2.0 N, and 3.0 N) the true forces?
8. Describe any sources of error in this experiment.

CONCLUSIONS . . .

5.5 Newton's Second Law of Motion

Newton's first law deals with situations in which the forces on an object are balanced, so no acceleration occurs. His second law deals with situations in which the external forces acting on an object are unbalanced (greater in one direction than in any other). The object therefore accelerates in the direction of the external unbalanced force. The rate of acceleration increases as the unbalanced force increases, but decreases as the mass of the object increases.

Newton's **second law of motion** summarizes these facts. It states:

The acceleration of an object varies directly as the external unbalanced force applied to it and inversely as its mass. The acceleration is in the same direction as the unbalanced force.

In mathematical form, we can state this law as follows:

$$\text{acceleration} = \frac{\text{unbalanced force}}{\text{mass}} \text{ or } \vec{a} = \frac{\vec{F}_{un}}{m}$$

This equation is often written in rearranged form:

$$\text{unbalanced force} = \text{mass} \times \text{acceleration or } \vec{F}_{un} = m\vec{a}$$

As stated earlier, the unit for force is the newton. We can use the second law equation to define the newton in terms of metric base units.

$$\vec{F}_{un} = m\vec{a}$$
$$\therefore N = kg \times m/s^2$$

Thus, we can define **one newton** as the unbalanced force required to give a 1.0 kg object an acceleration of 1.0 m/s².

Newton's second law of motion is an important law in physics. It affects all particles and objects everywhere in the universe. Because of its mathematical nature, the law is applied in many problems and questions.

SAMPLE PROBLEM 2
An unbalanced force of 48 N [W] is applied to a 4.0 kg cart. Calculate the cart's acceleration.

Solution

$$\vec{a} = \frac{\vec{F}_{un}}{m}$$

$$= \frac{48 \text{ N [W]}}{4.0 \text{ kg}} \quad \frac{\cancel{kg} \cdot m/s^2}{\cancel{kg}}$$

$$= 12 \text{ m/s}^2 \text{ [W]}$$

Therefore, the cart's acceleration is 12 m/s² [W].

MECHANICS

> **SAMPLE PROBLEM 3**
> A 2200 kg car, travelling at 25 m/s [S], comes to a stop in 10 s. Calculate (a) the car's acceleration and (b) the unbalanced force required to cause that acceleration.
>
> **Solution**
>
> (a) $\vec{a} = \dfrac{\vec{v}_f - \vec{v}_i}{t} = \dfrac{0 - 25 \text{ m/s [S]}}{10 \text{ s}}$
>
> $= -2.5 \text{ m/s}^2 \text{ [S]}$
>
> Thus, the car's acceleration is -2.5 m/s² [S], which is equivalent to 2.5 m/s² [N].
>
> (b) $\vec{F}_{un} = m\vec{a}$
> $= (2200 \text{ kg})(2.5 \text{ m/s}^2 \text{ [N]})$
> $= 5500 \text{ N [N]}$
>
> Therefore, the unbalanced force is 5.5×10^3 N [N].

Does Newton's second law agree with his first law of motion? According to the second law, $\vec{a} = \vec{F}_{un}/m$, so the acceleration is zero when the unbalanced force is zero; in other words, $\vec{a} = 0$ when the external forces are balanced. This is in exact agreement with the first law. In fact, the first law is simply a special case ($\vec{F}_{un} = 0$) of the second law of motion.

PRACTICE

14. Calculate the acceleration in each situation.
    (a) An unbalanced force of 15 N [W] is applied to a cyclist and bicycle having a total mass of 63 kg.
    (b) A bowler exerts an unbalanced force of 18 N [forward] on a 7.5 kg bowling ball.
    (c) An unbalanced force of 32 N [up] is applied to a 100 g model rocket.
15. Find the magnitude and direction of the unbalanced force in each situation.
    (a) A cannon gives a 5.0 kg shell a forward acceleration of 5.0×10^3 m/s² before it leaves the muzzle.
    (b) A 50 g arrow is given an acceleration of 2.5×10^3 m/s² [E].
    (c) A 500 passenger Boeing 747 jet (with a mass of 1.6×10^5 kg) undergoes an acceleration of 1.2 m/s² [S] along a runway.
16. Write an equation expressing the mass of an accelerated object in terms of its acceleration and the unbalanced force causing that acceleration.
17. In parts of outer space where the gravitational forces exerted by objects such as the earth, sun, and stars are practically zero, mass cannot easily be measured using a balance. However, it can be measured by means of an experiment using Newton's second law of motion. Calculate the mass of an object in such an experiment if an

unbalanced force of 87 N toward the front of the spacecraft gives it an acceleration of 150 cm/s² in that direction.
18. Derive an equation for unbalanced force in terms of mass, final velocity, initial velocity, and time.
19. Assume that during each pulse a mammalian heart accelerates 20 g of blood from 20 cm/s to 30 cm/s during a time interval of 0.10 s. Calculate the force (in newtons) exerted by the heart muscle on the blood.
20. The driver's handbook distributed by a provincial ministry of transportation states that the minimum safe distance between vehicles on a highway is the distance a vehicle can travel in 2.0 s at a constant speed. Assume that a 1200 kg car is travelling 72 km/h [S] when the truck ahead crashes into a northbound truck and stops suddenly.
    (a) If the car is at the required safe distance behind the truck, what is the separation distance?
    (b) If the average braking force exerted by the car is 6400 N [N], how long a time would the car take to stop?
    (c) Determine whether or not a collision would occur. (Assume that the driver's reaction time is an excellent 0.10 s.)

5.6 The Force of Gravity

We can apply Newton's second law of motion, in the form $\vec{a} = \vec{F}_{un}/m$, to calculate the acceleration due to gravity on the earth's surface. For example, if a 1.0 kg object is held up by a force scale, the scale reads 9.8 N. If this object is allowed to fall freely, the unbalanced force (gravity) acting on it is 9.8 N, and it will accelerate at a value of

$$\vec{a} = \frac{\vec{F}_{un}}{m} = \frac{9.8 \text{ N [down]}}{1.0 \text{ kg}}$$

$$= 9.8 \text{ m/s}^2 \text{ [down]}.$$

This value is, of course, the acceleration due to gravity ($\vec{g}$) discussed in Section 4.5 and Experiment 4B. Notice that the units N/kg are equivalent to m/s².

When an object is pulled downward by the force of gravity, it will accelerate downward unless that force is balanced by an upward force. We can use the value $\vec{g}$ = 9.8 m/s² to calculate the balancing force required to hold up an object. The equation is

$\vec{F}_w = m\vec{g}$, where $\vec{g}$ = 9.8 m/s² [down].

The $\vec{F}_w$ in this equation is called the weight of an object of mass m. **Weight** is the force of gravity pulling down on an object. Its magnitude is equal in magnitude to the force required to hold up an object or to raise an object without acceleration. See Figure 5-15. It is important, at least in physics, to remember the difference between weight and mass. Weight is the force of gravity measured in newtons. Mass is the quantity of matter measured in kilograms.

MECHANICS

Figure 5-15
Weight

(a) For an object undergoing free fall, the weight, F_w, is the unbalanced force which causes the object to accelerate.

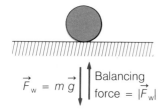

(b) When an object is supported by a surface, the weight of the object causes a downward force on the surface. This force is balanced by the upward force of the surface on the object.

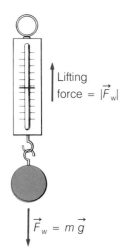

(c) The force required to raise an object at a constant speed is equal in magnitude to the object's weight.

From the equation $\vec{F}_w = m\vec{g}$, it is evident that the weight of an object of known mass changes if $\vec{g}$ changes. Although the average value of $\vec{g}$ on the earth to two significant digits is 9.8 m/s² [down], its value may change slightly depending on location. In general, the greater the distance from the centre of the earth, the lower is the value of $\vec{g}$. Thus, $\vec{g}$ is less at the top of a mountain than at the bottom of a valley. Also, $\vec{g}$ at the equator is somewhat less than at the North or South Pole, because the earth is slightly flattened at the Poles. Table 5-1 lists the values of $\vec{g}$ to four significant digits at various locations on the earth. (A more thorough analysis of the relationships involved in gravity can be found in Supplementary Topic 1 in this chapter.)

Table 5-1 Acceleration Due to Gravity at Various Locations on the Earth

Location	Latitude	Altitude (m)	$\vec{g}$ (m/s² [down])
Equator	0° [N]	0	9.780
North Pole	90° [N]	0	9.832
Java	6° [S]	7	9.782
Toronto	44° [N]	162	9.805
Brussels	51° [N]	102	9.811
Denver	40° [N]	1638	9.796

If you were to travel to the moon, you would find that the force of gravity there is only about 1/6 of that on the earth. As a result, the acceleration due to gravity on the moon is only 1.6 m/s² [down]. Accelerations due to gravity on different planets were listed in Table 4-2, Section 4.7.

SAMPLE PROBLEM 4
Calculate the weight of a fully-outfitted astronaut with a mass of 150 kg (a) on the earth and (b) on the moon.

Solution
(a) $\vec{F}_w = m\vec{g}$
$= (150 \text{ kg})(9.8 \text{ m/s}^2 \text{ [down]})$
$= 1470 \text{ N [down] or } 1.5 \times 10^3 \text{ N [down]}$

Therefore, the astronaut's weight on the earth is 1.5×10^3 N [down].

(b) $\vec{F}_w = m\vec{g}$
$= (150 \text{ kg})(1.6 \text{ m/s}^2 \text{ [down]})$
$= 240 \text{ N [down])}$

Thus, the astronaut's weight on the moon is 2.4×10^2 N [down]. Notice that, of course, the astronaut's mass does not change.

PRACTICE
21. (a) What is your mass in kilograms?
    (b) Calculate your own weight at sea level at (i) the equator and (ii) the North Pole.

22. (a) What is the weight of a 15 kg curling stone?
    (b) What force is required to raise the curling stone without acceleration?
23. To measure the weight of an object, is it proper to use a balance or a spring scale? Explain.
24. Calculate the weight of a 50 kg girl on the surface of (a) Venus and (b) Jupiter. Refer to Table 4-2.
25. Some science-fiction movies and television programs show people walking around spacecraft the way we walk around a room. How scientifically accurate is this representation?
26. A car of weight 1.8×10^4 N is accelerated uniformly from rest by an unbalanced force of 5.0×10^3 N [E] for 6.0 s. Calculate the following for the car:
    (a) its mass
    (b) its acceleration
    (c) the displacement from its original position

5.7 Newton's Third Law of Motion

Newton's first law of motion is descriptive and his second law mathematical. In both cases we consider the forces acting on only one object. When a force is applied *to* one object, however, it must be applied *by* a second object. This brings us to the third law, which considers forces acting in pairs on two objects.

Newton's **third law of motion**, often called the action-reaction law, states:

For every action force on an object there is an equal reaction force in the opposite direction on the object exerting the action force.

To illustrate the third law, imagine a boy jumping off the rear of a rowboat in a calm lake, as shown in Figure 5-16(a). He exerts a force against the boat; this force is obvious because the boat darts forward.

Figure 5-16
The Third Law of Motion

(a) *A swimmer exerts an action force on the boat, and the boat exerts a reaction force on the swimmer.*

MECHANICS

We call this the **action force**. The boy moves in a direction opposite to the boat because of the **reaction force** of the boat on him. The action and reaction forces are equal in size but opposite in direction, and act on different objects. See also Figure 5-16(b).

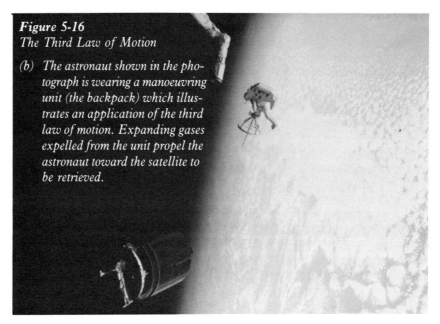

Figure 5-16
The Third Law of Motion

(b) The astronaut shown in the photograph is wearing a manoeuvring unit (the backpack) which illustrates an application of the third law of motion. Expanding gases expelled from the unit propel the astronaut toward the satellite to be retrieved.

Figure 5-17
An Apple in a Tree

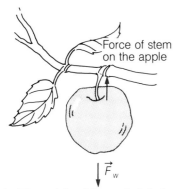

(a) *The weight of the apple is balanced by the upward force exerted by the stem. (This is not an action-reaction pair of forces.)*

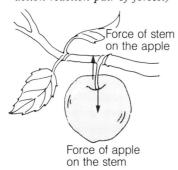

(b) *One action-reaction pair of forces exists where the stem and apple are attached.*

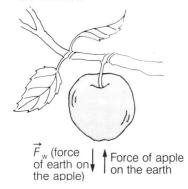

(c) *A second action-reaction pair exists between the apple and the earth.*

Do action-reaction forces exist on stationary objects? Yes, they do, but they might not seem as obvious as the example of the boy jumping off the boat. Consider, for instance, an apple hanging in a tree, as in Figure 5-17. The weight of the apple ($\vec{F}_w$) is balanced by the upward force of the stem holding the apple. However, these two forces act on the same object (the apple), so they are not an action-reaction pair. In fact, there are two action-reaction pairs in this example. One is the downward force of the apple on the stem, equal in size but opposite in direction to the force of the stem on the apple. The other is the downward force of earth's gravity, equal in size but opposite in direction to the upward force of the apple on the earth. Of course, if the stem breaks, the apple accelerates toward the earth because the earth is too large, relative to the apple, to accelerate significantly toward it.

The third law of motion has many interesting applications. As you read the following descriptions of some of them, remember there are always two objects to consider. One object exerts the *action force*; the other exerts the *reaction force*. In certain cases, one of the "objects" may be a gas such as air.

(a) When someone is swimming, the person's arms and legs exert an *action force* backward against the water. The water exerts a *reaction force* forward against the person's arms and legs, pushing his or her body forward.

(b) The propeller blades on a helicopter are designed to force air in one direction as the propeller spins rapidly. Thus, the *action force* is exerted downwards by the blades against the air. The *reaction force* is exerted by the air upward against the blades, sending the helicopter in a direction opposite to the motion of the air.

(c) A jet engine on an aircraft allows air to enter a large opening at the front of the engine. The engine compresses the air, heats it, then expels it rapidly out the rear (Figure 5-18). The *action force* is exerted by the engine backward on the expelled air. The *reaction force* is exerted by the expelled air forward on the engine, forcing the engine, and thus the entire airplane, in the opposite direction.

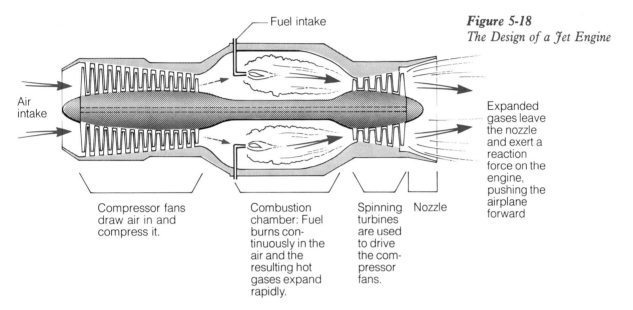

Figure 5-18
The Design of a Jet Engine

(d) A squid is a marine animal with a body size ranging from about 3 cm to 6 m. It propels itself by taking in water and expelling it in sudden spurts. The *action force* is applied by the squid backward on the discharged water. The *reaction force* of the expelled water sends the squid in the opposite direction.

PRACTICE

27. Explain each event described below in terms of Newton's third law of motion. In each case state what exerts the action force and what exerts the reaction force.
    (a) A space shuttle vehicle, like that shown in the photograph at the start of the chapter, is launched.
    (b) When a toy balloon is blown up and released, it flies erratically around the room.
    (c) A person with ordinary shoes is able to walk on a sidewalk.
    (d) A rocket accelerates in the vacuum of outer space.
28. What is meant by the term "whiplash" in an automobile accident? Explain how and why whiplash occurs.
29. (a) A certain string breaks when a force of 200 N is exerted on it. If two people pull on opposite ends of the string, each with a force of 150 N, will the string break? Explain.
    (b) Draw a force diagram of the situation in (a) showing all the action-reaction forces.

5.8 Applications of Forces

Several types of forces and their applications have already been mentioned in this chapter. Others will be described in this section, in the following discussions of gravitation and friction. Further applications of forces will be discussed later in the text.

Gravitation

The force of gravitational attraction exists between all particles of matter, everywhere in the universe. The size of the gravitational attraction between any two particles (or objects) depends on two important factors—the masses of the particles, and the distance between their centres. (Mathematical descriptions of these factors are found in Supplementary Topic 1.) The direction of gravitational attraction is along an imaginary line joining the centres of the two particles or objects.

One simple but very effective example of gravity is the use of plumb bobs in building construction and decoration. A *plumb bob* is a relatively large mass hung on a cord. To be sure that walls, door and window frames, or strips of wallpaper are vertical, a carpenter or interior decorator suspends a bob and checks whether the vertical lines are parallel to its cord. See Figure 5-19. To obtain a horizontal line, a 90° angle can be drawn from the vertical.

The force that keeps the moon in its orbit around the earth is the force of gravity. Not only does the earth exert an action force on the moon, but the moon also exerts an equal and opposite reaction force on the earth. This reaction force is evident in the formation of ocean tides on the earth. At any given time, ocean waters on two opposite sides of the earth are at high tide while the waters on the other sides are at low tide. See Figure 5-20. Since the earth rotates on its own axis once every 24 h and since the moon rises in the sky about 50 min later each day, the change of tides from one high tide to the next high tide occurs approximately every 12 h and 25 min. The tides in the Bay of Fundy in Nova Scotia are among the highest in the world. Figure 5-21 shows a photograph taken during low tide there.

Figure 5-19
Using a Plumb Bob

Figure 5-20
Ocean tides are caused by a complex set of forces involving the moon, the earth, the sun, and their motions. The diagram shows the result of these forces on an exaggerated layer of water. High tides result when water on the earth is made to bulge outward on the sides of the earth in line with the moon. The other sides of the earth experience low tide at this time.

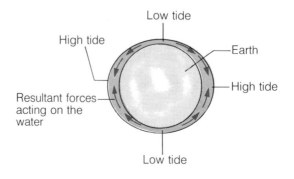

FORCE AND NEWTON'S LAWS OF MOTION

Figure 5-21
The difference between low and high tides along the coast of the Bay of Fundy is very evident in this photograph.

Like the moon, communication satellites orbit the earth. Canada is among the world's leaders in placing such satellites in a 24 h orbit directly above the equator. Each satellite must remain within a predetermined space so that messages can be sent from and received by communication dishes on the earth. Figure 5-22 shows the principle of satellite communication.

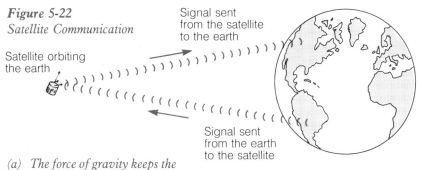

Figure 5-22
Satellite Communication

(a) *The force of gravity keeps the communication satellite in a stable orbit around the earth.*

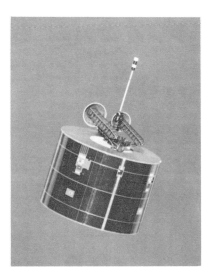

(b) *A typical communication satellite shown moments after it was deployed from a shuttle spacecraft.*

Of course, the earth and the other eight planets of the solar system are all satellites of the sun, held in orbit by the sun's gravity. So are the objects called *comets*. See Figure 5-23 on page 110. Probably the most famous of all comets is Halley's comet, which has a period of revolution around the sun of about 76 years. It is named after Edmund Halley (1656–1742). Halley, who lived in England, worked at certain times with Sir Isaac Newton. He charted the comet's path in 1682. The comet's appearance for several months in 1985–86 was predicted and anticipated by people all over the world.

The force of gravity exists throughout the universe. Our sun is but one ordinary star in our galaxy, the Milky Way, which consists of hundreds of billions of stars orbiting the galaxy's centre. And our galaxy is only one of millions of galaxies that make up the known universe and influence each other through the force of gravity.

MECHANICS

Figure 5-23
A comet is a dense heavenly body that has an elongated orbit around the sun. As a comet is attraced closer to the sun, the solar wind causes the comet's head to become brilliant, as seen in this photograph of the Mbros Comet, taken in 1957. The comet's tail can stretch for millions of kilometres. (The word "comet" is derived from the Greek word kometes, *which means "wearing long hair".)*

Figure 5-24
Starting and Kinetic Friction

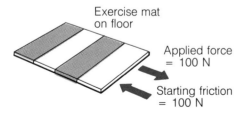

(a) *Starting friction must be overcome before an object begins moving.*

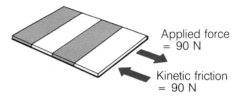

(b) *Kinetic friction must be overcome to maintain an object's motion. In general, kinetic friction between two surfaces is less than starting friction between the same surfaces.*

Friction

Friction has been defined and briefly discussed previously in this chapter, but it has been ignored in our calculations involving force and acceleration. In practical applications, however, it cannot be ignored.

Friction resists motion and acts in a direction opposite to the direction of motion. It occurs because of electrical forces at the surfaces of two objects in contact. No one would put on a pair of ice skates to try to glide along a concrete sidewalk. The friction between the sidewalk and the skate blades would prevent any skating.

One type of friction, called **static friction**, is the force that tends to prevent a stationary object from starting to move. ("Static" comes from the Greek word *statikos*, which means "causing to stand".) The maximum static friction is called the **starting friction**. It is the amount of force that must be overcome to start a stationary object moving. See Figure 5-24(a).

In certain circumstances static friction is useful; in others, it is not. A person trying to turn a stubborn lid on a jam jar appreciates the extra friction that comes with using a rubberized cloth between the lid and the hand. However, someone attempting to move a heavy filing cabinet across a floor does not appreciate static friction.

Once the force applied to an object overcomes the starting friction, the object begins moving. Then, moving or kinetic friction replaces static friction. **Kinetic friction** is the force that acts against an object's motion in a direction opposite to the direction of motion. See Figure 5-24(b).

Different types of kinetic friction have different names, depending on the situation. *Sliding friction* affects a toboggan; *rolling friction* affects a bicycle; and *fluid friction* affects a boat moving through water and an airplane flying through air.

Kinetic friction can be considered in solving motion problems. For example, assume that a parachutist of known mass is moving with a force of gravity of 630 N [downward] and a frictional air resistance of 450 N [upward]. The unbalanced force causing acceleration is 630 N − 450 N = 180 N [downward]. That force can be used to find the acceleration. Thus, the unbalanced force is the vector addition of the applied force and the frictional resistance. In equation form,

$$\text{unbalanced force} = \text{applied force} + \text{friction}$$
$$\text{or } \vec{F}_{un} = \vec{F}_{app} + \vec{F}_{fr}.$$

Because this equation involves vector quantities, direction must be taken into consideration. When you are analysing a problem involving force vectors, it is a good idea to draw a diagram showing the forces acting on an object. Such a diagram is shown in Figure 5-25 to illustrate the example of the parachutist.

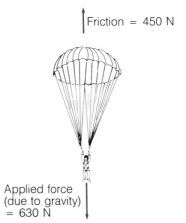

Figure 5-25
Forces Acting on a Parachutist

SAMPLE PROBLEM 5

A 60 g tennis ball is dropped from the ceiling and undergoes an average acceleration of 8.2 m/s² [down]. Calculate the average force of kinetic friction acting on the ball.

Solution
Let the downward direction of the acceleration be positive. The applied force is the weight ($\vec{F}_w$) of the ball, as shown in the diagram.

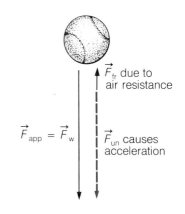

$$\vec{F}_{app} = \vec{F}_w = m\vec{g}$$
$$= (0.060 \text{ kg})(9.8 \text{ m/s}^2 \text{ [down]})$$
$$= 0.59 \text{ N [down]}$$
$$\vec{F}_{un} = m\vec{a}$$
$$= (0.060 \text{ kg})(8.2 \text{ m/s}^2 \text{ [down]})$$
$$= 0.49 \text{ N [down]}$$
$$\vec{F}_{fr} = \vec{F}_{un} - \vec{F}_{app}$$
$$= 0.49 \text{ N [down]} - 0.59 \text{ N [down]}$$
$$= -0.10 \text{ N [down] or } 0.10 \text{ N [up]}$$

∴ The average kinetic friction acting on the ball is 0.10 N [upward].

Throughout history, humankind has found ways to reduce kinetic friction. You read in Chapter 1 that 4500 years ago the Egyptians built enormous pyramids using huge stone blocks that were difficult to move by sliding. To move the blocks they placed logs beneath them and pushed them. By doing this, they were taking advantage of the fact that rolling friction is much less than sliding friction. A similar technique is used nowadays by fishermen on the coast of Senegal as they move their boats away from the water (Figure 5-26).

Modern technology uses the same principles as the Egyptians, though in a more sophisticated way. We try to reduce undesirable friction for many reasons. For instance, all machines have moving parts that experi-

Figure 5-26

PROFILE

H.W. Wevers, P.Eng.
Clinical Mechanics Group
Department of Mechanical Engineering
Queen's University

As a high school student in the Netherlands, I had always been interested in mathematics and science. However, I did not finally decide on what specific career I wished to follow until I was 18 years old. Like all 18- and 19-year-old males in the Netherlands, I was expected to serve in the army for two years. It was during this period of compulsory military service that I came to realize that my future lay in advanced engineering.

I did not want to limit my options by choosing a specialized field of study, so I opted for mechanical engineering. Later, I obtained an advanced degree in manufacturing engineering from the University of Delft. Upon graduation, I was hired as an engineer in a steel plant, but after only four years, I decided to leave the plant and return to the academic environment as a teacher of engineering.

In 1970 my young family and I immigrated to Canada where I had accepted a position as an Assistant Professor in Mechanical Engineering at Queen's University, Kingston, Ontario. Canada appealed to my wife and I because of its space, and the gentleness of its social system as reflected in good education, medical care, and social securities.

What I did not foresee was how dramatically the emphasis of my work would change in Canada. Looking back, I can see that the turning point came in 1974 when I befriended a colleague at Queen's who specialized in biomechanical engineering. His concerns about such problems as designing orthopaedic implants to help people with malfunctioning joints instantly captured my interest.

Eleven years later, I was appointed Professor of Mechanical Engineering as well as the Engineering Coordinator of the Clinical Mechanics Group at Queen's. This is a group of five senior researchers like myself supported by a group of approximately ten assistants and students.

Professor Wevers, on the left, working in his laboratory with a colleague, G. Saunders.

ence friction during operation. Friction can wear out the machines, reduce efficiency, and cause unwanted heat. (If you rub your hands together vigorously, you can feel the heat produced by friction.) Excess friction in machines can be overcome by making surfaces smooth, using materials with little friction, lubricating with grease or oil, and using bearings.

Bearings function on the principle of the rolling logs used by the Egyptians to move stones. A *bearing* is a device containing many rollers or balls that reduce friction while supporting a load. Bearings change sliding friction into rolling friction, reducing friction by up to 100 times.

Ways of reducing undesirable friction in other situations are also common. The wax applied to cross-country skis reduces sliding friction. A layer of air between a hovercraft and the water reduces fluid friction in a manner similar to the use of air pucks and the linear air track in a physics laboratory. A human joint is lubricated by *synovial fluid* between the layers of cartilage lining the joint. The amount of lubrication provided by synovial fluid increases when a person moves, giving an excel-

For the most part, we study ways in which we can help doctors rehabilitate patients who suffer from orthopaedic disabilities.

I specialize in joint prosthesis design, specifically, for the elbow and knee. In other words, I design implants to replace defective parts of an individual's joint. This is not as easy as it may at first sound. Joints are very complicated. To create the best prostheses, we have to develop models of human joints.

This is accomplished by a variety of methods. First, we take measurements of volunteer subjects. Then we compare these with X-ray photographs to get a clearer picture of how joints are built. Finally, we take joints from several embalmed cadavers and cut them into thin, 2 mm slices. Each slice is photographed in order to see the contours of the bone and cartilage that make up the joint. Then each photograph is digitized, or turned into a form that can be read by a computer. The computer takes the results of the digitized photographs and turns them into a three-dimensional image of the joint. Such an image can show us how the joint is shaped from any angle.

Using this three-dimensional model, we are able to help surgeons detect problems in the functioning of people's joints. It also helps us determine which of the hundreds of different prostheses that have been developed will be most suitable for use in a patient who requires a surgical implant.

This is by no means the only aspect of the Group's research. We have also invented new precision saw jigs which are useful during knee surgery. We have developed a unique weight-bearing standard X-ray frame for photographing the knee alignment, and we are studying new, longer lasting fixation techniques.

A vital part of our work involves brainstorming in the weekly meetings of the Group so that engineers, surgeons, and assistants can suggest new approaches to particular problems and plan further phases of research. This helps us develop new programs with other members of the faculty, with engineers and surgeons throughout the country.

Besides teaching, half of my time goes to research, doing the laboratory work with students and publishing the results. Some of our ideas have been patentable, and this has contributed to Canada's competitiveness in world trade.

I believe that in the future, medical technology will rely increasingly on artificial components to ensure that people can live long and useful lives. However, I feel it is vital that we do not forget the most important area of health care – prevention of disease through a healthy lifestyle.

lent example of the efficiency of the human body. In fact, our lubrication systems work so excellently that it is difficult for technologists to design artificial joints that function as well! See Figure 5-27.

Figure 5-27
Reducing Undesirable Friction

(a) *This hovercraft carries cars across the English Channel.*

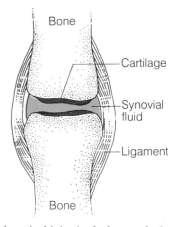

(b) *A typical joint in the human body*

MECHANICS

Figure 5-28
Tight-fitting Stones of a Large Inca Wall near Cuzco, Peru

It has been pointed out that, although friction is often undesirable, it can be useful. Consider, for example, a problem encountered by the Incas, who dominated a large portion of South America before the Europeans arrived in the sixteenth century. South America has many earthquake zones and, of course, buildings have a tendency to crumble during an earthquake. To help overcome this difficulty, the Inca stonemasons developed great skill in fitting building stones together very tightly so that a great deal of sliding friction would help hold their buildings together, even during an earthquake. Figure 5-28 shows an example of the skill of the Incas.

Now consider the modern technology involved in designing roads, bridges, automobile tires, athletic shoes, and surfaces of playing fields. Without friction, driving on highways and running on playing surfaces would not only be dangerous — it would be impossible! Friction between the moving parts and the stationary surfaces aids all types of acceleration: speeding up, slowing down, and changing direction. Engineers consider friction when they design treads for tires and athletic shoes, as well as surfaces for roads, bridges, and playing fields. Assume that the kinetic friction between rubber and ice is given a value of 1.0. Then the amount of kinetic friction between rubber and wet asphalt is 40. In other words, there is 40 times as much friction to help stop a car on wet asphalt as there is on a patch of ice. Table 5-2 compares the friction between rubber and common surfaces used on roads. All automobile drivers and athletes should be aware of the consequences of the reduction of helpful friction.

Table 5-2 Friction between Rubber and Other Surfaces

Materials	Friction (Compared to Rubber on Ice)
Rubber on ice	1
Rubber on wet asphalt	40
Rubber on dry asphalt	80
Rubber on wet concrete	60
Rubber on dry concrete	140

Figure 5-29
A Level

PRACTICE

30. What do you think happens to the gravitational attraction between two objects
    (a) when their masses increase?
    (b) when the distance between their centres increases?
31. The photograph in Figure 5-29 shows a device called a *level*, which is a common alternative to a plumb bob.
    (a) Describe how the level can be used to determine vertical and horizontal lines.
    (b) If the inside diameter of the hollow tube holding the liquid is not symmetrical, will the accuracy of the level be affected? Explain.
32. Assume that in a coastal village a low tide occurs at 4:20 h. Predict the times for the next two high tides and two low tides. (Your answers

will not necessarily coincide with the real-life situation because tides are also affected by local conditions.)

33. Research the meanings and causes of "spring tides" and "neap tides".
34. A communication satellite has small engines aboard that are used to keep the satellite within its proper space. Use at least one of Newton's laws to explain the use of the engines.
35. Determine the number of times Halley's comet has returned to intrigue earthlings since Edmund Halley made his discovery. Predict the year of its next passing.
36. Astronomers have used equations derived by Sir Isaac Newton to estimate that the Milky Way Galaxy has a mass of approximately 4×10^{41} kg. If the mass of the sun is 2×10^{30} kg, how many suns are there in our galaxy? (Assume for this question that all the suns have the same average mass and that the masses of any planets are negligible.)
37. Explain each of the following statements, taking into consideration the force of friction.
    (a) Streamlining is important in the transportation industry.
    (b) Friction is necessary to open a closed door that has a doorknob.
    (c) A highway sign reads, "Reduce speed on wet pavement."
    (d) Cars with front-wheel drives have distinct advantages over cars with rear-wheel drives.
    (e) Screwnails are useful for holding pieces of wood tightly together.
38. **Activity** Design and perform an experiment to determine how certain factors affect friction. You will need a force scale, a variety of surfaces, at least one wooden block, a cart, and masses to place on the block or cart. In your experiment, consider some or all of the following problems:
    (a) To what extent does the type of friction (starting, sliding, rolling) affect its magnitude?
    (b) Does an object's weight influence friction?
    (c) How does the contact area affect sliding friction?
    (d) Is sliding friction affected by the speed of motion?
    (e) How do the types of surface in contact affect friction?
39. For each situation given below, draw a diagram of the forces acting on the object, then answer the question. Consider that all forces mentioned are acting horizontally.
    (a) A butcher pulls on a freshly cleaned side of beef with a force of 2.2×10^2 N. The frictional resistance between the beef and the countertop is 2.1×10^2 N. What is the unbalanced force exerted on the beef?
    (b) An unbalanced force of 12 N [S] results when a force of 51 N [S] is applied to a box filled with books. What is the frictional resistance on the box?
    (c) Two students are exerting an applied horizontal force on a piano. The frictional resistance on the piano is 92 N [E] and the unbalanced force on it is 4 N [E]. What is the force on the piano applied by the students? Explain what is happening.

MECHANICS

40. A 60 kg sprinter accelerates from rest into a strong wind which exerts an average frictional resistance of 60 N. If the ground applies a forward force of 240 N on the sprinter's body, calculate the following:
    (a) the sprinter's unbalanced force
    (b) the sprinter's acceleration
    (c) the distance travelled in the first 2.0 s
41. A skydiver, initially undergoing a high acceleration, is approaching her terminal velocity. At a given instant the fluid friction (due to air resistance) on the diver and her diving gear is 680 N [up], and the unbalanced force causing acceleration is 40 N [down].
    (a) Calculate the applied force on the diver.
    (b) What is the cause of the applied force?
    (c) Find the mass of the diver and her gear.
42. Explain how you would solve each of the following problems. Relate your answer to friction.
    (a) A refrigerator door squeaks when opened or closed.
    (b) A small throw rug at the front entrance of a home slips easily on the hard floor.
    (c) A picture frame hung on a wall falls down because the nail holding it slips out of its hole.

Review Assignment

1. Name the type(s) of force responsible for each of the following:
   (a) A nickel is attracted to a special steel bar.
   (b) A coasting cyclist gradually comes to a stop.
   (c) An electron in a hydrogen atom travels in an orbit around a proton.
   (d) A meteor (or "shooting star") enters the earth's atmosphere.
   (e) The meteor in (d) begins to burn and give off light.
2. An airplane is travelling with constant speed, heading east at a certain altitude. What forces are acting on the airplane? Are the forces balanced?
3. A car has a fuel consumption of 7.2 L/100 km on an expressway and 9.5 L/100 km in the city. Explain why there is a difference.
4. **Activity** From a standing position, bend over, grab your ankles, and exert a large force to try to lift yourself off the floor. Use at least one of Newton's laws to explain what happens.
5. The scale used to draw the forces in Figure 5-30 is 1.0 cm = 2.0 N.
   (a) Find the unbalanced force acting on the cart.
   (b) If the cart's mass is 1.2 kg, what is its acceleration?

Figure 5-30

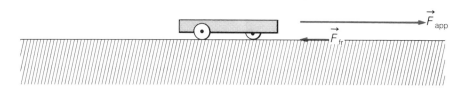

FORCE AND NEWTON'S LAWS OF MOTION

6. Use Newton's first law of motion to explain why it is necessary to push harder on a bicycle's pedals when accelerating than when coasting.
7. A tractor pulls forward on a moving plow with a force of 2.5×10^4 N, which is just large enough to overcome friction.
    (a) What are the action and reaction forces between the tractor and the plow?
    (b) Are these action and reaction forces balanced? Can there be any acceleration? Explain.
8. A punter kicks a 410 g football, giving it an acceleration of $|25\vec{g}|$ for 0.10 s.
    (a) What force is imparted to the ball?
    (b) Name and state the value of the reaction force.
9. (a) Calculate the earth's force of gravity on each of two steel balls of mass 5.0 kg and 10 kg.
    (b) If the force of gravity on the 10 kg ball is greater than on the other ball, why do the two balls accelerate at the same rate when dropped?
10. Where on the surface of the earth do you predict you would weigh the most? the least?
11. Just prior to a satellite recovery in the ocean, a parachute attached to a 150 kg satellite exerts a retarding force of 1100 N. Calculate the satellite's acceleration.
12. A vegetable vendor sets up a stall in an elevator of a tall building. She uses a spring scale to measure the weight of the vegetables in newtons. Under what conditions would it be advantageous
    (a) for you to buy from the vendor?
    (b) for the vendor to sell to you? Explain your answers.
13. One of the world's greatest jumpers is the flea. For a brief instant a flea can accelerate at a rate of about 10^3 m/s². What force would a 6.0×10^{-7} kg flea have to exert to produce this acceleration?
14. In an electronic tube an electron of mass 9.1×10^{-31} kg experiences an unbalanced force of 8.0×10^{-15} N over a distance of 2.0 cm.
    (a) Calculate the electron's acceleration.
    (b) Assuming it started from rest, how fast was the electron travelling at the end of the 2.0 cm motion?
15. A 1.2×10^4 kg truck is travelling south at 20 m/s.
    (a) What force is required to bring the truck to a stop in 300 m?
    (b) What is the cause of this force?
16. Calculate the acceleration of the cart shown in Figure 5-31, given the following assumptions:
    (a) There is no friction acting on the cart.
    (b) A frictional resistance of 2.0 N is acting on the cart.

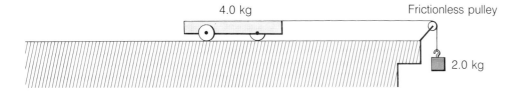

Figure 5-31

MECHANICS

17. You are standing on the edge of a frozen pond where friction is negligible. In the centre of the ice is a red circle 1.0 m in diameter. The Physics-for-Fun Society is offering a prize of a megadollar if you can apply all three of Newton's laws of motion to get to the red circle and stop there. Describe what you would do to win the prize.

18. A shuffleboard disk of mass 500 g accelerates forward under an applied force of 12 N.
    (a) If the frictional resistance is 8.0 N, find the disk's acceleration.
    (b) If the disk moves from rest for 0.20 s, how far does it travel while accelerating?

19. Each of the four wheels of a car pushes on the road with a force of 4.0×10^3 N [down]. The driving force produced by the engine on the car is 8.0×10^3 N [W]. The frictional resistance on the car is 6.0×10^3 N [E]. Calculate the following:
    (a) the mass of the car
    (b) the unbalanced force on the car
    (c) the car's acceleration

20. The photograph in Figure 5-32 shows a microscopic view of a surface that appears smooth to the unaided eye. Describe what the photograph reveals about friction.

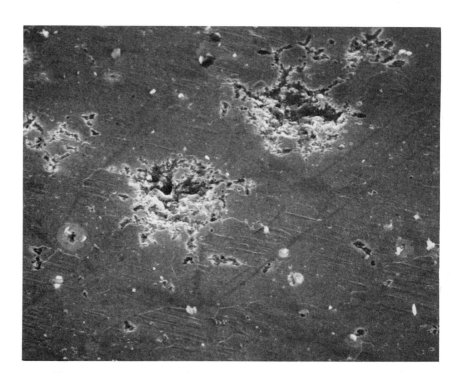

Figure 5-32
This photomicrograph of a polished metal surface was taken at a magnification of 250X.

21. Is friction desirable or undesirable when you are trying to tie a knot in a string? Explain.

22. A rocket sled weighing 2.0×10^4 N is gliding at 24 m/s [N] on ice where friction is negligible. Suddenly it passes over a rough patch 22 m long, which creates a frictional force of 6.0×10^3 N [S]. With what velocity does the sled leave the rough patch?

Supplementary Topic 1: Newton's Law of Universal Gravitation

One day, while sitting under an apple tree, Isaac Newton was struck on the head by a falling apple. Science historians relate that this seemingly unimportant event led to Newton's formation of a fundamental law of nature involving gravitational forces. The apple struck him on the head because it was pulled toward the earth by the force of gravity. Why then, Newton hypothesized, could it not be concluded that the moon stays in its orbit around the earth because it too "falls" toward the earth as it is influenced by the force of gravity?

After consulting known data about the moon's orbit around the earth and determining how the force of gravity depends on other variables, Newton devised the law of universal gravitation. Omitting directions, the relationships he discovered can be stated using these symbols:

F_G is the force of gravitational attraction between any two objects.
m_1 is the mass of one object.
m_2 is the mass of a second object.
d is the distance between the centres of the two objects.

If m_2 and d are constant, $F_G \propto m_1$ (direct variation).
If m_1 and d are constant, $F_G \propto m_2$ (direct variation).

If m_1 and m_2 are constant, $F_G \propto \dfrac{1}{d^2}$ (inverse square variation).

Thus, $F_G \propto \dfrac{m_1 m_2}{d^2}$ (joint variation).

$\therefore F_G = \dfrac{k m_1 m_2}{d^2}$, where k ≠ 0. See Figure 5-33.

Newton's **law of universal gravitation** therefore states:

> **The force of gravity between any two objects is directly proportional to the product of the masses of the objects and inversely proportional to the square of the distance between them. (The direction of the force extends along an imaginary straight line joining the centres of the two objects.)**

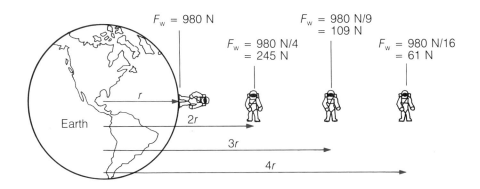

Figure 5-33
The force of gravity is inversely proportional to the square of the distance between two objects. In this case, the distance is measured between the centre of the earth and a fully outfitted astronaut whose mass is 100 kg.

MECHANICS

Did You Know?
The inverse square relationship applies to many phenomena, including electric forces and light intensity. For example, the intensity of light from a point source varies inversely as the square of the distance from the source.

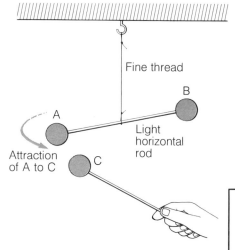

Figure 5-34
Cavendish's Experiment

The value of the constant, k, in Newton's gravitational equation is extremely small; Newton determined it by calculations. Verification of the value by direct experiment did not occur until over a century after Newton formulated the law. Then, the English physicist Henry Cavendish (1731–1810) performed an experiment using a set-up like that illustrated in Figure 5-34. When ball C was brought close to ball A, the force of gravitational attraction caused the double ball system (A and B) to rotate a measured amount. Since Cavendish had previously determined the force needed for a certain twist, he could now find the force of attraction between A and C. Using this technique, he calculated the value of the constant in Newton's equation, which now has the symbol G, to be 6.67×10^{-11} N·m²/kg².

Thus, Newton's law of universal gravitation in equation form is

$$F_G = \frac{Gm_1m_2}{d^2}$$

where G is 6.67×10^{-11} N·m²/kg²,
m_1 and m_2 are the masses of the objects measured in kilograms, and
d is the distance between the centres of the objects measured in metres.

SAMPLE PROBLEM 6
Calculate the force of gravitational attraction between two spherical boulders, each of mass 450 kg, whose centres are separated by a distance of 5.0 m.

Solution

$$F_G = \frac{Gm_1m_2}{d^2}$$

$$= \frac{6.67 \times 10^{-11} \frac{N \cdot m^2}{kg^2} \times 450 \text{ kg} \times 450 \text{ kg}}{(5.0 \text{ m})^2}$$

$$= 5.4 \times 10^{-7} \text{ N (Notice the cancellation of units.)}$$

∴ The force of attraction is 5.4×10^{-7} N, an extremely small force.

PRACTICE
43. Calculate the force of gravitational attraction between two spherical objects ($m_1 = 2.0 \times 10^3$ kg and $m_2 = 2.0 \times 10^4$ kg) separated by a distance of 4.0 m.
44. The mass of the earth can be calculated by knowing that the weight of an object (in newtons) is equal to the force of gravity between the object and the earth. Given that the radius of the earth is 6.4×10^6 m, determine its mass.
45. Find the force of attraction between the earth and the moon, using the following data:
    mass of the moon = 7.35×10^{22} kg
    mass of the earth = 5.98×10^{24} kg
    average distance from the earth to the moon = 3.84×10^8 m

Supplementary Topic 2: Impulse and Momentum

Consider the following two questions:

(a) If you were struck by a baseball, would you prefer a fast-moving or a slow-moving ball?

(b) If you were struck by a ball moving at 1.0 m/s, would you prefer the ball to be a table-tennis ball or a bowling ball of much larger mass?

The answers to these questions are obvious. What might not be obvious, however, is that the situations have in common an important physical quantity called "momentum". Likely you have heard the word before. In the first case, the baseballs have the same mass, but the faster baseball has a greater velocity and thus a greater momentum. As a result, it can do more damage. In the second case, the balls have the same velocity, but the bowling ball has a higher mass and thus a greater momentum, so it can do more damage.

From these examples, we can derive an equation for the momentum (symbol $\vec{p}$) of an object in terms of the object's velocity ($\vec{v}$) and mass (m).

When m is constant, $\vec{p} \propto \vec{v}$ (direct variation).
When $\vec{v}$ is constant, $\vec{p} \propto m$ (direct variation).
Thus, $\vec{p} \propto m\vec{v}$ (joint variation).
$\therefore \vec{p} = km\vec{v}$, where k $\neq$ 0.

If the momentum is measured in the appropriate SI units, then k = 1, and the equation for momentum is

$$\vec{p} = m\vec{v}.$$

Therefore, the **momentum** of an object is the product of its mass and velocity. Momentum is a vector quantity and has SI units of kilogram metres per second (kg·m/s). For example, a 2.0 kg ball moving at 3.0 m/s [E] has a momentum of 6.0 kg·m/s [E].

No doubt you can imagine how the momentum of an object of known mass can be changed. Since the object's mass does not change, its velocity must change. A changing velocity implies acceleration. From the second law of motion, acceleration is caused by an unbalanced force. Thus, a change in momentum ($\Delta\vec{p}$) results from an unbalanced force causing a change in velocity ($\Delta\vec{v}$). Therefore, $\Delta\vec{p} = m\Delta\vec{v}$. Combining these facts with Newton's second law gives us

$$\vec{F}_{un} = m\vec{a}, \text{ where } \vec{a} = \frac{\Delta\vec{v}}{t}$$

$$\therefore \vec{F}_{un} = \frac{m\Delta\vec{v}}{t}$$

and $\vec{F}_{un}t = m\Delta\vec{v}$

or $\vec{F}_{un}t = \Delta\vec{p}.$

MECHANICS

In this last equation, the product $\vec{F}_{un}t$ is called the **impulse**. Therefore, an impulse applied to an object equals the object's change in momentum. The impulse and the change in momentum are in the same direction.

SAMPLE PROBLEM 7
A 150 g baseball travelling at 24 m/s [W] is struck by a bat. It then travels at 40 m/s [E]. Calculate:
(a) the impulse imparted by the bat on the ball
(b) the time of contact if the average force exerted by the bat is 7.0×10^3 N [E]

Solution

(a) impulse $= \vec{F}_{un}t = m\Delta\vec{v}$
$= m(\vec{v}_f - \vec{v}_i)$
$= (0.15$ kg$)(40$ m/s [E] $- 24$ m/s [W])
$= (0.15$ kg$)[40$ m/s [E] $- (-24$ m/s [E])]
$= (0.15$ kg$)(64$ m/s [E])
$= 9.6$ kg·m/s [E]

∴ The impulse on the ball is 9.6 kg·m/s [E], or 9.6 N·s [E]. (Refer to Practice Question 46.)

(b) $\vec{F}_{un}t = m\Delta\vec{v}$ ∴ $t = \dfrac{m\Delta\vec{v}}{\vec{F}_{un}}$

$= \dfrac{9.6 \text{ N·s [E]}}{7.0 \times 10^3 \text{ N [E]}}$

$= 1.4 \times 10^{-3}$ s

Thus, the bat was in contact with the ball for 1.4×10^{-3} s.

Momentum is a useful quantity to use when solving certain types of physics problems, as you will learn in the questions and experiment that follow.

PRACTICE

46. Prove that N·s and kg·m/s are equivalent.
47. Calculate the momentum of the following:
    (a) a 0.42 kg football travelling at 24 m/s [E]
    (b) a 58 g tennis ball travelling at 50 m/s [N]
    (c) a 1500 kg car travelling at 72 km/h [W]
48. An unbalanced force of 2000 N [up] acts on a 500 kg rocket, causing its velocity to go from rest to 200 m/s [up]. How long does the impulse last?
49. An impulse of 3.2×10^3 N·s [W] is applied to a cannonball of unknown mass. If the ball started at rest and reached a velocity of 320 m/s [W], calculate its mass.
50. A tugboat of mass 8.0×10^5 kg is moving at 0.1 m/s [S].
    (a) What impulse is required to stop the tug?

(b) What unbalanced force must act on the tug to stop it in 2.0 min?
51. A soccer goalie kicks the ball with an unbalanced force of 1200 N for 8.0×10^{-3} s. If the 420 g ball starts from rest, how fast is it now travelling?

Experiment 5B: Momentum in Collisions

INTRODUCTION
In this experiment, as in several previous motion experiments, you must find the velocities of moving objects. The instructions are written for the recording timer technique. However, they can be adapted readily to other techniques.

You may find difficulty in creating the desired motion in this experiment. Read the instructions and practise the collisions before you begin taking measurements.

PURPOSE
To compare the momentum of two colliding objects before and after the collision when the objects
(a) stick together after the collision
(b) separate after the collision.

APPARATUS
three dynamics carts; adhesive substance (such as "Velcro"); two recording timers and related apparatus; beam balance

PROCEDURE
1. Check that the recording timers are functioning properly. Measure the mass of each cart.
2. Set up the apparatus as shown in Figure 5-35. Turn on the recording timers and give cart A a good push toward cart B. If the carts stick together and move together after the collision, determine the velocity of A before the collision and of AB after the collision.
3. Repeat the experiment with the third cart, C, atop A.
4. Set up the same apparatus, with the adhesive substance removed. (If cart A has a collision rod, aim it at B.) Turn on the recording timers and push A so that it collides with B. Determine the velocity of A before and after the collision and of B after the collision.
5. Repeat Step 4 with cart C atop A.

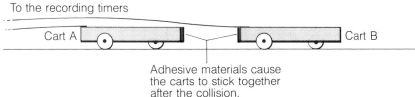

Figure 5-35
Set-up for Experiment 5B

ANALYSIS
1. For each collision, determine and compare the total momentum of the "system" of carts before and after the collision.
2. Explain the sources of error in this experiment.

CONCLUSIONS . . .

Supplementary Topic 3: The Law of Conservation of Momentum

There are many situations where objects collide or in other ways interact with one another. If we isolate the colliding objects, or if we neglect or eliminate the effects of external forces such as friction on the colliding objects, then we can conclude that momentum in a collision is conserved. This important concept is called the **law of conservation of momentum**:

> **The total momentum of all parts of a system before an interaction equals the total momentum after, if no external unbalanced force acts on the system.**

These words can be translated into mathematical form as follows:

initial momentum = final momentum
$$\vec{p}_i = \vec{p}_f$$

Now, for interacting objects A and B,

$$(\vec{p}_A + \vec{p}_B)_i = (\vec{p}_A + \vec{p}_B)_f$$
or $\quad m_A\vec{v}_{A_i} + m_B\vec{v}_{B_i} = m_A\vec{v}_{A_f} + m_B\vec{v}_{B_f}.$

This mathematical form of the law of conservation of momentum can be applied in numerous situations to solve motion problems.

SAMPLE PROBLEM 8

Car A, of mass 2.0×10^3 kg, is moving 9.0 m/s [E] and collides head-on with car B, mass 1.5×10^3 kg. Both cars come to a halt. Find the initial velocity of car B.

Solution

$$m_A\vec{v}_{A_i} + m_B\vec{v}_{B_i} = m_A\vec{v}_{A_f} + m_B\vec{v}_{B_f}$$

$$\therefore \vec{v}_{B_i} = \frac{m_A\vec{v}_{A_f} + m_B\vec{v}_{B_f} - m_A\vec{v}_{A_0}}{m_B}$$

But since $\vec{v}_{A_f} = \vec{v}_{B_f} = 0$, we have

$$\vec{v}_{B_i} = -\frac{m_A\vec{v}_{A_i}}{m_B}$$

$$= -\frac{2.0 \times 10^3 \text{ kg} \times 9.0 \text{ m/s [E]}}{1.5 \times 10^3 \text{ kg}}$$

$$= -12 \text{ m/s [E]}.$$

Therefore, car B's initial velocity is -12 m/s [E] or 12 m/s [W].

The law of conservation of momentum can also be applied when an explosion or separation sends parts of a system moving off in opposite

FORCE AND NEWTON'S LAWS OF MOTION

directions. For example, when a chemical explosion occurs in a rifle, the momentum given to the bullet as it leaves the muzzle in one direction is equal in magnitude to the momentum of the rifle as it recoils in the opposite direction. See Practice Question 56.

PRACTICE

52. **Activity** An interesting physics apparatus consists of steel balls suspended by strings and touching each other, as shown in Figure 5.36. Experiment with the apparatus by pulling aside one ball and releasing it. Then repeat, using two balls, then three, and so on. Explain what occurs using the momentum conservation law.
53. A girl is standing on a skateboard (total mass = 57 kg). A 60 kg boy throws a 500 g ball toward her with a velocity of 21 m/s [S]. The girl catches the ball. Neglecting any friction, determine the velocity with which the girl and the skateboard now move.
54. A neutron of mass 1.7×10^{-27} kg is moving at 2.7×10^3 m/s and collides head-on with a nitrogen atom at rest. If the mass of the nitrogen atom is 2.3×10^{-26} kg and if the two particles stick together after the collision, what is the final velocity of the particles?
55. An experiment is performed in a physics laboratory to find the mass of a stationary particle, B. A proton, A, of mass 1.7×10^{-27} kg, travelling at 4.0×10^6 m/s, strikes B and bounces straight back at a speed of 2.0×10^6 m/s. If B moves ahead at 1.0×10^6 m/s, calculate B's mass.
56. Two students, C and D, are facing each other at rest on "frictionless" roller skates. C has a mass of 80 kg and D has a mass of 50 kg. Now they push each other and D acquires a velocity of 4.4 m/s [W]. What velocity does C acquire?
57. A meteor crater is formed when a meteorite crashes into the earth. It is estimated that one meteorite had a mass of 10^{10} kg and was travelling at 10^4 m/s when it collided with the earth. Determine the earth's change of velocity as a result of the collision. (Assume that the earth's initial velocity was zero.)

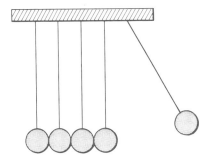

Figure 5-36

Key Objectives

Having completed this chapter, you should now be able to do the following:
1. List four forces that can act without contact.
2. State examples of mechanical forces.
3. State the SI unit of force.
4. Name and use the device commonly available to measure force.
5. Distinguish between balanced and unbalanced forces.
6. Calculate the unbalanced force acting on a single object with two or more parallel forces applied to it.
7. Define dynamics.
8. Appreciate why Galileo's explanations of motion differed from previously held explanations.

9. State Newton's first law of motion and describe examples illustrating and applying it.
10. Experimentally determine how the acceleration of an object depends on its mass and the unbalanced force applied to it.
11. State Newton's second law of motion in both sentence and equation form.
12. Apply the second law to the solution of motion problems.
13. Define the newton in terms of base metric units.
14. Distinguish between mass and weight.
15. Calculate the weight of an object of known mass, given the acceleration due to gravity.
16. Describe factors affecting the force of gravity.
17. State Newton's third law of motion and describe examples illustrating and applying it.
18. Draw force diagrams showing action-reaction pairs of forces acting on objects.
19. Describe applications of the force of gravity.
20. Define friction and describe types of friction.
21. List and describe factors that affect friction.
22. Given any two of applied force, unbalanced force, and frictional resistance, calculate the third quantity.
23. Describe examples of desirable and undesirable effects of friction.

Supplementary

24. State how the force of gravity between two objects depends on the masses of the objects and their distance of separation.
25. Apply Newton's law of universal gravitation to solve problems involving the force of gravity.
26. Define momentum and state the SI units used to measure it.
27. Given any two of momentum, mass, and velocity, calculate the third quantity.
28. Define impulse and state the SI units used to measure it.
29. Describe the relationship between momentum and impulse.
30. Solve problems involving momentum and impulse.
31. Perform a collision experiment to determine whether or not the momentum before a collision equals the momentum after the collision.
32. State the law of conservation of momentum in both sentence and equation form.
33. Apply the law of conservation of momentum to the solving of one-dimensional collision problems.

CHAPTER 6
Mechanical Energy and Power

Several principles of mechanics are applied in the demolition of this large structure.

In previous chapters, your study of mechanics focused on displacement, velocity, time, acceleration, and force. It is possible to combine these quantities to obtain other physical quantities. The photograph shows a device tackling a large job for its size. Many principles of physics are at work here. An engine operates a machine. A force displaces the heavy concrete "hammer" upward so the force of gravity can accelerate it into the old building. The hammer's energy works over a period of time to destroy the building. How can the quantities of motion be combined to study work, energy, and power?

MECHANICS

Figure 6-1
Examples of Energy

(a) *Chemical potential energy is released when fireworks explode over Ontario Place in Toronto, Ontario.*

(b) *Elastic potential energy is stored in this archer's stretched bow.*

(c) *Radiant energy from the sun is transformed into the thermal energy of air molecules in the production of wind. The wind can drive windmills like the one shown, which is used to pump water in a game park in Africa.*

6.1 The Importance of Energy

Energy is critical to our everyday lives. In ancient times, people consumed only the energy stored in the food they ate. That energy helped them perform work with their bodies. As civilization progressed, more energy was used. The further humans progressed, the greater was their consumption of energy. Today, therefore, the nations with the most advanced technology consume the most energy. That energy is used to heat homes in winter, cool homes in summer, manufacture necessities and luxuries, cook food, provide entertainment, transport people and goods — the list is endless.

Although a comprehensive definition of energy is difficult to find, a simple definition is that **energy** is the capacity to do work. "Energy" stems from the prefix *en* (which in Greek means "in") and the Greek word for "work", *ergos*.

Energy exists in many forms. Table 6-1 lists nine forms of energy, and Figure 6-1 shows some examples of them.

Table 6-1 Forms of Energy

Form of Energy	Comment
Radiant	Examples include visible light and X rays
Thermal	Results from the motion of particles
Electrical	Results from the forces of repulsion or attraction of charged particles
Nuclear	Stored in the nuclei of atoms
Sound	Allows us to hear vibrations
Chemical potential	Stored in materials such as food and fuel
Gravitational potential	Energy of position
Kinetic	Energy of motion
Elastic potential	Stored in stretched or compressed objects

The forms of energy listed in Table 6-1 are able to change from one to another. For example, in a microwave oven, electrical energy changes into radiant energy (microwaves) which is then transformed into thermal energy in the food being cooked. Undoubtedly you can give many other examples of energy changes.

PRACTICE

1. Which form(s) of energy listed in Table 6-1 can exist without matter?
2. Name at least one form of energy associated with each object in italics.
   (a) A *bonfire* roasts a marshmallow.
   (b) A *baseball* smashes a window.
   (c) A *solar collector* heats water for a swimming pool.
   (d) A stopwatch's *spring* is fully wound.
   (e) The *siren* of an ambulance warns of an emergency.

6.2 Using Force to Transfer Energy

Consider a situation in which some groceries must be lifted from the floor to a higher level, as in Figure 6-2. For the groceries to be raised, a force must be exerted over a certain distance. When a force is exerted to raise an object some distance, energy is transferred to the object. (Energy is also transferred to an object when an unbalanced force causes it to accelerate.)

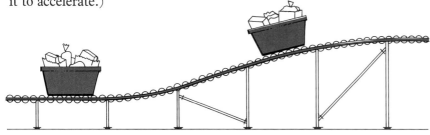

Figure 6-2
Energy is transferred to the box to lift it from the floor to the rollers.

MECHANICS

When energy is transferred to an object, we say that work has been done *on* the object. Similarly, when energy is transferred away from an object, work has been done *by* the object. Thus, **work** is a measure of the amount of energy transferred from one object to another. The amount of work (symbol W) depends on both the force ($\vec{F}$) and the displacement ($\vec{d}$) over which the force is exerted. Increasing either the force or the displacement increases the work.

In this text, the use of the equation for work is restricted to situations in which the force and displacement are parallel to one another. For this reason, we will omit the vector notation and write the following equation for work:

work = force × distance   or $W = Fd$

Since force is measured in newtons and displacement in metres, work is measured in newton metres (N·m). The newton metre is called the joule (J) in honour of James Joule (1818–1889), an English physicist who studied heat and electrical energy (Figure 6-3). Remember that the preferred SI units of metres, kilograms, and seconds must be used in order for work to be in joules.

When you are calculating the work done in lifting an object vertically, it may be necessary to find the force needed to lift the object, in other words, the object's weight. If the object's mass is known, we can calculate its weight using the equation $F_w = mg$, as discussed in Section 5.6.

Figure 6-3
James Joule (1818–1889)

SAMPLE PROBLEM 1

The mass of the box of groceries in Figure 6-2 is 20 kg, and the height of the rollers above the floor is 80 cm. Assuming that air resistance is negligible (*i.e.*, effectively zero), calculate the following:
(a) the force required to lift the box at a constant speed
(b) the work done by this force on the box in lifting it to the rollers

Solution

(a) $F_w = mg$
$= (20 \text{ kg})(9.8 \text{ m/s}^2)$
$= 196 \text{ N}$

∴ The force required is 2.0×10^2 N.

(b) $W = Fd$
$= (2.0 \times 10^2 \text{ N})(0.80 \text{ m})$
$= 1.6 \times 10^2 \text{ J}$

∴ The work done by the force on the box is 1.6×10^2 J.

In some situations, work is done *by* an object rather than *on* an object. The equation $W = Fd$ still applies, but the object doing the work loses, rather than gains, energy. See Sample Problem 3.

SAMPLE PROBLEM 2

A 2.0×10^2 N force parallel to the ramp is needed to push a loaded wheelbarrow at a constant speed up a ramp, as shown in the diagram. Calculate the work done by this force on the wheelbarrow, given that the ramp is 6.0 m long, and assuming that friction is negligible.

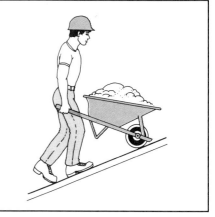

Solution
$W = Fd$
$ = (2.0 \times 10^2 \text{ N})(6.0 \text{ m})$
$ = 1.2 \times 10^3 \text{ J}$

Thus, the work done on the wheelbarrow is 1.2×10^3 J.

SAMPLE PROBLEM 3

A dog team is pulling a loaded sled across the snow with a total force of 150 N. The team and sled move with uniform velocity on a level surface.
(a) What does the 150 N force do?
(b) Calculate the work done by this force on the sled after the team and sled have travelled 1.0 km.

Solution
(a) The 150 N force is needed to balance the sliding friction of the sled in the snow and thus achieve uniform velocity.
(b) $W = Fd$
$ = (150 \text{ N})(1000 \text{ m})$
$ = 1.5 \times 10^5 \text{ J}$

Thus, the work done on the sled is 1.5×10^5 J. Most of this energy transferred to the sled changes to waste thermal energy, as shown by a slight increase in the temperature of the snow and the sled's runners.

PRACTICE
(Unless otherwise stated, assume that air resistance is negligible.)
3. A girl exerts a constant force of 20 N on a wagon for 3.2 m. (Assume that friction is negligible.)
   (a) How much work does she do on the wagon?
   (b) What type of motion does the wagon undergo?
4. A 150 g book is lifted from the floor to a shelf 2.0 m above. Calculate the following:
   (a) the force needed to lift the book at a constant speed
   (b) the work done by this force on the book in lifting it to the shelf
5. A boy pushes against a large maple tree with a force of 250 N. How much work does he do on the tree?
6. A 500 kg meteoroid is travelling through space far from any measurable force of gravity. If it travels at 100 m/s for 100 years, how much work is done on the meteoroid?

MECHANICS

7. A nurse holding a new-born 3.0 kg baby at a height of 1.2 m off the floor carries the baby 15 m at constant velocity along a hospital corridor. How much work has the nurse done on the baby?
8. Discuss the previous three questions in class, then write general conclusions regarding when work is or is not done *on* an object.
9. Calculate the amount of work done by a cherry picker in lifting you up a distance of 3.2 m at a constant speed.
10. An average force of 32 N is exerted on a box moving with uniform motion across a floor. The box moves a distance of 7.8 m along the level floor.
    (a) What does the 32 N force do?
    (b) How much work does the force do?
11. Express joules in terms of the base units of metres, kilograms, and seconds.
12. Rearrange the equation $W = Fd$ to express
    (a) F by itself and (b) d by itself.
13. A world champion weightlifter does 5.0×10^3 J of work in jerking a load from the floor to a height of 2.0 m. Calculate the following:
    (a) the average force exerted to lift the load
    (b) the mass of the load
14. An electric fork-lift truck is capable of doing a maximum of 4.0×10^5 J of work on a 4.5×10^3 kg load. To what height can the truck lift the load?

Experiment 6A: Work

INTRODUCTION
The ramp used in this experiment is actually a simple machine called an inclined plane. There are six simple machines, and analysing any one of them is useful to understand the operation of more complex machines.

PURPOSE
To measure the work done in pulling a cart up a ramp and to compare it to the work done in displacing the cart vertically.

APPARATUS
dynamics cart; ramp with a pulley at one end; pan; string; beam balance; set of masses from 1.0 g to 200 g (*Note*: A spring scale calibrated in

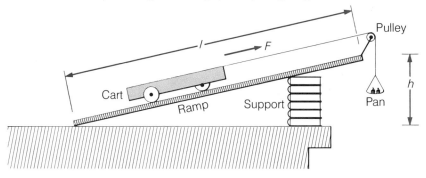

Figure 6-4
Set-up for Experiment 6A

newtons can be used as a substitute for the pulley-and-pan system to pull the cart up the ramp.)

PROCEDURE
1. Set up a table of observations based on the one shown, to a maximum of six trials.

	Trial 1	Trial 2 ...
Mass of cart, m (kg)		
Weight of cart, F_w (N)		
Length of ramp, l (m)		
Height of ramp, h (m)		
Force parallel to ramp, F (N)		
Input work, W_{in} (J)		
Output work, W_{out} (J)		

2. Measure the mass of the cart and calculate its weight, F_w.
3. Set up the apparatus as shown in Figure 6-4, so that the ramp is inclined at an angle of about 20° to the horizontal. Find the length and height of the ramp. Be sure the string is parallel to the ramp. Add masses to the pan until the cart moves without acceleration up the ramp.
4. Calculate the weight (F) of the masses on the pan. This is the force pulling on the cart. Next find the input work, the work done by F on the cart in moving it a distance l up the ramp ($W_{in} = Fl$).
5. Calculate the output work obtained by the ramp system in raising the cart a height h ($W_{out} = F_w h$).
6. Compare the input work (W_{in}) with the output work (W_{out}) and explain any differences.
7. Repeat the entire procedure using different slopes of the ramp and a constant cart mass.
8. Repeat Procedure Steps 2 to 6 using a constant slope but varying the cart mass.
9. Complete your table of data.

ANALYSIS
1. Describe the effect that each of the following had on the results of the experiment:
   (a) friction
   (b) changing the slope of the ramp
   (c) changing the mass of the cart
2. Name at least one advantage of moving an object up a ramp instead of lifting it vertically.
3. Since nature is not in the habit of giving us something for nothing, name at least one disadvantage of moving an object up a ramp instead of lifting it.
4. Describe how the results of this experiment could be used to determine the frictional resistance acting on the cart in each trial.

CONCLUSIONS . . .

MECHANICS

6.3 Gravitational Potential Energy

Suppose you are erecting a tent at an isolated park setting where the ground is harder than you expected. You don't have a hammer with you to pound the tent pegs into the ground, so you search for a solution to the problem. You grab a nearby rock, raise it above the peg, and pound the peg into the ground, as in Figure 6-5.

At the raised position, the rock has the potential to help you do some work on the peg. This potential arises from the fact that the force of gravity is pulling down on the rock. The type of energy possessed by an object because of its position is called **gravitational potential energy**, E_p. This potential energy can be used to do work on some object at a lower level.

In order to lift the rock in Figure 6-4 a height Δh, you would have to transfer energy $W = F\Delta h$ to it. Here, F is the force required to lift the rock without acceleration; it is equal in magnitude to the rock's weight. The transferred energy, or work, equals the change in the rock's potential energy. That is, $\Delta E_p = F\Delta h$, where ΔE_p is the potential energy of the rock raised a height h above the original level. Since the force F is the weight or force due to gravity ($F_w = mg$), we can now write the common equation for the change in gravitational potential energy:

$$\Delta E_p = F\Delta h = mg\Delta h, \text{ where } g = 9.8 \text{ m/s}^2$$

In SI, energy is measured in joules, mass in kilograms, and height (or displacement) in metres.

Often we are concerned about the potential energy relative to a particular **reference level**, the level to which the object may fall. Then the Δh in the potential energy equation is the height h of the object above the reference level. Thus, the equation for the potential energy of an object relative to a reference level is

$$E_p = mgh.$$

When answering questions on relative potential energy, it is important for you to indicate the reference level. In the rock and tent peg example, the rock has a greater potential energy relative to the ground than it has relative to the top of the peg.

Figure 6-5
Raising a rock gives it gravitational potential energy.

Did You Know?
Some animals are known to take advantage of gravitational potential energy. One example is the bearded vulture (or lammergeyer), the largest of all vultures. This bird, found in South Africa, is able to digest bones. Often its food consists of bones picked clean by other animals. When confronted with a bone too large to digest, the vulture carries the bone to a great height and drops it onto a rock. Then it circles down to scoop out the marrow with its gouge-shaped tongue.

SAMPLE PROBLEM 4
The 20 kg box of groceries in Figure 6-2 and Sample Problem 1 is 0.50 m above the level of the loading platform outside the store. Calculate the potential energy of the box relative to that level.

Solution
$E_p = mgh$
$ = (20 \text{ kg})(9.8 \text{ m/s}^2)(0.50 \text{ m})$
$ = 98 \text{ J}$
∴ The potential energy of the box relative to the platform is 98 J.

MECHANICAL ENERGY AND POWER

Figure 6-6
At the instant shown, the cars in this thrill ride have a large amount of gravitational potential energy.

Sample Problem 4 illustrates a useful application of potential energy. An object is raised to a position above a reference level, then gravity causes the object to accelerate down a sloped path. After accelerating, the object gains enough velocity to continue moving forward some distance. The grocery box, for instance, moves forward to the loading platform. Another example is shown in Figure 6-6, where a type of roller coaster is at a high position where its potential energy is greatest. Then gravity causes the coaster to accelerate downward, giving it enough speed to travel around the loop.

Figure 6-7 shows two more applications of potential energy. In (a), a pile driver is about to be lifted by a motor high above the pile. It will then have the potential energy to do the work of driving the pile into the ground. The pile will act as a support for a high-rise building.

In (b), the source of potential energy is provided by nature. A waterfall's potential energy can be used to drive turbines to produce hydroelectricity.

Figure 6-7
Applications of Gravitational Potential Energy

(a) Pile driver

(b) View of Victoria Falls in Zimbabwe, Africa

MECHANICS

PRACTICE

15. In April 1981, Arnold Boldt of Saskatchewan set a world high-jump record for disabled athletes in Rome, Italy, jumping to a height of 2.04 m. (At the age of three, Arnold had his right leg amputated above the knee after an accident.) Calculate Arnold's potential energy relative to the ground at the top of the record jump. (Assume that his mass at the time of the jump was 68 kg.)

16. A 500 g book is resting on a desk 60 cm high. Calculate the book's potential energy relative to
    (a) the desk top and (b) the floor.

17. Explain why the device shown in the photograph at the start of the chapter is an application of gravitational potential energy.

18. Assume that 850 J of work is done to raise an object from the ground to a higher position. What is the object's potential energy relative to the ground if
    (a) there is no friction in raising the object?
    (b) 30 J of work is removed by friction in raising the object?

19. Rearrange the equation $E_p = mgh$ to obtain an equation for (a) m; (b) g; and (c) h.

20. The elevation at the base of a ski hill is 350 m above sea level. A ski lift raises a skier (total mass = 72 kg, including equipment) to the top of the hill. If the skier's potential energy relative to the base of the hill is now 9.2×10^5 J, what is the elevation at the top of the hill?

21. The spiral shaft in a grain auger raises grain from a farmer's truck into a storage bin. Assume that the auger does 6.2×10^5 J of work on a certain amount of grain to raise it 4.2 m from the truck to the top of the bin. What is the total mass of the grain moved? Neglect friction.

22. A fully-dressed astronaut, weighing 1.2×10^3 N on earth, is about to jump down from a space capsule which has just landed safely on Planet X. The drop to the surface of X is 2.8 m, and the astronaut's potential energy relative to the surface is 1.1×10^3 J.
    (a) What is the acceleration due to gravity on Planet X?
    (b) How long does the jump take?
    (c) What is the astronaut's maximum velocity?

6.4 Kinetic Energy

A bowling ball resting on the floor has no energy of motion. One that is rolling along a bowling alley does have energy of motion. Energy due to the motion of an object is called **kinetic energy**, E_k. ("Kinetic", like the word "kinematics", stems from the Greek word *kinema*, which means "motion".)

To determine an equation for kinetic energy, we will use concepts from this chapter and Chapter 4. Assume that an object of mass m, travelling at a speed v_i, has an unbalanced force, F, exerted on it for a displacement Δd. The object will undergo an acceleration, a, to reach a

speed v_f. The work done on the object equals the change in its energy, so
$W = \Delta E = F \Delta d = ma \Delta d$.

From Chapter 4, we know that the acceleration (a) of the object is

$$a = \frac{\Delta v}{t} = \frac{v_f - v_i}{t}$$

and the displacement (Δd) during the acceleration is

$$\Delta d = (v_{ave})t = \frac{(v_f + v_i)t}{2}.$$

Substituting these equations into the equation for the energy change, we obtain

$$\begin{aligned}\Delta E &= ma\Delta d \\ &= m\frac{(v_f - v_i)}{t}\frac{(v_f + v_i)t}{2} \\ &= m\frac{(v_f^2 - v_i^2)}{2}\end{aligned}$$

or $\Delta E = \dfrac{mv_f^2}{2} - \dfrac{mv_i^2}{2}$.

To simplify this equation, let us assume that the object starts from rest, so $v_i = 0$, and the last term in the equation is zero. Then the change in energy equals the object's final kinetic energy. Thus, an object of mass m, travelling at a speed v, has a kinetic energy of

$$E_k = \frac{mv^2}{2}.$$

Again, in SI, energy is measured in joules, mass in kilograms, and speed in metres per second.

(*Note*: From Chapter 5 you will notice that both momentum and kinetic energy involve the two variables mass and speed. Momentum and kinetic energy are different quantities, but both are useful in solving a variety of physics problems.)

Did You Know?
Although the joule is the SI unit of energy, we often still hear of the heat calorie (cal), a former unit of heat, and the food calorie (C), a former unit of food energy. These energy units are related in the following way:
1.0 C = 1.0 × 10³ cal = 1.0 kcal
1.0 cal = 4.2 J
1.0 C = 4.2 × 10³ J = 4.2 kJ
Thus, a piece of apple pie rated at 400 C contains 1.68 × 10⁶ J, or 1.68 MJ, of chemical potential energy.

SAMPLE PROBLEM 5
Find the kinetic energy of a 6.0 kg bowling ball rolling at 5.0 m/s.

Solution

$$\begin{aligned}E_k &= \frac{mv^2}{2} \\ &= \frac{(6.0 \text{ kg})(5.0 \text{ m/s})^2}{2} \\ &= (3.0 \text{ kg})(25 \text{ m}^2/\text{s}^2) \\ &= 75 \text{ J}\end{aligned}$$

Thus, the kinetic energy of the bowling ball is 75 J.

MECHANICS

Figure 6-8
The ostrich has powerful legs which can carry it at high speed.

PRACTICE

23. Calculate the kinetic energy in each case.
    (a) A 7.2 kg shot leaves an athlete's hand during the shotput at a speed of 12 m/s.
    (b) A 140 kg ostrich is running at 14 m/s. (The ostrich is the fastest two-legged animal on earth.) See Figure 6-8.
24. Prove that the units for kinetic energy are equivalent to the units for work.
25. Starting with the equation $E_k = mv^2/2$, find an equation for
    (a) m and (b) v.
26. A softball travelling at 34 m/s has a kinetic energy of 98 J. What is its mass?
27. A 100 g cup falls from a kitchen shelf and shatters on the ceramic tile floor. Assume that the maximum kinetic energy obtained by the cup is 5.0 J and that air resistance is negligible.
    (a) What is the cup's maximum velocity?
    (b) What do you suppose happened to the 5.0 J of kinetic energy after the crash?
28. The energy available from the complete combustion of 1.0 L of gasoline is about 30 MJ. A certain car has a consumption rate of 10 L/(100 km) on the highway.
    (a) How many megajoules of energy does the car use in travelling 150 km?
    (b) What distance in metres can the car travel using 1.0 MJ of energy?

6.5 The Law of Conservation of Energy

As you have learned, energy can change from one form to another. Scientists say that when any such change occurs, energy is *conserved*. In other words:

When energy changes from one form to another, no energy is lost.

This statement is called the **law of conservation of energy**. (This law applies to all the forms of energy listed in Table 6-1.)

In this situation, the word "conservation" means that the total amount remains constant. This differs from the everyday usage of the expression "energy conservation" which refers to not wasting energy. Thus, you should distinguish between *conservation of energy* (the scientific law) and *conserving energy* (the wise thing to do).

Let us consider a practical application of the law of conservation of energy. Our example deals with the use of a pile driver, shown earlier in Figure 6-7(a).

Figure 6-9(a) illustrates the design of a pile driver. A hammer is lifted by an electric or gasoline engine (not shown) to a position above the pile. From there, the shaft guides the hammer as it falls and strikes the pile to knock it further into the ground to act as a structural support

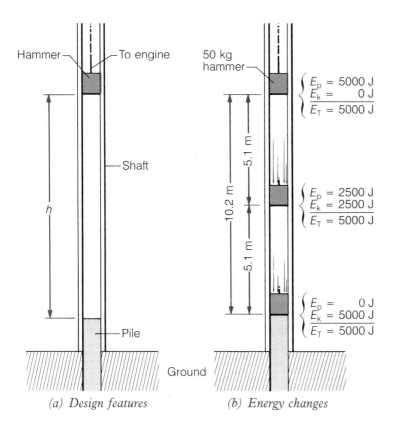

Figure 6-9
The Design and Energy Changes of a Pile Driver

for a building. We will analyse the energy changes occurring in these events, shown in Figure 6-9(b).

In hoisting the hammer from the level of the pile to the top position, the engine does the following work on the hammer:

$W = Fd$
$ = mgh$
$ = (50 \text{ kg})(9.8 \text{ m/s}^2)(10.2 \text{ m})$
$ = 5000 \text{ J}$

Notice that we are concerned here with the work done *on the hammer*. We will not consider the work required because of friction in the shaft or in the engine itself. Including these factors would involve an extra series of calculations, though the final conclusions would be much the same.

At the top position, then, the hammer has a potential energy ($E_p = mgh$) of 5000 J. Its kinetic energy is zero because its speed is zero. The hammer's total energy, $E_T = E_p + E_k$, is thus 5000 J.

When the hammer is released, it accelerates down the shaft at 9.8 m/s². (We assume that friction is negligible when the shaft is vertical.) At any position the potential energy and kinetic energy can be calculated. We will choose the half-way point, where the hammer is 5.1 m above the pile. Its potential energy there is

$E_p = mgh$
$ = (50 \text{ kg})(9.8 \text{ m/s}^2)(5.1 \text{ m})$
$ = 2500 \text{ J}$

To find the kinetic energy we must first find the hammer's speed after it has fallen 5.1 m. From Chapter 4,

$$v_f^2 = v_i^2 + 2ad$$
$$v = \sqrt{2ad} \quad (v_i = 0)$$
$$= \sqrt{2(9.8 \text{ m/s}^2)(5.1 \text{ m})}$$
$$= 10 \text{ m/s}.$$

Thus,

$$E_k = mv^2/2$$
$$= (50 \text{ kg})(10 \text{ m/s})^2/2$$
$$= 2500 \text{ J}.$$

Again, the total energy, $E_p + E_k$, is 5000 J.

Next, at the instant just prior to striking the pile, the hammer has zero potential energy ($h = 0$ in $E_p = mgh$) and a kinetic energy based on these calculations:

$$v = \sqrt{2ad}$$
$$= \sqrt{2(9.8 \text{ m/s}^2)(10.2 \text{ m})}$$
$$= 14.1 \text{ m/s}$$

Thus,

$$E_k = mv^2/2$$
$$= (50 \text{ kg})(14.1 \text{ m/s})^2/2$$
$$= 5000 \text{ J}.$$

Once again, the total energy is 5000 J.

Finally, the hammer strikes the pile. Its kinetic energy changes into other forms of energy such as sound, heat, and the kinetic energy of the pile as it is driven further into the ground.

This series of energy changes illustrates the law of conservation of energy. The work done on the object in raising it gives the object gravitational potential energy. This energy changes into kinetic energy and other forms of energy. Energy is not lost; it simply changes form.

The law of conservation of energy can be applied to other situations mentioned earlier in this chapter:

(a) raising a box of groceries to the rollers (Figure 6-2)
(b) using a rock to pound in a tent peg (Figure 6-5)
(c) operating a thrill ride (Figure 6-6)
(d) producing hydroelectric power at a waterfall (Figure 6-7(b))

The law can also be used to solve certain types of problems, as Sample Problem 6 and the Practice Questions illustrate.

PRACTICE

29. Use the law of conservation of energy to describe the energy changes that occur in the operation of a roller coaster at an amusement park.
30. A player spikes a 290 g volleyball, giving it 24 J of kinetic energy. How fast is the ball travelling after it is spiked?
31. A 90 kg kangaroo acquires 2.7 kJ of energy in jumping straight up. How high does this agile marsupial jump?

SAMPLE PROBLEM 6

As the water in a certain river approaches a 5.7 m vertical drop, its average speed is 5.1 m/s. For each kilogram of water in the river, determine the following:
(a) the kinetic energy at the top of the waterfall
(b) the gravitational potential energy at the top of the waterfall relative to the bottom
(c) the total energy at the bottom of the waterfall, not considering friction.)
(d) the speed at the bottom of the waterfall

Solution

(a) $E_k = mv^2/2$
$= (1.0 \text{ kg})(5.1 \text{ m/s})^2/2$
$= 13 \text{ J}$

∴ The kinetic energy of each kilogram of water at the top of the waterfall is 13 J.

(b) $E_p = mgh$
$= (1.0 \text{ kg})(9.8 \text{ N/kg})(5.7 \text{ m})$
$= 55.86 \text{ J}$

Thus, the potential energy of each kilogram of water relative to the bottom of the falls is 56 J.

(c) As the water falls, the potential energy changes to kinetic energy. By the law of conservation of energy, the total energy at the bottom of the falls is the sum of the initial kinetic energy and the kinetic energy gained due to the energy change. Thus, the total energy per kilogram at the bottom of the falls is 13 J + 56 J = 69 J.

(d) $v = \sqrt{\dfrac{2E_k}{m}}$
$= \sqrt{\dfrac{2(69 \text{ J})}{1.0 \text{ kg}}}$
$= 11.7 \text{ m/s}$

Therefore, the speed at the bottom of the waterfall is approximately 12 m/s.

32. A ball is dropped vertically from a height of 1.5 m; it bounces back to a height of 1.3 m. Does this violate the law of conservation of energy? Explain.
33. **Activity** Design and carry out an experiment to verify the law of conservation of energy for a metal mass which, starting from rest, falls a distance h vertically. You should find and compare the total energy ($E_T = E_p + E_k$) at three or four positions.
34. A 200 g pendulum bob is raised 20 cm above its rest position, as shown in Figure 6-10. The bob is released, and it reaches its maxi-

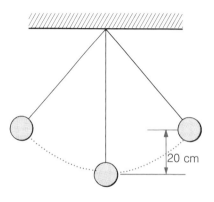

Figure 6-10

mum speed as it passes the rest position. Calculate its speed at that point. (*Hint*: Apply the law of conservation of energy.)

35. Study the interesting drawing by artist M.C. Escher shown in Figure 6.11.
    (a) What illusion is created in the artwork?
    (b) Does the water's perpetual motion defy the law of conservation of energy? Explain.

36. **Activity** If you performed Experiment 5B, refer to your observations and, for each collision, calculate the total kinetic energy of the system before and after each collision. Is there a general difference between the type of collision where the carts became joined together and the type where they bounced off each other?

37. For an object accelerating uniformly from rest, the speed attained after travelling a certain distance is $v = \sqrt{2ad}$. Substitute this equation into the equation for kinetic energy, $E_k = mv^2/2$. Explain the result.

Figure 6.11
Follow the path of the water in this drawing by M.C. Escher.

6.6 Power

It takes approximately 2000 J of work for an elevator to lift an average person one storey in a building. The amount of work remains the same whether the elevator lifts the person in 5.0 s or 30 s. If the work done is the same but the time changes, some other factor must change as well. That other factor is called "power".

Power (P) is the rate of doing work or transforming energy. Thus,

$$P = \frac{W}{t} \text{ or } P = \frac{\Delta E}{t}.$$

Since work and energy are measured in joules and time in seconds, power is measured in joules per second (J/s). This SI unit has the name watt (W) in honour of James Watt (1736–1819), a Scottish physicist who was the inventor of the first practical steam engine (Figure 6-12). Watts and kilowatts are commonly used to indicate the power used by electrical appliances, while megawatts are often used to indicate the power of electric generating stations.

Figure 6-12
James Watt (1736–1814)

SAMPLE PROBLEM 7

What is the power of a cyclist who transforms 2.7×10^4 J of energy in 3.0 min?

Solution

$$P = \frac{\Delta E}{t}$$

$$= \frac{2.7 \times 10^4 \text{ J}}{1.8 \times 10^2 \text{ s}}$$

$$= 150 \text{ W}$$

∴ The cyclist's power is 1.5×10^2 W.

MECHANICS

SAMPLE PROBLEM 8
A 52 kg student climbs 3.0 m up a ladder in 4.7 s. Calculate the student's
(a) potential energy at the top of the climb;
(b) power for the climb

Solution

(a) $E_p = mgh$
$= (52 \text{ kg})(9.8 \text{ m/s}^2)(3.0 \text{ m})$
$= 1.5 \times 10^3 \text{ J}$

∴ The potential energy is 1.5×10^3 J.

(b) $P = \Delta E/t$
$= (1.5 \times 10^3 \text{ J})/(4.7 \text{ s})$
$= 3.2 \times 10^2 \text{ W}$

Thus, the student's power is 3.2×10^2 W.

PRACTICE

38. Express watts in the base SI units of metres, kilograms, and seconds.
39. A fully outfitted mountain climber, complete with camping equipment, has a mass of 85 kg. If she climbs from an elevation of 2900 m to 3640 m in exactly one hour, what is her average power?
40. A 60 kg boy does 60 push-ups in 40 s. With each push-up he must lift an average of 70% of his mass a height of 40 cm off the floor. Calculate the following:
    (a) the work the boy does for each push-up, assuming he does work only when he pushes up
    (b) the total work done in 40 s
    (c) the power for this period
41. **Activity** Determine the maximum power of a student, such as yourself, running up a set of stairs. Safety considerations are important here. Only students wearing running shoes should try this experiment and, of course, they should be careful not to trip or pull their arm muscles while pulling on the rail. It may be interesting to compare the student power with that of an average horse, which can exert about 750 W of power for an entire working day. (This amount is called an old-fashioned "horsepower".)
42. **Activity** Design and carry out an experiment to determine the power of a student performing a variety of activities. Examples include lifting weights in the exercise room, climbing a rope in the gymnasium, lifting books, or doing push-ups.
43. Rearrange the equation $P = E/t$ to obtain an equation for (a) E and (b) t.
44. The power rating of the world's largest wind generator is 3.0 MW. How long would it take such a generator to produce 1.0×10^{12} J, the amount of energy needed to launch a rocket?
45. An elevator motor develops 32 kW of power while it lifts the elevator 24 m at a constant speed. If the elevator's mass is 2200 kg, including the passengers, how long does the motion take?

46. The nuclear generating station located at Pickering, Ontario, one of the largest in the world, is rated at 2160 MW. How much energy, in megajoules, can this station produce in one day?
47. A water pump rated at 2.0 kW can raise 50 kg of water per minute at a constant speed from a lake to the top of a storage tank. How high is the tank above the lake? Assume that all the energy from the pump goes into raising the height of the water.

Review Assignment

1. In physics, what is the relationship between energy and work?
2. A black bear's greatest enemy is the grizzly bear. To escape a grizzly attack, a black bear does what its enemy cannot do—it climbs a tree whose trunk has a small diameter. Calculate the work done by a 140 kg black bear in climbing 18 m up a tree.
3. Compare the amount of work you would do in climbing a vertical rope with the work done in climbing a stairway inclined at 45° if both activities get you 6.0 m higher.
4. A golf ball is given 115 J of energy by a club that exerts a force over a distance of 4.5 cm while the club and the ball are in contact.
   (a) Calculate the average force exerted by the club on the ball.
   (b) If the ball's mass is 47 g, find its average acceleration.
   (c) What speed does the club impart to the ball?
5. An average force of 220 N is required to keep a motorcycle moving at uniform velocity.
   (a) How far will one megajoule of energy take the motorcycle?
   (b) What happens to the energy consumed?
6. A hoist in an automobile service centre raises a 1600 kg car 1.8 m off the floor.
   (a) What is the car's potential energy relative to the floor?
   (b) How much work did the hoist do on the car to raise it? Neglect friction.
7. A roast of beef waiting to be taken out of a refrigerator's freezer compartment has a potential energy of 35 J relative to the floor. If the roast is 1.7 m above the floor, what is the mass of the roast?
8. A 50 kg diver has 1470 J of potential energy relative to the water when standing on the edge of a diving board. How high is the board above the water?
9. A group of winter enthusiasts returning from the ski slopes is travelling at 90 km/h along a highway. A pair of ski boots having a total mass of 2.8 kg has been placed on the shelf of the rear window.
   (a) What is the kinetic energy of the pair of boots?
   (b) What happens to that energy if the driver must suddenly stop the car?
10. What happens to an object's kinetic energy when its speed doubles? triples?

MECHANICS

11. A discus travelling at 20 m/s has 330 J of kinetic energy. Find the mass of the discus.
12. An archer nocks a 200 g arrow on her bowstring. Then she exerts an average force of 110 N to draw the string back 0.60 m. Assume that friction is negligible.
    (a) What speed does the bow give to the arrow?
    (b) If the arrow is shot vertically upwards, how high will it rise?
13. An interesting and practical feature of the Montréal subway system is that, in some cases, the level of the station is higher than the level of the adjacent tunnel, as Figure 6-13 demonstrates. Explain the advantages of this design. Take into consideration such concepts as force, acceleration, work, potential energy, and kinetic energy.

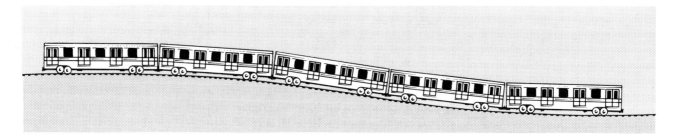

Figure 6-13
The Montréal Subway System

14. Calculate the power rating of a lightbulb that transforms 1.5×10^4 J of energy per minute.
15. Some people perform difficult tasks to raise money for charity. For example, walking up the stairs in Toronto's CN Tower helps both charity and personal fitness. Assume that for every 1.0 J of usable energy there are 3.0 J of waste energy (heat) in climbing stairs. If a 70 kg man climbs the 342 m height 10 times in 4.0 h, calculate the following:
    (a) the work he does on each trip up the stairs
    (b) the total work for the 10 upward trips
    (c) the power rating for the upward trips (including waste)
16. How much energy is transformed by a 1200 W electric kettle during 5.0 min of operation?
17. Scientists are researching methods of powering spaceships for future interplanetary voyages. One serious possibility is the use of an "ion drive" propulsion system in which electric forces eject charged particles (ions) rearward, accelerating the spaceship forward. Such ships might reach speeds up to 45 km/s relative to the earth. If a 4.0 W ion drive operates continuously for one week, how much work is done? Express your answer in megajoules.
18. A human who is swimming consumes energy at a rate of about 600 J/s.
    (a) What is the source of energy consumed?
    (b) How long does it take the swimmer to consume one megajoule of energy?
    (c) How long does it take the swimmer to consume an amount of energy equivalent to that stored in one kilogram of coal (30 MJ)? Express this answer in hours.

19. An alternative unit to the joule or megajoule is the kilowatt hour (kW·h), which is still used in many parts of Canada to measure electrical energy. One kilowatt hour is equivalent to one kilowatt of power used for one hour. Prove that 1.0 kW·h = 3.6 MJ.
20. The world record for the pole vault is about 5.8 m. To achieve this feat, an athlete must raise his centre of mass about 4.6 m off the ground to clear the bar. How fast must the vaulter be running prior to take-off to be able to get over the bar? (Assume that all his kinetic energy changes to potential energy at the top of the jump.)
21. Throughout the year, the average power received from the sun in the densely populated regions of Canada is about 150 W/m^2.
    (a) Estimate, then calculate, the average yearly amount of energy received by a roof with a surface area of 200 m^2.
    (b) At a cost of 3.0¢/MJ (the average cost of electrical energy), how much is the energy in (a) worth?
    (c) Estimate the average yearly amount of energy received by your province.

Key Objectives

Having completed this chapter, you should now be able to do the following:
1. Recognize and list examples of the various forms of energy.
2. Describe the conditions under which work is done on an object.
3. Given any two of work done, force, and displacement, calculate the value of the third quantity.
4. Write the SI unit for work in base units.
5. Determine experimentally the amount of work done in raising an object to a given height above a reference level.
6. Define gravitational potential energy.
7. Given any three of potential energy, mass, acceleration due to gravity, and height, calculate the value of the fourth quantity.
8. Describe applications of gravitational potential energy.
9. Define kinetic energy.
10. Prove that the work done on an object in changing its velocity from zero to some value equals the change in the object's kinetic energy.
11. Given any two of kinetic energy, mass, and velocity, find the value of the third quantity.
12. State the law of conservation of energy and give examples illustrating it.
13. Define power, and write the SI unit used to measure it in base units.
14. Given any two of power, work (or change of energy), and time, find the value of the third quantity.

MECHANICS

CHAPTER 7
Energy in a Modern World

This vertical-axis wind generator is part of Canada's continuing effort to develop energy resources.

Our standard of living is better now than at any time in the past. We have advanced medical facilities, the comfort of temperature-controlled buildings, ample fresh or frozen food, clothing to suit our variable climate, complex transportation systems, the assistance of numerous applications of electricity, and many interesting sports and leisure activities. For all these advantages to exist, energy is required. Unfortunately, there are problems we must overcome if we wish to continue using energy to maintain or improve our standard of living. The photograph shows the installation of a Canadian-designed vertical-axis wind generator with a potential power output of 500 kW of electricity. This type of wind generator, located in Prince Edward Island, may help solve some of our energy problems. What are those problems, and how can we solve them?

7.1 The Importance of Energy

Before you begin reading the next paragraph, estimate an answer to this question: How much energy (in joules) do you think you have been responsible for consuming in the past 24 h? Remember that it takes 9.8 J of energy change (in the form of work) to raise a 1 kg object a vertical distance of 1 m; but your body is not very efficient, so for every joule of energy you transform, you body consumes 4 J, 5 J, maybe even 10 J. When you have completed your estimate, read on.

If you estimated about 10^3 J (1 kJ), you would be wise to go back to Chapter 6 and study it again. It takes about 200 J to raise your body from a sitting to standing position and about 2 kJ to climb a set of stairs. If you estimated about 10^7 J (10 MJ), you've forgotten something. Just your food intake in a day should contain about 10 MJ of energy!

Keep in mind that you also have consumed a large amount of other energy directly: energy used to cook your food, provide light and heat at home and school, and perhaps provide transportation. Indirectly you have consumed energy used to make your clothes, books, radio, furniture, and so on, as well as energy to build and light your streets, take away your sewage and garbage—the list is becoming longer. So, if you estimated that you are responsible for consuming about 10^9 J (1 GJ or 1000 MJ), you were close to the daily energy consumption of the average Canadian.

Thus, the energy we consume (about 1000 MJ per person per day) is about 100 times as much as the amount we need to survive (about 10 MJ). Of course, much of the consumed energy is beyond our control, but we should be aware of the entire energy situation. Consider, for example, Figure 7-1. Are we, as individuals, responsible for the energy consumed to illuminate vacant offices at night?

Energy exists in many forms, as you could see in Table 6-1. The forms of energy studied in the previous chapter were gravitational potential energy and kinetic energy. Other important forms of energy will be studied throughout this text. But energy is not only important in physics; it is significant in all walks of life. Can you describe how energy relates to each area of interest listed below?

industry	economics
technology	communication
travel	agriculture
leisure	medicine
politics	scientific research

We consume energy for countless uses. Energy does work for us and makes our lives more comfortable and more productive. But politicians,

Figure 7-1
The Skyline of Toronto, Ontario, after Business Hours

economists, and scientists are concerned about the problems associated with our appetite for energy. They are searching for answers to such questions as these:

- Where will we obtain the energy resources we need in the future?
- How will we make wise use of energy when we convert it from one form, such as chemical, to another, such as electrical, to do work for us?
- How can we protect our environment from the polluting effects of obtaining and using energy?

Numerous other possible questions come to mind when you look at the titles of energy-related articles from magazines and newspapers, such as those shown in Figure 7-2.

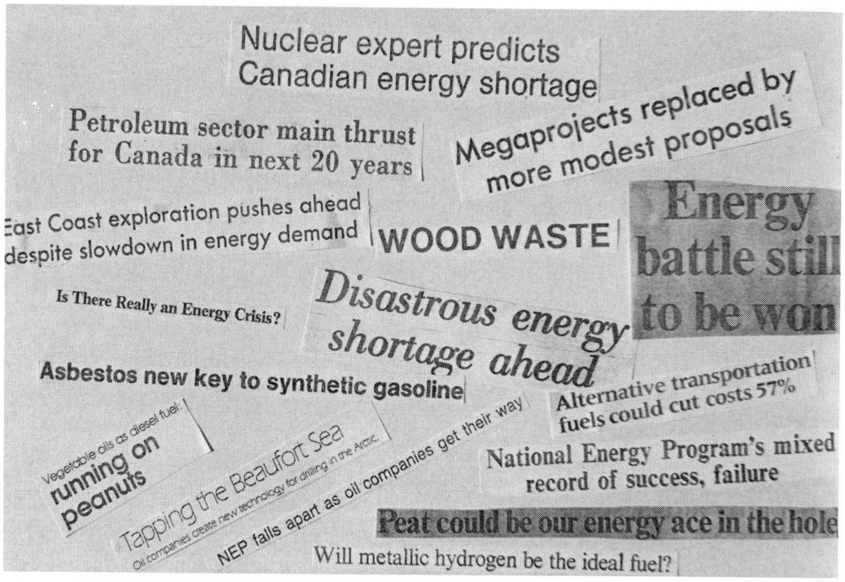

Figure 7-2
Do these energy-related headlines exhibit positive or negative reporting?

This chapter is intended to give you background information, from the physics point of view, which you can use to become aware of the "good" and "bad" aspects of energy use in our society.

PRACTICE

1. Discuss in class how energy relates to each walk of life listed in the middle of this section. Could some other human activities be included in the list?
2. Describe the difference between an energy source and an energy conversion technique. Give examples.
3. **Activity** Collect (or list the titles of) current energy-related newspaper and magazine articles. Classify them into at least two categories. Depending on the articles you find, you might try one of these sets of categories:
   (a) optimistic, pessimistic
   (b) problems, solutions
   (c) political, scientific
   (d) past, present, future
   (e) domestic, foreign

   Discuss your collection in class.

7.2 The Consumption of Energy

It is interesting to compare our current consumption of energy (1000 MJ/person/day) with that of people in different eras. As civilization progressed, the amount of energy consumed per person increased remarkably (see Table 7-1). Simultaneously, the world's population has grown, so the net effect is that we are now consuming a vast amount of energy.

Table 7-1 Daily Average Energy Consumption per Person

Lifestyle	Era	Conditions	Energy Consumption (MJ/d)
Primitive	Pre-Stone Age	Survival; no use of fire	10
Seasonally nomadic	Stone Age	Energy from wood fires for cooking and heating	22
Agricultural	Medieval	Energy from domesticated animals, water, wind, and coal	100
Industrial	19th Century	Energy mainly from coal to run industries and steam engines	300
Technological	Present	Energy from fossil fuels and nuclear sources used for electricity, transportation, industry, agriculture, and so on	1000 (in Canada)

Table 7-1 indicates that an average person living in primitive conditions consumes only about 10 MJ of energy per day. Learning how that value is computed will help you learn about your own body's energy consumption. All living things require energy to sustain life. The rate at which energy is consumed per unit mass is called the **metabolic rate**. Since the rate of energy consumption is called power ($P = \Delta E/t$), the equation for metabolic rate is

$$\text{M.R.} = \frac{P}{m}.$$

In SI, metabolic rate is measured in watts per kilogram (W/kg).

The average metabolic rate for humans during various activities is given in Table 7-2. To calculate primitive man's daily consumption we will make two assumptions — first, let us assume that the average mass per person was 50 kg; second, let us assume that the average metabolic rate for a 24 h period was the same as the rate for humans today, about 2.3 W/kg. Then

$$P = \text{M.R.} \times m$$
$$= 2.3 \frac{\text{W}}{\text{kg}} \times 50 \text{ kg}$$
$$= 115 \text{ W}.$$

Therefore, the estimated energy used in one day is

$$E = Pt$$
$$= (115 \frac{\text{J}}{\text{s}})(24 \text{ h})(3600 \frac{\text{s}}{\text{h}})$$
$$= 9.9 \times 10^6 \text{ J or about 10 MJ.}$$

We can apply the information in Table 7-2 to familiar situations, as the following Sample Problem shows.

SAMPLE PROBLEM 1
A girl with a mass of 50 kg swims for 20 min. Calculate (a) her power rating and (b) the energy she has consumed.

Solution

(a) $P = \text{M.R.} \times m$
$$= (10 \frac{\text{W}}{\text{kg}})(50 \text{ kg})$$
$$= 500 \text{ W}$$

$\therefore$ Her power rating is 5.0×10^2 W.

(b) $E = Pt$
$$= (500 \text{ W})(20 \text{ min})(60 \frac{\text{s}}{\text{min}})$$
$$= 6.0 \times 10^5 \text{ W or 0.60 MJ}$$

Thus, the energy consumed is 0.60 MJ.

Table 7-2 Approximate Metabolic Rates of Humans

Activity	Metabolic Rate (W/kg)
sleeping	1.0
sitting	1.4
standing	2.5
walking	4.2
bicycling	7.5
swimming	10
running	17

In many developing countries, the average energy consumption is certainly more than that during primitive times, but seldom as much as during medieval times. Refer to Table 7-1. In other words, a large portion of the world's population today consumes energy at a rate less than 100 MJ/person/day. (Can you think of reasons why that value is so much less than the Canadian average?) Thus, developing countries are not the culprits of excess energy consumption. It is the industrialized, highly technological nations that consume the greatest amount of energy. In fact, about 80% of the energy consumed in the world each year is used by the industrialized countries. Embarrassingly, North Americans are the people with the greatest demand for energy. Both the United States and Canada consume much more than their share of the world's energy.

Now let us look at one of the most enlightening and frightening aspects of energy consumption. Frequently we hear news reports indicating the rate of growth *per annum* of some factors in our society. Using "a" to represent *annum* or "year", we have these examples:

- The population increase is 1.2%/a.
- The inflation rate is 5%/a.
- Consumption of oil for heating is decreasing at 3%/a.
- Postal rates might rise 12%/a.

To discover the impact of growth rates, we will use 8%/a as an example. Assume, for the sake of simple calculations, that a person starts working at the age of 20 for an annual salary of $20 000. At a rate of increase of salary of 8%/a, what do you expect this salary will be upon retirement at age 65? After you make a guess, look at the values in Table 7-3.

Table 7-3 Effects upon Salary of a Growth Rate of 8%/a

Age	Salary	Age	Salary
20	$20 000	28	$ 37 000
21	21 600 ($20 000 × 1.08)	29	40 000
22	23 300 ($21 600 × 1.08)	38	80 000
23	25 200 *etc.*	47	160 000
24	27 200	56	320 000
25	29 400	65	640 000
26	31 700		
27	34 300		

Table 7-3 shows that at the seemingly low growth rate of 8%/a, the original value is doubled in 9 years, and after 45 years the value is greater by a factor of 32 times! Now, this may appear marvellous in the case of salaries. But what about energy consumption? If our energy consumption were to increase at 8%/a, after the average person's period of work expectancy (45 years), our energy use would be 32 times greater! You can now appreciate why scientists are, and politicians should be, concerned about our growth rate of energy consumption.

Another interesting fact emerges from Table 7-3. At a growth rate of 8%/a, the time required for an amount to double, called the **doubling time**, is 9 years. The product of the two numbers is 8%/a × 9 a = 72%. This value can be used to estimate the doubling time for particular growth rates. For example, you have seen that at 8%/a, the doubling time is 72% ÷ 8%/a = 9 a. At a growth rate of 10%/a, the doubling time is approximately 72% ÷ 10%/a = 7.2 a. Thus,

$$\text{doubling time} = \frac{72\%}{\text{growth rate}} \text{ (an approximation).}$$

This equation gives a relatively accurate result at low rates of growth, but it becomes inaccurate at high rates.

SAMPLE PROBLEM 2
By what factor will Canada's energy consumption increase in the next 24 a at a growth rate of (a) 3%/a and (b) 6%/a?

Solution

(a) $\text{doubling time} = \dfrac{72\%}{\text{growth rate}}$

$= \dfrac{72\%}{3\%/a}$

$= 24 \text{ a}$

∴ After 24 years energy consumption will be 2 times its present value.

(b) $\text{doubling time} = \dfrac{72\%}{\text{growth rate}}$

$= \dfrac{72\%}{6\%/a}$

$= 12 \text{ a}$

∴ After 24 years energy consumption will be 4 times its present value.

As you can see from Sample Problem 2, the growth rate of our energy consumption, which is currently estimated to be 3%/a, should be of major concern to all Canadians. Unfortunately, it is not the only energy-related problem, as you will learn in the next section.

PRACTICE

4. A 65 kg athlete runs a 100 m dash in 11 s. What is the athlete's (a) power and (b) energy consumed?

5. Calculate your own power and energy consumed during an average night of sleep.

6. An 80 kg man has a power of 720 W while shovelling snow. What is the man's metabolic rate for this activity?

7. **Activity** Research and report on current statistics for consumption of energy in Canada and around the world.

8. **Activity** Determine the electrical energy consumption in your home for a two-week period. Refer to Section 19.4 for details on how to measure electrical energy. Then persuade your family to conserve energy as much as possible and repeat the calculations for the next two weeks.

9. **Activity** This activity is best done in groups of four to six students. The main pieces of apparatus are a stopwatch, an eyedropper, and a large bucket with a capacity of at least 15 L. You will also need a variety of beakers and graduated cylinders. [Semi-log graph paper is required for the optional exercise in (d).] Before you begin the activity, estimate the answers to (a) and (c) below.

    (a) Determine the amount of time needed to fill the bucket if you add one drop of water to the empty bucket, then cause the amount of water in the bucket to double every 10 s. (If at first you are unsuccessful, try again.)
    (b) Relate your observations to the concepts of doubling time and growth.
    (c) Calculate the number of drops of water in the bucket after each 10 s interval. What is the total number of drops in the filled bucket?
    (d) Use semi-log graph paper to plot a graph of the number of drops (along the logarithmic scale) as a function of time (along the horizontal scale). What do you conclude?

10. The world population is growing at 1.7%/a. How long will it take for it to double? What will its total numbers be at that time?

11. If our oil reserves are dwindling at a rate of 6%/a, what percentage of our current supply will remain after 24 a?

12. What is meant by a population growth rate of zero? Do you believe that all countries in the world should aim for such a rate? Explain.

13. At a postal rate increase of 12%/a, what would be the cost of mailing a first-class letter in (a) 12 a and (b) 60 a?

7.3 Non-renewable Resources

An **energy resource** is a raw material, obtained from nature, which can be used to do work. A resource is called **renewable** if it renews itself in the normal human lifespan. All other resources are called **non-renewable**. The latter are the subject of this section.

Figure 7-3 illustrates Canada's main sources of energy. Approximately 11% of our energy consumption originates from water power (at waterfalls, for example). This resource is renewable. Almost all of the remaining energy is taken from non-renewable resources — crude oil, natural gas, and coal, which are fossil fuels, and uranium.

Fossil fuels make up the largest portion of non-renewable energy resources. Fossil energy begins as radiant energy from the sun that is absorbed by plants. The plants use the energy to manufacture carbohydrates which store energy. Most of that energy is used during the lifetime of the plants, but some remains after death. If the plants become buried, they do not decay; rather, they are compressed into various new forms. Their energy (chemical potential energy) can be extracted later. Because it takes millions of years for plant life to become useful fuel, once we have consumed the fossil fuels currently available, they will be gone forever.

Fossil fuels are composed mainly of carbons and hydrocarbons. **Hydrocarbons** are compounds containing only carbon and hydrogen. (They differ from carbohydrates such as starch and sugar, which contain oxygen as well as carbon and hydrogen.) Hydrocarbons are found in the solid, liquid, or gaseous state. Table 7-4 lists the main categories of fossil fuels currently mined in Canada and their energy content.

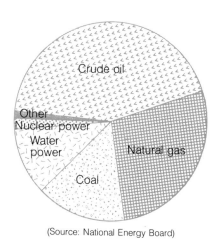

Figure 7-3
Canada's Sources of Energy

Table 7-4 Major Fossil Fuels

Fuel	State	Composition	Energy Content
coal	solid	70% C	28 MJ/kg
lignite	solid	30% C	12 MJ/kg
gasoline	liquid	varies	44 MJ/kg
kerosene	liquid	varies	43 MJ/kg
methane	gas	CH_4	49 MJ/kg
ethane	gas	C_2H_6	44 MJ/kg
propane	gas	C_3H_8	43 MJ/kg

(*Note:* All values are approximate.)

Canada is fortunate in that it happens to be one of the most resource-rich nations in the world. But just how much of the earth's non-renewable resources do we have, and how long will our supplies last? Table 7-5 answers these questions for oil, natural gas, coal, and uranium.

Fossil fuels are recovered from the earth in raw form. Then they are refined and may be used either directly (in transportation vehicles, industries, homes, and buildings) or indirectly. When a fuel is used indirectly, its stored energy must be converted to another form of energy.

Table 7-5 Canada's Non-renewable Resources on a World-wide Scale

Energy Resource	Approximate Portion of World's Supply	Estimated Time Remaining before Depletion of World Supply
oil	1%	less than 100 a
natural gas	3%	less than 200 a
coal	1%	about 500 a
uranium	20%	more than 1000 a

This is accomplished by means of devices called **energy converters**. One of the most useful and important energy converters in Canada is the electric generator. All across the country various forms of energy are converted into electrical energy using generators. Fossil fuels are often used to heat water to produce steam, which in turn drives huge electric generators. Figure 7-4 shows the basic method of using fuel to generate electricity. (More details of electricity generation are discussed in Chapters 21 and 23.)

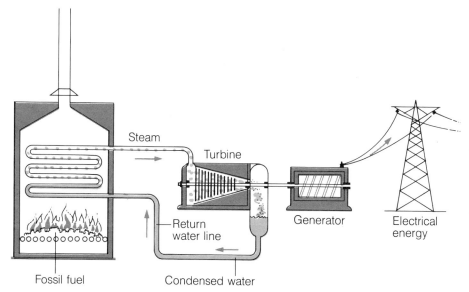

Figure 7-4
The drawing shows the use of fossil fuel to generate electricity. Chemical energy stored in the fuel changes to heat. The heat boils water which changes to steam. The steam, under pressure, forces the turbine to spin. The generator, connected to the turbine, changes the mechanical energy of spinning into electrical energy.

Uranium is an important source of energy in nuclear generation of electricity. Uranium undergoes **nuclear fission**, in which the nucleus (core) of each atom splits up and, in doing so, releases a relatively large amount of energy which heats water. Thus, for electrical energy production, uranium serves the same function as fossil fuels. (Fission is discussed in greater detail in Chapter 23.)

The fossil fuels already mentioned are relatively easy to recover from the earth. However, as the supply of oil and natural gas diminishes, less conventional fossil fuels, which are more difficult to recover, will become important. Two sources of such fuels are located in Western Canada's tar sands and Eastern Canada's oil shale rock.

MECHANICS

Figure 7-5
Surface mining in the Athabasca Tar Sands is performed by large machines like the one seen in the lower right-hand corner of this photograph. The bitumen extracted is refined in the operation seen in the background.

Canada's tar sands, located predominantly in Northern Alberta, may become extremely valuable in the future. The largest known petroleum accumulation in the world is the deposit called the Athabasca Tar Sands. In fact, the Athabasca Tar Sands are the site of the largest mining operation in the world. This all sounds impressive, but there are numerous problems to overcome before the tar sands can be mined efficiently.

Tar sands are composed of a mixture of sand grains, water, and a thick tar called **bitumen**. Only about 10% of the bitumen can be surface mined. The remainder lies beneath the surface, down to a depth as great as 600 m. Currently, surface mining is carried out using huge, expensive machines, shown in Figure 7-5. Methods of extracting tar from below the surface use up to 50% of the energy recovered. This is obviously a waste of energy, so much research and development are being carried out to improve the techniques.

As we mentioned, Eastern Canada offers another source of unconventional fossil fuel. In New Brunswick (Albert County) and Nova Scotia (Pictou County), oil shale rock has trapped a very thick, almost solid oil called **kerogen**. This material is mined by undergound explosions followed by an injection of steam or hot air. The process heats the shale and vaporizes the oil for recovery. Kerogen is difficult and expensive to mine, but as energy prices rise and other resources become depleted, this resource may become important.

PRACTICE

14. Assume that Canadians consume about 10^{11} kg of methane *per annum*. Use the information in Table 7-4 to determine how much energy is available in that amount of methane.
15. Canadians consume about 2×10^8 kg of crude oil per day. If the energy content of oil is 43 MJ/kg, how much energy do we obtain from crude oil each day?

16. Describe examples of energy converters other than an electrical generator.
17. The giant tar sands excavator can move 4.5×10^7 kg of oil sand per day. How much work is done just in lifting that amount of oil sand a vertical height of 10 m?

7.4 Renewable Energy Resources

Fossil fuels and fissionable materials, such as uranium, will not last many more centuries. But renewable energy resources are alternatives which can supply a seemingly endless amount of energy. Our challenge is to develop means of converting the available renewable energy into usable energy.

Many of the world's renewable energy resources currently being used or researched are briefly described in this section. As you read each description, try to categorize the resource as being available either locally or non-locally.

Solar energy, radiant energy from the sun, can be used to produce small amounts of electrical energy when it strikes *photovoltaic* cells. These cells are used in satellites and such instruments as calculators. Solar energy can also be used to heat buildings and swimming pools directly. The expression **passive solar heating** refers to designing and building a structure to take best advantage of the sun's energy at all times of the year. The sun's rays enter such a building in winter but not in summer, as illustrated in Figure 7-6. Passive solar heating is much less expensive

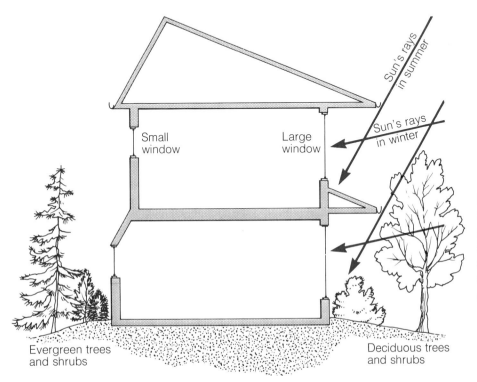

Figure 7-6
The diagram illustrates the basic features of a home with passive solar heating. Other features may include carpets that absorb light energy in winter, and window shutters that are closed at night to prevent heat loss.

MECHANICS

Figure 7-7
Active Solar Heating

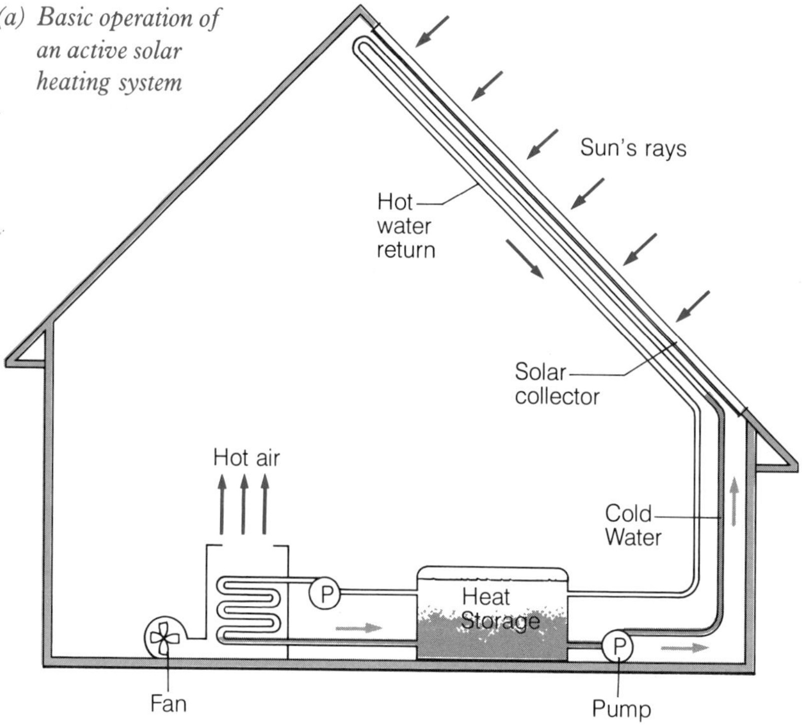

(a) Basic operation of an active solar heating system

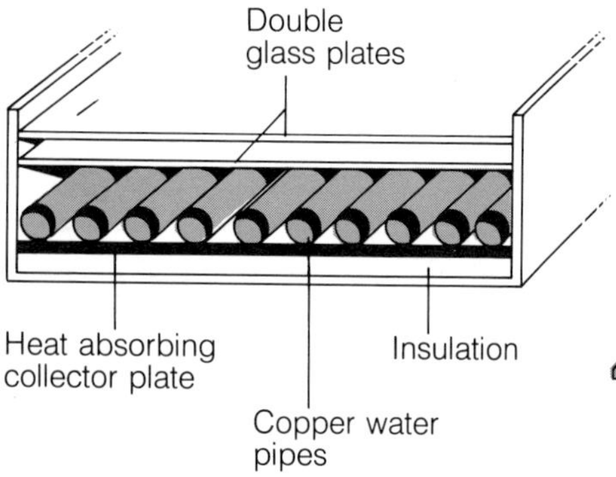

(b) An example of solar collector design

(c) This experimental home in Regina, Saskatchewan, was built to test features of good insulation and active and passive solar heating.

to install than active solar heating. An **active solar system** (Figure 7-7) absorbs the sun's energy and converts it into other forms of energy, such as electricity. For instance, an array of solar cells placed on a slanted south-facing wall or roof can convert light energy into electrical energy.

Hydraulic energy comes indirectly from solar energy. The sun's radiant energy strikes water on the earth. The water evaporates, rises, condenses into clouds, then falls as rain. The rain gathers in rivers and lakes and has gravitational potential energy at the top of a dam or waterfall. That energy can then be changed into another form, such as electricity, which is useful.

Wind energy, again caused indirectly by solar energy, is a distinct possibility as an energy source in those areas of Canada where wind is common throughout the year. Wind generators can change the kinetic energy of the wind into clean, non-polluting electrical energy, or into energy for pumping water. The photograph at the start of the chapter shows a wind generator designed in Canada. Figure 7-8(a) shows the estimated wind energy available in Canada; and Figure 7-8(b) shows how wind energy is transformed into electrical energy in a vertical-axis wind generator.

Tidal energy is a possible energy resource in regions where ocean tides are large. It is one of the few resources not resulting from the sun's radiation. It occurs because of gravitational forces of the moon and sun on the earth. To obtain electrical energy from tidal action, a dam must first be built across the mouth of a river that empties into the ocean. The gates of the dam are opened when the tide rolls in. The moving water spins turbines which produce electricity. When the tide stops rising, the gates are closed until low tide approaches. Then the gates are opened and the trapped water rushes out past the turbines,

ENERGY IN A MODERN WORLD

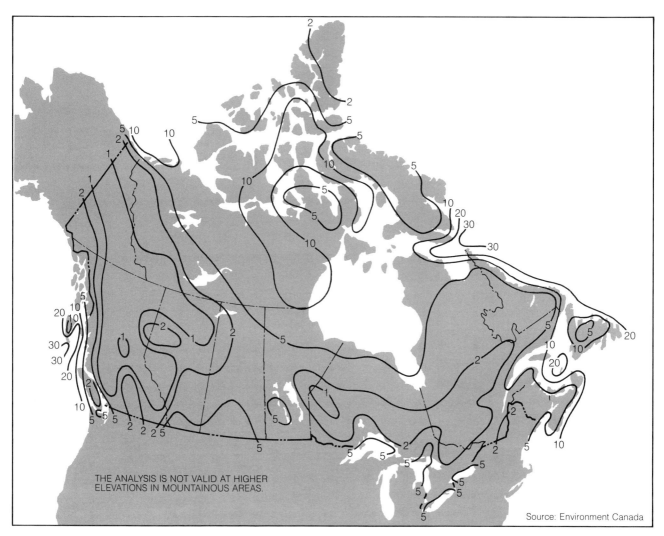

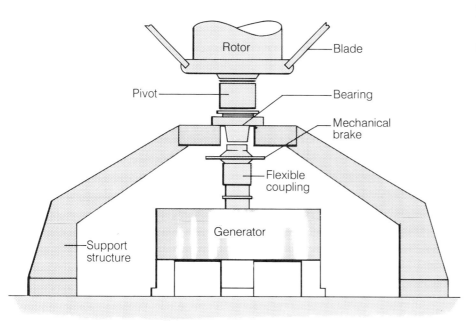

Figure 7-8
Wind Energy

(a) This map shows the mean annual wind energy density available in Canada during the period 1967–1976. The units of measurement are $GJ/m^2/a$ at an elevation of 10 m.

(b) This diagram shows the structure of the base of the vertical-axis wind generator shown in the photograph at the start of the chapter.

MECHANICS

Figure 7-9
Canada's first tidal generating station takes advantage of the high tides in the Bay of Fundy.

once again producing electricity. A major advantage of this system is that it does not produce either air pollution or thermal pollution. Its disadvantages are that it is difficult to produce a supply of electricity whenever it is most needed; as well, it is hard to judge how the construction of the dam will affect the tides and the local ecology.

Two tidal-energy electrical plants have been in operation for several years. One is in Russia; the other, in France. North America's first tidal-energy generating station began operation in the mid 1980s. Built on an existing causeway, it is located at Annapolis Royal, Nova Scotia, which is linked to the Bay of Fundy system which has among the highest tides (15 m) in the world. Refer to Figure 7-9. Other possible sites in Canada are at Ungava Bay in Northern Québec, Frobisher Bay and Cumberland Sound on Baffin Island, and Jervis and Sechelt Inlets near Vancouver.

Waves on large bodies of water gain their energy from the wind and thus, indirectly, from the sun. Apparatus can be designed to convert the up-and-down wave action into electricity. Figure 7-10 shows one design of a wave-power generator.

Figure 7-10
The drawing shows a device for converting the energy in waves into electrical energy. As a crest passes by, water flows into the upper chamber. When the following trough passes by, the water in the upper chamber forces the inlet flaps to close. Thus the water drops to the lower chamber and on its way causes the turbine to spin. The water exits through the flaps in the lower chamber.

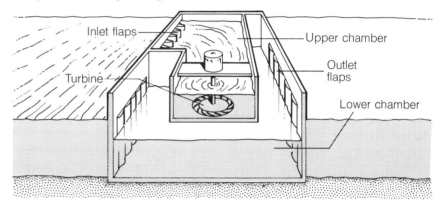

Biomass energy is the chemical potential energy stored in plants and animal wastes. Again, this energy comes indirectly from the sun. Burning wood is a common source of such energy. Wood is used both in home fireplaces and woodstoves, and in large industries that burn the leftover products of the forestry industry. Of course, good planning must be carried out to ensure that the trees are replanted — otherwise, this resource cannot be called renewable.

Numerous schemes are being developed to use forms of biomass other than wood. One proposal is to burn trash to produce heat. Another is the fermentation of sugar molecules in grain by bacteria to produce methane and ethanol (grain alcohol). A mixture of one part of this alcohol in nine parts of gasoline can be used to run automobile engines. This mixture, called gasohol, is being used in various parts of Canada and to a great extent in Brazil, where it is sugarcane, not grain, that is fermented.

One further biomass scheme has interesting possibilities. Certain plants produce not only carbohydrates but also hydrocarbons. An example of such a plant is the rubber tree, which produces *latex*. Research is underway to develop the use of fast-growing trees and shrubs that produce hydrocarbons directly, and thus require much less processing than carbohydrates before being used as fuel.

Geothermal energy is thermal energy or heat taken from the earth. It results from radioactive decay (the nuclear fission of elements in rocks). This enormous resource increases the earth's subsurface temperature an average of 25°C with each kilometre of depth. Hot springs and geysers spew forth hot water and steam from within the earth's crust. They can be used directly to heat homes and generate electricity. See Figure 7-11. However, most of the thermal energy contained underground does not find its way to the surface, so methods for its extraction are being researched. For example, if the rocks are hot and dry, certainly no water

Figure 7-11
This thermal generating station is located in California, U.S.A.

or steam will come to the surface. Still, there is a technique for utilizing this heat. First, two holes are drilled deep into the ground a set distance apart. Water is poured down one hole and gains energy as it seeps through the hot, porous rocks. Then the water rises up the other hole. The circulating water runs turbines to produce electricity. Geothermal energy is plentiful in the former volcanic regions of British Columbia and the Yukon Territories, as well as in the Western Canada sedimentary basin in the Prairie provinces.

Nuclear fusion is the process in which nuclei of the atoms of light elements join together at extremely high temperatures to become larger nuclei. (Notice that this process differs from nuclear fission, in which the nuclei of heavy elements split apart.) With each fusion reaction, some mass is lost; it changes into a relatively large amount of energy. Fusion is the energy source for the sun and stars. On the earth, the process has been possible in physics laboratories since the 1920s.

Two main problems must be overcome in order to use fusion to generate electricity. The first is producing temperatures as high as hundreds of millions of degrees, which are needed to begin the fusion reaction. The second is confining the reacting materials, so that fusion may continue. Research is currently progressing in the use of magnetic fields and lasers to solve both problems. (A more detailed explanation of fusion is given in Chapter 23.)

Hydrogen, one of the most abundant substances on the earth, is used, in certain forms, as a common fuel to operate fusion reactors. Nuclear fusion has certain advantages, one of them being a potentially limitless supply of fuel from the world's oceans. Another is that it produces much less radioactive waste than nuclear fission, so it is more desirable from the environmental point of view.

The **atmosphere** can also be used as a source of heat. An electric **heat pump**, for instance, operates to heat a home in winter and cool it in summer. It works on the principles that when a substance changes from a liquid to a vapour, energy is absorbed (that is, evaporation requires heat), and when a substance changes from a vapour to a liquid, energy is given off (that is, condensation gives off heat).

The substance that circulates in a heat pump system is called a *refrigerant*. It flows in one direction in summer and the opposite direction in winter. The refrigerant evaporates inside the home in summer, absorbing heat. It evaporates outside in winter, again absorbing heat. Figure 7-12 illustrates the basic operation of a heat pump in both seasons.

Did You Know?
The principle of operation of a heat pump (air-to-air heat extraction) can be applied to a system in which the outside source of heat is the earth, rather than the air. A closed-loop piping system is buried underground, where the temperature is relatively constant. This type of system requires less energy to operate than a regular heat pump.

PRACTICE
18. List as many renewable energy resources originating from the sun's radiant energy as you can.
19. Which renewable energy resources do not originate from the sun's radiant energy?
20. Starting with the sun, trace the energy transformations that occur in order to cook a roast in an electric oven. The electricity comes from a hydraulic generating system.
21. Which alternate energy resource described in this section is most likely to be developed in your area? Explain why.

22. Compare the operation of a refrigerator to the operation of a heat pump.
23. Choose a specific renewable energy resource, and research its current availability and experimental development.

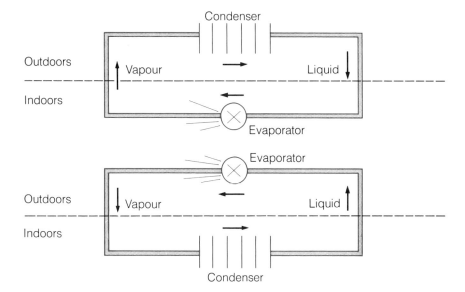

Figure 7-12
The Basic Operation of a Heat Pump

(a) *In summer, evaporation occurs indoors to absorb heat; condensation occurs outdoors to give away the heat.*

(b) *In winter, evaporation occurs outdoors to absorb heat; condensation occurs indoors to give away the heat.*

7.5 Advantages of Using Energy

This topic, like any of the others in this chapter, could fill an entire book. We will consider a few obvious advantages of using energy, then take a closer look at an example related to physics.

Of course, we need energy to survive. Our bodies cannot use the sun's energy directly — nor can other animal's. Plants convert the sun's energy into a form we can utilize. Food animals such as cattle and sheep eat plants and we, in turn, eat both plants and these animals.

But our lifestyle requires more than survival. Without energy, we would not be able to take advantage of the following:

- Comfort: climate control in homes, offices, and cars
- Communication: round-the-clock television, telephone, and radio
- Leisure: electronic games, audio and visual systems, three-dimensional movies, athletic facilities
- Travel: safe, comfortable, and fast cars, boats, trains, and planes
- Information: news broadcasts, printed materials, computerized storage and retrieval systems
- Health: medical research and diagnostic systems, a much longer life expectancy than the world average
- Working conditions: use of safe and efficient machinery

Likely you could add many more applications that reflect the advantages of using energy. Let us look in detail at a modern application relating directly to physics.

Humans have always wanted to learn more about themselves and their surroundings. Scientists perform research to discover more about nature so they and other humans can better understand and utilize it. Much of this research would be impossible without large amounts of energy input.

Particle physicists, who study the tiniest known particles in the universe, perform research using sophisticated apparatus that requires much energy. Figure 7-13 shows a particle accelerator. In this device, electrical energy is fed into the coils of powerful electromagnets. The forces of electromagnetism accelerate the particles being researched up to speeds approaching the speed of light. When collisions occur between the fast-moving particles, the physicists obtain valuable information which helps them determine the secrets of previously unknown realms.

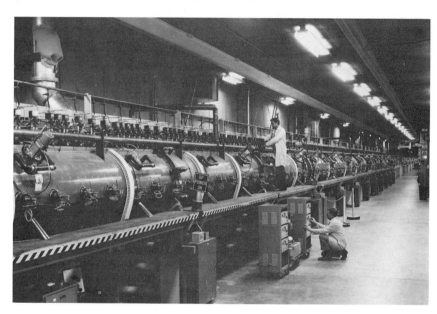

Figure 7-13
This photograph shows part of an accelerator used to increase the speed of protons up to 85% of the speed of light. Analysing the action of these high-energy protons helps scientists develop a better understanding of the universe.

PRACTICE
24. Discuss some advantages of energy consumption, other than those mentioned in this section, that relate directly to physics or some other areas of science.

7.6 Disadvantages of Using Energy

Nature is selfish — it never gives something for nothing. We live in a modern technological society with leisure time, conveniences, and gadgets undreamed of only a few generations ago. Most unfortunately, however, there is a price we must pay for all the advantages.

The law of conservation of energy states that energy is neither created nor destroyed; it simply changes form. Where, then, does all the energy go when we consume such vast quantities of fuel? If you follow the flow of energy from the source to the final result, you come to one conclusion — it almost always ends up as thermal energy. A large percentage of our energy resources goes toward refining raw materials, overcoming friction, and producing thermal pollution. This is, of course, wasteful.

Besides thermal pollution, there are many detrimental effects on the environment that result from our use of energy. Table 7-6 lists several sources of energy-related problems, as well as some immediate or long-term effects of those problems.

Table 7-6 Environmental Effects of Energy Use

Source of Problem	Effect of Problem
oil spills	death of plant and animal life
hydro dams	loss of land; alteration of ecological system
electric generating sites	loss of land
electric transmission lines	loss of land
liquid or gas pipelines	loss of land; disruption of animal migration patterns
sewage	lethal chemicals in water system
excess sound	psychological effect on humans and animals
high-energy radiation (X rays)	health hazard
thermal energy	alteration of sea life and atmosphere
chemical emissions	threat to all forms of life including humans; deterioration of building materials such as stone and brick
microwaves	potential health hazard
aerosol cans	reduction of ozone, which absorbs harmful radiation from the sun, in upper atmosphere
nuclear generation of electricity	long-term storage of radioactive waste

The list in Table 7-6 is frightening. Some of the problems are controversial; there is still disagreement about their seriousness. Some are unavoidable, and we must learn to accept them if we wish to maintain our lifestyle. Still other problems are, to some extent, avoidable, and it is the responsibility of individuals and society in general to work at eliminating them. What can be done to alleviate energy-related problems is discussed in the next two sections.

PRACTICE
25. Starting with the chemical energy stored in fuel for a car engine, describe how the energy eventually becomes thermal energy.
26. Categorize the problems listed in Table 7-6 under these three headings: Avoidable; Unavoidable; and Controversial. Discuss your lists in class.

27. One potential disadvantage of energy use is our extreme dependence on energy. It was not discussed in this section, but it could be a serious problem in an emergency situation. Describe the major consequences on Canadians of the following emergencies.
   (a) All the world's countries that normally export oil declare a 90 day oil embargo, during which they export no oil.
   (b) Part of a cross-Canada natural gas pipeline explodes and cannot be repaired for several days.
   (c) An electrical blackout occurs across your province during the two coldest days of the winter.

7.7 Society's Responsibilities to Itself

In this chapter we have seen both sides of a very important concern. Energy use has many advantages, but it also causes serious problems in our modern world. The advantages will continue to grow and, if we are wise in designing applications, they will remain advantageous. But what solutions does society have for the problems of a limited supply of convenient energy and the environmental pollution resulting from the use of energy?

The answer to this question will be considered under three headings: Improving the Efficiency of Energy Use; Pollution Control; and Energy Conservation.

Improving the Efficiency of Energy Use

We all rely heavily on electricity. The production of most of our electricity creates much unwanted pollution and requires vast amounts of non-renewable resources. To improve the efficiency of energy production, some of the thermal energy created during electricity generation could be used as an alternate source of energy. Doing this would reduce not only thermal pollution, but also the need for other resources for heating purposes. The process of producing electricity and using the resulting thermal energy for heating is called **cogeneration**. Cogeneration is used mainly in industrial plants located near generating stations. It is likely to become more important in the future.

Another way to improve the efficiency of generating electricity is to learn how to produce electricity for **local consumption**, so that the energy does not have to be transferred long distances using huge transmission lines. Sources of energy for localized power include renewable resources such as solar, hydraulic, and wind energy.

Yet another means of improving efficiency is the use of an energy converter called a **fuel cell**. This device has attracted much attention after its use in the space program. In a fuel cell, chemical potential energy is changed directly into electrical energy. The chemical fuel used in a fuel cell is usually hydrogen gas. The hydrogen combines with oxygen chemically in the presence of a third chemical called a *catalyst*. The result is the production of water and an electric current. (Refer to Figure 7-14.) Because the energy turns directly from chemical into electri-

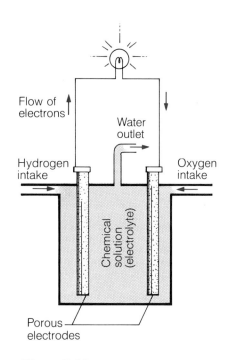

Figure 7-14
The Fuel Cell

cal energy, the fuel cell is much more efficient than electric generating stations. It can operate at a relatively low temperature, so it emits fewer pollutants. Furthermore, it has few moving parts, so it is quiet and easy to maintain. Electricity is taken from the fuel cell in a manner similar to the process for a dry cell.

Pollution Control

Governments in North America have made improvements in the standards required for control of polluting emissions from engines. For example, the "catalytic converter" is a device placed in the exhaust system of all new automobiles. Its interior is coated with a thin layer of platinum. This versatile metal turns poisonous gases that flow past it into water and carbon dioxide. One problem with the converters is that they lose their efficiency after the vehicle has been driven about 75 000 km. Therefore, in order to control pollution, drivers of older cars should replace their converters.

Politicians and scientists have also promoted other methods of reducing pollution from engines. One important step is the increasing use of lead-free gasoline, which does not create the poisonous lead compounds produced from leaded fuels. Another step is the design of more efficient automobile engines. Ordinary internal combustion engines are not efficient, because the heat to run them comes from a sequence of explosions in the engine. Newer engines (such as the *Stirling engine*) burn the fuel continuously and are thus more efficient and less prone to polluting.

Of course, much pollution occurs in industry and in the heating of buildings. Many methods, too numerous to mention here, are being researched to help reduce pollution.

Energy Conservation

Conserving energy is indeed a vital goal for all members of our society. Governments have provided incentives for people to improve home insulation and to replace old, inefficient furnaces with new ones that use cleaner, more plentiful sources of energy. They also promote the use of active and passive solar systems.

In transportation, governments support reduced speed limits, car pools, public transporation, and lanes on city roads restricted to bus use only. Table 7-7 makes it clear that a person driving alone in a car consumes a relatively large amount of energy for the distance travelled.

Governments also sponsor tests on electric consumer products to determine how energy-efficient they are. A booklet containing the research results is available, and every appliance sold in Canada must be approved with an energy guide ("Energuide") sticker indicating its energy consumption rate. See Figure 7-15.

Our democratic society will not dictate the number of children allowed in each family. But in some countries, especially where over-population and energy supply are serious problems, each family might be limited to a single child. (Remember that at a population rate of increase of 2.0%, it takes only 36 years to double a country's population.)

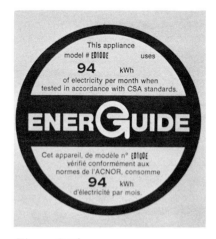

Figure 7-15
Energuide Label

MECHANICS

Table 7-7 A Comparison of Forms of Surface Transportation

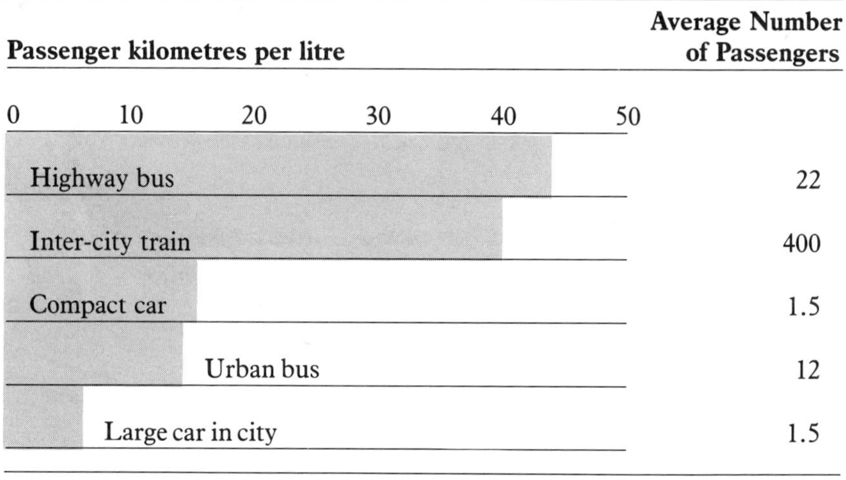

Passenger kilometres per litre	Average Number of Passengers
Highway bus (≈45)	22
Inter-city train (≈40)	400
Compact car (≈15)	1.5
Urban bus (≈12)	12
Large car in city (≈9)	1.5

PRACTICE

28. Use at least one of the following questions to initiate a project, debate, or class discussion.
    (a) Is cogeneration of electricity possible in your region?
    (b) Is generation of electricity for local consumption possible in your area?
    (c) Which type of gasoline, leaded or lead-free, should be more heavily taxed? How does your answer compare with the existing situation?
    (d) How could school officials improve energy consumption in your school?
    (e) Should Canadian families be restricted to one child?
    (f) If you were in control of time zones in Canada, would you advise the use of standard time, daylight savings time, or a combination? Explain your reasons.
    (g) Should highway express lanes in and near large cities be reserved during rush hours for cars with two or more people?
    (h) Which technique would be more effective at conserving automobile fuel — fuel rationing or higher taxes?
    (i) Should consumers pay more or less when they increase the rate of electrical energy they consume? What is the present pricing policy in your province?

7.8 Our Personal Responsibilities in Solving Energy Problems

Most of us are somewhat selfish: we like our lives to be care-free and happy. To help achieve this ideal state, we must all share responsibility for reducing pollution and conserving energy.

Our first duty is to become aware of energy problems and their solutions. We should express interest in energy-related reports on news broadcasts and in magazines. Also, we should discuss information about energy found in texts and other references.

Our next task is more specific: we should try to conserve energy and reduce pollution at home. There are many ways to accomplish this. Consider the following questions:

- Do you use more hot water than necessary to take a bath or shower?
- Do you leave the refrigerator door open while you decide what you want to eat?
- Do you keep your home quite cool in the winter and wear a sweater?
- If you have a fireplace, does most of the heat it produces go up the chimney? See Figure 7-16.
- Is your home properly insulated?
- Do you practise proper maintenance of the furnace and other appliances in your home?
- Do you leave lights and electric appliances on when they are not in use?
- Are you aware of which types of lights and appliances are more efficient than others?
- When you use an appliance such as a toaster, clothes washer, or clothes dryer, do you make maximum use of its energy?
- Do you tend to consume goods contained in throw-away packages?

Figure 7-16
In a fireplace heat circulator, cool air enters through the lower vent and passes through a duct adjacent to the fire. The hot air rises and is discharged through the upper vent or passes through ducts to other rooms.

Besides conserving energy in our homes, we should conserve it outside as well. Consider these questions:

- When you travel short distances, do you usually walk or take a car?
- Do you take part in entertainment and sports activities which are large energy consumers such as water skiing behind a motorboat? Are there substitute activities that would consume less energy but still benefit you as much – wind surfing, for example?
- Do you keep informed of ways in which you can help conserve energy?

These are just a few of the many important questions about energy we can ask ourselves. The answers will help determine the fate of future generations who will, no doubt, wish to be able to consume energy, and to do so in a pollution-free environment.

PRACTICE
29. Describe ways in which you can conserve energy in your own home.
30. Ice cubes that remain in the freezer compartment of a frost-free refrigerator gradually disappear. Why does this occur? Is letting ice cubes do this an efficient use of energy? Explain.

Review Assignment

1. Discuss reasons why Canada's rate of energy consumption is much higher than the world average.

2. What is the metabolic rate of a 60 kg person who exerts a power of 900 W running up a stairway?
3. Assume that your metabolic rate while bicycling is 7.5 W/kg.
   (a) What is your power?
   (b) How much energy do you consume during 1.0 h of this activity?
4. A girl has a power of 520 W while swimming. Determine her mass, using the average metabolic rate for swimming in Table 7-2.
5. If the cost of natural gas increases 3.6%/a, by what factor will the price have increased after
   (a) 20 a; (b) 40 a; and (c) 100 a?
6. Assume that our federal government wants to be sure that 50 years will elapse before the number of cars in Canada doubles. What growth rate per annum should the government advocate?
7. Explain why scientists and politicians should understand growth rate when dealing with energy use.
8. What is a fossil fuel? What are its main components?
9. Should Canada export any of its natural resources? Discuss.
10. Contrast and compare the use of falling water with the use of fossil fuels for generating electricity.
11. State the main advantage and main disadvantage of using each of the following non-renewable energy resources:
    (a) oil   (d) uranium
    (b) natural gas   (e) tar sands
    (c) coal   (f) kerogen
12. Assume that the world's annual energy consumption is 3×10^{20} J and that Canadians consume an average of 1000 MJ/d. To answer, you will need to know the approximate population of Canada and the world.
    (a) How much energy does each Canadian consume *per annum*?
    (b) How much energy does all of Canada consume *per annum*?
    (c) What percentage of the world's population lives in Canada?
    (d) What percentage of the annual world energy consumption does Canada consume?
    (e) Calculate the ratio of the answer in (d) to the answer in (c). What do you conclude?
13. State the main advantage and main disadvantage of each of the following renewable energy resources:
    (a) solar energy   (f) biomass
    (b) hydraulic energy   (g) geothermal energy
    (c) wind   (h) nuclear fusion
    (d) tides   (i) the atmosphere
    (e) waves
14. Research and report on alternate sources of energy not mentioned in this text.
15. Describe long-term objectives that governments in Canada should pursue to ensure that we have a plentiful supply of low-pollution energy in the future.
16. State immediate objectives that all of us can pursue to help alleviate the problems of energy use in our country.

Supplementary Topic: The Physics of Heat

Heat, Temperature, and Expansion

We stated earlier that much of the energy we consume is eventually transformed into thermal energy. Furthermore, we are surrounded by the use and effects of heat—thermostats control furnaces, large bodies of water help moderate the climate of certain regions, wind is generated by thermal effects, and the weather influences the clothes we wear. Thus, heat plays a significant role in our lives.

In order to discuss the effects of heat, it is necessary to define thermal energy, heat, and temperature. **Thermal energy** is the total kinetic and potential energies of the atoms or molecules of a substance. It depends on the mass, temperature, nature, and state of the substance. **Heat** is a measure of the energy transferred from a hot body to a colder one. **Temperature** is a measure of the average kinetic energy of the atoms or molecules of a substance. It increases if the motion of the particles increases.

Consider, for example, 1.0 kg of water at 50°C and 60 kg of water at 50°C. The samples have the same temperature, but the 60 kg sample contains much more thermal energy. Consider also 1.0 kg of ice at 0°C and 1.0 kg of water at 0°C. The temperatures are the same because the average kinetic energy of the molecules is the same. However, the water molecules have more potential energy than the ice particles. This added potential energy, gained during the melting process, means that the water at 0°C has a higher thermal energy than the ice at 0°C.

Measuring temperature is important in the study of heat. The **thermometer** is a device used to measure temperature. One common type of thermometer is made of a long, thin, transparent tube containing mercury or alcohol. As the temperature rises, the liquid expands more than the glass and rises up the tube. Mercury and alcohol are useful in thermometers because of their freezing and boiling temperatures. Mercury freezes at −39°C and boils at 357°C. Alcohol freezes at −117°C and boils at 79°C.

The temperature scale in everyday use is the Celsius scale, invented by the Swede Anders Celsius (1701–1744). He based his scale on the common substance, water. He called the freezing temperature of pure water 0°C and the boiling temperature 100°C. Then he divided the scale into 100 equal parts or degrees. Of course, temperatures may go below 0°C or above 100°C.

Another temperature scale, used by scientists, is the Kelvin scale, invented by William Thompson Kelvin, a British scientist (1824–1907). One Kelvin degree equals one Celsius degree; however, the two scales have different zero points. Zero on the Kelvin scale is the coldest temperature possible, known as **absolute zero**. It is the same as −273°C. Thus, water freezes at 273 K, which is read 273 kelvins. To change

Did You Know?
Room temperature is about 20°C and body temperature is about 37°C. The hottest weather temperature ever recorded was 58°C, in Libya, and the coldest was −88°C, in Antarctica.

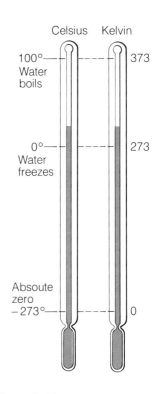

Figure 7-17
The Celsius and Kelvin Temperature Scales

MECHANICS

Figure 7-18
These hot air balloons are drifting over a golf course near the Bow River in Calgary, Alberta.

Figure 7-19
The volume of a given mass of water depends on the temperature of the water.

Did You Know?
Very cold temperatures are useful to scientists, because much can be learned about matter when it is at a low temperature. For example, if the gas helium is cooled to −269°C (or 4 K), it becomes a liquid. It can then be used to conduct such experiments as determining the properties of some metals, called **superconductors**, which lose all resistance to the flow of electric current at low temperatures.

kelvins to degrees Celsius, subtract 273, and to change degrees Celsius to kelvins, add 273. See Figure 7-17.

In general, heating a substance causes it to expand, and cooling a substance causes it to contract. When a substance is heated, its molecules gain energy and move more quickly. The molecules collide more, pushing each other apart and making the substance expand. The opposite occurs when a substance loses heat.

Gases expand readily when heated. In a hot-air balloon, for example, an energy source at the base of the balloon heats the air. The hot air in the balloon expands and becomes less dense than the surrounding cool air, causing the balloon to rise (Figure 7-18).

Most liquids expand when heated and contract when cooled. For instance, you have seen that mercury and alcohol, which expand uniformly, are useful in thermometers. Water, however, is different from other liquids. As water cools from 100°C to 4°C it contracts, so that at 4°C its density is greatest because the molecules are closest together. Then, as water cools from 4°C to 0°C, it expands. During the cold winter months this unusual behaviour is very fortunate. It means that in a large body of water, such as a lake, the water at the bottom does not drop below about 4°C. The colder, less dense water rises to the top, and that is where it freezes. The layer of ice on the top of the lake protects the water below from the cold air. Thus, even in winter, northern lakes do not entirely freeze. The graph in Figure 7-19 shows how the volume of water changes at temperatures near 4°C.

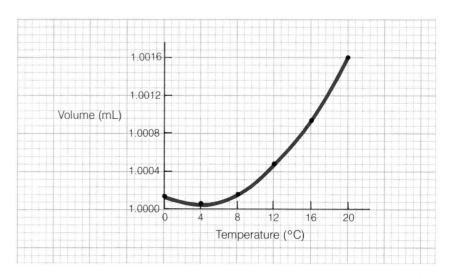

Solids are also subject to expansion and contraction. For example, the gaps between sections of railway tracks are large in winter and small in summer. See Figure 7-20.

The amount of expansion a substance undergoes as its temperature increases depends on the nature of the substance. For example, the type of glass with the trademark "Pyrex" expands very little when heated, so it does not break as easily at high temperatures as ordinary glass. To compare substances, we can measure the **coefficient of linear expansion**, which is the ratio of the expansion (ΔL) to the length (L) of a solid

Figure 7-20
The spaces between sections of railway tracks allow for expansion of the metal in hot weather.

sample, per degree Celsius. Using the symbol α (alpha) for this coefficient, we have

$$\alpha = \frac{\Delta L}{L \Delta T}.$$

The units for L and ΔL are the same, so they cancel. The unit for the coefficient of linear expansion is therefore $1/°C$ or $°C^{-1}$. Table 7-8 lists the coefficients of linear expansion of common solids.

SAMPLE PROBLEM 3
A 1.3 m rod of aluminum, initially at room temperature (20°C), increases its length by 2.6 mm as it is heated to 100°C. Determine the coefficient of linear expansion of aluminum.

Solution

$$\alpha = \frac{\Delta L}{L \Delta T}$$

$$= \frac{2.6 \times 10^{-3} \text{ m}}{1.3 \text{ m} \times 80°C}$$

$$= 2.5 \times 10^{-5} °C^{-1}$$

∴ The coefficient of linear expansion is $2.5 \times 10^{-5} °C^{-1}$.

Table 7-8 Coefficients of Linear Expansion at 20°C

Material	Coefficient of Linear Expansion ($°C^{-1}$)
aluminum	2.5×10^{-5}
brass	1.9×10^{-5}
iron or steel	1.2×10^{-5}
glass (Pyrex)	3.0×10^{-5}
glass (ordinary)	9.0×10^{-5}

PRACTICE

31. Distinguish between thermal energy and temperature.
32. Do you think it would be possible for a substance to have no thermal energy whatsoever? Explain.
33. Which type of thermometer, alcohol or mercury, would you use in the Arctic in the winter? Why?
34. Compare the size of a Kelvin degree with the size of a Celsius degree.
35. State the following temperatures in kelvins:
    (a) boiling point of water
    (b) room temperature
    (c) body temperature
36. Convert the following to degrees Celsius:
    (a) 393 K   (b) 473 K   (c) 164 K
37. If a glass bottle of pop is placed in a freezer, it might burst. Why?
38. Hot water can be used to loosen a metal screw cap on a glass jar. Explain how this is possible.
39. A material called "Invar" expands 0.90 mm for each kilometre of length with a temperature increase of 1.0°C. Aluminum expands 25 mm under the same conditions.
    (a) Which material would make a more accurate measuring tape? Why?
    (b) Calculate the coefficient of linear expansion of Invar.
40. **Activity** Design and perform an experiment to determine the coefficient of linear expansion (α) of various metal rods. The procedure you use depends on the type of apparatus available.
41. A *bimetallic strip*, also called a *compound bar*, consists of two strips of different metals attached lengthwise. It can be used as a thermostat to control such devices as furnaces and air conditioners. The bimetallic strip shown in Figure 7-21 is at room temperature.
    (a) Will the bar bend toward A or B when it is heated? when it is cooled?
    (b) In this case, should the control device (C) be connected to a furnace or an air conditioner? Explain.

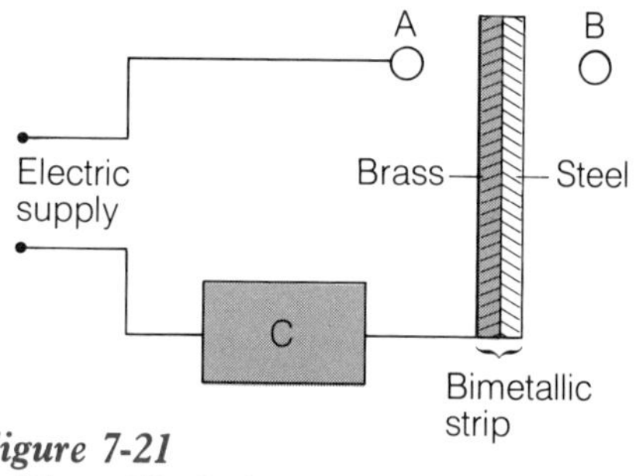

CAUTION:
When experimenting with heat, exercise care. Avoid loose hair or clothing, and wear safety goggles and an apron.

Figure 7-21
A Bimetallic Strip

Transfer of Heat

The transfer of heat from one body to another can cause either a temperature change or a change of state. We will begin this section by considering temperature changes.

Different substances require different amounts of energy to increase the temperature of a given mass of the substance. This occurs because different substances have different capacities to hold heat. For example, water holds heat better than steel. Therefore, water is said to have a higher specific heat capacity than steel. The word "specific" is used to indicate that we are considering an equal mass of each substance. (In preferred SI units, the mass is 1.0 kg.) Thus, **specific heat capacity**, c, is a measure of the amount of energy needed to raise the temperature of 1.0 kg of a substance by 1.0°C. It is measured in joules per kilogram degree Celsius, J/(kg·°C).

The English scientist James Joule performed original investigations to determine the specific heat capacities of various substances. He discovered, for instance, that it takes 4.2×10^3 J of energy to raise the temperature of 1.0 kg of water by 1.0°C:

$c_W = 4.2 \times 10^3 \frac{J}{kg \cdot °C}$, where c_W is the specific heat capacity of water

This value also means that 1.0 kg of water gives up 4.2×10^3 J of energy when its temperature drops by 1.0°C.

The heat, E_H, gained or lost by a body is directly proportional to the mass, m of the body, its specific heat capacity, c, and the amount of the temperature change, ΔT. The equation relating these factors is

$E_H = mc\Delta T$.

SAMPLE PROBLEM 4

How much heat is needed to raise the temperature of 2.2 kg of water from 20°C to the boiling point?

Solution

$$E_H = mc\Delta T$$
$$= (2.2 \text{ kg})(4.2 \times 10^3 \frac{J}{kg \cdot °C})(80°C)$$
$$= 7.4 \times 10^5 \text{ J}$$

∴ The heat required is 7.4×10^5 J or 0.74 MJ.

The specific heat capacities of different substances are shown in Table 7-9.

Table 7-9 Specific Heat Capacities of Common Substances

Substance	Specific Heat Capacity $\left(\frac{J}{kg \cdot °C}\right)$	Substance	Specific Heat Capacity $\left(\frac{J}{kg \cdot °C}\right)$
glass	8.4×10^2	water	4.2×10^3
iron	4.5×10^2	alcohol	2.5×10^3
brass	3.8×10^2	ice	2.1×10^3
silver	2.4×10^2	steam	2.1×10^3
lead	1.3×10^2	aluminum	9.2×10^2

When heat is transferred from one body to another, it always flows from the hotter body to the colder one. The amount of heat transferred obeys the **principle of heat exchange**, which states:

> When heat is transferred from one body to another, the amount of heat lost by the hot body equals the amount of heat gained by the cold body.

MECHANICS

In equation form, this law is written

$$[E_H]_{lost} = [E_H]_{gained}$$
$$\text{or} \quad [mc\Delta T]_{lost} = [mc\Delta T]_{gained}.$$

SAMPLE PROBLEM 5
A 200 g piece of steel at 350°C is submerged in 300 g of water at 10°C to be cooled quickly. Determine the final temperature of the steel and the water.

Solution
Let the final temperature be T. Then, for steel, $[\Delta T]_{lost} = 350°C - T$; for water, $[\Delta T]_{gained} = T - 10°C$.

$$[mc\Delta T]_{lost} = [mc\Delta T]_{gained}$$

$$(0.20 \text{ kg} \times 4.5 \times 10^2 \tfrac{J}{kg \cdot °C})(350°C - T) = (0.30 \text{ kg} \times 4.2 \times 10^3 \tfrac{J}{kg \cdot °C})(T - 10°C)$$

These units cancel.

$$31\,500°C - 90\,T = 1260\,T - 12\,600°C$$

$$44\,100°C = 1350\,T$$

$$\therefore T = 32.7°C$$

Thus, the final temperature of the steel and water is about 33°C.

Figure 7-22
A Temperature-Time Graph for a Substance That Is First Heated and Then Cooled

Next, let us consider what happens during changes of state. When heat is added to a substance at either the melting point or the boiling point, the added heat goes to separating the molecules farther from one another, thereby giving them greater potential energy. Thus, the temperature remains constant during a change of state. This fact is illustrated on the temperature-time graph in Figure 7-22.

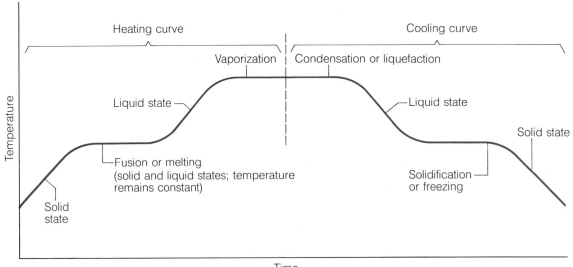

The amount of heat needed to change a unit mass of a substance from a solid to a liquid is called the **specific latent heat of fusion** (l_f). Again, the word "specific" denotes a unit mass, in this case 1.0 kg. The word "latent" means "hidden", which refers to the fact that adding heat to a substance during a change of state does not change the kinetic energy of the molecules; in other words, the temperature remains constant.

Similarly, the amount of heat needed to change a unit mass of a substance from a liquid to a gas is called the **specific latent heat of vaporization** (l_v). The temperature remains constant as the substance vaporizes.

Table 7-10 lists the specific latent heats of fusion and vaporization of five substances. Notice that the units are joules per kilogram (J/kg). Notice also that vaporization requires more heat per kilogram than fusion does.

Table 7-10 Specific Latent Heats of Fusion and Vaporization

Substance	Latent Heat of Fusion (J/kg)	Latent Heat of Vaporization (J/kg)
alcohol	1.1×10^5	8.6×10^5
iron	2.5×10^5	6.3×10^6
lead	2.3×10^4	8.7×10^5
silver	1.1×10^5	2.3×10^6
water	3.3×10^5	2.3×10^6

The amount of heat needed to melt (or vaporize) a substance is directly proportional to the specific heat of fusion (or the specific heat of vaporization) and the mass of the substance. Thus,

$$E_H = l_f m \quad \text{and} \quad E_H = l_v m.$$

Here, it is understood that in the first case E_H is the heat required to melt a known mass, and in the second case E_H is the heat required to vaporize a known mass.

SAMPLE PROBLEM 6

Determine the amount of heat needed to change 8.4 kg of solid silver into liquid silver.

Solution

$$E_H = l_f m$$
$$= 1.1 \times 10^5 \, \frac{J}{kg} \times 8.4 \, kg$$
$$= 9.2 \times 10^5 \, J$$

∴ The amount of heat needed is 9.2×10^5 J.

From Table 7-10, water requires 2.3×10^6 J/kg or 2.3 MJ/kg to evaporate. This explains why a person feels cool when first stepping out of a

MECHANICS

lake or ocean on a hot day. As the moisture evaporates from the body, heat is taken away both from the body and from the surrounding air. Sweating involves a similar process helping to cool the body.

Finally, water's high specific heats of fusion and vaporization help maintain a moderate climate in regions near large bodies of water. Areas by the Atlantic and Pacific coasts experience more moderate climates than areas on the Prairies. On the dry Prairies, there are few large bodies of water to absorb heat in the summer or give it off in the winter.

PRACTICE

42. Calculate the amount of heat needed to raise the temperature of the following:
    (a) 8.4 kg of water by 6.0°C
    (b) 2.1 kg of alcohol by 32°C
    (c) 6.4 kg of lead from 12°C to 39°C
43. Determine the heat lost when
    (a) 3.7 kg of water cools from 31°C to 24°C
    (b) a 540 g piece of silver cools from 78°C to 14°C
    (c) a 2.4 kg chunk of ice cools from $-13°C$ to $-19°C$
44. Rearrange the equation $E_H = mc\Delta T$ to obtain an equation for
    (a) c (b) m (c) ΔT
45. An electric immersion heater, rated at 500 W, changes the temperature of 5.0 kg of a liquid from 32°C to 42°C in 100 s. Find the following:
    (a) the energy given to the liquid ($E = Pt$)
    (b) the specific heat capacity of the liquid
46. Determine how much brass can be heated from 20°C to 32°C using one megajoule of energy.
47. A 2.5 kg pane of glass, initially at 41°C, loses 4.2×10^4 J of heat. What is the new temperature of the glass?
48. **Activity** Perform an experiment to determine the relationship between the quantity of heat (E_H) transferred to some water and the product of the mass, specific heat capacity, and change of temperature of the water ($mc\Delta T$). Figure 7-23 shows the apparatus to be used. The quantity of heat can be found by multiplying the power rating (in watts) of the immersion heater by the time (in seconds) that the heater warms the water; i.e., $E_H = Pt$.
49. In an experiment to determine the specific heat capacity of a metal sample, a student quickly transfers a 70 g bar of metal M from boiling water into 45 g of water at 16°C. The highest temperature reached by the metal and water together is 28°C. Determine the following:
    (a) the specific heat capacity of metal M
    (b) the identity of metal M
50. **Activity** Apply the principle of heat exchange in designing and performing an experiment to determine the specific heat capacity of various metals. Figure 7-24 shows how to obtain an initial temperature of 100°C for the metal samples. Not shown in the diagram is the apparatus to find mass.

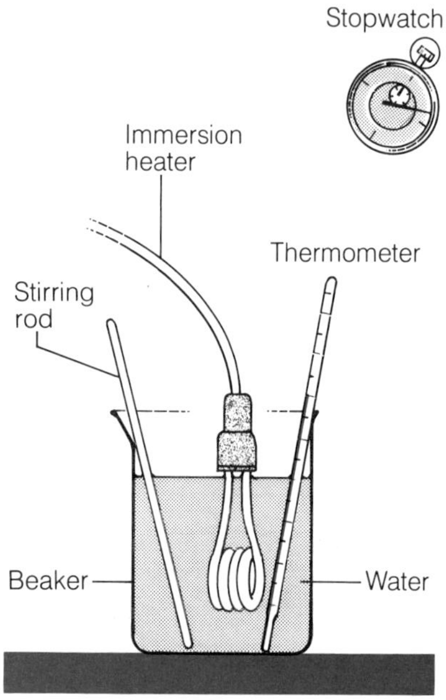

Figure 7-23

CAUTION:
Be sure that the electrical connections to the immersion heater are in safe condition and that the heating coil remains submerged in the water. Also avoid loose hair and clothing, and wear safety goggles and an apron.

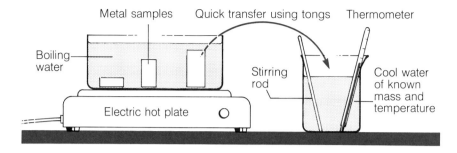

Figure 7-24

CAUTION:
Safety goggles are advised.

51. **Activity** Design and perform an experiment to determine the specific latent heat of fusion of ice. One way to do this is to add ice of known mass to high-temperature water of known mass, and apply the principle of heat transfer as shown in this equation:

    $[E_H]_{\text{lost by hot water}} = [E_H]_{\text{fusion}} + [E_H]_{\text{gained by melted ice}}$
    $[mc\Delta T]_{\text{hot water}} = [ml_F]_{\text{ice}} + [mc\Delta T]_{\text{melted ice}}$

    Here, the only unknown is l_F.

52. **Activity** Perform an experiment to determine the specific latent heat of vaporization of water. Figure 7-25 gives clues about how to use an immersion heater of known power to obtain data. You will have to find the time (in seconds) to boil away some of the water (*e.g.*, 50 g or 0.050 kg). Then you can apply the equation:

    $E_H = Pt = ml_V$

    where l_V is the only unknown.

53. How much heat is required to
    (a) change 15 kg of liquid silver into gaseous silver at the boiling temperature?
    (b) change 15 kg of solid iron into liquid iron at the melting temperature?

54. Calculate the specific latent heat of fusion if 2.1×10^6 J of energy are needed to melt 8.4 kg of a substance at a constant temperature.

55. Calculate the specific latent heat of vaporization of a substance if 8.4×10^5 J of energy are needed to boil away 0.40 kg of a substance at a constant temperature.

56. Just as their ancestors did thousands of years ago, nomads who cross the hot African desert by camel use porous goatskin bags to keep their water supplies cool. Modern-day travellers in extremely hot areas apply the same principle when they try to prevent their films from heat damage by using a wet towel to keep their film supplies cool. Explain the principle in both situations.

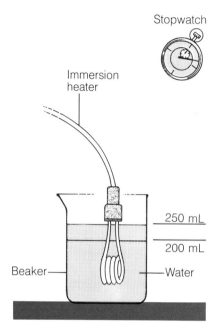

Figure 7-25

CAUTION:
Safety goggles are advised.

CAUTION:
Be sure that the electrical connections to the immersion heater are in safe condition and that the heating coil remains submerged. Avoid loose hair and clothing, and wear safety goggles and an apron.

Key Objectives

Having completed this chapter, you should now be able to do the following:
1. Appreciate the complexity of energy supply and use in our society.
2. Compare our rate of energy consumption with the rates of other people in the world or in the past.

3. Define metabolic rate, and recognize the SI units used to measure it.
4. Given any two of metabolic rate, power, and mass, calculate the third quantity.
5. Given the power of a device and the time used, find the total amount of energy consumed.
6. Appreciate why growth rate is an important quantity for anyone concerned with energy supply and demand.
7. Given one of the growth rate or the doubling time of a variable, find the other quantity.
8. Distinguish between renewable and non-renewable resources.
9. List Canada's main non-renewable energy resources, and describe their origin and composition.
10. Describe how fossil fuel energy is converted into electrical energy.
11. List and describe Canada's most important renewable energy resources.
12. Distinguish between active and passive solar heating systems.
13. Distinguish between nuclear fission and nuclear fusion.
14. Describe the operation of a heat pump.
15. Discuss major advantages and disadvantages of using energy.
16. Describe ways in which both society and individuals can utilize alternate sources of energy, control pollution, and conserve energy.
17. Define cogeneration and describe how it can be implemented.
18. Describe the basic operation of a fuel cell.

Supplementary

19. Distinguish thermal energy, heat, and temperature.
20. Describe the origin of the Celsius temperature scale.
21. Convert temperature in degrees Celsius to kelvins, and *vice versa*.
22. Explain the effects of the fact that water behaves differently from most substances when cooled.
23. Define coefficient of linear expansion, and state the unit used to measure it.
24. Describe how to determine experimentally the coefficient of linear expansion of a metal.
25. Describe applications of heat expansion.
26. Define specific heat capacity and state its SI unit of measurement.
27. Calculate the amount of heat gained or lost by a substance that experiences a temperature change.
28. Apply the principle of heat exchange to determine the specific heat capacity of substances.
29. Determine experimentally the specific heat capacity of metals.
30. Define specific latent heat of fusion, and determine it experimentally for ice.
31. Define specific latent heat of vaporization, and determine it experimentally for water.
32. Describe applications of the transfer of heat.

UNIT III
Fluids

The information in this Unit will be especially useful if you plan a career in medicine, sports, engineering, meteorology, environmental science, boat building, deep-sea diving, food production, wine-making, fire-fighting, dry cleaning, navigation, aeronautics, or transportation.

FLUIDS

CHAPTER 8
Fluids at Rest

Canal locks, like the one shown here in Scotland, enable boats to travel between water systems at different elevations.

A fluid is a substance which flows and takes the shape of its container. Water and air are two very important fluids, but there are many others. After you learn about forces and pressures in fluids at rest, you will be able to answer the following questions: Can it be true that a little Dutch boy could hold back the North Sea by putting his finger into a leaking hole in a dike? How is fluid pressure applied in the fields of transportation, sports, food production, and health care? Why are liquid drops spherical? And how many applications of fluids at rest can be seen in the photograph of the canal locks in Scotland?

8.1 General Properties of Fluids

You have learned that the three common states of matter are solid, liquid, and gas. (A fourth state, *plasma*, exists only at extremely high temperatures. It is described in Chapter 23.) Liquids and gases have two properties that allow them to be called fluids. First, they can flow; second, they take the shape of their container.

These two properties are easily shown for a common liquid such as water. Water flows readily when it is poured from one container into another, and it takes the shape of a container, even one that is open at the top.

Gases also take the shape of their container, although the container must be closed. For example, if a flask filled with air is held under water, as in Figure 8-1(a), it is evident that the air occupies the entire flask. One way to demonstrate that a gas can be poured is shown in Figure 8-1(b). Why does the gas (air) flow upward in this case?

In most cases, it is easy to determine whether or not a substance is a fluid. However, there are exceptions. For example, butter at about room temperature (20°C) is neither a rigid solid nor a readily-flowing liquid. Glass is even more complex. Images seen through old panes of glass often appear distorted because the glass has become thicker near the bottom. One possible explanation of this phenomenon is that glass flows, but its rate of flow is extremely slow.

An interesting property of liquids can be demonstrated with two containers that are connected, like those in Figure 8-2. When one container is raised above the other, the liquid flows toward the lower level until both levels are equal. The fact that liquids in connected vessels reach a common level has several applications, one of which was used by the Egyptians over 4500 years ago. In constructing their pyramids, the Egyp-

Figure 8-1
Gases have properties of fluids.

(a) *Air takes the shape of its container. Since air is invisible, this property is not easily observed. However, it is evident when the container, with the air sealed inside, is inverted in water and the seal is removed.*

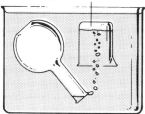

(b) *Air can be "poured" from one container into another.*

Figure 8-2
Water in Connected Vessels

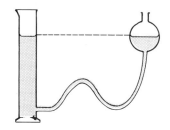

Figure 8-3
Before the construction of this large pyramid began, the ground was made level by applying the principle that water in connected vessels, in this case, trenches, reaches the same level. The same principle is used in the operation of canal locks like those shown in the photograph at the start of the chapter.

FLUIDS

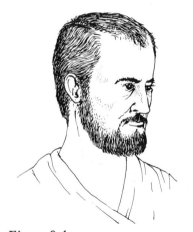

Figure 8-4
Archimedes (about 287 B.C.–212 B.C.)

tians realized the necessity of having a level base. To achieve an even base, they began by marking off the outside dimensions of the pyramid, which were as large as 250 m on each side! Then they dug a set of trenches parallel to one another and deep enough to hold water, followed by another set of trenches perpendicular to the first set. The ground probably resembled a large checkerboard. The workers filled the trenches with water which, because it was level, could be used as a guide for making the ground level. Finally, they drained the water away and began constructing the pyramid. See Figure 8-3.

Other properties of fluids, especially liquids, were also discovered by the ancients. In Greece, the brilliant scientist Archimedes (Figure 8-4) led the way in applying the properties of fluids. A well-known story tells of how Archimedes discovered the solution to a problem put to him by King Hiero. The King wanted to know whether his crown was made of the amount of gold he had paid for, or if some lead or silver had been dishonestly added. Melting the crown would provide the answer, but was not a sensible solution. The problem bothered Archimedes until one day in a public bath near his home, he stepped into a tub filled with water. As the water displaced by his body spilled onto the floor, he realized how to solve the problem. Using a piece of pure gold equal in mass to the King's crown, he could find out whether the pure gold displaced the same amount of water as the crown. He was so excited by this discovery that he jumped out of the tub and ran naked through the streets to his home yelling "Eureka!" ("I've found it!"). Later, he proved that the King had been cheated.

An important property of fluids (and of solids as well) is the characteristic called density, the mass per unit volume of a substance. If the mass and volume of a sample of a substance are known, the density can be found using the equation $D = m/V$. The preferred SI unit for density is kilograms per cubic metre (kg/m^3), although other units can be used. For example, a density of 870 kg/m^3 is equivalent to 870 g/L and 0.87 g/mL for a fluid, and 870 g/dm^3 and 0.87 g/cm^3 for a solid.

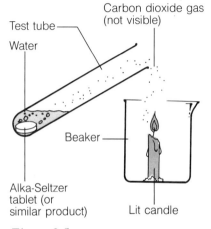

Figure 8-5

PRACTICE

1. Which of the following materials are fluids? Explain your reasoning.

   (a) steam
   (b) glycerin
   (c) sand
   (d) liquid nitrogen

2. Explain how the demonstration shown in Figure 8-5 verifies that a gas (in this case, carbon dioxide) has the properties of a fluid.

3. In each case determine the density and express your answer in the preferred SI unit.

   (a) 0.50 m^3 of glycerin has a mass of 630 kg.
   (b) The mass of 10 L of ethyl alcohol is 7900 g.
   (c) 75 mL of sea water has a mass of 77 g.

8.2 Pressure

A person wearing a pair of skis or snowshoes stays on top of the snow, while a person of equal mass wearing only boots sinks into the snow. The two people exert an equal force on the snow. However, the difference lies in the pressure each exerts on the snow. With skis or snowshoes, the pressure on any part of the snow is less because the force is spread out over a large surface area.

Pressure is therefore the force applied per unit area. In equation form,

$$p = \frac{F}{A}$$

Here, F is the force measured in newtons; A is the area measured in square metres, perpendicular to the force; and p is the pressure in newtons per square metre. The SI unit for pressure has been given the name pascal (Pa) in honour of a French scientist, Blaise Pascal, 1623–1662, who contributed greatly to our knowledge of fluids, as well as to other fields of learning.

A pressure of 1.0 N/m² or 1.0 Pa is very small. It is approximately the pressure exerted by a dollar bill resting flat on a table. Scientists prefer to use the more practical unit of kilopascals (1.0 kPa = 1000 Pa). The pressure a chair exerts upwards against you when you are sitting is approximately 3 kPa or 4 kPa.

SAMPLE PROBLEM 1

A wooden box having a weight of 1.2×10^3 N measures 60 cm long by 50 cm wide by 20 cm high. Calculate the pressure the box exerts on the floor when it is resting upright.

Solution

First, the area of the box in contact with the floor is

$$A = lw$$
$$= 0.60 \text{ m} \times 0.50 \text{ m}$$
$$= 0.30 \text{ m}^2$$

Now, $p = \dfrac{F}{A}$

$$= \frac{1.2 \times 10^3 \text{ N}}{0.30 \text{ m}^2}$$

$$= 4.0 \times 10^3 \text{ Pa or } 4.0 \text{ kPa}$$

∴ The box exerts a pressure of 4.0 kPa on the floor.

An interesting example demonstrating that pressure depends on area is a "bed of nails". Imagine the sharp end of a nail protruding upward from a board lying on the floor. A person would not want to stand on

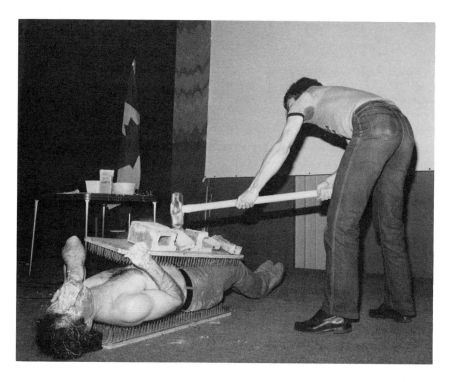

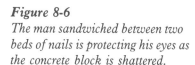

Figure 8-6
The man sandwiched between two beds of nails is protecting his eyes as the concrete block is shattered.

that nail in bare feet, because the pressure on a small surface would be so great that the nail would easily pierce the skin. Now consider a person lying on a bed of nails, shown in Figure 8-6. If the nails are spaced appropriately, about 2.0 cm apart, the force of the person lying down is spread out over the surface area provided by hundreds of nails, so it is unlikely that any one nail would break through the skin.

The examples in this section and the Practice Questions that follow do not relate directly to fluids. However, they provide the basis for understanding pressure in fluids, studied next.

PRACTICE

4. **Activity** Place your left forearm flat on a desk or table. With the palm of your right hand, press down as hard as you can on your left forearm near the elbow. Now try to exert the same force downward using only the tip of the index finger of your right hand. Explain what you feel and why.

5. Explain each of the following statements, considering especially the difference between force and pressure.
   (a) An all-terrain vehicle has wide wheels.
   (b) If boxing gloves were allowed to be smaller, the sport of boxing would be more dangerous than it already is.
   (c) It is much easier to break a walnut by hand using two walnuts pressed against one another than it is using a single walnut in the hand.

6. Calculate the pressure in each case:
   (a) $F = 420$ N, $A = 0.50$ m^2
   (b) $F = 6.2 \times 10^4$ N, $A = 2.0$ m^2
   (c) $F = 360$ N, $A = 3.0 \times 10^{-2}$ m^2

7. Calculate the pressure applied on the floor by the toe of a ballet dancer's shoe when she balances, briefly, on that toe. Assume that the dancer weighs 500 N and the surface area of her toe is 2.0×10^{-4} m².
8. Calculate the pressure exerted on the ground by a circus elephant balancing on two feet. Assume that the elephant's mass is 5.0×10^3 kg and that the area of each foot is 1.0×10^{-2} m². (Be sure to change the mass to weight.) Compare the pressure exerted by the elephant to the pressure exerted by the ballet dancer in Question 7 above.
9. Rewrite the equation $p = F/A$ to express
   (a) F by itself and (b) A by itself.
10. Calculate the unknown quantities:
    (a) $p = 2.5$ Pa, $A = 0.22$ m², $F = ?$
    (b) $p = 8.0 \times 10^4$ Pa, $F = 6.4 \times 10^2$ N, $A = ?$
11. Assume that the air pressure in a bicycle tire is 400 kPa higher than the air pressure outside the tire and is spread over an area of 0.20 m². What total force acts on the inside of the tire?
12. Suppose that the ground in a playground can withstand a pressure of 1.4×10^4 Pa, and you are asked to design a sandbox which can rest on the ground without sinking into it. If the weight of the sand and sandbox is 4.9×10^4 N, what is the minimum recommended surface area of the bottom of the box?

8.3 Atmospheric Pressure

One major reason our earth can support life as we know it is that it has an atmosphere. Our atmosphere, which is composed of the fluid, air, is piled up layer upon layer, each layer pressing down on the one below. The result is a pressure called **atmospheric pressure**, which is greatest near the surface of the earth because of the weight of all the air above.

Normally we do not notice atmospheric pressure, because the pressure inside our bodies balances the pressure outside. However, our ears are sensitive to changes in atmospheric pressure. Likely you have experienced a "pop" in your ears when your elevation above ground level changes rapidly. This may happen when you are riding in an airplane during take-off or landing, in an elevator in a tall building, or in a car on a mountain highway. The pop results when the pressure difference on either side of the eardrum is suddenly altered. Swallowing helps the pressure difference return to normal.

Atmospheric pressure, like any other pressure, is measured in pascals or kilopascals. (Other units are often used for medical and other purposes.) The lowest atmospheric pressure on record is 87.7 kPa, on the island of Guam in the South Pacific. The highest pressure on record is 108.38 kPa, measured in Siberia, U.S.S.R. Standard atmospheric pressure, used by scientists for comparison purposes, is 101.3 kPa. This is the average atmospheric pressure at sea level, and is commonly called one atmosphere. Table 8-1 indicates the average atmospheric pressures at elevations above sea level.

FLUIDS

Figure 8-7
Using Atmospheric Pressure

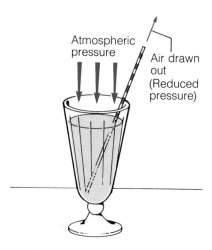

(a) A drinking straw

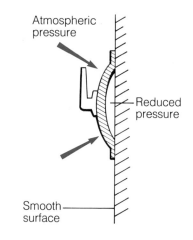

(b) A rubber sucker

CAUTION:
Be sure the burning paper falls to the bottom of the flask.

Table 8-1 Average Pressures at Various Elevations

Elevation (m)	Pressure (kPa)
0	101.3
1 500	85.0
3 000	70.0
5 500	50.0
9 000	30.0
12 500	20.0

Analysing the changes in atmospheric pressure at a certain locality is useful in forecasting the weather. A constant pressure indicates little, if any, change in the weather, while a changing pressure indicates a probable alteration in the weather pattern. Generally, decreasing atmospheric pressure signals the approach of a possible storm; increasing pressure means the approach of fair weather.

Besides weather forecasting, there are other everyday applications of atmospheric pressure. The use of the drinking straw, which was patented in 1888, is an example. Sucking air out of a straw reduces the pressure inside the straw. The pressure on the surface of the liquid, which is due to the atmosphere, is then greater than the pressure in the straw. It forces the liquid up the straw. See Figure 8-7(a). A rubber sucker, shown in Figure 8-7(b), works in a similar way. The sucker is moistened and pressed on a smooth, flat surface. The air pressure between the sucker and the surface is thereby reduced, and atmospheric pressure forces it against the surface. Such devices are used in industry to lift metal sheets and in homes to support towel holders. Other devices that operate on the same principle are the syringe and the medicine dropper (or eye-dropper).

PRACTICE

13. **Activity** This activity can be performed either individually or as a class demonstration using a flask (with a capacity of at least 500 mL), a piece of paper (about 10 cm by 20 cm), and a match. Roll the paper loosely so it will fit into the mouth of the flask. Light one end of the paper with a match and insert the burning paper into the flask. Place the palm of your hand tightly onto the mouth of the flask and, after a few minutes, lift your hand up gently. Describe and explain what happens.
14. In the CN Tower in Toronto, Ontario, the elevator takes only 60 s to rise to the main observation deck, 342 m above the ground. If a person's eardrums are forced to curve during the ascent, would they curve inward or outward? Explain. (The curving is what causes the ears to "pop".)
15. The pressure outside a jet aircraft in flight is 25.0 kPa. What is the elevation of the craft? Refer to Table 8-1. What is done to ensure the comfort of the passengers and crew?
16. A certain weather report indicates that in 6.0 h the atmospheric pressure has fallen from 100.7 kPa to 100.1 kPa. What prediction can be made about the weather?

17. Suppose that at sea level a student is able to suck water up a straw to a height of 110 cm. If the same student tried this experiment with the same straw at the top of a high mountain, would the water rise to a level higher or lower than 110 cm? Explain.

8.4 Measurement of Pressure

Measuring Atmospheric Pressure

The instrument that measures atmospheric pressure directly is called a **barometer**. It was invented in Italy around 1640 by Evangelista Torricelli (1608–1647). Torricelli, who was a student of Galileo Galilei, made the first barometer by filling a long glass tube with water so no air could get into it. The tube, closed at the top and open at the bottom, was placed in a water-filled container, as shown in Figure 8-8. The atmospheric pressure was great enough to hold about 10.3 m of water in the tube. Since a glass tube over 10 m long was impractical for a science laboratory, Torricelli tried the same experiment using liquid mercury, which has a density about 13.6 times that of water. He found that the tube needed to be only about 76 cm (0.76 m) long.

Figure 8-8
The diagram shows an artist's conception of how Torricelli might have experimented with a water-filled barometer. He discovered that atmospheric pressure can support a column of water over 10 m in height.

Torricelli's type of barometer, especially the kind using mercury, is still used today. However, it is not as convenient as the **aneroid barometer**. (The word "aneroid" means "without liquid".) An aneroid barometer consists of an enclosed container having thin metal walls which are sensitive to pressure changes (Figure 8-9). A needle attached to the container indicates the atmospheric pressure. Such a barometer is used both as a weather gauge and as an **altimeter**, a device that measures the altitude of an airplane above sea level.

Measuring Gauge Pressure

Values for pressures other than atmospheric pressure are often required for both gases and liquids. Devices that serve this function do not measure the true or absolute pressure; rather, they measure the **gauge pres-**

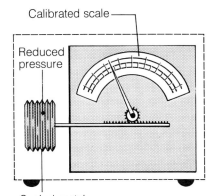

Figure 8-9
Aneroid Barometer

sure, the difference between the absolute pressure and the atmospheric pressure. Thus,

gauge pressure = absolute pressure − atmospheric pressure
or absolute pressure = gauge pressure + atmospheric pressure

For example, if air is added to a bicycle tire until the gauge pressure is 400 kPa, and the atmospheric pressure is 100 kPa, then the absolute pressure in the tire is 500 kPa.

One type of instrument used to measure gauge pressure is a **manometer**. A simple U-shaped manometer, which consists of two pieces of glass tubing (each about 30 cm long) connected by a rubber tube, is shown in Figure 8-10(a). Water is added to a depth of about 15 cm from the bottom of the instrument. If the pressure on one side of the manometer increases, as in Figure 8-10(b), the water level in that side falls, while the water level in the other side rises. The tubes can be calibrated using a reference manometer, or the gauge pressure can be found using the derivation that follows.

Figure 8-10
A Manometer

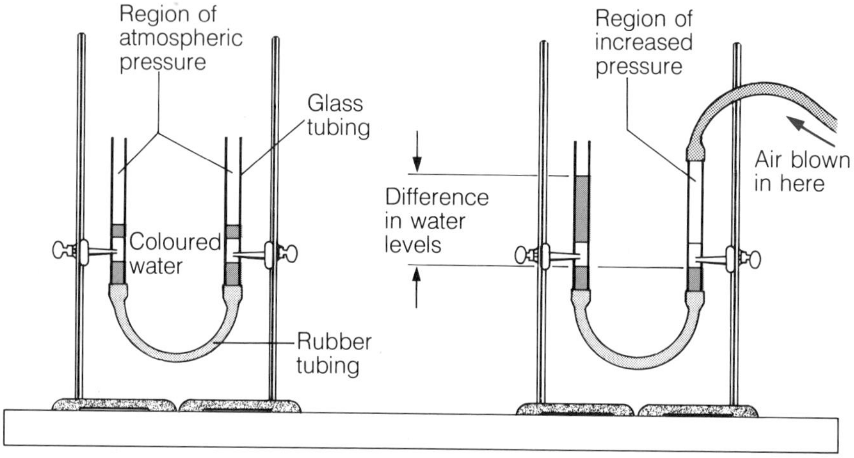

(a) Making a manometer *(b) Using a manometer*

The pressure at the bottom of a column of liquid of height h is given by $p = F/A$. From Chapter 5, the force of gravity on the liquid is $F = mg$, where $g = 9.8$ N/kg. Thus,

$$p = \frac{mg}{A}.$$

However, since the density of the liquid is m/V, its mass can be expressed as $m = DV$. Therefore, the pressure is

$$p = \frac{DVg}{A}.$$

Finally, since $V = Ah$, the pressure is

$$p = Dhg.$$

Here, D is the density of the liquid in kilograms per cubic metre and h is the height of the column in metres. In the case of the U-shaped manometer, h is the difference in height between the two levels of liquid. (Notice that the equation $p = Dhg$ applies only to liquids which maintain a relatively constant density with increasing depth. The equation is not applicable to gases, because their density changes when the depth changes.)

SAMPLE PROBLEM 2

An air pump is connected to one tube of a manometer containing water (D = 1000 kg/m³). Pressure is increased until the difference in water levels is 15 cm. What is the pressure difference?

Solution

$p = Dhg$

$= 1000 \frac{\text{kg}}{\text{m}^3} \times 0.15 \text{ m} \times 9.8 \frac{\text{N}}{\text{kg}}$

$= 1470 \text{ N/m}^2$

$= 1500$ Pa or 1.5 kPa

Thus, the pressure difference is 1.5 kPa.

Figure 8-11
Because it has such a long neck, the giraffe needs a specially adapted circulatory system.

Measuring Blood Pressure

The human heart pumps blood under pressure to all parts of the body. The amount of pressure exerted by the heart is easy to measure and is an important indicator of a person's general health.

To measure blood pressure, a gauge is wrapped around the upper arm at the same level as the heart. See Figure 8-12. Two measurements are made: first, the maximum blood pressure, called the **systolic pressure**; then, the pressure when the heart is resting, called the **diastolic pressure**.

A hand pump is used to increase the pressure in the air bag above the systolic level. Then the pressure is gradually reduced, as the doctor or nurse listens with a stethoscope to the flow of blood in the main artery of the arm. When the systolic pressure is reached, a tapping sound is heard as the blood spurts through the artery. When the pressure has decreased to the diastolic level, the sound disappears.

Did You Know?
The fact that the pressure of liquids increases with depth is not a problem for the giraffe, because it has a large heart and a complex system of valves and blood vessels in its brain. Without these features, the giraffe would faint when raising its head suddenly or bleed when lowering it. See Figure 8-11.

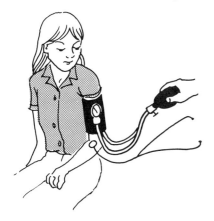

Figure 8-12
Measuring Blood Pressure

Normal blood pressures, which are gauge pressures, range from about 16 kPa (systolic) down to about 11 kPa (diastolic). They are usually written as a ratio (16/11). Anyone with high blood pressure, for example, above 19/12, requires medical attention. In practice, blood pressure is measured in terms of millimetres of mercury (mm Hg). In these units, the ratio of normal blood pressures is 120/80.

PRACTICE

18. The best suction pump in the world can pull liquid mercury in an open tube no higher than about 76 cm. Why?
19. State an advantage of each of the following:
    (a) a mercury barometer compared with a water barometer
    (b) an aneroid barometer compared with a mercury barometer
20. The gauge pressure of a car tire is 205 kPa, and the atmospheric pressure is 101 kPa. What is the absolute pressure of the tire?
21. If the atmospheric pressure is 102 kPa and the absolute pressure of a bicycle tire produced by an air pump is 463 kPa, what is the gauge pressure of the tire?
22. A vacuum pump sucks air out of one side of a manometer that uses mercury ($D = 1.36 \times 10^4$ kg/m^3). The other side of the manometer is open to the atmosphere. The resulting difference in mercury levels is 50 cm.
    (a) Calculate the difference in pressures between the two sides of the manometer.
    (b) If the atmospheric pressure is 101 kPa, what is the absolute pressure on the vacuum pump side of the manometer?
23. **Activity** Design and perform an experiment to determine how the pressure beneath the surface of a liquid depends on the following factors:
    (a) the direction (up, down, sideways)
    (b) the depth beneath the surface
    (c) the size of the container holding the liquid
    (d) the density of the liquid

A U-shaped manometer containing water can be used to measure gauge pressure, as Figure 8-13 shows. The rubber diaphragm connected to the thistle tube should be stretched tightly so it is sensitive to pressure changes. Some possible liquids for the activity include water, alcohol, and glycerin.

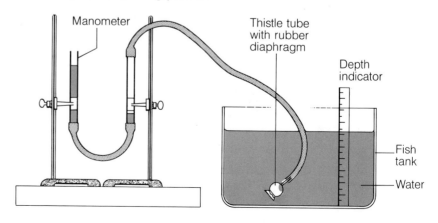

Figure 8-13

8.5 Pascal's Law

An important property of liquid pressure was discovered by the French scientist Blaise Pascal, who was mentioned earlier. His discovery, called **Pascal's law**, is based on the fact that liquids cannot be (easily) compressed. It states:

Pressure applied to an enclosed liquid is transmitted equally to every part of the liquid and to the walls of the container.

Pascal applied his law in the design of a useful device called the **hydraulic press**. (The word "hydraulic" means "operating by the force of a liquid".) Figure 8-14 illustrates how such a press works. A small downward force applied to the small movable piston can produce a large upward force on the large movable piston. According to Pascal's law, the pressure (p_s) on the small piston equals the pressure (p_L) on the large piston. Thus, since $p = F/A$, we can write

$$\frac{F_s}{A_s} = \frac{F_L}{A_L} \quad \text{which means that} \quad F_L = F_s \cdot \frac{A_L}{A_s}.$$

Here, "s" means small and "L" means large. This equation can be used to solve for any one of the variables, given the other three.

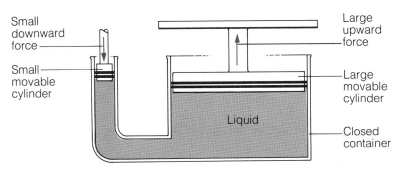

Figure 8-14
The Hydraulic Press

SAMPLE PROBLEM 3

Assume that for the hydraulic press shown in Figure 8-14 a force of 220 N is applied to the small piston, which has a surface area of 0.12 m². If the surface area of the large piston is 3.0 m², what total force is exerted on the large piston?

Solution

$$F_L = \frac{F_s A_L}{A_s}$$

$$= \frac{220 \text{ N} \times 3.0 \text{ m}^2}{0.12 \text{ m}^2}$$

$$= 5500 \text{ N or } 5.5 \times 10^3 \text{ N}$$

∴ The total force exerted on the large piston is 5.5×10^3 N.

FLUIDS

The principle of the hydraulic press has several important applications. In automobile service stations, a *hydraulic jack* is used to hoist a car. A relatively small force applied to the small piston can lift a car perched on the platform attached to the large piston. A hydraulic fork-lift truck operates in a similar manner. The *hydraulic brakes* on automobiles (Figure 8-15) also apply Pascal's law. When the brake pedal is pushed, the pressure on the brake fluid in the master cylinder is increased. This pressure is transmitted through the fluid to eight other cylinders, two on each wheel. The increased pressure on the brake pads exerts a frictional resistance on the wheel disks, slowing down the car.

Figure 8-15
Hydraulic Brake System

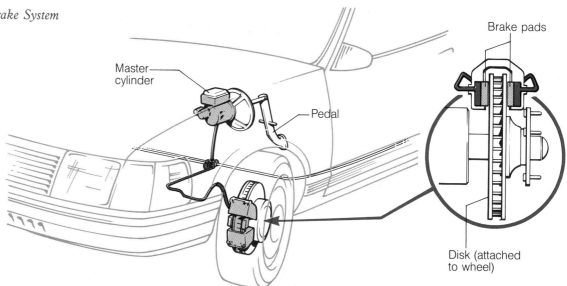

PRACTICE

24. Both gases and liquids are fluids. Why is it not possible to use a gas to operate a hydraulic press?

25. Starting with the equation $\dfrac{F_s}{A_s} = \dfrac{F_L}{A_L}$, write an equation for each of the following:
    (a) F_s
    (b) F_L
    (c) A_s
    (d) A_L

26. Determine the unknown quantities:

F_s (N)	A_s (m²)	F_L (N)	A_L (m²)
?	0.25	1800	1.5
65	?	520	0.64
540	0.15	?	1.2
1.4×10^2	2.0×10^{-2}	7.7×10^3	?

27. In an automobile service centre, the hydraulic jack can exert a maximum force on the small piston of 2.2×10^3 N. The surface area of the small piston is 0.10 m² and the surface area of the large piston is 2.0 m². Calculate
    (a) the maximum force which can be used to lift a car
    (b) the mass of that car.

28. In a hydraulic brake system, a force of 25 N is exerted on the piston in the master cylinder. This piston has a surface area of 5.0 cm². What force can thus be exerted on each brake piston having an area of 100 cm²?

Figure 8-16
Illustrating Buoyancy Exerted by a Hand

8.6 Buoyancy in Fluids

Anyone who has tried to lift a rock out of water realizes that the rock appears to be lighter until it breaks through the surface of the water. The force that pushes upward on objects in fluids, causing the objects to seem lighter, is called **buoyancy**. Buoyancy helps hold a swimmer up in water and weather balloons up in air. Because buoyancy is a force, it is measured in newtons.

A demonstration to illustrate buoyancy is shown in Figure 8-16. In Figure 8-16(a), a mass is hung on a force scale; Figure 8-16(b) shows the forces acting on the mass. In Figure 8-16(c), the hand is exerting a type of buoyant force, so the upward force required to balance the mass is now less, as Figure 8-16(d) shows.

To learn why buoyancy exists in fluids, recall the relationship derived for a liquid in a manometer (Section 8.4). That relationship, $p = Dhg$, indicates that as the depth or height (h) of a liquid increases, the pressure also increases. If this fact is applied to an object submerged in a liquid, such as the block shown in Figure 8-17, it is seen that the pressure at the bottom surface is greater than the pressure at the top surface. The difference in pressure at the two levels results in a difference of total force, with the upward force being greater than the downward force. The amount by which the upward force resulting from the liquid exceeds the downward force resulting from the liquid is the buoyant force (or force of buoyancy).

Figure 8-17
The Cause of Buoyancy in Liquids

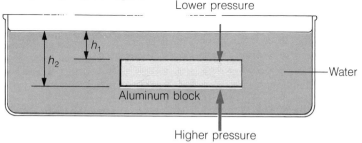

(a) *The object's weight is balanced by the spring scale.*

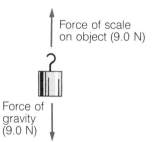

(b) *Two forces act on the object.*

(c) *The hand exerts a buoyant force of 3.0 N.*

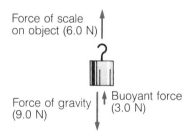

(d) *Three forces act on the object.*

SAMPLE PROBLEM 4
A man who weighs 700 N lies in water and experiences a buoyant force of 680 N. What force is required to prevent the man from sinking?

Solution
Only 20 N (700 N − 680 N) is required.

SAMPLE PROBLEM 5

The weight of a wooden block in air is 80 N. When placed in water, the block floats easily. What is the buoyant force of the water on the block when the block is floating?

Solution
The buoyant force must be 80 N, enough to support the block.

PRACTICE

29. A cottager is trying to remove a 100 kg rock from the water near the shore of a lake. The buoyant force on the rock is 400 N. Calculate
    (a) the weight of the rock in air
    (b) the force required to lift the rock in the water.

30. How does the weight of an object that floats compare to the magnitude of the buoyant force on it when it is floating?

Experiment 8A: Measuring Buoyant Force

INTRODUCTION

In the previous section you learned the definition of and reason for buoyancy. In this experiment, you will learn how to determine the buoyant force on objects that either sink or float in a fluid.

All forces in this experiment must be measured in newtons. Two facts are needed to aid in the calculations. First, the force of gravity or weight of an object can be found using $F = mg$, where $g = 9.8$ m/s^2 or 9.8 N/kg. Second, the density of water is 1000 kg/m^3, which is equivalent to 1.0 g/mL.

PURPOSE

To determine the relationship between the buoyant force on an object in water and the weight of the water it displaces.

APPARATUS

beam balance; force scale (in newtons); overflow can; graduated cylinder; 2 different hooked metal masses; 2 blocks of wood of different mass that can fit into the overflow can

PROCEDURE

1. Set up a table for observations and calculations based on the one shown. You will use a minimum of four objects. The last two rows are optional. See Analysis Question 4.
2. Find the mass of the larger metal object and calculate its weight.
3. Fill the overflow can with water and let the excess water drip away.
4. With the graduated cylinder ready to catch the overflowing water

(Figure 8-18), lower the metal object into the water. Find the following:
   (a) the volume of water displaced
   (b) the apparent weight of the submerged object
5. Repeat Procedure Steps 2 to 4 using the smaller metal object, and enter the data in your table of observations.
6. Now repeat the experiment using the two blocks of wood which must be lowered *gently* into the water. *Do not* submerge the wood.

Table of Observations and Calculations

object	large metal mass
mass of object (kg)	
weight of object (N)	
volume of water displaced (mL)	
apparent weight of object in water (N)	
buoyant force (N)	
weight of displaced water (N)	
density of metal object (g/cm³)	
(weight of metal object)/(weight of displaced water)	

(*Note:* The last two rows are optional.)

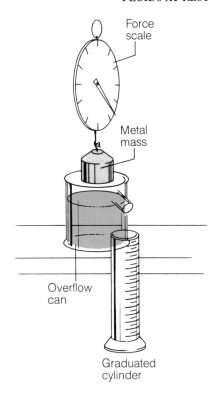

Figure 8-18
Set-up for Experiment 8A

ANALYSIS
1. Calculate the force of buoyancy on each object placed into the water. (For objects which sink, this is the difference between the weight of the object in air and the apparent weight in water.) Enter the values in your chart.
2. Calculate the weight of the displaced water for each object.
3. In each case, compare the buoyant force on the object in water with the weight of the water that the object displaces.
4. Calculate the density of each metal object in grams per cubic centimetre. Compare the *magnitude* of that density to this ratio:

$$\frac{\text{weight of object in air}}{\text{weight of water displaced by object}}$$

(This calculation provides a means of finding the relative density of an object that sinks. Relative density will be defined in Section 8.8.)

CONCLUSIONS . . .

QUESTIONS
1. How would the results of this experiment be affected if a liquid other than water were used? If possible, verify your answer experimentally using a liquid such as alcohol or glycerin.
2. A metal block of mass 8.1 kg measures 20 cm by 15 cm by 10 cm. When the block is placed in water, calculate
   (a) the volume of the water displaced
   (b) the weight of the water displaced
   (c) the apparent weight of the block in water.
3. Calculate the density of the block in Question 2 above, using
   (a) $D = m/V$ and (b) the ratio named in Analysis Question 4.

8.7 Archimedes' Principle

As you learned in Experiment 8A, an object placed in a fluid displaces some of the fluid. The weight of the displaced fluid is equal to the apparent loss of weight of the object placed in the fluid. The apparent loss of weight results from the buoyant force of the fluid on the object. This relationship is summarized in a statement called **Archimedes' principle** in honour of the Greek scientist who discovered it:

> **The buoyant force on an object in a fluid is equal to the weight of the fluid displaced by that object.**

This principle applies whether the object sinks or floats in the fluid. For an object that sinks, the volume of the displaced fluid equals the volume of the object. For an object that floats, the volume of the displaced fluid equals the volume of the portion of the object beneath the surface of the fluid. These concepts are applied in Sample Problems 6 and 7.

SAMPLE PROBLEM 6

A 500 g piece of metal with a volume of 180 cm³ is submerged in water. What is the apparent weight of the metal in water?

Solution

The apparent weight of the metal is its weight in air (F_w) minus the buoyant force (F_B):

$$F_w = mg$$
$$= 0.50 \text{ kg} \times 9.8 \text{ m/s}^2$$
$$= 4.9 \text{ N}$$

According to Archimedes' principle, the buoyant force equals the weight of the 180 cm³ of displaced water. Since the density of water is 1.0 g/cm³, the mass of 180 cm³ of water is 180 g, or 0.18 kg (from $m = DV$). Then, the buoyant force is

$$F_B = mg$$
$$= 0.18 \text{ kg} \times 9.8 \text{ m/s}^2$$
$$= 1.8 \text{ N}.$$

∴ The apparent weight of the metal is 4.9 N − 1.8 N = 3.1 N.

PRACTICE

31. A piece of metal with a mass of 5.4 kg and a volume of 6.0×10^2 cm³ is suspended from a force scale. Determine the reading on the scale when the metal is
    (a) in air
    (b) half-submerged in water (density = 1.0 g/mL)

SAMPLE PROBLEM 7
A piece of wood placed into a beaker of alcohol displaces 58 mL of the liquid. If the density of alcohol is 790 kg/m³ (or 0.79 g/mL), what is the buoyant force acting on the wood?

Solution
The mass of the displaced alcohol is

$$m = DV$$
$$= 0.79 \text{ g/mL} \times 58 \text{ mL}$$
$$= 46 \text{ g or } 0.046 \text{ kg}.$$

The buoyant force is

$$F_B = mg$$
$$= 0.046 \text{ kg} \times 9.8 \text{ m/s}^2$$
$$= 0.45 \text{ N}.$$

∴ The buoyant force on the wood is 0.45 N.

    (c) fully submerged in water
    (d) fully submerged in gasoline (density 0.69 g/mL).
32. An astronaut on the moon (where $g = 1.6$ m/s²) places a moon rock of mass 1.4 kg into a graduated beaker of water. The water rises from the 740 mL level to the 1300 mL level. Determine the buoyant force on the rock.
33. Repeat Practice Question 32, assuming that the same experiment was performed on the earth.
34. A rectangularly-shaped barge has a cross-sectional area of 2200 m² at the water line. When fully loaded, its bottom is 3.8 m into the water.
    (a) Determine the buoyant force acting on the barge.
    (b) What is the mass of the barge and its load?
35. A car ferry has a uniform cross-sectional area of 1400 m² and operates in water having a density of 1000 kg/m³. How far will the ferry sink if 50 cars with an average mass of 1400 kg each are taken aboard?

8.8 Applications of Buoyancy

Long before Archimedes discovered how buoyant force relates to the weight of the fluid displaced by an object, people were applying the principle of buoyancy in the building of ships. Their wooden ships floated easily, despite the fact that they had never heard of Archimedes' ideas. However, such ships were small and weak and would not be able to serve the purposes for which ships are currently used.

    Today's ships can be made large and strong using metal construction. Since metals are more dense than water, why does a ship float so easily, even when loaded with cargo? The answer lies in the shape of the ship.

FLUIDS

Did You Know?
Over 300 years ago, a Swedish warship called *Wasa* sank in the Stockholm Harbour. In 1961 the ship was refloated using a technique involving buoyancy. Four large buoyancy tanks were filled with water, taken down to the sunken ship, and attached firmly to it. Then the water was pumped out of the buoyancy tanks, making them much less dense than the surrounding water. The resulting buoyant force caused the ship to rise.

The ship must be built so that its weight is less than the weight of the maximum volume of water it can displace. To accomplish this, the ship is built containing a large volume of air. The average density of the air, metal, and cargo combination is less than the density of water, so the ship floats. This principle also applies to concrete boats, one of which is pictured in the photograph in Figure 8-19, taken along the Grand Canal in Wuxi, China.

Another common application of buoyancy is in the use of hydrometers. A **hydrometer** is a long, hollow tube weighted at the bottom so it can float upright in liquids. The fraction of a hydrometer submerged in a liquid is equal to the ratio of the density of the hydrometer to the density of the liquid. For example, a hydrometer with a density of 700 kg/m^3 placed in water will sink until 70% of it is under the surface. This fact allows the hydrometer to be calibrated.

Hydrometers are usually calibrated to indicate the ratio of the density of a liquid to the density of water. This ratio is called the **relative density*** of the liquid being compared with water. Since relative density is a ratio, it has no units. (The optional exercise in Experiment 8A indicated how to determine the relative density of an object that sinks in water by finding the ratio of the weight of the object in air to the weight of the water displaced by the object.) Table 8-2 lists the relative densities of several common materials.

Table 8-2 Relative Densities of Common Materials

Material	State	Relative Density	
hydrogen	gas (0°C)	0.000 089	
helium	gas (0°C)	0.000 178	— at atmospheric pressure
air	gas (0°C)	0.001 29	
cork	solid	0.240	
pine (wood)	solid	0.40 — 0.50	
oak (wood)	solid	0.70 — 0.90	
ethyl alcohol	liquid	0.79	
ice	solid (0°C)	0.92	
water	liquid (4°C)	1.00	
salt water	liquid (4°C)	1.03 (depends on salt content)	
glycerin	liquid	1.26	
aluminum	solid	2.70	
iron	solid	7.86	
copper	solid	8.95	
mercury	liquid	13.6	
gold	solid	19.3	

(*Note:* Values listed are at 20° C unless otherwise stated.)

Hydrometers are used to check the relative densities of battery water and antifreeze in automobiles. They also serve useful functions in the production of such liquids as syrup, milk, perfume, wine, and oil byproducts.

---

*Relative density was formerly called specific gravity.

Figure 8-19
The weight of this Chinese concrete boat is less than the maximum weight of water it can displace. Such boats are used to transport goods along canals, where waves are never a problem.

Applications of buoyancy are also found in sea life. For example, many fish can alter their density by controlling the amount of gas in a sack called the *swim bladder*. Doing this allows them to ascend and descend in the water in their constant search for food. (The pressure of sea water increases with depth, so a fish would have to decrease the amount of gas in its bladder to go deeper.)

The principle of buoyancy is also applied in gases such as air, but only for extremely light objects. Helium balloons, for instance, are used both for fun and for sending weather-watching equipment aloft.

PRACTICE

36. Would a ship be able to carry more cargo in fresh water or ocean water? Explain.
37. A 4.5 kg piece of pure aluminum is moulded into the shape of a boat with a capacity of 6.0×10^{-3} m³. Determine whether or not the boat will float in pure water.
38. A uniformly-shaped wooden dowel is placed in water in a graduated cylinder. The dowel is 24 cm long, and 6.0 cm of it remain above the surface. Determine the relative density of the dowel.
39. **Activity** Design and perform an experiment for determining the relative density of a variety of substances by floating them in water.
40. **Activity** Use a hydrometer to determine the relative density of a variety of liquids.
41. State whether the following substances will float in ethyl alcohol. Refer to Table 8-2.
    (a) pine wood (b) ice (c) aluminum
42. A piece of solid copper that measures 10 cm by 8.0 cm by 5.0 cm is placed in liquid mercury. (The relative densities of both substances are found in Table 8-2.)
    (a) What fraction of the copper will sink beneath the surface of the mercury?
    (b) Determine the buoyant force of the mercury on the copper.

Did You Know?
Every 12 h, hundreds of weather stations around the world release helium-filled balloons into the atmosphere. These balloons weigh less than the air they displace, and the resulting buoyant force lifts them upward. Each balloon carries a package of instruments called a **radiosonde**. The instruments collect data about temperature, pressure, and humidity and transmit the data back to the earth. The balloons are also tracked by radar to give information about wind velocity.

FLUIDS

8.9 Interaction of Particles of Liquids

Thus far in the study of fluids at rest you have learned how the forces of gravity and buoyancy relate to liquids and gases, as well as how force can be transmitted through a liquid. From the microscopic point of view, there are other forces affecting the action and properties of liquids.

Consider what happens if a droplet of water comes in contact with a piece of tissue paper. The water spreads out very quickly and becomes absorbed by the tissue paper. Evidently the water particles are attracted to the paper particles and spread out between them. The force involved is called **adhesion**, the force of attraction between particles that are not alike. Adhesion depends on the nature of the particles in contact with one another.

If a droplet of water is placed onto a shiny, smooth surface such as a laboratory bench, it does not become absorbed by the surface. Rather, it appears to form a flattened ball, with the particles of water attracting one another more than the smooth surface. In this case, the force is called **cohesion**, the force of attraction between particles that are alike. Like adhesion, cohesion depends on the nature of the particles involved.

Figure 8-20(a) shows a photograph of droplets of dew on a plant. Notice the spherical shape of the tiny droplets. This shape is a result of the cohesion of the particles at the surface of the liquid. **Surface tension** is the special name given to the cohesion of surface particles. Figure 8-20(b) illustrates, in a non-mathematical fashion, why the force of cohesion causes the surface particles to be drawn toward the centre of the sphere, in effect creating the spherical shape.

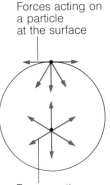

Figure 8-20
Surface Tension

(a) *The spherical shape of the water droplets on this plant is caused by surface tension.*

(b) *Comparing the force of cohesion of particles in a droplet: For each particle at the surface of the droplet, the net force is toward the centre of the sphere. For each particle beneath the surface, the net force is zero.*

The Activities in the Practice Questions that follow will help you understand more about the factors affecting the forces of adhesion, cohesion, and surface tension. Applications of these forces are discussed in the final section of this chapter.

PRACTICE

43. For which products listed below would a large adhesion to water particles be beneficial? Explain your answers.
    (a) cloth towel
    (b) facial tissue
    (c) floor polish
    (d) garden soil

44. **Activity** Design and perform an experiment to compare the forces of adhesion and cohesion for various combinations of liquids and solids. You will need a separate eye-dropper for each type of liquid used.

45. **Activity** Predict the result of this activity, then verify your prediction experimentally. Shape a copper wire into a loop and tie a thread loosely across the loop, as shown in Figure 8-21. Dip the loop into a liquid soap solution, then use a pin to break the soap film above the thread. Use a diagram and the concept of cohesion to explain the result.

46. **Activity** Add water and then alcohol carefully to a glass beaker so they form a visible boundary. Finally, add a few drops of olive oil. Describe what occurs, and why.

47. **Activity** This activity is useful for illustrating the surface tension of water.
    (a) Add water to a *clean*, medium-sized beaker until it is about 3/4 full. Use a paper clip made into a handle, as shown in Figure 8-22, to lower a second paper clip gently onto the water surface. Describe the water surface where the clip is resting. Explain why the clip, whose density is greater than the density of water, does not sink.

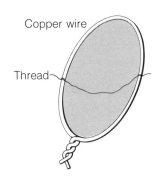

Figure 8-21

Figure 8-22

(b) Put a small amount of liquid soap onto the tip of one finger and, as you observe carefully, touch your finger to the surface of the water as far from the paper clip as possible. Describe what happens.

FLUIDS

Figure 8-23
The Formation of a Meniscus in a Glass Container

(a) Mercury creates a convex meniscus.

(b) Water creates a concave meniscus.

Figure 8-24
A water strider takes advantage of surface tension.

8.10 Applications of the Interaction of Liquid Particles

The forces of adhesion, cohesion, and surface tension are evident in many instances in the science laboratory and our everyday lives.

Often, students working with liquid chemicals notice a curved shape where the liquid touches its container. This curved shape is called a **meniscus** after a Greek word meaning "crescent moon". Consider two types of liquids placed in graduated cylinders made of glass. A liquid composed of particles that have low adhesion for glass but high cohesion for each other creates a convex meniscus. Mercury is an example of this type of liquid, as Figure 8-23(a) reveals. A liquid composed of particles that display high adhesion for glass creates a concave meniscus. This is the case with water, as you can see in Figure 8-23(b).

In Figure 8-23(b), the water appears to be crawling up the surface of the graduated cylinder. This crawling action becomes even more noticeable if the cylinder or tube is small in diameter. Such a tube is called a **capillary tube**. The rising of a liquid up a capillary tube resulting from the adhesion between the particles of the liquid and the particles of the tube is called **capillary action**. In nature, capillary action is one mechanism responsible for moving water and nutrients from the ground through the stems of plants or trees to the leaves. A plant's capillary tubes are extremely narrow, with diameters ranging from about 0.02 mm to 0.6 mm. In very tall plants, such as the trees growing in the rain forests near the Pacific Ocean, capillary action plays a minor role in transporting water to the leaves. Atmospheric pressure is important, but remember that it can hold water to just over 10 m. The greatest effect results from the cohesion of water particles that are drawn up the tree to replace the water which continually evaporates from the leaves.

Surface tension results in other benefits. In nature, for example, a water strider can move on water without piercing the surface. The surface tension prevents the strider's thin legs from sinking into the water. See Figure 8-24. In industry, surface tension is an important factor in the development of devices in which a special liquid changes to a solid when an electric potential (voltage) is applied. The fluid (technically known as *electro-rheological fluid*) is a mixture of oil, water, and tiny solid particles which have the same density as the oil and are strongly attracted to water. When a direct current (DC) voltage is applied to the liquid, the solid particles line up and are held together by surface tension, creating a solid substance. This unique material has potential use in hydraulic devices such as automatic transmissions, power steering, and anti-skid brake systems in automobiles, as well as in artificial human limbs.

Surface tension can also be a problem. When dirty or greasy clothes are being washed in water, the cohesion of the water particles prevents the water from breaking through the dirt or grease and getting between the fibres of the clothes. The surface tension can be reduced by adding soap to the water, allowing the water to seep into the places where it is needed.

PRACTICE

48. Discuss each of the following applications in terms of concepts described in this section.
    (a) blotting paper
    (b) a wick in an oil lamp
49. Why do the fibres of an artist's paint brush tend to adhere to one another? Is this a benefit or a hindrance to the artist?

Review Assignment

1. Define
   (a) fluid   (b) pressure   (c) pascal (Pa).
2. (a) Estimate the total mass (in kilograms) of air in your science classroom.
   (b) Estimate the total volume (in cubic metres) of the air, then use that estimate and the density of air (1.29 kg/m^3) to calculate the mass of the air. How good was your estimate in (a) above?
3. Assume that a man's weight is 780 N.
   (a) What pressure does he apply to the floor when he stands on one foot? The area of the sole of his shoe is 0.020 m^2.
   (b) If the man now stands on a snowshoe whose area is 0.20 m^2, how much pressure does he apply to the snow?
   (c) Explain the advantage of using snowshoes to walk on snow.
4. An absolute pressure of 300 kPa inside a car tire is exerted over a surface area of 1.2 m^2. Calculate the total resulting force on the inside of the tire.
5. A piece of styrofoam can withstand 27 kPa of pressure without being crushed. A 6000 N box is to be placed on the styrofoam.
   (a) Calculate the minimum area of the bottom of the box needed to prevent crushing the styrofoam.
   (b) If the box is placed on its side (area = 0.10 m^2), will the styrofoam be crushed? Explain.
6. Explain what causes atmospheric pressure.
7. Explain the action of a siphon, illustrated in Figure 8-25.
8. **Activity** Certain drinks for individual consumption are available in rectangular cartons. Obtain one of these drinks and a straw that fits snugly into the carton. Drink the contents, then continue sucking on the straw. Explain what happens. Remove your mouth from the straw, and again explain what happens. (This procedure can be repeated until the apparatus wears out.)
9. Would it be possible to drink from a straw on the moon, where there is no atmosphere? Explain.
10. A room that measures 10 m by 8.0 m has an atmospheric pressure at the base of 100 kPa.
    (a) Determine the total force exerted by the air on the floor of the room.
    (b) Calculate the mass of a column of water with the same cross-sectional area that would cause the same force on the floor.

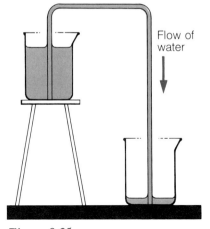

Figure 8-25

FLUIDS

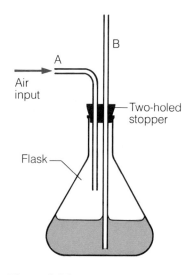

Figure 8-26

Figure 8-28

11. Estimate the atmospheric pressure at the top of Canada's highest mountain, Mount Logan, which has an elevation of 6050 m. Refer to Table 8-1.
12. A household barometer indicates that the air pressure is increasing. What is the general weather forecast?
13. Figure 8-26 shows a device in which a person blows into tube A, causing water to rise up tube B.
    (a) If the water in B rises 20 cm above the water in the flask, what is the difference in pressure between the two water levels?
    (b) If the atmospheric pressure is 102 kPa, what is the absolute pressure in the flask?
14. A patient visits a doctor when the atmospheric pressure is 101 kPa. The patient's absolute systolic pressure is 115 kPa; the absolute diastolic pressure is 110 kPa.
    (a) What are the corresponding blood pressures registered on the doctor's pressure gauge?
    (b) Is the patient's blood pressure normal?
15. State the effect of each of the following on the pressure beneath the surface of a liquid:
    (a) decreasing the depth beneath the surface
    (b) increasing the density of the liquid
    (c) changing the direction of a manometer diaphragm from facing downward to facing upward
    (d) going from a large lake to a swimming pool at the same depth
16. Explain the feasibility of the story of the Dutch boy who held back the North Sea by putting his finger into a leaking hole in a dike. (*Hint*: The hole in the dike was not far beneath the top surface of the sea.)
17. Figure 8-27 shows the construction of a dam. Explain the design of the dam.

Figure 8-27

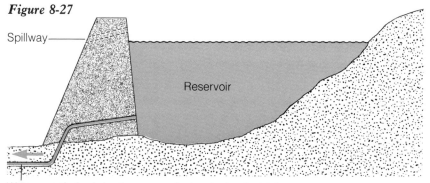

18. A swimmer may experience a popping sensation in the ears when diving to a depth of more than about 1 m. Explain why this happens.
19. Figure 8-28 is a photograph of a silo used to store grain. How does the spacing of the horizontal braces change with the distance from the top of the silo? Explain why this design is used.
20. Calculate the gauge pressure at a depth of 2.0 m beneath the surface of (a) water and (b) mercury ($D = 13\ 600$ kg/m^3).

21. A *bathyscaphe* is an underwater vessel used for researching the ocean floor. The small glass porthole at the bottom of the bathyscaphe must withstand extremely high pressures. Calculate the gauge pressure at a depth of 11.6 km, which was reached in 1960. (Assume that the density of salt water is 1030 kg/m^3.)
22. Describe how Pascal's law is applied to the operation of a hydraulic press.
23. The principle of the hydraulic press is used to operate a dentist's chair. Assume that the pressure everywhere in the liquid is 15.0 kPa. If the small cylinder has an area of 8.0×10^{-3} m^2, and the large cylinder has an area of 7.0×10^{-2} m^2, calculate the force on each cylinder.
24. In a hydraulic jack, a force of 150 N is applied to a cylinder with an area of 6.6 cm^2. How much force is exerted on the larger cylinder, whose area is 33 cm^2?
25. In a hydraulic press, a small input force can be used to obtain a large output force. Is nature giving us something for nothing? Explain your answer.
26. What is the cause of buoyancy in a fluid?
27. The mass of a human brain is about 1.5 kg. If the fluid around the brain exerts a buoyant force of 14.2 N, what is the force exerted by the skull on the brain to hold it in place?
28. The measurements listed below were taken during an experiment involving Archimedes' principle.

    mass of crown = 1.5 kg
    volume of water displaced by the crown = 120 mL

    Calculate
    (a) the weight of the crown in air
    (b) the buoyant force acting on the crown when immersed in water
    (c) the apparent weight of the crown in water
    (d) the relative density of the crown. (Is the crown pure gold?)
29. The weight of an object in air is 4.7 N. The object is placed into water in an overflow can, and 470 mL of water overflow.
    (a) What is the weight of the displaced water?
    (b) What is the buoyant force acting on the object?
    (c) Does the object float in water?
30. An ice cube is placed in water in a drinking glass. What happens to the water level after the cube has melted?
31. How can a hydrometer be used to determine the strength of a car's antifreeze? (*Hint*: The density of pure antifreeze, ethylene glycol, is less than the density of water.)
32. **Activity** Obtain a hydrometer used to measure the strength of a car battery. Describe its operation.
33. Distinguish between cohesion and adhesion.
34. Why does hair cling together when it is wet?
35. Figure 8-29 shows a soap film in the shape of a rectangle. Draw the shape of the soap film that remains after the film has been broken on the side of the thread indicated.

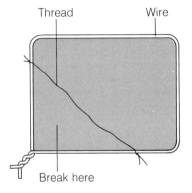

Figure 8-29

36. What factors affect the action of a liquid in a capillary tube?
37. In the past, the various applications of fluids in our society were generally taken for granted. Water supplies for home and industrial use seemed limitless; clean air was assumed to be the rule, not the exception; and the average citizen was not concerned about potential problems associated with the use of fluids.
    (a) Discuss specific ways in which our society has abused various applications of liquids and gases.
    (b) Describe steps currently being taken to reduce the effects of such abuse.

Key Objectives

Having completed this chapter, you should now be able to do the following:
1. Define and recognize examples of a fluid.
2. Define pressure, and state its SI unit of measurement.
3. Convert pascals to kilopascals and *vice versa*.
4. Given any two of pressure, force, and area, determine the third quantity.
5. Explain the cause of atmospheric pressure, and describe evidence that it exists.
6. State the value of standard atmospheric pressure in kilopascals.
7. Describe the ways in which atmospheric pressure affects our daily lives.
8. Describe the construction and use of instruments that measure pressure.
9. Distinguish between gauge pressure and absolute pressure.
10. Recognize the average blood pressures of a normal, healthy heart.
11. State the factors determining the pressure beneath the surface of a liquid.
12. State Pascal's law, and describe how it is applied in common hydraulic devices.
13. Given any three of the four areas and forces in a hydraulic press, calculate the fourth quantity.
14. Describe the cause of buoyancy in fluids.
15. Experimentally determine Archimedes' principle for objects that either sink or float in a liquid.
16. Apply Archimedes' principle in calculating buoyancy and density.
17. Define relative density.
18. Describe applications of buoyancy in fluids.
19. Define the forces of adhesion, cohesion, and surface tension, and describe how they relate to practical applications.

CHAPTER 9
Fluids in Motion

This fan drives air in a wind tunnel used to study the streamlining of automobiles.

Moving fluids are an important and practical part of our lives. One type of fluid motion occurs when a fluid such as water or natural gas moves through a container such as a pipe. The other type occurs when an object like a baseball or a submarine moves through a fluid such as air or water.

In this chapter, you will study the concepts of viscosity, turbulence, and streamlining, as well as the effects of changes in a fluid's speed. After studying these characteristics of fluids in motion, you will be able to answer the following questions. How are wind tunnels like the one shown in the photograph, which have air driven through them by huge fans, used to research the streamlining of vehicles? How does streamlining affect the performance of such vehicles as cars, trucks, airplanes, submarines, and racing bicycles? How does moving air affect various sports, among them baseball, football, golf, sailing, and skiing? Finally, what applications of moving fluids are found in and around the home?

9.1 Viscosity and Turbulence

Figure 9-1
Laminar Flow in Fluids (The length of each vector represents the magnitude of the velocity at that point.)

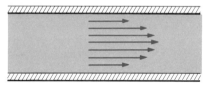

(a) Water in a pipe

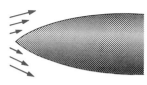

(b) Air around a cone

The motion of a fluid is influenced by forces that result when particles of the moving fluid interact with one another and their surroundings. Interactions within a fluid are caused by cohesive forces and particle collisions. Interactions with the surroundings are caused by adhesive forces.

As a fluid flows, there is internal friction caused by the cohesive forces between molecules. This internal friction is called **viscosity**. A fluid with a high viscosity, such as liquid honey, has a large amount of internal resistance and does not flow readily. A fluid with a low viscosity, such as water, has low internal resistance and flows easily. Viscosity depends not only on the nature of the fluid, but also on the fluid's temperature. As the temperature increases, the viscosity of a liquid generally decreases, and the viscosity of a gas generally increases.

Fluid particles also interact with their surroundings. For example, as water flows through a pipe, the water molecules closest to the walls of the pipe experience a frictional resistance which reduces their speed to virtually zero. Measurements show that the water speed changes from a minimum at the wall of the pipe to a maximum at the centre. If the speed of a fluid is slow and the adjacent layers flow smoothly over one another, the flow is known as **laminar flow**. It is illustrated for water in a pipe in Figure 9-1(a). Laminar flow can also occur when a fluid such as air passes around a smooth object, as shown in Figure 9-1(b).

In most situations involving moving fluids, laminar flow is difficult to achieve. Rather, as the fluid flows through or past an object, the flow can become irregular. The result is whirls called **eddies**, shown in Figure 9-2. Eddies cause a disturbance called **turbulence**, which resists the fluid's motion. Some of the kinetic energy of the fluid is converted into heat and sound, and thus is wasted. The likelihood of turbulence occurring increases as the velocity of the fluid relative to its surroundings increases.

Figure 9-2
Turbulence Caused by Eddies

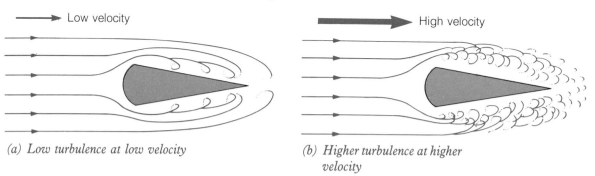

(a) Low turbulence at low velocity

(b) Higher turbulence at higher velocity

The problem of turbulence in liquids moving in tubes or pipes can be reduced in various ways. In one interesting technique, small amounts of liquid plastic are injected into the system. This has been tried successfully in the over-worked sewage system in London, England. The plastic particles mix with the sewage particles, reducing the liquid's viscosity and adhesion to the sewer pipe and walls and thereby making it easier for the pumps to transfer the sewage. A similar method can be used to reduce the turbulence of water being ejected from fire hoses, allowing the water-jet to stream farther. This is advantageous, especially in fighting fires in tall buildings. In the human body, liquid plastic can be added to the bloodstream of a person with blood-flow restrictions. Doing this helps reduce turbulence in the blood, thus lessening the chances of a blood stoppage.

Another problem associated with turbulence is found in the design of high-rise buildings in urban areas. The tall, box-like structures tend to direct fast-moving air from near the top of the building, where the winds are greatest, down toward the bottom. At the street level of a poorly designed building, gusts of wind can have a devastating effect on an unsuspecting pedestrian. To help overcome this problem, scientists build models of a proposed structure and its surroundings and send wind through the models in a wind tunnel. After analysing the observed problems, changes are made and the models re-tested. Changes that are commonly made include adding overhangs, trees, and shrubs, and changing the size of the lower part of the building.

One extensive wind study is being done by the National Research Council (NRC) in Ottawa, Ontario. Scientists built a wind tunnel model of the nation's capital in order to accumulate large quantities of data on wind patterns throughout the entire city. A model of any new development is added to the original model, and its effects are tested before the development is approved. See Figure 9-3.

Figure 9-3
Model of the City of Ottawa in NRC's Low-Speed Aerodynamics Laboratory

Figure 9-4
Toronto's City Hall

An interesting example of the turbulence caused by tall buildings occurred at the City Hall in Toronto, Ontario. Figure 9-4 shows a photograph of this relatively isolated structure. Wind tunnel testing of the design revealed that a circular podium, the two-storey structure in the middle of the photograph, was necessary to reduce the wind gusts at street level. Several years after the City Hall was built, the city decided to add a jogging path atop the podium, probably without realizing how gusty the winds there were. Unfortunately, on one occasion, a gust of

FLUIDS

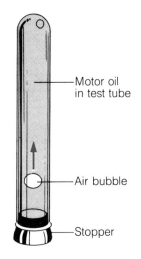

Figure 9-5

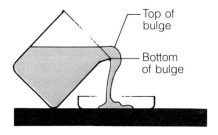

Figure 9-6

Figure 9-7

wind blew part of the track over the side of the podium, injuring a family on the track.

Turbulence is also a major concern for the transportation industry. Methods of studying and reducing turbulence are discussed in the next section.

PRACTICE

1. Based on your past experience, discuss whether the liquids named below have a high or low viscosity.
   (a) skim milk   (c) whipping cream
   (b) liquid honey   (d) methyl alcohol
2. Describe what you think is meant by the following:
   (a) As slow as molasses in January. (Molasses is the syrup made from sugar cane.)
   (b) Blood runs thicker than water.
3. **Activity** Observe the effect of a temperature change on the viscosity of various grades of motor oil (*e.g.*, SAE 20, SAE 50, and SAE 10W-40). You can do this by recording the time it takes an air bubble to travel through the oil in a stoppered test tube which is flipped over after being placed in a cold water bath and then a hot water bath. See Figure 9-5.
4. When pouring liquid honey or syrup from a bottle, as in Figure 9-6, compare the speeds of the top and bottom of the bulge where the syrup or honey leaves the jar. How does this pattern relate to laminar flow?
5. Common sense tells us that two flags blowing in the wind within close proximity of one another should face the same direction. However, as you can see in Figure 9-7, this is not always the case. Here, the wind is blowing from right to left, around the building. Using a diagram, explain why the flags are facing opposite directions.

9.2 Streamlining

Turbulence around an object moving through a fluid is a problem observed both in nature and in the transportation industry. The main technique used to reduce this type of turbulence is called streamlining. **Streamlining** is the process of reducing turbulence by altering the design of an object that moves rapidly through a fluid.

To study an example of the effect of streamlining, refer to Figure 9-8. Predict what will happen to the flame of the candle when a person blows toward it, as indicated in each case. Then verify your prediction experimentally and explain what occurs.

Fish, birds, and many animals which need to move quickly in water or air provide excellent examples of streamlining. Scientists study animal streamlining closely and try to apply their findings to technology. The transportation industry in particular devotes much research to trying to improve the streamlining of cars, trucks, motorcycles, trains, boats, submarines, airplanes, spacecraft, and other vehicles. Streamlining often enhances the appearance of a vehicle, but more importantly, it improves safety and reduces fuel consumption.

Streamlining is an experimental science, and the best way to research it is in large wind tunnels and water tanks. (Water tanks serve the same function as wind tunnels, described earlier.) The photograph at the start of this chapter shows a fan that propels air in a wind tunnel used to research the streamlining of automobiles. Figure 9-9 shows that the fan directs air along a tunnel, around two corners, then through a smaller tunnel. As the air moves into the smaller tunnel it accelerates, reaching speeds up to 100 km/h, and flows past the automobile being tested. It then returns to the fan to be recirculated. Researchers view the action from behind an adjacent glass wall (shown in Figure 9-9), and analyse the turbulence around the model. Pressure-sensitive beams, electronic sensors, drops of coloured water, small flags, and plumes of smoke are among the various ways in which turbulence can be detected.

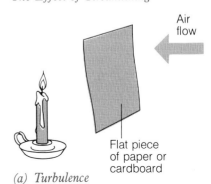

Figure 9-8
The Effect of Streamlining

(a) Turbulence

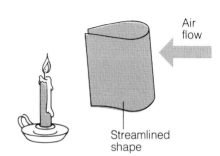

(b) Simple streamlining

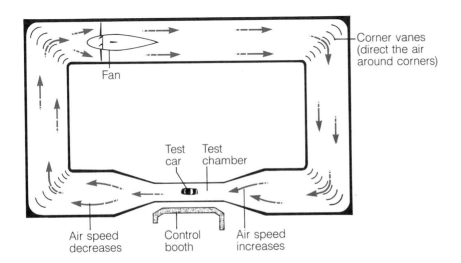

Figure 9-9
The drawing shows a typical wind tunnel arrangement used to analyse the streamlining of automobiles.

Figure 9-11
Bicycle Designs and Their Drag Coefficients

(a) *Commuter bicycle, upright position,* $C_d = 1.1$

(b) *Aerodynamic components, crouched position,* $C_d = 0.83$

(c) *Partial fairing, crouched position,* $C_d = 0.70$

(d) *Closely following another bicycle (drafting),* $C_d = 0.50$

(e) *Vector single (three wheels), complete fairing,* $C_d = 0.10$

Figure 9-10
Some Streamlining Features on an Automobile

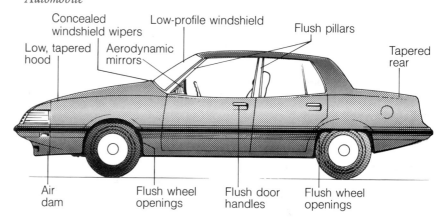

In order to communicate the results of their research in a simple fashion, scientists determine a number, called the **drag coefficient** (symbol, C_d), for each vehicle tested. To understand the range of values of this coefficient, consider the following two extremes. For a highly streamlined airplane wing, $C_d = 0.050$. For an open parachute, which is designed for maximum drag, $C_d = 1.35$. Most other C_d values lie between these extremes.

About 60 years ago, when cars were not streamlined and gasoline was much less expensive, the average C_d for cars was about 0.70. Today, the average value has dropped to less than 0.40, although some test models have C_d values reported to be as low as 0.15. Figure 9-10 shows some of the features used to reduce turbulence around a car.

A bicycle, together with its rider, has a drag coefficient as high as 1.1 when the rider is sitting in an upright position. However, when the rider is in the crouched, racing position on a streamlined racing bicycle, the C_d is reduced to about 0.83. The C_d can be further reduced by following closely behind another cyclist. The lower air resistance in this case allows a C_d as low as 0.50. Some cycling enthusiasts nowadays use a lightweight shielding, called *fairing*, which creates a streamlined envelope around cycle and cyclist and reduces the C_d to an amazing 0.10. See Figure 9-11.

Researchers have also found interesting ways of reducing the turbulence that limits the speed of submarines travelling underwater. For example, to reduce the adhesion of water particles to a submarine's hull, compressed air is forced out from a thin layer between the hull and its porous outer skin. Millions of air bubbles then pass along the submarine, preventing adhesion and thus reducing the drag.

Another way to reduce drag around a submarine is to take in some of the water the submarine is passing through and expel it, under pressure, from the rear. What law of motion is being applied here?

A third method of drag reduction for submarines, which at first seems surprising, applies the principles used by certain sharks. It has long been assumed that the best means of reducing turbulence is to have perfectly smooth surfaces and hidden joints. However, nature has provided a clue that this is not necessarily so. Sharks are obviously well-adapted to moving through water with reduced drag. A microscopic view of the skin of certain fast-moving sharks reveals that the skin has tiny grooves parallel to the flow of water. Similarly, a thin plastic coating with fine grooves applied to the surface of a submarine has the effect of reducing turbulence and increasing the maximum speed. See Figure 9-12.

Did You Know?
In the game of golf, the ball was originally smooth. When it was discovered that a ball with scratches travelled farther than a smooth ball, the surface was designed with dimples. Experiments verify that a person who can drive a dimpled ball over 200 m can drive a smooth ball of equal mass only about 50 m! As a smooth ball travels through the air, laminar flow produces a high pressure at the front of the ball and a low pressure at the rear, causing a large drag. As a dimpled ball travels through the air, however, turbulence causes the pressure difference between front and rear to be minimal, thereby reducing drag.

Figure 9-12
Using Grooves to Improve Streamlining

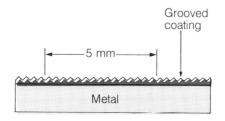

(a) The diagram illustrates the grooved nature of shark skin. The patch of skin shown is magnified to about 3000 times the actual size.

(b) A thin plastic coating with three grooves per millimetre reduces the drag of a metal surface passing through water.

Some of the discoveries applied to submarines can also be used for ordinary ships and boats, as well as for airplanes.

PRACTICE
6. Discuss the types of features used on each of the following types of vehicles to reduce drag.
   (a) trailer-trucks
   (b) spacecraft
   (c) motorcycles
   (d) trains

9.3 Bernoulli's Principle

The speed of a moving fluid has an effect on the pressure exerted by the fluid. Consider water flowing under pressure through a pipe having the shape illustrated in Figure 9-13. As the water flows from the wide section to the narrow section, its speed increases. This effect is seen in a river which flows slowly at its wider regions, but speeds up when it passes through a narrow gorge. The effect can also be verified experimentally, as you will learn if you perform the Activity in Practice Question 8.

In Figure 9-13, the water particles must accelerate as they travel from region A into region B. Acceleration is caused by an unbalanced force, but what is its source in this case? The answer lies in the pressure difference between the two regions. The pressure (or force per unit area) must be greater in region A than in region B in order to accelerate the molecules as they pass into B.

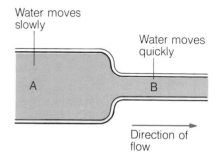

Figure 9-13
The speed of water flowing under pressure in a pipe depends on the pipe's diameter.

FLUIDS

These concepts were analysed in detail by the Swiss scientist Daniel Bernoulli (1700–1782). His conclusions became known as **Bernoulli's principle**, which states:

Where the speed of a fluid is low, the pressure is high, and where the speed of a fluid is high, the pressure is low.

A device used to demonstrate pressure differences in a fluid at various speeds is shown in Figure 9-14. The same conclusions can be drawn from viewing this apparatus. (In the diagram, the top part of the apparatus is drawn horizontally so that the effect of gravity need not be considered.)

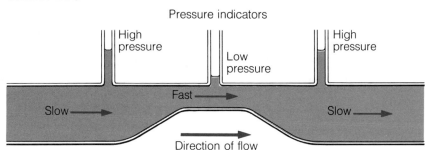

Figure 9-14
The pressure of the water depends on its speed.

The situation of water flowing in a pipe is just one example of Bernoulli's principle. Now consider an example in which a ball spins as air blows by it (Figure 9-15). As the ball spins, it exerts a dragging force on the air near its surface, because of the interaction of the particles there. This causes the speed of the air above the ball (at A) to be greater than the speed below the ball (at B). According to Bernoulli's principle, where the speed of the air is greater, the pressure is less. Thus, the upward pressure at B exceeds the downward pressure at A, resulting in a net upward pressure on the ball.

The behaviour of a spinning ball in *moving* air is demonstrated in Figure 9-16. The air is coming from a fan or from a hose connected to a vacuum cleaner exhaust. The ball chosen may be a table-tennis ball, a styrofoam ball, or a tennis ball, depending on the strength of the air flow. As the air's path is slowly changed to a small angle from the vertical, the ball begins to spin in such a way that an upward pressure prevents it from falling.

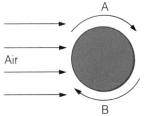

Figure 9-15
A Spinning Ball in the Path of Moving Air

Figure 9-16
A Ball in Moving Air

(a) Vertical path of air

(b) Air path at an angle to the vertical

PRACTICE

7. **Activity** Obtain a piece of paper about 10 cm by 20 cm. Hold the short edge of the paper above your mouth and blow under the paper to try to lift it. Now place the paper below your mouth and blow again. Explain your observations.

8. **Activity** Design and perform an experiment to measure the linear speed of water leaving a horizontal hose of known diameter. (*Hint*: If you collect a measured volume of water for a certain amount of time, then divide that value by the area of the nozzle, you will obtain the speed: $(cm^3/s)/cm^2 = cm/s$.) If possible, extend the experiment to find the speed of the water-jet when the original fluid pressure is constant but the area of the end of the nozzle is halved.

9. **Activity** Predict what will happen when a person blows air between two empty pop cans arranged as in Figure 9-17. Verify your prediction experimentally, then explain the results. (*Hint*: Recall some of the information from the previous chapter.)
10. Explain the following statements in terms of Bernoulli's principle.
    (a) As a convertible car with its top up cruises along a highway, the top bulges outward.
    (b) A fire in a fireplace draws better when the wind is blowing outside.

Figure 9-17

9.4 Applications of Bernoulli's Principle

A *paint sprayer*, shown in Figure 9-18, is one device that applies Bernoulli's principle. Air from a pump moves rapidly across the top end of a tube, reducing the pressure in the tube. Atmospheric pressure forces the paint up the tube to be mixed with the flowing air, creating a spray.

Airplane wings, illustrated in Figure 9-19, are designed to direct air to flow a longer distance above the wing than below. As a result, the air above the wing travels at an increased speed, producing a region of reduced pressure. The pressure below the wing, which is therefore greater, exerts an upward force, or *lift*, on the wing.

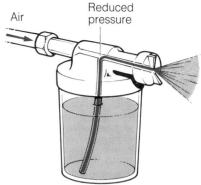

Figure 9-18
A Paint Sprayer

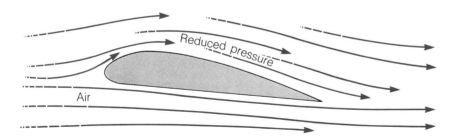

Figure 9-19
An Airplane Wing

Some modern *skis* used in ski-jumping competitions have flexible rubber extensions at their rear. As the skier glides through the air, the air above the skis travels faster than the air below. The result is a higher pressure below the skis than above, which creates an upward lift on the skis. See Figure 9-20.

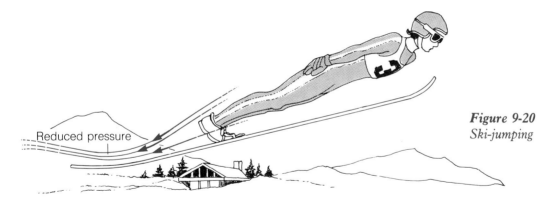

Figure 9-20
Ski-jumping

FLUIDS

Figure 9-21
An Automobile Carburetor

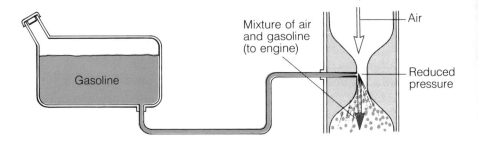

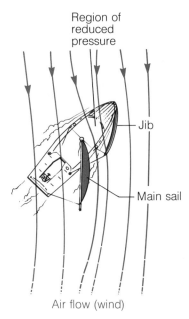

Figure 9-22
Sailing into the Wind

Figure 9-23
Throwing a Curve Ball (as viewed from above)

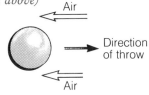   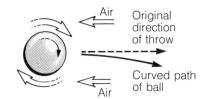

A *carburetor* in a car is yet another device applying Bernoulli's principle. It has a barrel in which air flow controls the amount of gasoline sent to the engine. Figure 9-21 shows air flowing by the gasoline intake. The fast-moving air has a reduced pressure, so the gasoline, which is under atmospheric pressure, is forced into the carburetor. There it mixes with the air and goes to the engine.

A *sailboat* can sail into the wind by having its jib and main sail set so that the air speed between them is greater than that behind the main sail. Then the pressure behind the main sail is greater than ahead of it, forcing the boat forward. See Figure 9-22.

Throwing a curve in baseball is also an application of Bernoulli's principle! In Figure 9-23, a ball is thrown to the right. This means that, relative to the ball, the air is moving to the left. The ball is thrown with a clockwise spin which causes air to be dragged along with the ball. To the left of the moving ball the speed of the air is slow, so the pressure is high. The ball is forced to curve to the right, following the path shown in Figure 9-23(c).

(a) *No spin on a thrown ball; motion is in a straight line.*

(b) *Spinning ball; air is dragged around near the surface of the ball.*

(c) *Ball thrown with a spin; the pressure to the left of the ball is greater than the pressure to the right of the ball.*

Fluidic control devices are the final application of Bernoulli's principle which we shall examine. Used to control mechanical systems, they are in some cases replacing computers. Fine layers of metallic chips, each with a different set of openings, are stacked one on top of the other, as in Figure 9-24. High-pressure fluid such as oil, steam, or air flows through the stack, taking a path determined by the design as well as by external conditions. Such a device can be used to control something as simple as a car's windshield washing system. More complex fluidic devices have applications in the control of the flight of jet aircraft, which normally use computers for control purposes. In the latter case, a fluidic device has the advantage that its components will not be damaged by lightning, which is a major concern in using computer systems in aircraft.

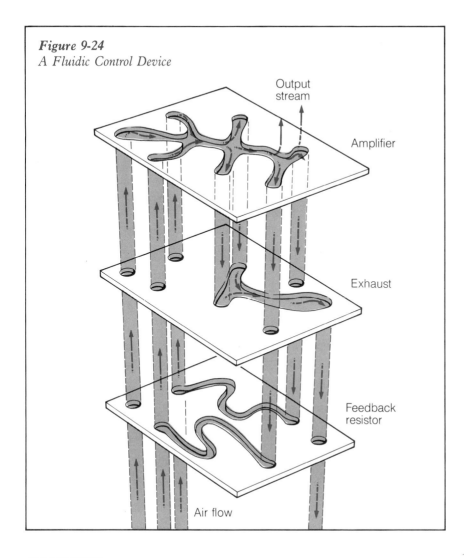

Figure 9-24
A Fluidic Control Device

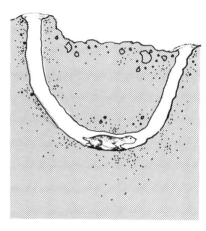

Figure 9-25

Figure 9-26

PRACTICE

11. Animals such as prairie dogs and gophers, which live underground, require sufficient air circulation in their burrows. To solve the problem, these creatures make one burrow entrance higher than the other, as shown in Figure 9-25. Explain how this design helps increase air circulation.
12. A baseball (viewed from above) is thrown as indicated by the dashes in Figure 9-26. If the ball is spinning counter-clockwise, determine the approximate direction of the path of the ball. Use diagrams in your explanation.

Review Assignment

1. A steel ball is falling in glycerin and reaches a terminal velocity. Will the terminal velocity be greater if the glycerin is at 20°C or at 60°C? Explain.

FLUIDS

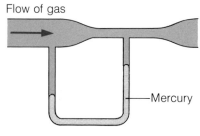

Figure 9-27

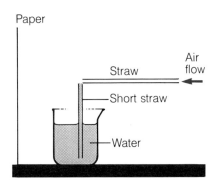

Figure 9-28

Figure 9-29

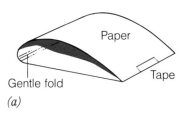
(a)

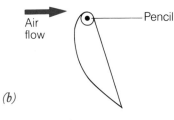
(b)

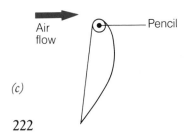
(c)

2. Why are pumping stations required at regular intervals along the cross-Canada natural gas pipeline?
3. What reasonable value of the drag coefficient would you be satisfied with if you were buying a car? a bicycle? Explain your answer.
4. Name an animal which you think has a relatively low drag coefficient. Describe features of the animal that help it reduce drag.
5. It takes 30 s to fill a 2.0 L container with water from a hose having a diameter of 1.0 cm. Assuming that the hose is held horizontally, what is the speed of the water being ejected from the hose?
6. Figure 9-27 shows a device called a **venturi flowmeter**, used to measure the speed of gas flowing through a tube. How does its design relate to Bernoulli's principle?
7. An accumulation of ice on an airplane's wings can be dangerous, even if the ice has little mass compared to the plane's mass. Explain the danger.
8. Research the meanings of the terms "slice" and "hook" in golfing. What causes each of these, and what should a golfer do to try to prevent them?
9. **Activity** Set up the apparatus illustrated in Figure 9-28. Predict what will happen when a person blows through the horizontal straw as indicated. Verify your prediction experimentally, then relate your observations to a practical application.
10. **Activity** Obtain a piece of paper about 10 cm by 30 cm and fold it in the shape of an airplane wing, as in Figure 9-29(a). Tape the ends of the paper together. Hold the middle of the wing with a pencil, as shown in (b), and blow across the wing as indicated. Repeat the procedure for the situation shown in (c). Explain what you observe.
11. **Activity** Design and carry out a competition that requires the application of at least one principle studied in this Unit on fluids.

Key Objectives

Having completed this chapter, you should now be able to do the following:
1. Define viscosity, and state how the viscosity of liquids and gases depends on the temperature.
2. Define laminar flow.
3. Describe the meaning and cause of turbulence.
4. Describe methods of reducing turbulence in liquids transported in tubes, as well as in areas surrounding tall buildings.
5. Define streamlining, and describe examples of streamlining in transportation.
6. Given the drag coefficient of a vehicle, recognize whether or not the vehicle is highly streamlined.
7. State the relationship between the pressure of a moving fluid and its speed.
8. Describe applications of Bernoulli's principle in liquids and gases.

UNIT IV
Waves, Sound, and Music

The information in this Unit will be especially useful if you plan a career in acoustical engineering or technology, music, architecture, radio, television, the recording industry, environmental (sound) pollution, or medicine.

WAVES, SOUND, AND MUSIC

CHAPTER 10
Vibrations and Waves

The vibrations seen on the 850 m centre span of this bridge eventually caused the bridge to collapse.

Vibrations are common occurrences. A swing with a child on it vibrates to and fro. A guitar string vibrates far more quickly, producing sound. Electrons in an alternating electric current (AC) in North America commonly vibrate sixty times each second. The wings of an insect also vibrate, causing a buzzing sound. These vibrations, and many others, set up waves that travel through material or space. The following questions about vibrations and waves will be answered in this chapter: What types are there, and how are they produced? What is the function of a wave? How do waves act under different circumstances? And why did the bridge shown in the photograph undergo such large vibrations?

VIBRATIONS AND WAVES

10.1 Vibrations

A **vibration** is the periodic or repeated motion of a particle or mechanical system. There are three types of vibration which we will discuss.

A **transverse vibration** occurs when an object vibrates perpendicularly to its axis at the normal rest position. An example is a child swinging on a swing.

A **longitudinal vibration** occurs when an object vibrates parallel to its axis at the rest position. An example is a coil spring supporting a vehicle.

A **torsional vibration** occurs when an object twists around its axis at the rest position. An example occurs when a string supporting an object is twisted, causing the object to turn or vibrate around and back. The three kinds of vibrations are illustrated in Figure 10-1.

Figure 10-2
The Pendulum

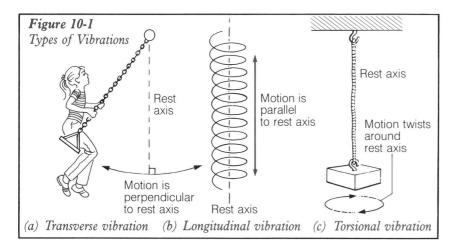

(a) Transverse vibration (b) Longitudinal vibration (c) Torsional vibration

Figure 10-1
Types of Vibrations

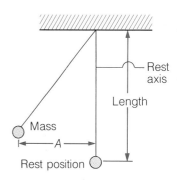

(a) Definitions

To study transverse vibrations we will use a pendulum, shown in Figure 10-2. Figure 10-2(a) illustrates the mass and rest axis of a pendulum, as well as two quantities, length and amplitude, which can be measured. The **length** of a pendulum is the distance from its suspension point to the centre of the mass. The **amplitude** (A) is the maximum horizontal displacement of the mass from its normal rest position. Figures 10-2(b) and (c) show that a **cycle** is one complete vibration of the pendulum.

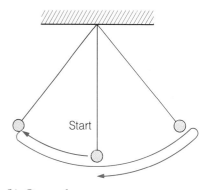

(b) One cycle

SAMPLE PROBLEM 1
A child is swinging on a swing with a constant amplitude of 1.2 m. What total distance does the child move through horizontally in 3 cycles?

Solution
In one cycle the child moves 4×1.2 m = 4.8 m.
Therefore, in 3 cycles the child moves 3×4.8 m = 14.4 m.

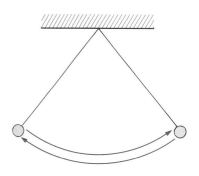

(c) One cycle is equivalent to one complete vibration.

WAVES, SOUND, AND MUSIC

Figure 10-3
The Spring

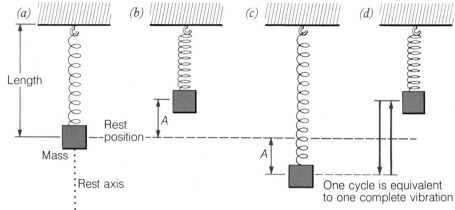

Figure 10-4
Torsional vibrations occur on an "anniversary" clock.

To study longitudinal vibrations we will consider a mass hung on the end of a spring, shown in Figure 10-3. Figure 10-3(a) illustrates the spring and mass at rest. In Figure 10-3(b), the mass has been raised to an amplitude (A), its maximum displacement from the rest position. Then the mass is released; it drops to its lowest position in Figure 10-3(c). The cycle is complete when the mass returns to its highest position, as in Figure 10-3(d).

Torsional vibrations can be viewed on the type of clock shown in Figure 10-4. The central shaft rotates first in one direction, then in the opposite direction, repeating this motion in uniform periods of time.

PRACTICE

1. State the type of vibration in each case.
   (a) A diving board vibrates momentarily after a diver jumps off.
   (b) A woodpecker's beak pecks a tree trunk.
   (c) The shock absorbers on a motorcycle vibrate as the bike travels over a rough road.
2. For the pendulum shown in Figure 10-5, state the following:
   (a) the type of vibration
   (b) the amplitude
   (c) the total distance the mass moves through horizontally in 5 cycles

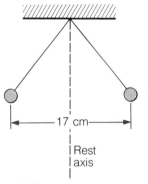

Figure 10-5

Figure 10-6

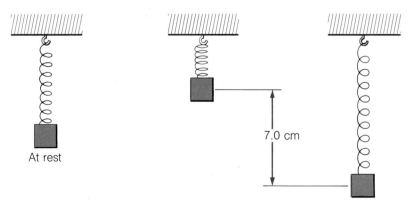

3. The diagrams in Figure 10-6 show a mass at rest on a spring and then vibrating. State
   (a) the type of vibration
   (b) the amplitude
   (c) the total distance the mass moves through in 3.5 cycles.

10.2 Frequency and Period of Vibration

A vibrating object has a frequency and a period of vibration. The **frequency** (f) is the number of cycles that occur in a specified amount of time.

$$f = \frac{\text{number of cycles}}{\text{total time}}$$

The SI unit of frequency is cycles per second, or hertz (Hz). This unit is named after the German physicist Heinrich Hertz (1857-1894), whose brilliant experiments with radio waves laid the foundation for the radio, television, and radar of today.

The **period** (T) is the time required for one cycle of vibration.

$$T = \frac{\text{total time}}{\text{number of cycles}}$$

The SI unit of period is seconds per cycle, or simply seconds (s).

SAMPLE PROBLEM 2

A mass hung on a spring vibrates vertically 15 times in 12 s. Calculate the (a) frequency and (b) period of the vibration.

Solution

(a) $f = \dfrac{\text{number of cycles}}{\text{total time}}$

$= \dfrac{15 \text{ cycles}}{12 \text{ s}}$

$= 1.2 \text{ cycles/s}$

∴ The frequency is 1.2 Hz.

(b) $T = \dfrac{\text{total time}}{\text{number of cycles}}$

$= \dfrac{12 \text{ s}}{15 \text{ cycles}}$

$= 0.80 \text{ s/cycle}$

∴ The period is 0.80 s.

Since frequency is measured in cycles per second and period in seconds per cycle, they are reciprocals of each other. Thus,

$$T = \frac{1}{f} \text{ and } f = \frac{1}{T}.$$

SAMPLE PROBLEM 3

What is the period of vibration of a 60 Hz recording timer?

Solution

$$T = \frac{1}{f}$$

$$= \frac{1}{60} \text{ Hz}$$

$$= 0.017 \text{ or } 1.7 \times 10^{-2} \text{ s}$$

∴ The period of vibration is 1.7×10^{-2} s.

SAMPLE PROBLEM 4

A certain piano string has a period of 9.1×10^{-3} s. What is its frequency?

Solution

$$f = \frac{1}{T}$$

$$= \frac{1}{9.1 \times 10^{-3} \text{ s}}$$

$$= 110 \text{ Hz}$$

∴ The frequency is 1.1×10^2 Hz.

PRACTICE

4. Calculate the frequency in hertz and the period in seconds for each situation described below.
   (a) A movie projector displays 1800 frames each minute.
   (b) A South African bird, the horned sungem, has the fastest wing-beat of any bird, at 1800 beats in 20 s.
   (c) Most butterflies beat their wings between 460 and 640 times per minute. (In this case, find a range for the answer.)
5. Calculate the period of vibration in seconds if the frequency is
   (a) 0.17 Hz; (b) 4.0 MHz; (c) 1.2×10^{-5} Hz.
6. Calculate the frequency in hertz of an object that vibrates with a period of
   (a) 0.010 s; (b) 2.0×10^{-8} s; (c) 1.0 d (one day).

Experiment 10A: The Pendulum

INTRODUCTION
A pendulum swings with a regular period, so it is useful as a device for measuring time. In fact, Galileo Galilei made the first pendulum clock in 1581. He used his pulse to discover the regular period of vibration of a lamp hanging in a church in Pisa, Italy. Then he performed laboratory experiments similar to this one, in order to determine the factors affecting the period and frequency of a swinging pendulum.

Parts of the Analysis of this experiment relate to the concepts of mechanics described earlier in the text. The law of conservation of mechanical energy is useful for calculating the maximum velocity of the pendulum mass. Work is done on the mass to raise it to its highest position, where it has maximum gravitational potential energy. When the mass is released, the potential energy becomes kinetic energy, which is at its maximum at the bottom of the swing. The velocity can be found by using the equation developed in Section 6.4.

PURPOSE
To determine the relationship between the frequency of a pendulum and its amplitude, mass, and length.

APPARATUS
retort stand; clamp; split rubber stopper; string; stopwatch; metre stick; metal masses (50 g, 100 g, and 200 g)

PROCEDURE
1. Predict, with an explanation, what you think will happen to the frequency of the pendulum in the following cases:
   (a) The mass increases, but the length and amplitude remain constant.
   (b) The amplitude increases, but the mass and length remain constant.
   (c) The length increases, but the mass and amplitude remain constant.

2. Set up a table of data based on Table 10-1.

Table 10-1 Observations for Experiment 10A

Length (cm)	Mass (g)	Amplitude (cm)	Time for 20 cycles (s)	Frequency (Hz)
100	200	10		

3. Obtain a string about 110 cm long and attach a 200 g mass to one end. Place the other end of the string into the split rubber stopper (Figure 10-7), adjust the pendulum length to 100 cm, and clamp the rubber stopper firmly. Remember that the length is measured to the centre of the mass.

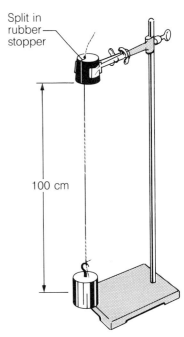

Figure 10-7
Support for the Pendulum String

WAVES, SOUND, AND MUSIC

4. Give the pendulum an amplitude of 10 cm and measure the time taken for 20 complete cycles. Repeat once or twice for accuracy, then calculate the frequency. Enter the data in your observation table.
5. Use amplitudes of 10 cm, 20 cm, and 30 cm to determine the relationship between the frequency and the amplitude of a pendulum. (If a computer program is available to verify your results, use it before proceeding.) Tabulate your data.
6. Determine the relationship between the frequency and the mass of a pendulum, using a length of 100 cm and an amplitude of 10 cm. Use masses of 50 g, 100 g, and 200 g, and be sure to measure the length to the middle of each mass. Tabulate the data.
7. Determine the relationship between the frequency and the length of a pendulum using an amplitude of 10 cm and a constant mass. Use lengths of 100 cm, 80 cm, 60 cm, 40 cm, and 20 cm. Tabulate your data.

ANALYSIS

1. With frequency as the dependent variable, plot graphs of frequency versus
   (a) amplitude, for a length of 100 cm and a constant mass
   (b) mass, for a length of 100 cm and a constant amplitude
   (c) length, for a constant amplitude and a constant mass. Relate each graph to the experiment.
2. For each of the different lengths, calculate the period of vibration. Plot a graph of period as a function of the length of the pendulum. Replot the graph to try to obtain a straight line. (*Hint:* You can either square the values of the period or find the square roots of the values of the length.)
3. Calculate the maximum speed of the 100 g pendulum mass when it has a length of 100 cm and an amplitude of 50 cm. (*Hint:* You can apply the law of conservation of mechanical energy to this problem. The maximum vertical displacement of the mass above its rest position can be found by means of a scale diagram, by actual measurement using the pendulum, by applying the Pythagorean theorem, or by using trigonometry.)

CONCLUSIONS . . .

10.3 Transfer of Energy

Energy can be transferred from one place to another by several methods involving matter. Mechanical energy, such as the kinetic energy of a hammer, is transferred by exerting a force on an object. Heat may be transferred through particles by means of conduction and convection. A further means of transferring energy is the action of **waves**, which are disturbances caused by vibrations.

There are many types of waves, including water waves. Water waves

VIBRATIONS AND WAVES

Figure 10-8
The village in the photograph is located on the west coast of Portugal.

on the surfaces of oceans and lakes are created by wind forces. The energy from water waves can wear away rocks along shorelines, producing interesting formations like those in Figure 10-8, as well as small pebbles and grains of sand.

Waves on water are easily seen, as are waves on ropes and coiled springs. For this reason, waves on ropes, springs, and water are looked at first here. The study of invisible waves, in particular, sound waves, will follow.

Another type of energy transfer by means of waves is called **radiation**. It is different from the other types mentioned above because it does not require matter. Heat, for example, may be transferred by radiation. Visible light energy and invisible radio waves and X rays are other examples of radiation. You will learn more about such waves in later chapters.

PRACTICE

7. Assume that someone is sitting near the entrance of a home across the street from where you are seated. List several methods you could use to attract the person's attention. State which methods involve waves.

Experiment 10B: Pulses on a Coiled Spring

INTRODUCTION

A **pulse** (symbol ⌒ or ⌣) is simply half a wave (symbol ⌒⌣). The knowledge gained by studying pulses on a spring can be applied to all types of waves. Thus, the conclusions to this experiment are important to the study of waves.

WAVES, SOUND, AND MUSIC

The distances indicated in the Procedure instructions are intended for short springs. If long springs are used, the distances should be increased.

PURPOSE
(a) To study the action of pulses moving along a coiled spring.
(b) To study fixed-end and open-end reflection.

APPARATUS
coiled spring (such as a Slinky toy); piece of masking tape; piece of paper; stopwatch; metre stick; piece of string at least 4 m long

CAUTION:
Do not overstretch the springs, and do not release a fully stretched spring.

PROCEDURE
1. Attach the masking tape to a coil near the middle of the spring. Stretch the spring along a smooth surface (the floor) to a length of 2.0 m. With one end of the spring held rigidly, use a rapid sideways jerk at the other end to produce a transverse pulse. See Figure 10-9(a). Describe the motion of the particles of the spring. (*Hint:* Watch the tape attached to the spring.)
2. With the same set-up as in Procedure Step 1, use a rapid forward push to produce a longtitudinal pulse along the spring. Refer to Figure 10-9(b). Again describe the motion of the particles of the spring.
3. Stand a folded piece of paper on the floor close to the middle of the spring, as in Figure 10-9(c). Use the energy transferred by a transverse pulse to knock the paper over. Describe where the energy came from and how it was transmitted to the paper.
4. Hold one end of the spring rigid and send a transverse pulse toward it. Does the pulse that reflects off this fixed end return on the same side of the rest axis as the original or incident pulse?
5. Tie a piece of string at least 4 m in length to one end of the spring. Send a transverse pulse toward the string, as shown in Figure 10-9(d). Does the pulse that reflects off this open end of the spring return on the same side as the incident pulse? (Notice that the open end is not truly "open" because a string is attached. However, it is a good approximation.)
6. Remove the string, then stretch the spring to an appropriate length (*e.g.*, 2.0 m). Measure the time taken for a transverse pulse to travel from one end of the spring to the other. Repeat the measurement several times for accuracy while trying to keep the amplitude constant. Then find the time taken for the same type of pulse to travel from one end of the spring to the other and back again. Does the reflected pulse take the same time to travel the spring's length as the incident pulse?
7. Determine whether the time taken for a transverse pulse to travel from one end to the other and back again depends on the amplitude of the pulse.
8. Predict the relationship between the speed of the pulse and the stretch of the spring. Check your prediction experimentally by stretching the spring to various lengths and finding the time for a

transverse pulse to travel down and back again. (Apply the following equation: average speed = total distance travelled ÷ time.)
9. If different types of springs are available, compare the speed of a transverse pulse along each of them.

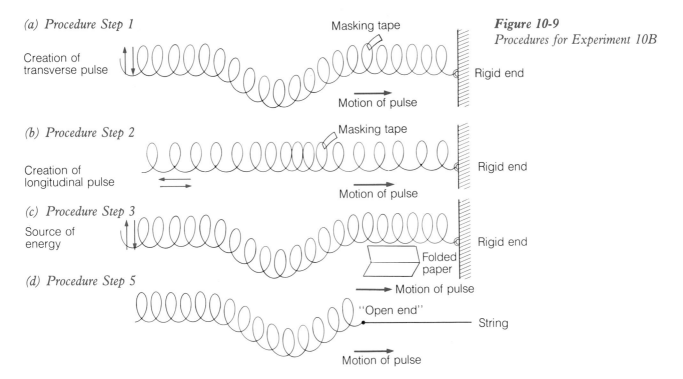

Figure 10-9
Procedures for Experiment 10B

CONCLUSIONS...

QUESTIONS
1. Based on the observations in this experiment, discuss whether the following statements are true or false:
   (a) Energy may move from one end of a spring to the other.
   (b) When energy is transferred from one end of a spring to the other, the particles of the spring are also transferred.
2. State what happens to the speed of a pulse in a material under the following circumstances:
   (a) The condition of the material changes. (For instance, stretching a spring changes its condition.)
   (b) The amplitude of the pulse increases.
   (c) The pulse is reflected off one end of the material.
3. Does the speed of a pulse depend on the material through which the pulse travels? Explain.
4. A reflected pulse which is on the same side as the incident pulse is said to be **in phase** with the incident pulse. A reflected pulse on the opposite side of the incident pulse is **out of phase** with it. Is the reflected pulse in phase or out of phase for (a) fixed-end reflection and (b) open-end reflection?

WAVES, SOUND, AND MUSIC

10.4 Periodic Waves

Periodic waves are produced by a source vibrating at some constant frequency. Both periodic transverse and periodic longitudinal waves may be set up on a coiled spring.

Figure 10-10
Transverse Waves on a Rope

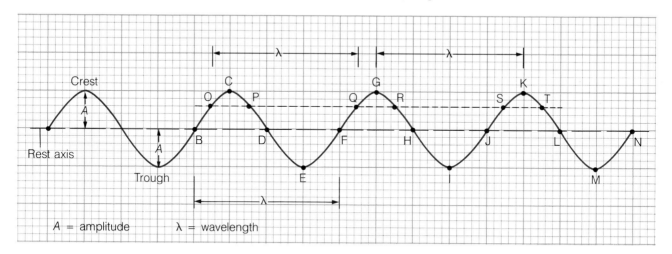

Figures 10-10 and 10-11 respectively show the important parts of transverse and longitudinal waves. A transverse wave consists of a **crest** above the rest axis and a **trough** below it. A longitudinal wave consists of a compression, where the particles are close together, and a **rarefaction**, where the particles are *rarefied* or spread apart. The **amplitude**

Figure 10-11
Longitudinal Waves on a Spring

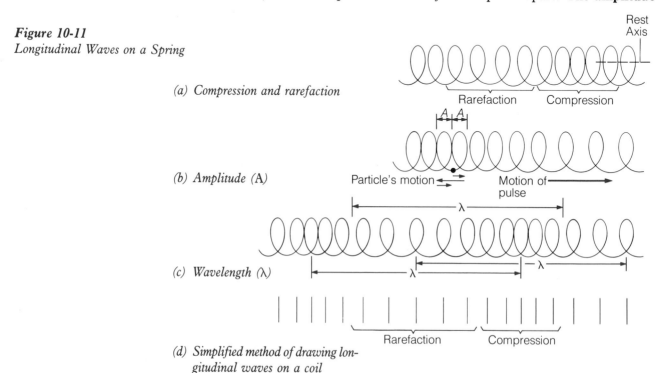

(a) Compression and rarefaction

(b) Amplitude (A)

(c) Wavelength (λ)

(d) Simplified method of drawing longitudinal waves on a coil

(A) for both is the maximum displacement from the rest axis or rest position. The **wavelength** is the distance between any two consecutive points which are in phase. For example, in Figure 10-10, points C, G, and K are located on a similar part of the wave, so they are said to be in phase with one another. Thus, one wavelength is the distance from C to G or from G to K. Points B, F, and J are also in phase with each other, so the distance from B to F or from F to J is also one wavelength. Points O, Q, and S are in phase also. Can you see the other sets of in-phase points? The symbol for wavelength is λ, *lambda*, which is the letter "l" in the Greek alphabet.

Figure 10-11(d) shows a technique that may be used to draw longitudinal waves on a coil. You may wish to use it yourself when drawing such waves.

Did You Know?
A **tidal bore** is a moving crest of water which carries the ocean tide upstream along some rivers. The tidal bore on the Severn River in Britain, for example, is large enough to allow surfers to ride upstream for several kilometres. In Canada, the largest tidal bore occurs on the Petitcodiac River in New Brunswick, which is part of the Bay of Fundy system. This crest is approximately 1 m in height, and is followed for over 2 h by a gradual rise in the river.

SAMPLE PROBLEM 5
Draw a periodic transverse wave consisting of two wavelengths with $A = 1.0$ cm and $\lambda = 2.0$ cm.

Solution
Draw the rest axis, PQ, then draw two light lines 1.0 cm above and below PQ, as shown. Label a starting point, B, then mark the points wherever the wave will cross the rest axis (*i.e.*, at D, F, H, and J). Between B and D, mark the top of the crest, C, and mark all other crests and troughs in a similar fashion. Finally, draw a smooth curve joining the outlined points.

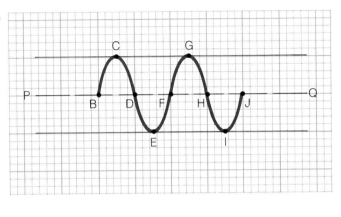

Periodic waves are, of course, caused by vibrations. A vibrating source of waves has both a frequency and a period, as you learned in Section 10.2. If the frequency of the vibrating source increases, the wavelength of the periodic wave decreases. This fact can be demonstrated on a rope, a coiled spring, or a wave machine.

PRACTICE
8. Measure the amplitude and wavelength of the periodic transverse wave in Figure 10-10.
9. Measure the wavelength of the periodic longitudinal wave in Figure 10-11(c).
10. Draw a periodic wave consisting of two complete wavelengths, each with $\lambda = 4.0$ cm, for (a) a transverse wave (use $A = 0.5$ cm) and (b) a longitudinal wave.
11. What is the relationship between the wavelength of periodic waves and their frequency?

10.5 The Universal Wave Equation

You learned in Experiment 10B that the speed of a pulse or wave depends on the material and the condition of the material through which the wave is travelling. For a pulse on a spring, the speed can be found using the equation speed = distance/time. For a periodic wave, however, another equation should be derived.

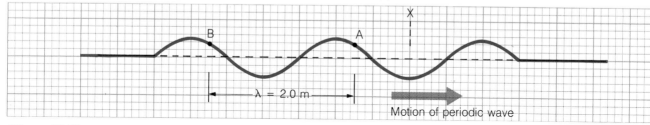

Figure 10-12
Finding the Speed of a Wave

Consider Figure 10-12, which shows a periodic transverse wave, of $\lambda = 2.0$ m, travelling to the right past an observer at X. To determine the speed of the wave, the observer starts timing the motion when point A passes by and stops it when point B passes by. The measured time is equivalent to the period (T) of vibration of the source causing the wave. In T seconds, the wave has advanced a distance of one wavelength. Therefore, since $v = d/t$, we have

$$v = \frac{\lambda}{T}$$

which indicates the speed of a wave in terms of wavelength and period.

Now, since $f = 1/T$, the equation $v = \lambda/T$ or $v = \frac{1}{T} \times \lambda$ can be written

$$v = f\lambda.$$

This equation, called the **universal wave equation**, indicates the speed of a wave in terms of wavelength and frequency.

SAMPLE PROBLEM 6
A wave machine, vibrating at a frequency of 4.0 Hz, makes water waves of wavelength 2.5 m. What is the speed of the water waves?

Solution

$$v = f\lambda$$
$$= 4.0 \text{ Hz} \times 2.5 \text{ m}$$
$$= 4.0 \, \frac{\text{cycles}}{\text{s}} \times \frac{2.5 \text{ m}}{\text{cycle}} \quad \text{(See note below.)}$$
$$= 10.0 \text{ m/s}$$

∴ The speed of the waves is 10 m/s.
(*Note:* If frequency is stated in cycles per second and wavelength in metres per cycle, the cycles divide out, leaving metres per second, the correct unit of speed.)

VIBRATIONS AND WAVES

SAMPLE PROBLEM 7
A periodic source produces a wave of λ = 3.2 cm every 0.50 s. Calculate the speed of the waves.

Solution

$$v = \frac{\lambda}{T} = \frac{3.2 \text{ cm}}{0.50 \text{ s}}$$

$$= 6.4 \text{ cm/s}$$

∴ The speed of the waves is 6.4 cm/s.

PRACTICE

12. In each case, calculate the speed of the waves; express the answer in metres per second.
    (a) $f = 18$ Hz, $\lambda = 2.7$ m
    (b) $f = 2.1 \times 10^4$ Hz, $\lambda = 2.0 \times 10^5$ cm
    (c) $T = 4.5 \times 10^{-4}$ s, $\lambda = 9.0 \times 10^4$ m
    (d) $T = 2.0$ ms, $\lambda = 3.4$ km
13. Write an equation for each of the following:
    (a) f in terms of v and λ
    (b) T in terms of v and λ
    (c) λ in terms of v and f
    (d) λ in terms of v and T
14. Calculate the unknown quantities in the table below.

	Speed (m/s)	Frequency (Hz)	Period (s)	Wavelength (m)
(a)	3.0×10^5	?	////	1.5×10^2
(b)	0.60	////	?	0.12
(c)	850	25	////	?
(d)	2.0×10^8	////	4.2×10^{-7}	?

15. A 17 cm sound wave is moving at 3.4×10^2 m/s. Calculate
    (a) the frequency of the sound.
    (b) the period of vibration of the source of the sound.
16. A wave machine creates 7.5 Hz waves in a water tank 45 cm long. If each wave takes 0.50 s to travel the length of the tank, calculate the wavelength of the waves.

10.6 Interference of Pulses

You have studied the action of a single pulse travelling along a spring. What happens if a pulse moving in one direction meets a pulse moving in the opposite direction? The pulses interfere with each other for an instant. The interference of transverse pulses in one dimen-

Figure 10-13
Interference of Pulses

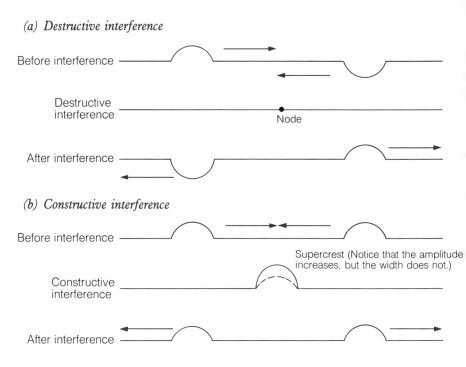

sion can be observed using a spring or wave machine. Interference also occurs for longitudinal pulses and torsional pulses, but it is not as easily demonstrated.

Two types of interference, constructive and destructive, can occur. For transverse pulses, **destructive interference** occurs when a crest meets a trough. If the crest and trough have equal amplitude and shape, their amplitudes cancel each other for an instant. Then the crest and trough continue in their original directions, as shown in Figure 10-13(a). For longitudinal pulses, destructive interference occurs when a compression meets a rarefaction.

Constructive interference occurs when pulses build each other up, resulting in a larger amplitude. This occurs for transverse pulses when a crest meets a crest, causing a **supercrest**, or a trough meets a trough, causing a **supertrough**. Figure 10-13(b) shows a supercrest. Constructive interference also occurs for longitudinal pulses.

In Figure 10-13, the interference in each case is shown at the instant that the pulses overlap or become superimposed on one another. If we call amplitudes on one side of the rest axis positive, and amplitudes on the opposite side negative, then the superimposed amplitude is simply the addition of the individual amplitudes. For example, amplitudes of +1.0 cm and −1.0 cm are added to produce a zero amplitude. The concept of amplitude addition is summarized in the **principle of superposition**, which states:

> **The resulting amplitude of two interfering pulses is the algebraic sum of the amplitudes of the individual pulses.**

This principle is especially useful for finding the resulting pattern when pulses which are unequal in size or shape interfere with one another. It applies only to amplitudes that are reasonable in size, because large pulses

VIBRATIONS AND WAVES

which interfere may cause distortion of the medium through which they are travelling.

To learn how to apply the principle of superposition in one dimension, refer to Figure 10-14, where coloured arrows are used to show amplitudes. In Figure 10-14(a) two straight-line pulses are added; in Figure 10-14(b) two curved-line pulses are added.

Figure 10-14
Applying the Principle of Superposition

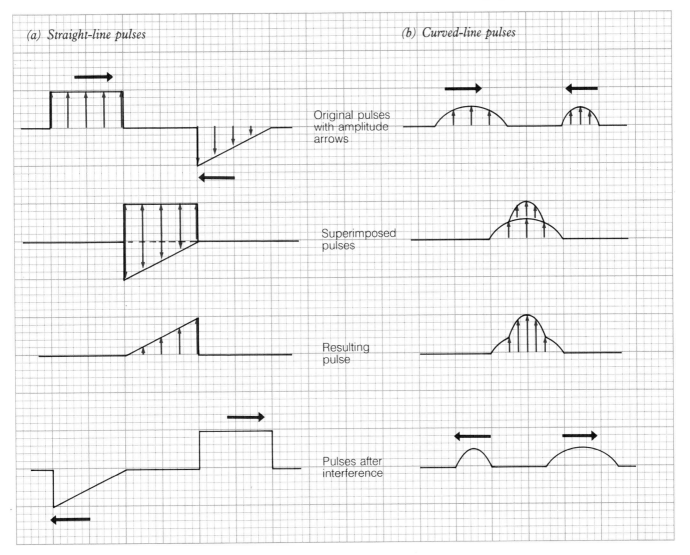

PRACTICE

17. State whether the interference is constructive or destructive in each case.
    (a) A large crest meets a small trough.
    (b) A supertrough is formed.
    (c) A small compression meets a large compression.
18. Use the principle of superposition to determine the resulting pulse when the pulses shown in Figure 10-15 are superimposed on each other. (The point of overlap should be at the horizontal midpoints of the pulses.)

WAVES, SOUND, AND MUSIC

Figure 10-15

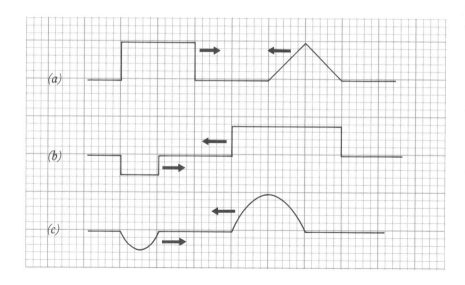

10.7 Viewing Waves on Water

Studying pulses on coiled springs is a good way to begin the topic of waves. However, a spring has only one dimension – length. A surface of water has two dimensions, length and width, so studying water waves is the next step. This will help you understand sound waves, which travel in air in all three dimensions, length, width, and depth.

A **ripple tank** like that in Figure 10-16(a) is a device used in science laboratories to study waves in water. It is a raised, shallow tank with a glass bottom. For most experiments the tank is level and contains water to a depth of about 10 mm. A light source held by a stand above the water allows the transverse water waves to be seen easily. Each crest

Figure 10-16
The Ripple Tank

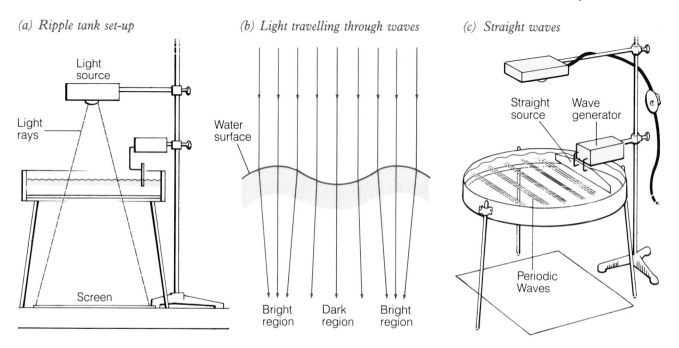

(a) Ripple tank set-up

(b) Light travelling through waves

(c) Straight waves

240

VIBRATIONS AND WAVES

acts like a magnifying glass, focusing the light to a bright region below the tank. Each trough spreads the light out, producing a dark region. The bright and dark regions appear on the screen, as in Figure 10-16(b).

In a ripple tank, periodic straight waves are produced by a motor connected to a straight bar, shown in Figure 10-16(c).

Experiment 10 C: The Diffraction of Water Waves

INTRODUCTION
In this experiment you will look at the interaction of periodic waves with obstacles. When waves interact with an obstacle, their motion is affected by what is called "diffraction". **Diffraction** is the bending effect on a wave as it passes through an opening or by an obstacle. What you learn in this experiment can be applied to sound waves in the next chapter.

PURPOSE
To observe diffraction of waves through openings and around obstacles, and to determine how diffraction depends on wavelength.

APPARATUS
ripple tank and related apparatus; ripple tank motor; retorts and clamps; wax barriers; a pencil or dowel

PROCEDURE
1. Place water in the ripple tank to a depth of about 10 mm. Make sure the tank is level.
2. Set up the motor to provide a source of straight waves. Place a straight wax barrier about 3 cm or 4 cm in front of the source, as in Figure 10-17(a). Operate the motor at a fairly high frequency (short wavelength), and draw a diagram of the resulting diffraction pattern around the edge of the barrier.
3. Gradually reduce the frequency used in Procedure Step 2 and describe the effect on the pattern. Draw a diagram of the diffraction pattern at a low frequency (long wavelength).
4. Predict the shape of the diffraction pattern when waves of short and long wavelength interact with barriers, as shown in Figures 10-17(b) and (c). Experiment to check your predictions.
5. Empty the tank and dry it completely.

CONCLUSIONS . . .

QUESTIONS
1. Draw a diagram showing the likely diffraction pattern when straight waves with λ = 1.0 cm interact with the barrier opening shown in Figure 10-18.
2. Which type of sound waves, high frequency or low frequency, do you predict will diffract better around corners? Explain your answer.

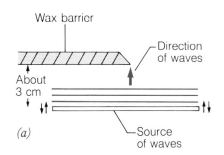

Figure 10-17
Experiment 10C: Barrier Arrangements

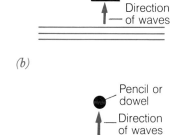

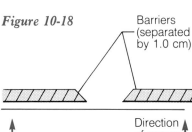

Figure 10-18

10.8 Mechanical Resonance

Any object that vibrates will do so with its largest amplitude if it is vibrating at its own natural frequency. This natural or **resonant frequency** is the frequency at which a vibration occurs most easily. It arises when a small, repeated force causes a relatively large vibration. When the vibration is mechanical, the natural vibration is called *mechanical resonance*. If you understand the examples of mechanical resonance that follow, you should be able to understand the resonance of sound waves, discussed in the next two chapters.

A pendulum has its own resonant frequency (recall Experiment 10A). A playground swing, which acts like a long pendulum, therefore also has its own resonant frequency. If you are pushing someone on a swing, the amplitude of vibration can be built up by pushing at the correct instant in each cycle. In other words, the frequency of the repeated force equals the resonant frequency of the swing in order to generate a large amplitude.

If a car is stuck in snow, it can be rocked back and forth at the resonant frequency of the system. This motion builds up the amplitude, helping to get the car out of the snow.

Another example of resonant frequency was discovered by military leaders in previous centuries. If the soldiers in an army marched across a small bridge in unison, the amplitude of vibration of the bridge could build up. If the frequency of the soldiers' steps was near the resonant frequency of the bridge, the vibration could cause the bridge to collapse. To prevent this, the soldiers were told to "break step" as they crossed bridges.

A spectacular example of mechanical resonance was the disaster that caused the collapse of a bridge in the State of Washington in 1940. The Tacoma Narrows Bridge was suspended by huge cables across a valley. Shortly after its completion it was observed to be unstable. On a windy day four months after its official opening, the bridge began vibrating at its resonant frequency. At first the bridge vibrated as a transverse wave. Then one of the suspension cables came loose, and the entire 850 m centre span of the bridge started to vibrate torsionally. The vibrations were so great that the bridge collapsed! See Figure 10-19 and page 224.

The human body also has resonant frequencies. Experiments have shown that the entire body has a mechanical resonant frequency of about 6 Hz, the head of between 13 Hz and 20 Hz, and the eyes of between 35 Hz and 75 Hz. Large amplitude vibrations at any of these frequencies could irritate or even damage parts of the body. In occupations such as transportation and road construction, efforts are made to reduce the effects of mechanical vibrations on the human body.

Radio and television provide our final example of resonance. When you tune a radio or television, you are actually adjusting the frequency of vibration of particles in the receiver so that they resonate with the frequency of a particular signal from a radio or television station.

Did You Know?
Have you ever tried to carry water in a flat dish as you walk? The reason this is difficult to do without spilling the water is that the resonant frequency of the water slopping back and forth is close to the frequency of walking. Thus, the amplitude of the wave in the dish becomes relatively large.

VIBRATIONS AND WAVES

Figure 10-19
The Tacoma Narrows Bridge

(a) The centre span of the bridge is shown vibrating torsionally before collapse.

(b) The vibrations have caused the bridge to collapse.

(c) This photograph shows the rebuilt bridge. What structural changes have been made to the newer bridge?

PRACTICE

19. Describe examples of mechanical resonance other than those examined in this section.

10.9 Standing Waves – A Special Case of Both Interference and Resonance

If periodic transverse waves of equal length and amplitude travel in opposite directions on a spring, rope, or wave machine, a pattern called a "standing wave" pattern can be set up. It has positions of zero amplitude called **nodes** which remain in the same position. This explains why the formation is called a **standing wave pattern**.

To understand why a standing wave pattern is an example of interference, consider Figure 10-20. In each of (a) – (e) the solid black wave is travelling to the right, the broken black wave is travelling to the left, and the coloured wave is the resultant wave, which is found using the

Did You Know?
Tidal action in the world's oceans and seas can set up standing wave patterns with very long wavelengths. A few places on the earth, including Tahiti in the South Pacific Ocean, have no noticeable tide because they lie on a tidal node.

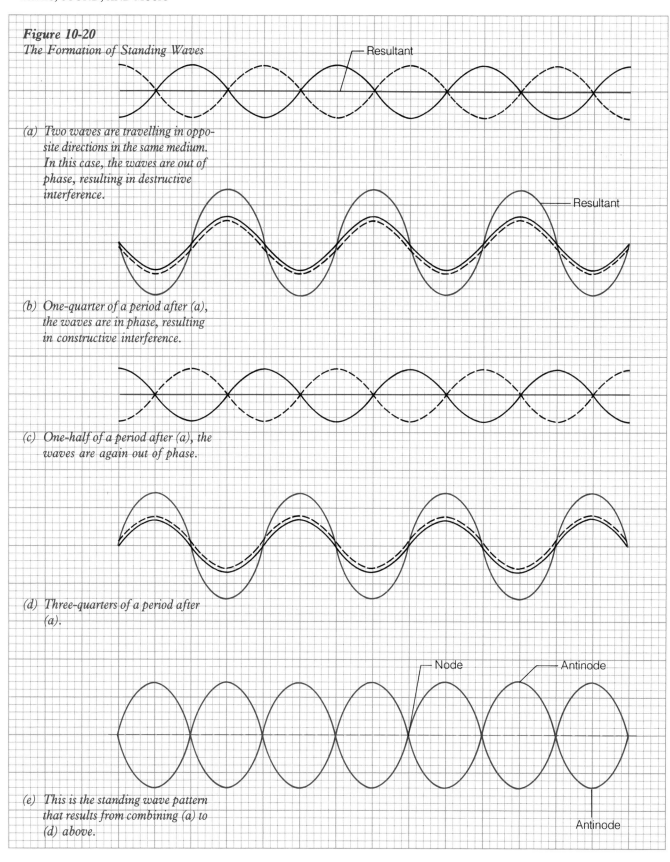

Figure 10-20
The Formation of Standing Waves

(a) Two waves are travelling in opposite directions in the same medium. In this case, the waves are out of phase, resulting in destructive interference.

(b) One-quarter of a period after (a), the waves are in phase, resulting in constructive interference.

(c) One-half of a period after (a), the waves are again out of phase.

(d) Three-quarters of a period after (a).

(e) This is the standing wave pattern that results from combining (a) to (d) above.

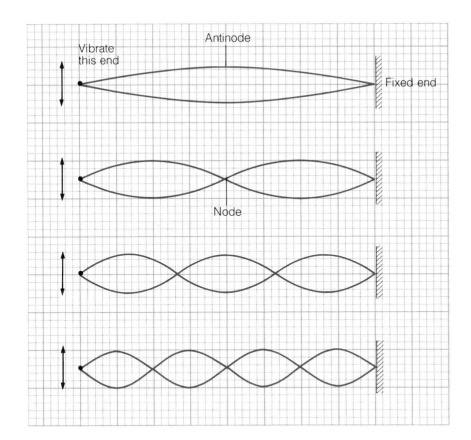

Figure 10-21
Producing Standing Waves

(a) Low frequency, long wavelength, zero nodes between ends

(b) Higher frequency, shorter wavelength, one node between ends

(c) Two nodes between ends

(d) Three nodes between ends

principle of superposition. Figure 10-20(a) shows the waves at one instant; (b) is $\frac{1}{4}$ period later; (c) is $\frac{1}{2}$ period after (a); and (d) is $\frac{3}{4}$ period after (a). Constructive interference occurs in (a) and (c); destructive interference occurs in (b) and (d). The resulting standing wave pattern is shown in Figure 10-20(e). Notice that a position of maximum amplitude on a standing wave is called an **antinode**.

To understand why a standing wave pattern is an example of resonance, you should set up such a pattern in class. Tie one end of a long rope to a rigid support, and send periodic waves toward the rigid end where fixed-end reflection will occur. Discover the frequency that results in a single antinode between the ends of the rope, as shown in Figure 10-21(a). This is the lowest resonant frequency for the system, the **fundamental frequency**. Now try to double the frequency to obtain two antinodes between the ends, as in Figure 10-21(b). At the correct frequency, the pattern is easily maintained. Continue increasing the frequency until you achieve three antinodes, four antinodes, and so on, as shown in Figures 10-21(c) and (d). In each case, the standing wave pattern is possible only at a specific resonant frequency which is a whole-number multiple of the fundamental frequency.

It is evident in Figure 10-21 that the distance from one node to the next in a standing wave pattern is half of the wavelength that produced the pattern, or $\frac{1}{2}\lambda$. The distance between the centres of adjacent antinodes is also $\frac{1}{2}\lambda$. These facts are applied in the Sample Problem 8 on page 246.

WAVES, SOUND, AND MUSIC

SAMPLE PROBLEM 8
A standing wave pattern is produced on a 6.0 m rope using a 5.0 Hz source. If there are three antinodes between the ends, what is the speed of the waves that produced the pattern?

Solution
As shown in the diagram,

$$1.5 \lambda = 6.0 \text{ m}$$
$$\therefore \lambda = 4.0 \text{ m}.$$

Then,
$$v = f\lambda$$
$$= 5.0 \text{ Hz} \times 4.0 \text{ m}$$
$$= 20 \text{ m/s}.$$

$\therefore$ The speed of the waves is 20 m/s.

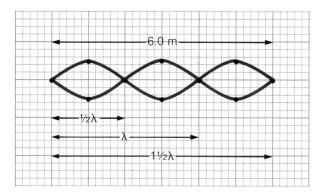

The information about waves in this chapter will be applied in the chapters about sound and music that follow.

PRACTICE
20. Draw a scale diagram of a standing wave pattern of waves on an 8.0 m rope with four antinodes between the ends. What is the wavelength of the waves that produced the pattern?
21. The speed of a wave on a certain 4.0 m rope is 3.2 m/s. What frequency of vibration is needed to produce a standing wave pattern with
    (a) 1 antinode; (b) 2 antinodes; and (c) 4 antinodes?

Review Assignment

1. State the type of vibration in each case:
   (a) A tree sways in the wind.
   (b) A sewing-machine needle moves up and down.
2. The mass of a long pendulum moves a total horizontal distance of 14 cm in one cycle. What is its amplitude of vibration?
3. Calculate the period in seconds of each of these motions:
   (a) A pulse beats 25 times in 15 s.
   (b) A woman shovels snow at a rate of 15 shovelsful per minute.
   (c) A turntable rotates at 45 rpm (revolutions per minute).
4. Calculate the frequency of the following:
   (a) A tuning fork vibrates 88 times in 0.20 s.
   (b) A recording timer produces 3600 dots in one minute.
   (c) A turntable rotates at $33\frac{1}{3}$ rpm.
5. The world record for pogo jumping is over 122 000 times in 15 h and 26 min.
   (a) What type of vibration is occurring in the spring of the pogo stick?

(b) Calculate the average period of vibration.
(c) Calculate the average frequency of vibration.
6. State the relationship (if any exists) between these pairs of variables:
   (a) period and frequency of a vibration
   (b) frequency and wavelength of a periodic wave
   (c) amplitude and speed of a pulse
   (d) period and length of a pendulum
   (e) mass and frequency of a pendulum
   (f) wavelength and period of a periodic wave
7. The frequency of certain microwaves is 2.0×10^{10} Hz. What is their period?
8. A sound wave in a steel rail has a period of 4.0×10^{-3} s. What is the wave's frequency?
9. Prove that one hertz is the reciprocal of one second.
10. A metronome, an inverted pendulum which produces ticks at various frequencies, is used mainly by musicians. The frequency can be altered by moving the mass on the metronome closer to or farther from the pivot point. What should a music student do to increase the metronome's frequency of vibration?
11. Each diagram in Figure 10-22 shows an incident pulse travelling toward one end of a rope. Draw a diagram showing the reflected pulse in each case.
12. A **tsunami** is a fast-moving surface wave that travels on an ocean after an underwater earthquake or volcanic eruption. In the deep ocean the wavelength might be over 250 km, the amplitude only about 5 m, and the speed up to 800 km/h. The wave might pass under a ship and not even be noticed, but it can strike a shore with an amplitude of perhaps 30 m and do severe damage. How is it possible that a wave which is seemingly harmless at sea can do such damage onshore?
13. Describe the factor affecting the speed of a wave in a given material.
14. A sonar signal (sound wave) of 500 Hz travels through water with a wavelength of 3.0 m. What is the speed of the signal in water?
15. The energy of earthquakes is transmitted by waves. Besides surface waves (tsunamis), there are **body waves** which move through the depths of the earth. The two types of body waves are called **primary (P)** waves, which are longitudinal, and **secondary (S)** waves, which are transverse. Both P and S waves travel in solids, but only P waves travel in fluids. P waves travel at about 8.0 km/s; S waves at about 4.5 km/s. Assume that a certain earthquake creates P and S waves of wavelength 200 km.
   (a) Calculate the period and frequency of the P waves.
   (b) Calculate the period and frequency of the S waves.
   (c) Speculate on how P and S waves have been used to determine that much of the earth's inner core is molten.
16. An electronic sound generator emits a note of frequency 512 Hz. If the speed of the sound in air is 344 m/s, what is the wavelength of the note?
17. Ocean waves that measure 12 m from crest to adjacent crest pass by a fixed point every 2.0 s. What is the speed of the waves?

Figure 10-22

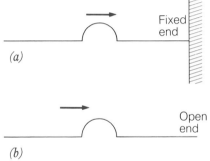

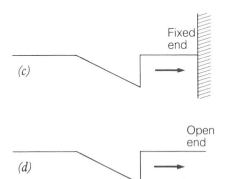

WAVES, SOUND, AND MUSIC

18. Certain radio waves which travel at 3.0×10^8 m/s in air have a wavelength of 6.0×10^4 m. What is the period of these radio waves?
19. A spring is stretched 5.0 m along the floor. Periodic transverse waves of period 0.25 s are generated from one end toward the other end, which is held rigidly. Each wave returns to the source after 0.50 s.
    (a) What is the speed of the waves along the spring?
    (b) What is the wavelength of the waves?
    (c) If a standing wave pattern is produced, how many nodes and antinodes occur between the ends of the spring?
20. Apply the principle of superposition to draw the resultant shape when each of the sets of pulses shown in Figure 10-23 interferes. (Draw the diagrams so that the horizontal midpoints of the pulses coincide.)

Figure 10-23

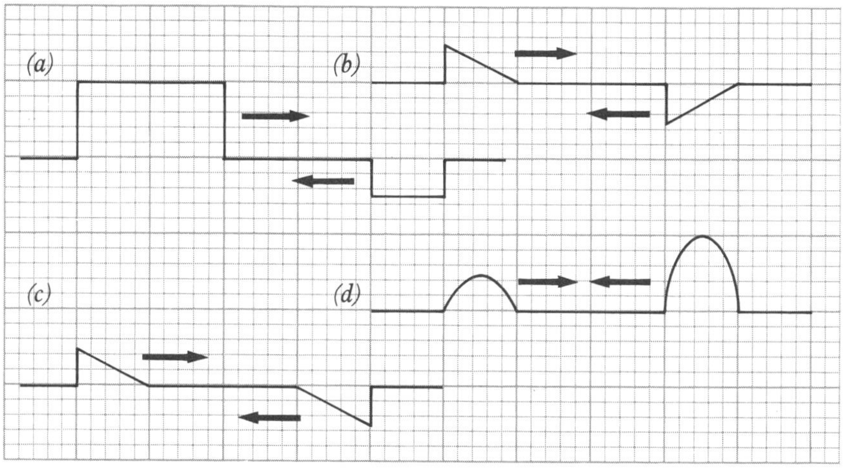

21. Straight water waves of low frequency are sent toward an opening between two barriers.
    (a) What happens to the waves as they pass through the opening?
    (b) Describe what occurs as the frequency is gradually increased.
22. Describe in your own words the meaning of mechanical resonance.
23. **Activity** Set up the apparatus shown in Figure 10-24, with the retort stands about 50 cm apart. Connect the stands with a tight string, and suspend two pendulums of equal length (about 30 cm) and equal mass (50 g).
    (a) Predict what will happen in (b) and (c) below, then check your predictions experimentally.
    (b) Set one pendulum swinging perpendicularly to the horizontal string. Watch carefully for several minutes, then explain your observations.
    (c) Shorten one of the pendulums to about 20 cm, and repeat (b) above.
24. A 6.0 m rope is used to produce standing waves. Draw a diagram of the standing wave pattern produced by waves having a wavelength of (a) 12 m; (b) 6.0 m; and (c) 3.0 m.
25. The distance between nodes of a standing wave is 40 cm, and the frequency of the source producing the wave by reflection is 880 Hz. Find the speed of the wave in metres per second.

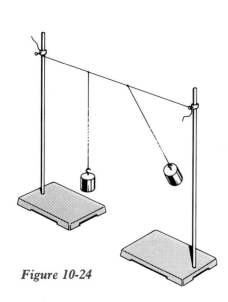

Figure 10-24

Key Objectives

After having completed this chapter, you should be able to do the following:
1. Define and recognize transverse, longitudinal, and torsional vibrations.
2. Define, for a vibration, cycle and amplitude.
3. Define period and frequency, and state their units.
4. Given one of period or frequency, calculate the other quantity.
5. Name the factors affecting the frequency of a simple pendulum.
6. Recognize the shape of a pulse, a single wave, and a periodic wave.
7. Describe what the speed of a wave depends on.
8. Distinguish between fixed-end and open-end reflection.
9. Recognize these parts of a transverse wave: rest axis, crest, trough, wavelength, and amplitude.
10. Recognize these parts of a longitudinal wave: rest position, compression, rarefaction, wavelength, and amplitude.
11. State the relationship between the frequency of a periodic wave and its wavelength.
12. Derive the universal wave equation.
13. Given any two of frequency, wavelength, and speed, calculate the third quantity.
14. Given any two of period, wavelength, and speed, calculate the third quantity.
15. Describe how constructive interference and destructive interference occur in pulses and waves.
16. Apply the principle of superposition to determine the resultant shape of interfering pulses and waves.
17. Define diffraction of waves, and state how diffraction around obstacles depends on wavelength.
18. State the meaning of mechanical resonance, and describe examples.
19. Describe how a standing wave pattern is produced by transverse waves.
20. Draw a standing wave pattern of transverse waves, and determine the wavelength of the waves that produce the pattern.

WAVES, SOUND, AND MUSIC

CHAPTER 11
Sound Energy and Hearing

The energy of amplified sound waves having the resonant frequency of this wine glass causes the glass to shatter.

Imagine you are asked to describe sound and hearing to someone who has been unable to hear since birth. Perhaps you would begin by describing sounds heard when you were a baby, sounds such as a mother's soothing voice. Then you might try to give word pictures of sounds of voices, music, radio, television, animals, machines, and so on. Next you might describe how sound is produced and transmitted. You could rely on the visual examples of vibrations and waves discussed in the previous chapter or on a photograph which shows a wine glass being shattered by the amplified sound of a human voice. Finally, you could describe the physical functions of the human ear. What are the characteristics of sound waves, and how does the human ear receive sound energy and send it to the brain for interpretation?

SOUND ENERGY AND HEARING

11.1 Production and Transmission of Sound Energy

The sounds we hear are described in many ways. Leaves rustle, lions roar, babies cry, birds chirp, corks pop, orchestras crescendo – the list is long. The energy that produces all these types of sounds originates from vibrating objects.

Some vibrations that make sound are visible. If you pluck a guitar string or strike a low-frequency tuning fork, the actual vibrations of the object can be seen. Similarly, if you watch the low-frequency woofer of a loudspeaker system, you can see it vibrating. See Figure 11-1.

There are vibrations which produce sound that are not visible, however. When you speak, for instance, parts of your throat vibrate. When you make a whistling sound by blowing over an empty pop bottle, the air molecules in the bottle vibrate to produce sound.

Nevertheless, it is possible to say that *all* sound energy, both visible and invisible, is produced by vibrations. The definitions and concepts related to mechanical vibrations studied in Chapter 10 also apply to vibrations that cause sound.

Once sound energy is produced, it must be transmitted through some material to the listener. That is, sound, unlike light energy, cannot travel through a vacuum. To illustrate this phenomenon, we can place an electric bell inside a bell jar that is connected to a vacuum pump. The bell produces sound when a vibrating arm strikes a gong. As the vacuum pump sucks the air out of the jar, sound from the bell becomes increasingly difficult to hear. When most of the air is removed, no sound can be heard, although the arm is still seen to be vibrating. See Figure 11-2.

Air is the most common material that transmits sound energy to our ears. Sound travels through air by means of longitudinal waves: there are compressions where the air molecules are close together, and rarefactions where the air molecules are spread apart. Figure 11-3 shows compressions and rarefactions coming from a drum membrane.

Figure 11-1
Visible Vibrations

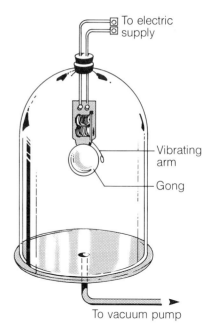

Figure 11-2
A Bell in a Vacuum

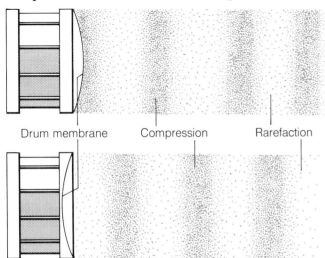

Figure 11-3
The Production of Longitudinal Sound Waves by a Drum Membrane

251

WAVES, SOUND, AND MUSIC

Figure 11-4
The Tuning Fork and Longitudinal Waves

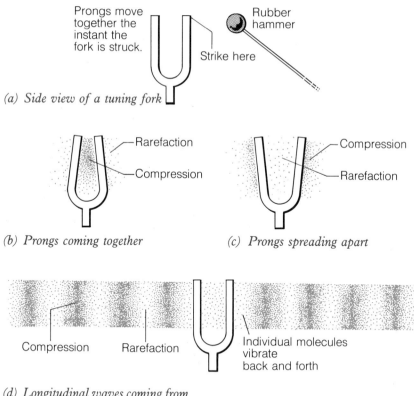

In a physics laboratory, a tuning fork is often used as a source of sound energy. A tuning fork has two prongs, joined rigidly at one end. It is constructed so that when one prong is struck from the side with a rubber hammer, both prongs move together, as Figure 11-4(a) demonstrates. This causes a compression of the air molecules between the prongs and, at the same instant, rarefactions outside the prongs. See Figure 11-4(b). At the next instant, shown in Figure 11-4(c), the prongs spread apart, causing a rarefaction between the prongs and compressions outside them. The air molecules vibrate back and forth at the same frequency as the tuning fork. They transfer the sound energy from the source to the listener by colliding with each other, as Figure 11-4(d) illustrates.

PRACTICE

1. State what vibrates to produce sound from the following:
   (a) a banjo (c) a drum
   (b) a coach's whistle (d) a crackling fire
2. **Activity** Strike a low-frequency tuning fork, and touch the prongs to the surface of water in a beaker. Describe what happens and why.
3. Assume that you are speaking to a friend located about one metre from you. When you speak, air molecules near you are set into vibration. Do those molecules reach your friend's ears? Explain your answer.

CAUTION:
Be careful not to touch the sides of the glass beaker with the vibrating tuning fork.

4. An explosion occurs at an oil refinery, and several seconds later, windows in an adjacent neighbourhood shatter. Why do the windows shatter?
5. There is no air on the moon. How do you think astronauts on the moon can hear each other speak?
6. Measure the wavelengths of the longitudinal waves in Figure 11-4(d).
7. **Activity** View and describe a demonstration of a rotating Crova's disk (Figure 11-5).

Figure 11-5
Crova's Disk

11.2 The Speed of Sound in Air

During a thunder storm, a stroke of lightning causes thunder, but the lightning is always seen before the thunder is heard. A similar phenomenon occurs at track meets, where the timer of a 100 m sprint stands at the finish line and watches for a puff of smoke from the starter's pistol. Shortly after the smoke is seen, the sound of the gun is heard. Light travels extremely fast (3.0×10^8 m/s in air); sound travels far more slowly.

You learned in Chapter 10 that the speed of a wave in a given material depends on the condition of the material. The speed of sound waves in air depends on the temperature of the air. At 0°C, the speed of sound in air at atmospheric pressure is 332 m/s. As the air temperature increases, the speed increases, because the air molecules move more rapidly. For every rise in temperature of one Celsius degree, the speed of sound in air increases by about 0.6 m/s. Thus, at 1°C, the approximate speed of sound is (332 + 0.6) m/s = 332.6 m/s. In general, the speed of sound in air can be found using the equation

$$v = 332 \text{ m/s} + 0.6 \frac{\text{m/s}}{°C} T.$$

Here, T is the air temperature in degrees Celsius. If the temperature drops below 0°C, the speed of sound in air is naturally less than 332 m/s.

SAMPLE PROBLEM 1
Calculate the speed of sound in air when the temperature is 16°C.

Solution

$v = 332 \text{ m/s} + 0.6 \frac{\text{m/s}}{°C} \times T$

$= 332 \text{ m/s} + 0.6 \frac{\text{m/s}}{°C} \times 16°C$

$= 332 \text{ m/s} + 9.6 \text{ m/s}$

$= 341.6 \text{ m/s}$

∴ The speed of sound is about 342 m/s.

WAVES, SOUND, AND MUSIC

PRACTICE

8. A pistol is used at the starting line to begin a 500 m race along a straight track. At the finish line, a puff of smoke is seen, and 1.5 s later the sound is heard. What is the speed of the sound?
9. Assume that, during a storm, thunder is heard 8.0 s after lightning is seen. If the speed of sound in air is 350 m/s, what is the distance from the lightning to the observer?
10. At a speed of 340 m/s, how long does it take sound to travel 1.0 km?
11. Calculate the speed of sound when the air temperature is
    (a) 7°C; (b) 23°C; and (c) −4°C.
12. **Activity** Perform an experiment outside to determine the speed of sound in air. You will need a loud source of sound such as two pieces of hardwood to be struck together, a stopwatch, and a thermometer. Stand a known distance of at least 150 m from a wall off which sound echoes distinctly. Find the time for a sound you produce to bounce off the wall and return to you. Then calculate the speed of the sound. Find the air temperature, and use it to determine the theoretical or "accepted" value of the speed. Then calculate the percent error of your experimental calculation.
13. A hiker standing on one side of a canyon uses principles of physics to estimate the width of the canyon. The hiker strikes a dead branch against a boulder and discovers that the echo of the sound off the far wall of the canyon returns in 1.8 s. Assuming that the air temperature is 10°C, how wide is the canyon?
14. Rearrange the equation $v = 332 \text{ m/s} + 0.6 \frac{\text{m/s}}{°\text{C}} T$ to express T by itself.
15. Calculate the air temperature when sound travels in the air at
    (a) 348 m/s and (b) 320 m/s.
16. A 380 Hz tuning fork is struck in a room where the air temperature is 22°C.
    (a) What is the speed of sound in the room?
    (b) What is the wavelength of the sound from the tuning fork?

11.3 The Speed of Sound in Various Materials

Children at play sometimes discover that sound travels very easily along a metal fence. Swimmers notice that they can hear a distant motor boat better with their ears under the water than in the air. In both cases, sound is travelling in a material other than air.

Sound travels most rapidly in certain solids, less rapidly in many liquids, and quite slowly in most gases. Thus, the speed of the sound depends not only on the temperature of the material, as discussed in the previous section, but also on the characteristic properties of the material. Table 11-1 lists the speed of sound in several materials.

SOUND ENERGY AND HEARING

Table 11-1 The Speed of Sound in Common Materials

State	Material	Speed (m/s)
solid	aluminum	5100
	glass	5030
	steel	5030
	maple wood	4110
	bone (human)	4040
	pine wood	3320
solid/liquid	brain	1530
liquid	fresh water	1500
	sea water	1470 (depends on salt content)
	alcohol	1240
gas (at atmospheric pressure)	hydrogen	1270 (at 0°C)
	helium	927 (at 20°C)
	nitrogen	350 (at 20°C)
	air	332 (at 0°C)
	oxygen	317 (at 0°C)
	carbon dioxide	258 (at 0°C)

Scientists use knowledge of the speed of sound in various materials to search for oil and minerals, study the structure of the earth, and locate objects beneath the surface of the sea, among many other applications. (This topic will be discussed further in Section 11.7.)

PRACTICE

17. **Activity** Strike a tuning fork with a rubber hammer, hold the fork at arm's length, and listen to the sound produced. Now strike the tuning fork again with the same force, and touch its base to a desk or a laboratory bench. What do you observe?
18. Compare the time it takes sound to travel 600 m in steel with the time to travel the same distance in air at 0°C.
19. A twenty-one gun salute is about to be given by a Navy ship anchored 3000 m from a person who is swimming near the shore. The swimmer sees a puff of smoke from the gun, quickly pops her head under water, and listens. In 2.0 s she hears the sound of the gun; she then lifts her head above water. In another 6.6 s, she hears the sound from the same shot coming through the air. Find the speed of the sound in (a) the water and (b) the air.

Experiment 11A: Interference of Sound Waves

INTRODUCTION
You read in Chapter 10 that, for longitudinal waves, destructive interference occurs when a compression overlaps a rarefaction. Constructive interference occurs when a compression overlaps a compression or a rarefaction overlaps a rarefaction.

WAVES, SOUND, AND MUSIC

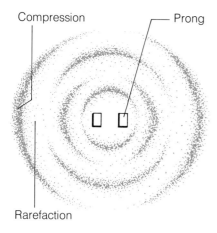

Figure 11-6
Top View of Waves from a Tuning Fork

In this experiment you will observe effects of interference in two distinct situations. One involves a single frequency; the other, two different frequencies.

In Section 11.1, the vibrations of a tuning fork were described for a fork seen from the side. Figure 11-6 shows longitudinal waves spreading out from a tuning fork viewed from the top. This diagram will help explain the production of the interference pattern near a tuning fork.

When two sounds of slightly different frequencies are heard together, an interference pattern consisting of loud and soft sounds occurs. This pattern is called the production of **beats**. Once you have heard beats, you will find them easy to recognize.

PURPOSE
To observe and explain interference patterns of sound in air.

APPARATUS
one unmounted tuning fork; two identical tuning forks mounted on resonance boxes; rubber hammer or stopper; two strong elastic bands

PROCEDURE
1. Strike a tuning fork with a rubber hammer or on a rubber stopper. Hold the fork vertically near your ear and *slowly* rotate it. Listen carefully for loud and soft sounds, and have your partner help you locate their exact positions. Repeat until you are certain of the results, then draw a diagram of the top view of the tuning fork showing the positions of the loud and soft sounds.
2. Place two mounted tuning forks close to and facing each other. Wrap an elastic band tightly around a prong of one of them, as shown in Figure 11-7. Sound the two forks together and describe the resulting sound. Repeat the experiment, using two elastic bands on the same prong. Finally, remove the bands and try the procedure a third time.

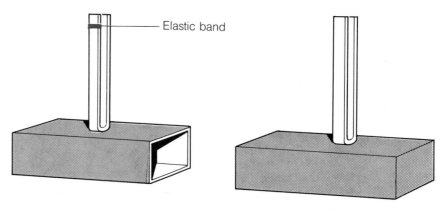

Figure 11-7
Mounted Tuning Forks

ANALYSIS
1. Use a diagram similar to the one in the Introduction to this experiment to illustrate how interference occurs near a single tuning fork. (*Hint*: What type of interference occurs where a compression interacts with a rarefaction?)

2. Figure 11-8 shows the top view of a vibrating tuning fork with one prong positioned in a slot cut into a piece of cardboard (the solid lines). Predict the relative loudness of the sounds along lines A, B, C, D, and E, and explain your reasoning. You may try to verify your predictions experimentally.
3. Do you think that the frequency of a tuning fork increases or decreases when elastic bands are added to a prong? Explain your answer.
4. Predict a relationship between the frequency of beats produced and the frequencies of the sources producing the beats.
5. Explain how this experiment verifies that sound energy travels by means of waves.

CONCLUSIONS . . .

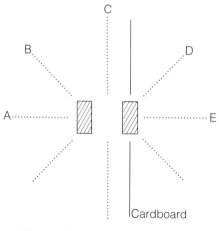

Figure 11-8

11.4 Beat Frequency

The loud and soft sequence of sounds called beats can be distinguished most easily when the frequencies of the two sources producing the beats are almost equal. If the frequencies are almost equal, the wavelengths are also nearly the same. We can use this fact to draw diagrams to show the production of beats. Two transverse waves represent the longitudinal sound waves; they are "added" according to the principle of superposition. The resulting interference pattern illustrates the reason for the loud and soft sounds of beats. See Figure 11-9.

Beat frequency, measured in hertz, is the number of beats heard per second. It is found by subtracting the lower frequency from the higher frequency. For example, if a 256 Hz tuning fork is heard with a 250 Hz note from an audio frequency generator, the beat frequency is 256 Hz − 250 Hz = 6 Hz.

Figure 11-9
Applying the Principle of Superposition to the Production of Beats

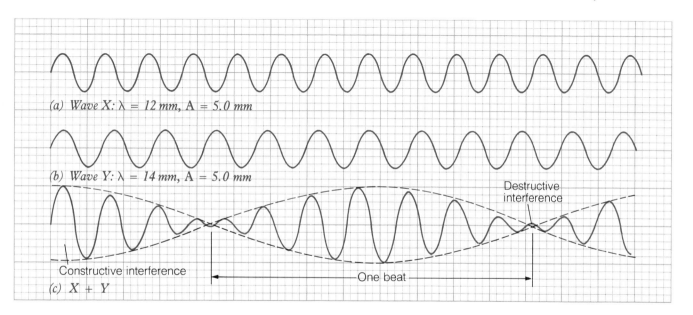

WAVES, SOUND, AND MUSIC

> **SAMPLE PROBLEM 2**
> Two 384 Hz tuning forks are sounded together, and no beats are heard. Then a metal clip is attached to a prong of one fork, and again the forks are sounded together. This time a beat frequency of 4 Hz is heard. What is the new frequency of the fork with the clip?
>
> **Solution**
> The new frequency must be either 4 Hz higher or 4 Hz lower than 384 Hz. Adding extra mass to a tuning fork reduces the resonant frequency, so the frequency must be 4 Hz lower. Thus, the new frequency is 380 Hz.

One application of beats arises in the tuning of a musical instrument such as a piano. A note on the piano is sounded with a tuning fork of corresponding frequency. The tension in the piano strings can be adjusted until no beats are heard.

PRACTICE

20. Use the principle of superposition to "add" these two waves together on a piece of graph paper turned sideways:
    Wave A $\lambda = 4.0$ cm; $A = 1.0$ cm; 5 wavelengths
    Wave B $\lambda = 5.0$ cm; $A = 1.0$ cm; 4 wavelengths
    Describe how the resulting pattern relates to the production of beats.
21. State the beat frequency when the following pairs of frequencies are heard together:
    (a) 202 Hz, 200 Hz   (b) 341 Hz, 347 Hz   (c) 1003 Hz, 998 Hz
22. Four tuning forks, with frequencies of 512 Hz, 518 Hz, 505 Hz, and 503 Hz, are available. What are all the possible beat frequencies when any two forks are sounded together?
23. A 440 Hz tuning fork is sounded together with the A-string on a guitar, and a beat frequency of 3 Hz is heard. Then an elastic band is wrapped tightly around one prong of the tuning fork, and a new beat frequency of 2 Hz is heard. Determine the frequency of the guitar string.

11.5 Resonance in Sound

Likely you have heard the high-pitched squeal produced when a person runs a moist finger around the lip of a long-stemmed glass. The frequency of such a sound is called the resonant frequency, as you learned in the previous chapter. The resonant frequency of a long-stemmed glass can be changed by putting water into it. Also, if a nearby sound having the same frequency as the resonant frequency is loud enough, the glass may shatter, as the photograph at the start of this chapter illustrates.

Sound resonance, like mechanical resonance, occurs when a small force produces a relatively large natural vibration. For example, when a tuning fork is struck, it vibrates at its own resonant frequency. A short tuning fork has a high frequency and a long tuning fork has a low frequency.

Two identical tuning forks can be used to show how easily an object vibrates at its own resonant frequency. Obtain two identical low-frequency tuning forks. Strike one fork and hold it close to the other, as in Figure 11-10. After about 15 s, stop the first fork from vibrating and listen to the second one. It is vibrating because of the influence of the first fork, which has the same resonant frequency. Energy is transferred easily through nearby air molecules which are vibrating at the frequency of the forks. If the forks had different resonant frequencies, the transfer of energy would not occur easily.

When one object begins to vibrate because of energy received from a second vibrating object with the same resonant frequency, the vibration is called a **sympathetic vibration**. The demonstration of a sympathetic vibration described above is more effective if the tuning forks are of equal frequency and are mounted on resonance boxes.

Sympathetic vibrations at resonant frequencies can also be shown if a piano is available. Depress the right (sustaining) pedal to free all the strings in the piano. Sing a certain note loudly into the piano and listen for the sound of the strings that vibrate in sympathy with your voice.

In some cases, it is necessary to prevent an object from vibrating with maximum amplitude at its own resonant frequency or frequencies. A good loudspeaker, for example, is designed in such a way that its resonant frequencies, which are many, are controlled so that none is very dominant. Otherwise, the sound of some frequencies would be louder than that of other frequencies.

Resonance of sound in musical instruments will be studied in greater detail in Chapter 12.

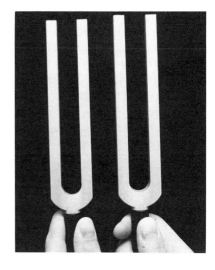

Figure 11-10
Demonstrating Resonance in Sound with Tuning Forks

PRACTICE

24. (a) What happens to the resonant frequency of a source of sound energy when the length of the source increases?
    (b) On what facts do you base your answer in (a)?
25. Describe how the concepts of sympathetic vibrations in sound relate to the Activity in Review Assignment Question 23 in Chapter 10.

11.6 Hearing and the Human Ear

Hearing is an important sense that we often take for granted. Understanding what and how we hear helps us appreciate and care for this precious sense.

Our ears are very sensitive organs which react to a wide range of frequencies. With an audio frequency generator connected to an appropriate loudspeaker, your teacher can check the frequency range of hearing of everyone in the class. All the frequencies you can hear make up your **audible range**. Most students have an audible range of about 25 Hz to 20 kHz. The ear is most sensitive to frequencies between about

PROFILE

Hans Kunov, Ph.D., P.Eng.
Professor of Electrical Engineering and Otolaryngology
Institute of Biomedical Engineering, University of Toronto
Silverman Hearing Research Laboratory, Mount Sinai Hospital

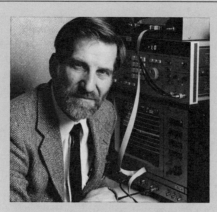

In the days when I was still in elementary school in Denmark, most young boys were fascinated by radio, much as many children are fascinated by robotics today. I was attracted to the mystery and sophistication of the technology and begged my mother to let me spend some money saved "for my education" on a correspondence course in radio technology. She finally gave in, and I spent many Sundays by myself learning the secrets of electronics.

There was nobody I could turn to for extra help when I was young. My father had died when I was quite young, and my mother never remarried. Like most women of that time, she had not been educated in the technological areas and, thus, could not assist me with my special interests. I had a physics teacher who was also of little help. I recall once asking him a question about the pronunciation of a particular unit. He withdrew with many excuses before I even had told him exactly what I wanted. I remember thinking that he must have been concerned that he might appear to lack knowledge.

I am mentioning the problems I had because I think sometimes students feel, when they see or read about adults who are successful, that these people were probably always great students and always knew just what they wanted to do. From knowing my own experiences and those of so many of my colleagues, I can say this just isn't so. We come from extremely varied backgrounds, and, earlier in our lives, had many concerns about our future. For example, I worried a great deal about what I should do in life and, even today, I do not think of myself as extraordinary in any way. In my case, what did happen as a result of my difficulty finding someone to help me with my hobbies and school work was that I learned, early in life, a very important skill – how to be self-reliant. Learning how to help yourself is a difficult but very important step toward leading a satisfying adult life. I also found that I learned to work hard which is, I guess, a part of being self-reliant.

During my high school years, I further developed my radio hobby and obtained an amateur radio licence. I bought an old transmitter and receiver, did some work on the set, and spent many happy hours in my room at the Morse key developing friendships thousands of kilometres away.

After briefly toying with the idea of studying medicine or literature at university, I finally did as everyone had always thought I might. I chose electrical engineering. At university, I found that having radio as a hobby was not as interesting any longer. Now, such matters had become part of my daily work, and I took up new interests. My main concern, of course, was my studies. I worked very hard, especially during the first two years. It paid off handsomely, academically as well as economically: I could virtually survive on scholarships, making it

1000 Hz and 5000 Hz. People who expose their ears to loud sounds for long periods of time may lower their audible range. In general, the aging process also decreases the audible range in the higher frequencies.

The human ear is complex, but a short description of it will help you understand how we hear. The human ear consists of three parts – the outer ear, middle ear, and inner ear – which are illustrated in Figure 11-11(a).

The **outer ear** consists of the **pinna**, the external portion that we can see, and the **auditory canal**. The pinna funnels the longitudinal sound waves into the auditory canal, which in turn directs the waves to the easier for my mother who did not have to support me.

During one summer when I was still an undergraduate, I stumbled over Norbert Wiener's book *Cybernetics* and it inspired me immensely. I saw the potential of a marriage between engineering and biology, and I decided that I wanted to be part of this. After my graduation, I found a professor who was interested in biomedical electronics. I worked with him on my Ph.D. which involved developing mathematical and electronic models of nerve cells.

After obtaining my Ph.D., I spent one last year in Denmark working on a project in which we developed a method of measuring pressure in the human gastro-intestinal tract through a tiny radio-pill which was swallowed by the patient. The project worked well, but we did have some interference from marine and air traffic. The press got hold of this, and for a few weeks various cartoonists had a field day drawing patients' stomachs communicating with fishermen and pilots!

My first permanent position was the professorship I hold to this day. My wife and I came to Canada in 1967 when I joined the Institute of Biomedical Engineering at the University of Toronto. At first, I took up work on lasers and ultrasound, in addition to the nerve models I had worked on for my Ph.D. I also became involved in work on muscle stimulation. Later, however, I returned to some older interests centred on hearing and acoustics; and that is where I currently concentrate my energy.

One major interest outside my research, I have found to be very special to me. The loss of my father early in my life and the difficulty I felt of not having a father or a male influence in my life made me feel I would like to serve that role for someone else, so I very much enjoy the time I spend as a Big Brother. (It also keeps me away from my laboratory for at least one day a week, both in mind and body.)

At present, my work involves the development of a speech-processing scheme for special implantable hearing aids and other hearing-assistive devices. I am currently trying to develop an accurate model of the acoustic behaviour of the head, ear, ear canal, and middle ear. The objectives of this work are threefold. First, I want to create new scientific knowledge about acoustics and the hearing process. Second, I am interested in developing diagnostic and therapeutic procedures to help those with hearing disabilities. Finally, I hope to develop practical instruments which can be produced by Canadian manufacturers.

The path which scientists follow may look straight and logical in hindsight, but there are hundreds of decision-points along the way. I often wonder how different my career would have been if I had been inspired by different people. Science for me is a never-ending search for knowledge. There appear to be no natural boundaries to science, as each new advance opens up new possibilities. Being part of the process requires long hours of work, dedication, and occasionally sacrifice. But the sense of freedom, the exhilaration of discovery, and the satisfaction of making a useful or at least recognized contribution far outweigh the negative sides.

If our country is to remain in the forefront culturally, technologically, and economically, it is absolutely imperative that a number of the best young minds dedicate themselves to scientific research. It is important not to forget that the best and most useful scientific work comes from people who have allowed themselves to be educated beyond their narrow disciplines.

eardrum. The **eardrum** connects the outer ear to the middle ear. It consists of a membrane about the thickness of a human hair which reacts to soundwaves by vibrating. The eardrum vibrates surprisingly small distances: the range of amplitude of eardrum vibration is from about 10^{-11} to 10^{-7} m.

The **middle ear** consists of three tiny bones called the **hammer**, **anvil**, and **stirrup** because of their shapes. The vibrating eardrum transfers its energy to these tiny bones. The bones act like a system of levers to transfer energy to the **oval window**, which joins the middle ear to the inner ear.

WAVES, SOUND, AND MUSIC

Figure 11-11
The Human Ear

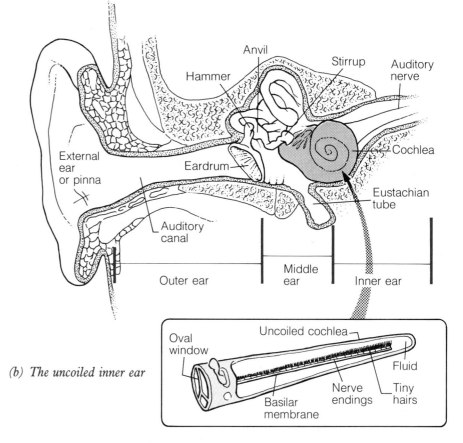

(a) The three main parts of the human ear

(b) The uncoiled inner ear

Did You Know?
At 3.6 mm in length, the stirrup, also called the *stapes*, is the smallest bone in the human body.

Did You Know?
The most common surgical outpatient operation in North America is the placement of a ventilating tube through the patient's eardrum to provide a pathway for air. This operation becomes necessary when the eustachian tube fails to open, a condition often caused by defects or an infection.

The **eustachian tube**, shown entering the middle ear in Figure 11-11(a), joins the throat to the middle ear. As you know, your ears "pop" when the air pressure is equalized during an ascent or descent in elevators or aircraft. It is the eustachian tube that allows air pressure to become equal on both sides of the eardrum.

The **inner ear** consists of a coiled, fluid-filled tube called the **cochlea** (from the Greek word *kochlias*, which means "snail"). The cochlea is shown in its correct position in Figure 11-11(a) and uncoiled in Figure 11-11(b). Incoming signals from the oval window stimulate a part of the cochlea called the **basilar membrane** into vibrating. ("Basilar" means "at the base".) The portion of this membrane closest to the oval window responds to high-frequency sounds, and the portion at the tip of the coil responds to low-frequency sounds. Connected to the basilar membrane are about 30 000 nerve endings which transform the vibrational energy into electrical impulses that are sent through the **auditory nerve** to the brain for interpretation. The details of this last step in the hearing process are not yet well understood.

Approximately one person in twenty in North America is either deaf or hard of hearing. Deafness may occur if signals cannot travel through the auditory nerve to the brain. There is no cure for such deafness. Deafness may also be caused by damage to the eardrum or the middle ear. This problem may be solved by an operation or by the use of a hearing aid. A hearing aid transmits energy through the skull to the inner ear. The inner ear then acts in the normal manner to send the sound to the brain. Yet another type of deafness is caused by an impaired cochlea. Researchers are working on improving a tiny electronic device that can be implanted into the cochlea, with the hope that someday this device will be capable of overcoming the problem of an imperfect cochlea.

11.7 Infrasonics, Ultrasonics, and Echo Finding

You have read that the average human audible range is from about 25 Hz to 20 kHz. Frequencies lower than 25 Hz are called **infrasonic**. (*Infra* is a Latin word meaning "lower than".) If we could hear frequencies lower than 25 Hz, we would often be bothered by sounds in and around us. You can imagine, for example, how annoying it would be to hear the sound of muscles every time they were activated!

Frequencies higher than 20 kHz are called **ultrasonic**. (*Ultra* is a Latin word meaning "higher than".) Dogs' ears are sensitive to frequencies that are higher than we can hear, so dog whistles are made to produce ultrasonic frequencies. Dogs are not the only animals that hear ultrasonic sounds; the audible ranges of several creatures are listed in Table 11-2.

Table 11-2 Audible Ranges

Animal	Audible Range (Hz)
human	25 to 20 000
dog	15 to 50 000
cat	60 to 65 000
bat	1000 to 120 000
porpoise	150 to 150 000
robin	250 to 210 000

Did You Know?
Recently, scientists have begun to study the sounds produced by muscles in action. They have discovered that the sound produced by muscles in the human body have a frequency of about 20 Hz. They have also found that the amplitude of the sound is proportional to the load the muscle is lifting. This fact helps scientists research the development and usefulness of muscles for various activities, including sports.

Some animals navigate and hunt using ultrasonics. Some types of bats, for instance, emit high-frequency sounds that are reflected off objects. The reflected sounds return to the bat and allow it to judge what is nearby. This explains how such bats can navigate in darkness.

The reflection of ultrasonic sounds in water is applied in a process called **echo finding**. It is used to determine the depth of water below a ship or to locate reefs, submarines, or schools of fish. Ultrasonic sounds are sent out in pulses from the ship, as shown in Figure 11-12. They are reflected off the object and return to the ship. An instrument measures the time taken for the signal to return, then the equation speed = distance/time ($v = d/t$) is used to find either the speed of the sound or the distance it travelled. This method of detection is also called **sonar**, which stands for **so**und **n**avigation **a**nd **r**anging.

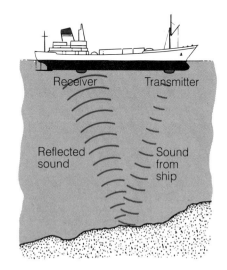

Figure 11-12
Echo Finding by Means of Ultrasonics

WAVES, SOUND, AND MUSIC

> **SAMPLE PROBLEM 3**
> A ship is anchored where the depth of water is 120 m. An ultrasonic signal is sent to the bottom of the lake and returns in 0.16 s. What is the speed of the sound in water?
>
> **Solution**
> The distance travelled by the sound is 240 m.
>
> $v = \dfrac{d}{t}$
>
> $= \dfrac{240 \text{ m}}{0.16 \text{ s}}$
>
> $= 1500 \text{ m/s}$
>
> ∴ The speed of the sound is 1.5×10^3 m/s.

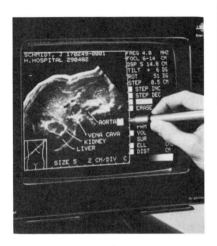

Figure 11-13
Ultrasonic sound is used to analyse body organs. One mode of viewing the results of the analysis is to display the scan on a monitor, as shown. In this case, the scan of a kidney is being traced with a light pen which will yield useful information.

Ultrasonic sounds have several other applications. In industry they help locate flaws in welded joints; they are also used to drill small holes in glass and steel and to clean electronic parts of watches and other instruments. The medical uses include the detection of brain damage and certain cancers, the study of the growth of infants in the womb, and the making of glasses for the blind (Figure 11-13). For medical applications, ultrasonic sounds are less dangerous than high-energy X rays. In the home, they are used in television remote control units.

PRACTICE
26. Ultrasonic sound is sent from a submarine to the ocean floor 360 m below. The sound is reflected to the submarine in 0.50 s. Find the speed of the sound in water.
27. The speed of sound in a freshwater lake is 1500 m/s. Ultrasonic sound sent from the surface of the water to the bottom of the lake returns in 0.20 s. How deep is the lake?

11.8 The Doppler Effect and Supersonic Speeds

As a racing car streaks past an observer by the side of the track, the observer detects changes of frequency of the sound from the car. As the car approaches, the sound becomes higher in frequency. Then, at the instant the car passes the observer, the frequency drops noticeably. The apparent changing frequency of sound due to an object's motion is called the **Doppler effect**. It is named after Christian Doppler, 1803–1853, an Austrian physicist and mathematician who first analysed the phenomenon. Likely you have heard this effect from train whistles, car horns, or sirens on fire trucks, ambulances, or police cruisers.

SOUND ENERGY AND HEARING

To understand why the Doppler effect occurs, look at Figure 11-14. Figure 11-14(a) shows sound waves travelling outwards from a stationary source. Figure 11-14(b) shows the source of sound waves travelling to the left. As the waves approach observer A, they are closer together than they would be if the source were not moving. Thus, observer A hears a sound of higher frequency. Observer B, however, hears a sound of lower frequency because the source is travelling away, producing sounds of longer wavelength. A similar effect occurs when the source of sound is stationary and the observer is moving toward or away from it.

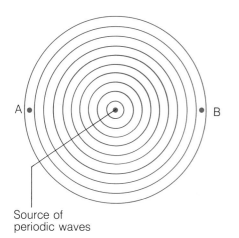

Figure 11-14
The Doppler Effect

(a) The source is stationary. Both observers, A and B, hear the same frequency of sound.

(b) The source is moving to the left. Observer A hears a higher frequency and Observer B hears a lower frequency than the observers in (a).

The Doppler effect has several applications. Astronomers use the Doppler effect of light waves to estimate the speed of faraway stars and galaxies relative to that of our solar system. Doctors use the effect at ultrasonic frequencies to determine the rate of blood flow and to view the heartbeat of unborn babies. Police take advantage of the Doppler effect when they reflect radar waves off an approaching car to determine its speed. See Figure 11-15 on page 266. A similar technique is used to track the path of satellites circling the earth.

The effect shown in Figure 11-14(b) applies if the source of sound is travelling at **subsonic speeds**, in other words, at speeds less than the speed of sound in air. When the speed of the source equals the speed of sound in air at that location, the speed is given the name Mach 1. The **Mach number** of a source of sound is the ratio of the speed of the source to the speed of sound in air at that location. Thus, at 0°C near the surface of the earth, Mach 2 is 2 × 332 m/s = 664 m/s. Mach number is named after Ernst Mach (1838–1916), an Austrian physicist and philosopher.

As an airplane approaches a speed of Mach 1, the air waves it produces pile up, creating a high pressure region called the **sound barrier**.

WAVES, SOUND, AND MUSIC

Figure 11-15
The Doppler effect is applied in the use of radar to determine the speed of vehicles on a highway. The radar system in a moving police cruiser can determine the speed of a car ahead or behind, travelling in the same direction or the opposite direction.

(a) Same direction

(b) Opposite direction

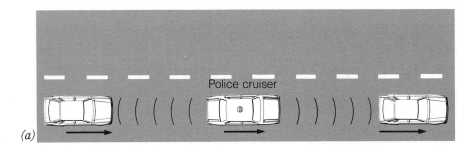

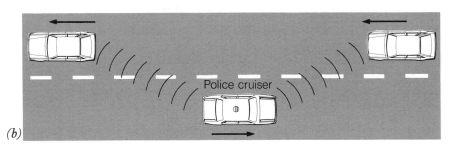

To exceed the speed of sound, extra thrust is needed until the aircraft "breaks through" the sound barrier. Only specially constructed aircraft can withstand the vibrations caused in breaking through the sound barrier to reach **supersonic speeds**, speeds greater than Mach 1. The record high speed of a fixed-wing aircraft is approximately Mach 25; it is held by the space shuttle *Columbia* and was achieved where the speed of sound was about 300 m/s.

SAMPLE PROBLEM 4
What is the speed of an airplane travelling at Mach 2.5 where the speed of sound is 320 m/s?

Solution

Mach number = v_{source}/v_{sound}
$\therefore v_{source}$ = (Mach number)(v_{sound})
   = 2.5 (320 m/s)
   = 800 m/s

Thus, the speed of the airplane is 8.0×10^2 m/s.

When an object travels at supersonic speeds, it produces what is known as a **shock wave**. The source of sound actually overtakes its own waves, causing the compressions to overlap one another. The overlapping waves form a shock wave cone that spreads outward from the source. This is illustrated in Figure 11-16. If a shock wave from a supersonic airplane reaches the ground before its energy is weakened, it generates a **sonic boom**, a thunder-like noise that can do much damage to buildings and ears. To prevent serious sonic booms, supersonic aircraft are required to travel at very high altitudes.

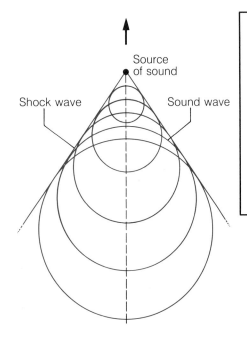

Figure 11-16
Shock Wave

PRACTICE

28. State what happens to the apparent frequency of a sound source in the following situations:
    (a) The listener is stationary and the source is approaching.
    (b) The listener is stationary and the source is receding.
    (c) The source is stationary and the listener is approaching.
    (d) The source is stationary and the listener is receding.
29. Assume that the speed of sound at a certain altitude is 300 m/s. Calculate the speed of an airplane that is travelling at (a) Mach 0.70 and (b) Mach 4.2.
30. A plane travelling at 260 m/s has a speed of Mach 0.80. What is the speed of sound in the air?
31. Discuss the advantages and disadvantages of supersonic air travel (a) near large cities; (b) over rural areas; and (c) over the oceans.
32. Assume that you are approaching a blowing whistle which is emitting a 510 Hz sound. The speed of the sound in the air is 340 m/s, and your speed relative to the whistle is 25 m/s.
    (a) What is the wavelength of the sound from the whistle?
    (b) What is your speed relative to the sound when you are approaching the whistle? going away from the whistle?
    (c) Using the wavelength you found in (a), calculate the frequencies of the two sound waves in (b).

Did You Know?
The first person to sail across the Atlantic Ocean alone in a helium balloon reported that the shock waves from supersonic aircraft slammed the balloon without warning, producing a sound resembling a huge dynamite explosion.

Review Assignment

1. Are these statements true?
   (a) All sound is produced by vibrating objects.
   (b) All vibrating objects produce sound.
   Explain your answers.
2. What vibrates to produce the sound originating from each of the following?
   (a) an acoustic guitar
   (b) an electric doorbell
   (c) a stereo system
3. Describe how sound energy is transferred from its source to the listener through air.
4. Why does sound energy not travel in a vacuum?
5. At a baseball game, a physics student with a stopwatch sits behind the centre-field fence marked 136 m. He starts the watch when he sees the bat hit the ball and stops the watch when he hears the resulting sound. The time observed is 0.40 s.
   (a) How fast is the sound energy travelling?
   (b) What is the air temperature?
6. A 200 m dash along a straight track was timed at 21.1 s by a timer who used the flash from the starter's pistol to start the stopwatch. If the air temperature was 30°C, what would the time have been if the timer had started the watch upon hearing the sound of the gun?

7. Thunder is heard 3.0 s after the lightning that caused it is seen. If the air temperature is 14°C, how far away is the lightning?
8. A marching drummer strikes the drum every 1.0 s. At what distance from the drummer will a listener hear the drum at the instant when the drumstick is farthest from the drum? The air temperature is 20°C.
9. Why does sound travel faster in air when the temperature increases?
10. How long does it take sound to travel 2.0 km in (a) aluminum and (b) hydrogen at 0°C? (Refer to Table 11-1.)
11. Killer whales make sounds so loud that hydrophones can pick them up from a distance of about 8 km. Assume that a killer whale travelling at 50 km/h is approaching an observer using a hydrophone. How long after the sound is heard will the whale take to reach the observer from an initial distance of 8.0 km in sea water (Table 11-1.)
12. State the conditions necessary for sound interference that is (a) constructive and (b) destructive.
13. Describe the sound heard when beats are produced.
14. What beat frequencies are possible when these tuning forks are used in pairs: 256 Hz, 259 Hz, and 251 Hz?
15. Describe two examples of sound resonance.
16. Define and give an example of a sympathetic vibration.
17. State the meaning of the following:
    (a) human audible range
    (b) ultrasonic sounds
    (c) infrasonic sounds
18. What is your own audible range?
19. Describe how a bat uses sound energy to locate its next meal.
20. Assume that an eardrum vibrates with an amplitude of 2.0×10^{-10} m when listening to a 3.0 kHz sound. Through what total distance will the eardrum vibrate in one minute?
21. A pulse is sent from a ship to the floor of the ocean 420 m below the ship; 0.60 s later the reflected pulse is received at the ship. What is the speed of the sound in the water?
22. Ultrasonic sound is used to locate a school of fish. The speed of sound in the ocean is 1.45 km/s, and the reflection of the sound reaches the ship 0.12 s after it is sent. How far is the school of fish from the ship?
23. What happens to the frequency of sound heard by the listener in these situations?
    (a) The source and listener are moving closer together.
    (b) The source and listener are moving apart.
24. At 18°C and atmospheric pressure near the surface of the earth, what is the speed of a plane travelling at Mach 0.90?
25. At a certain altitude the speed of sound is 1066 km/h. Jet A is travelling at Mach 1.2 [W] and jet B is travelling at Mach 2.0 [E] at a safe distance from A. Both velocities are relative to the ground.
    (a) What is the velocity of each jet relative to the ground in kilometres per hour?
    (b) What is the velocity of one jet relative to the other?
26. Why is it difficult for an aircraft to break through the sound barrier?

27. Estimate how long it would take the space shuttle *Columbia*, travelling at Mach 25 where the speed of sound is 296 m/s, to travel once around the earth at an altitude of 3.5×10^2 km. Now calculate the answer, knowing that the earth's radius is 6.4×10^3 km.

Key Objectives

Having completed this chapter, you should now be able to do the following:
1. State what produces sound energy.
2. Describe how sound energy is transmitted.
3. Find either the speed of sound in air or the air temperature, given one quantity.
4. Describe how the speed of sound in air can be found experimentally.
5. Given any two of the speed of sound in a material, the distance travelled, and the time, determine the third quantity.
6. Explain the interference of sound waves near a tuning fork.
7. Explain the condition required to hear beats, and use the principle of superposition to illustrate your explanation.
8. Calculate the beat frequency when sounds of two slightly different frequencies are heard together.
9. Describe how beats can be used to tune a musical instrument.
10. Describe examples of resonance in sound.
11. Explain and give examples of sympathetic vibrations.
12. Define audible range, and state its average value for human hearing.
13. Explain how a human ear receives sound and transmits it to the brain.
14. Define infrasonic and ultrasonic frequencies.
15. Describe applications of ultrasonic frequencies.
16. Describe the Doppler effect and explain its cause.
17. Define Mach number, subsonic speed, and supersonic speed.
18. Given any two of Mach number, speed of the source, and speed of the sound in air, calculate the third quantity.
19. Explain what causes the sound barrier.
20. Describe the formation and effect of a shock wave.
21. Recognize applications of the Doppler effect.

WAVES, SOUND, AND MUSIC

CHAPTER 12
Music, Musical Instruments, and Acoustics

The Roy Thomson Hall, home of the Toronto Symphony Orchestra, has modern design features to provide a balance of reflection and absorption of sound.

How would you describe the difference between noise and music? You could say that noise is sound that is unpleasant and annoying, and music is sound that is pleasant and harmonious. However, such a description of the difference between the two depends largely on individual judgement. But there is also a scientific difference. In this chapter, you will apply the knowledge of vibrations, waves, and sound you gained from the previous two chapters to the study of music and the analysis of the scientific difference between music and noise. Among the important questions you will learn to answer are these: What are the characteristics of musical sounds? How do musical instruments produce pleasant sounds? How does the study of acoustics help improve the sound in auditoriums such as the one shown in the photograph?

12.1 "Seeing" Sound

The study of music and musical instruments is unique among topics studied in physics. There are two main problems associated with it. First, there is an immense variety of tastes regarding what is beautiful music. Second, any two people, even if they have the same tastes in music, may have different abilities to hear the subtle differences among sounds, depending on their individual training and sensitivity to sounds. If you have a "musical ear", you will easily be able to distinguish the characteristics of musical sounds. If you do not, you should try all the more diligently to concentrate on the sounds you hear as you study music and musical instruments.

Whether or not you have a musical ear, your study of this chapter will be made objective by means of a set-up that allows you to "see" sound at the same time as you hear it. Such a set-up is a combination of a microphone and an oscilloscope.

An **oscilloscope** is an electronic instrument that receives signals from a source and displays them as traces on a television-like screen. Figure 12-1 shows an example of longitudinal sound waves that have been transformed into electrical signals on an oscilloscope. The oscilloscope has several control knobs which can be adjusted to give it sensitivity to a wide range of frequencies, amplitudes, and shapes.

Figure 12-1
Using an Oscilloscope to "See" Sound

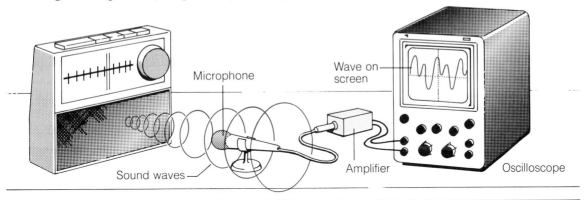

Figure 12-2 illustrates the shapes of traces that appear on an oscilloscope screen when various sounds are displayed. Figure 12-2(b) shows that noise waves are not smooth or regular. The "oooooo" sound in Figure 12-2(c) is the result of a high falsetto sound.

Figure 12-2
Sound Traces on an Oscilloscope Screen

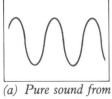

(a) Pure sound from a tuning fork

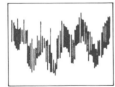

(b) Random noise

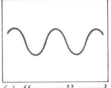

(c) "oooooo" sound

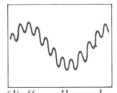

(d) "eeeee" sound

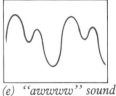

(e) "awwww" sound

The next three sections discuss in detail the three main characteristics of musical sounds – pitch, loudness, and quality. Each of these characteristics depends not only on the source of the musical sound but also on the listener. Thus, they are called *subjective characteristics*. Oscilloscope displays of the sound waves from musical instruments will help you analyse these characteristics in a scientific, objective way.

PRACTICE

1. Discuss in class the similarities and differences between an oscilloscope, a television, and a microcomputer monitor.
2. Predict the shape of an oscilloscope trace of each of the following sounds, and draw a diagram of the shape:
   (a) a person whistling
   (b) an impact sound such as a clap

12.2 Pitch and Musical Scales

If you are near a pond on a summer evening, you might hear crickets chirping and bullfrogs croaking. The sounds are easily distinguished: cricket sounds have a high pitch; bullfrog sounds have a low pitch.

A pitch wheel, shown in Figure 12-3, is a device that demonstrates the relationship between pitch and frequency of vibration. It consists of a set of three or four wheels of equal diameter, each with a different number of teeth. As the device is spun by an electric motor, a piece of paper is held up to each wheel in turn. The paper vibrates with the lowest frequency when it touches the wheel with the fewest teeth. This action produces the sound of lowest pitch. The more teeth a wheel has, the higher the pitch of the sound produced. Thus, we find that *pitch increases as frequency increases*.

As the pitch of a sound changes, the waveform also changes. This can be illustrated using an audio frequency generator connected to an oscilloscope (Figure 12-4). It is clear that *as pitch increases, wavelength decreases*.

Although there is an objective relationship between pitch and frequency or wavelength, the actual pitch a person hears depends on other factors, including the observer, the complexity of the sound, and, to a certain extent, the loudness of the sound. Thus, pitch is a subjective characteristic.

Most musical tunes consist, at any one instant, of more than a single sound. In general, two or more sounds are *harmonious* if their frequencies are in a simple ratio. (Recall from Chapter 1 that Pythagoras discovered this over 2000 years ago.) Pleasant, harmonious pairs of sounds have high **consonance**. Unpleasant, unharmonious pairs of sounds have high **dissonance**, or low consonance. Table 12-1 lists some ratios of frequencies in order of decreasing consonance. The interval called the **unison** is simply a set of sounds of the same frequency. The interval called an **octave** has sounds with double the frequency of the sounds in another octave. For example, a 200 Hz sound is one octave above a

Figure 12-3
The Pitch Wheel

Figure 12-4
Oscilloscope Traces of Pitch and Wavelength

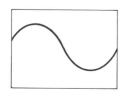

(a) *Low frequency, low pitch, long wavelength*

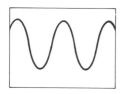

(b) *High frequency, high pitch, short wavelength*

100 Hz sound. The names of other intervals in Table 12-1 are given for the interest of musicians.

Table 12-1 Consonance and Frequency Ratios

	Interval	Ratio of Frequencies
	Unison	1/1
	Octave	2/1
Decreasing	Fifth	3/2
consonance	Fourth	4/3
↓	Major third	5/4
	Major sixth	5/3
	Minor third	6/5
	Minor sixth	8/5

We will study two common musical scales, which are sets of tones with increasing or decreasing frequencies. One scale, which we will call the **scientific scale**, is used in many science laboratories. Its standard frequency, based on the number 2^8, is 256 Hz. The standard is multiplied by simple ratios, such as 3/2, to give the entire scale. The calculations for one octave of the scientific scale are shown in Table 12-2. Tuning forks in laboratories are often labelled with the frequencies of this scale.

Table 12-2 The Scientific Musical Scale

Note	Ratio of Frequencies	Frequency (Hz)	Interval
Middle C	1/1	256.0	standard
D	9/8	288.0	major whole tone (second)
E	5/4	320.0	major third
F	4/3	341.3	fourth
G	3/2	384.0	fifth
A	5/3	426.7	major sixth
B	15/8	480.0	major seventh
High C	2/1	512.0	octave (eighth)

The notes G and C in Table 12-2 sound pleasant together because their frequencies are in the ratio 3/2. Another pleasant-sounding pair is B and E, because 480/320 is 3/2.

The other musical scale, which we will call the **musicians' scale**, has a standard frequency of 440 Hz on fixed frequency instruments such as the piano. That frequency is the note A above middle C on the piano. An octave below has a frequency of 220 Hz, and an octave above has a frequency of 880 Hz. Figure 12-5 on page 274 shows parts of the piano scale, including the notes, their frequencies, and their staff notations. This is the scale used in tuning most musical instruments.

The calculations for the frequencies of the musicians' scale are based on two main facts.

(a) A note one octave above another is double the frequency of the other.

Did You Know?
Archaeologists in China have discovered sets of musical bells made there about 2400 years ago. The bells are based on a twelve-tone system, with one bell having a frequency (256.4 Hz) almost the same as the standard frequency of the scientific scale.

WAVES, SOUND, AND MUSIC

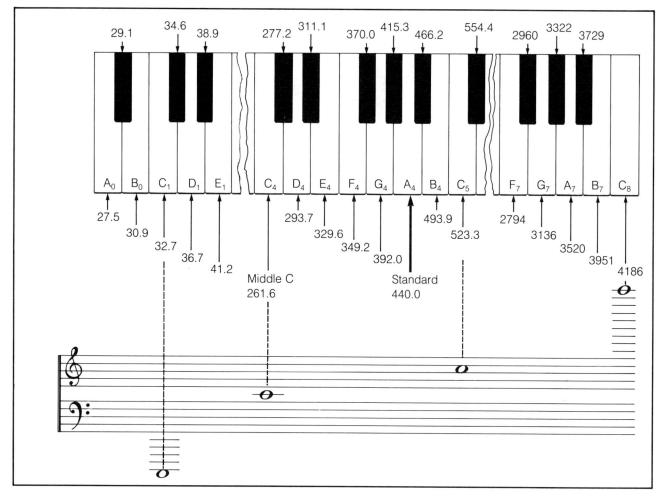

Figure 12-5
The Musicians' Scale, Illustrated on a Piano Keyboard

Did You Know?
The lowest note in classical music, D_2 or 73.4 Hz, is found in a composition by Mozart. Some singers have been able to sing sounds with frequencies lower than the lowest note on the piano, A_0, while other singers could sing above the highest note, C_8. However, these extreme frequencies have no musical value.

(b) There are exactly twelve intervals per octave. On a piano keyboard, for example, there are five black keys and seven white keys per octave.

Thus, the musicians' scale uses a number (approximately 1.0595) that, when multiplied by itself 12 times, gives the number 2. In other words, $\sqrt[12]{2} = 1.0595$. We will use the symbol "a" to denote this special number.

Starting with the standard frequency of 440 Hz, we multiply and divide 440 by a, then by a^2, then by a^3, and so on. This procedure gives new notes and is repeated until the entire musical scale is calculated. Calculations for one octave are shown in Table 12-3.

In Table 12-3, one example of a pair of notes that sounds harmonious is E_5 and A_4. The ratio of their frequencies is very close to 3/2.

The standard frequency A = 440 Hz is used in many countries throughout the world. To allow the public to check for that frequency, government agencies broadcast it by telephone and radio. For example, if you dial the telephone number 1-303-499-7111 (at your own expense!) you will be given the standard time in Greenwich, England, and a standard frequency broadcast (440 Hz) at intervals of exactly 1.0 s.

Table 12-3 The Musicians' (Equitempered) Scale

Note	Calculation (Using a = 1.0595)	Frequency (Hz)	Approx. Ratio	Interval
A_4	standard	440.00	1/1	standard
$A\#, Bb$	$440 \times a$	466.16	16/15	semitone
B_4	$440 \times a^2$	493.88	9/8	major whole tone, minor whole tone
C_5	$440 \times a^3$	523.25	6/5	minor third
$C\#, Db$	$440 \times a^4$	554.37	5/4	major third
D_5	$440 \times a^5$	587.33	4/3	fourth
$D\#, Eb$	$440 \times a^6$	622.25	45/32	augmented fourth, diminished fifth
E_5	$440 \times a^7$	659.26	3/2	fifth
F_5	$440 \times a^8$	698.46	8/5	minor sixth
$F\#, Gb$	$440 \times a^9$	739.99	5/3	major sixth
G_5	$440 \times a^{10}$	783.99	9/5	minor seventh
$G\#, Ab$	$440 \times a^{11}$	830.61	15/8	major seventh
A_5	$440 \times a^{12}$	880.00	2/1	octave (eighth)

Did You Know?
The musicians' scale, also called the *equitempered scale*, was standardized relatively recently, in 1953. It is still not accepted everywhere in the Western world. The advantage of an equitempered scale is that it allows musicians the freedom of control either to change from one key to another within one piece of music, or to play a variable-frequency instrument such as a trumpet along with a fixed-frequency instrument such as a piano. This flexibility is possible because all whole tones are equal in size and two semitones are exactly equal to one whole tone. (A semitone is an interval of $a = \sqrt[12]{2}$ and a whole tone is a^2.)

PRACTICE

3. On what does the pitch of a musical sound depend?
4. A portable electric saw is used to cut through a wooden log. What happens to the pitch of the sound from the saw as the blade cuts further into the log? Why?
5. State the frequency of a note one octave above notes with these frequencies:

    (a) 210 Hz   (b) 318 Hz   (c) 590 Hz

6. State the frequency of a note one octave below notes with these frequencies:

    (a) 310 Hz   (b) 684 Hz   (c) 1.2×10^4 Hz

7. State the frequency of a note two octaves above notes with these frequencies:

    (a) 300 Hz   (b) 512 Hz

8. For each pair of notes listed, determine the ratio of their frequencies as a fraction, and state whether their consonance when sounded together is high or low.

    (a) 750 Hz, 500 Hz   (c) 2000 Hz, 1000 Hz
    (b) 1000 Hz, 800 Hz   (d) 820 Hz, 800 Hz

9. Calculate the frequency of B_5 and C_6 on the musicians' scale.
10. Calculate a seven-interval (eight-note) scientific scale with 300 Hz as the standard frequency. Use the ratios indicated in Table 12-2.
11. Set up a two-octave musicians' scale based on five intervals per octave and a standard frequency of 500 Hz. ($\sqrt[5]{2} = 1.1486$) Give a name or symbol to each note.

12.3 Intensity and Loudness of Sounds

In physical terms, there is a difference in loudness between a soft whisper and the roar of nearby thunder. However, the loudness of a sound you hear is a subjective evaluation that depends on several factors, including the objective quantity known as "intensity".

Sound intensity is a measure of the power of a sound per unit area. (Since power is the rate at which energy is received, intensity can also be defined as energy/time/area.) Its SI unit is watts per square metre, or W/m². A sound intensity of 1.0 W/m², at a frequency of 1000 Hz, is called the **threshold of pain**. That is, an intensity above 1.0 W/m² would be painful to your ears. The lowest sound intensity a normal ear can hear, at 1000 Hz, is 4×10^{-12} W/m², a value called the **threshold of hearing**. This value is valid for a teenager with normal hearing. (Formerly, the threshold of hearing was defined as 1×10^{-12} W/m².)

Thus, the threshold of pain has an intensity on the order of 10^{12} times greater than the threshold of hearing. Since values in this measurement scale are awkward for everyday usage, a different scale, a logarithmic scale, is used to indicate sound intensity levels. In this more convenient scale, 10^{-12} W/m² is defined as an **intensity level** of zero bels or zero decibels. The decibel is more common than the bel (1.0 dB = 10^{-1} B), so henceforth sound intensity level will be stated in decibels in this text, while sound intensity will be stated in watts per square metre. The unit bel (B) is named after Alexander Graham Bell (Figure 12-6), who invented the telephone. Bell was born in Scotland in 1847. He worked in Canada and the United States, and died in Nova Scotia in 1922.

Table 12-4 Intensities of Sounds

Intensity (W/m²)	Intensity Level (dB)	Ratio of Intensity to the Intensity at 0 dB	Example
10^{-12}	0	1	Formerly defined as the threshold of hearing
10^{-11}	10	10	Empty church on quiet street
10^{-10}	20	100	Average whisper, at 1 m
10^{-9}	30	1000	Library reading room
10^{-8}	40	10 000	Inside a car with engine on
10^{-7}	50	100 000	Quiet restaurant
10^{-6}	60	1000 000	Conversation, at 1 m
10^{-5}	70	10 000 000	Machinery in factory
10^{-4}	80	100 000 000	Noisy street corner
10^{-3}	90	1000 000 000	Loud hi-fi in average room
10^{-2}	100	10 000 000 000	Rock concert
10^{-1}	110	100 000 000 000	Jet taking off, at 60 m
10^{0}	120	1000 000 000 000	Threshold of pain

In Table 12-4, the first column lists the intensities of sounds in the SI unit of watts per square metre. The second column lists the sound inten-

Figure 12-6
Alexander Graham Bell (1847–1922) is shown making the famous phone call inaugurating the line linking Chicago and New York City.

sity levels in decibels, and the third column compares these levels to the level at the threshold of hearing. The final column indicates examples of sounds having the intensity levels named.

The loudness of the sounds we perceive relates to the intensity of the sound. However, the two measures are not the same because the human ear does not respond to all frequencies equally. Figure 12-7 shows a graph of sound level as a function of frequency. From the graph, it is

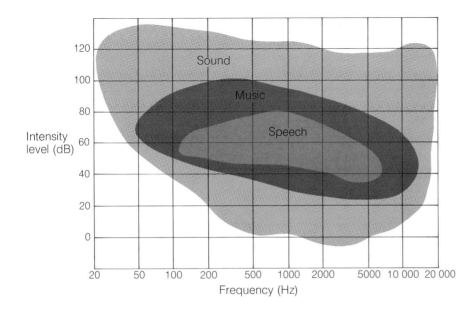

Figure 12-7
Sensitivity of the Average Human Ear to Different Frequencies

WAVES, SOUND, AND MUSIC

evident that the average human ear is most sensitive to sound frequencies between about 1000 Hz and 5000 Hz. Lower frequencies must have a higher sound level or intensity in order to be heard.

Listening to loud sounds for long periods of time can cause hearing loss. We use the term **noise pollution** to describe the effect of excess noise in our modern society. Governments try to prevent noise pollution by setting standards of noise levels on streets and in workplaces. Ear protection must be provided for people who work where the intensity level is greater than about 80 dB (Figure 12-8).

PRACTICE

12. On what factors does the loudness of a sound depend?
13. State the number of times the intensity level of the first sound in the pairs below exceeds the second.
    (a) 4 B, 3 B   (c) 8 B, 6 B
    (b) 100 dB, 70 dB   (d) 120 dB, 80 dB
14. A 50 dB sound is increased in intensity level by a factor of 10^4. What is its new intensity level?
15. Using the graph in Figure 12-7, state the approximate threshold of hearing in decibels at the following frequencies:
    (a) 50 Hz   (b) 1000 Hz
16. State the approximate threshold of pain in decibels at the following frequencies:
    (a) 100 Hz   (b) 2000 Hz

Figure 12-8
Ground personnel working near airplanes must have ear protection.

12.4 Quality of Musical Sounds

Assume that you hear a 440 Hz note equally loudly from a piano, a violin, and a trumpet. The three sounds have equal pitch and equal loudness. However, they sound different because each has a unique quality.

A novice music student will create sound of poor quality on a musical instrument, whereas an experienced player using the same instrument will produce sound of high quality. A small portable radio produces sound of poor quality compared to that from an expensive sound system. Thus, the quality of musical sounds can differ greatly.

The **quality** of a sound refers to how musical or pleasing the sound is to the human ear. Subjectively, quality depends on many factors. Objectively, however, it depends on the shape of the sound waves, which in turn depends on the harmonic structure of the sound waves. **Harmonics** are the **fundamental frequency** of a musical sound as well as the frequencies which are whole number multiples of the fundamental. The quality of a sound depends on the number of harmonics that make up the sound, and their relative loudness.

For example, assume that an instrument produces a fundamental frequency (f), also called the **first harmonic**, of 100 Hz, and that it has a second harmonic of $2f$ (or 200 Hz) with 80% of the amplitude of f.

Figure 12-9 shows that these sounds can be "added" using the principle of superposition to obtain a more complex waveform.

The quality of musical sounds is discussed in more detail for various types of musical instruments later in the chapter.

Figure 12-9
Adding the First and Second Harmonics to Change the Quality of a Sound

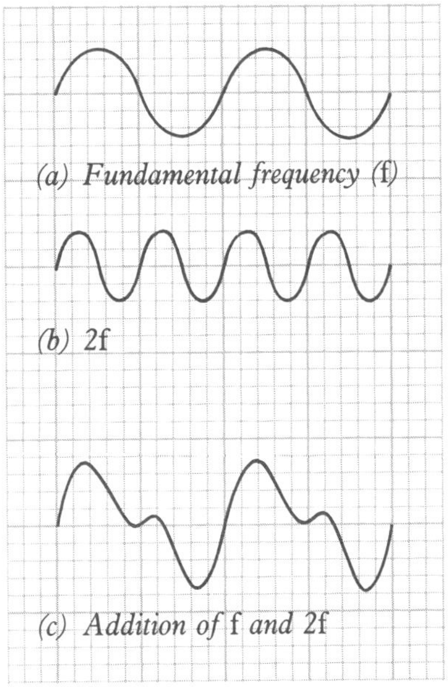

PRACTICE

17. **Activity** If the equipment is available to your class, view the oscilloscope traces of varying qualities of musical sounds as you listen to them. Draw diagrams of the waveforms you observe.
18. The note A (f = 440 Hz) is struck on a piano. Determine the frequency of the following:
    (a) the second harmonic
    (b) the third harmonic
    (c) the fifth harmonic
19. Use the principle of superposition to add the following waves, then describe how the resulting waveform relates to the quality of musical sound.
    Wave X: λ = 6.0 cm, A = 2.0 cm (Draw 1 wavelength.)
    Wave Y: λ = 2.0 cm, A = 1.0 cm (Draw 3 wavelengths to coincide with wave X.)

Experiment 12A: The Frequency of Vibrating Strings

INTRODUCTION
The use of stringed instruments originated thousands of years ago. Today, stringed instruments such as the guitar, piano, and violin, are still very popular. The frequency produced by the strings or wires of such instruments depends on the following variables, which you will control in this experiment:

- the length of the string
- the diameter of the string
- the tension or tightness of the string
- the density of material in the string

The most important apparatus used in this experiment is called a *sonometer*. It is a device with a variety of diameters and densities of strings and a means of adjusting both the length and the tension of each string. One way of controlling the tension is changing the mass suspended from the end of the string as it loops over the end of the sonometer. Another way is adjusting the peg around which the string is wound. This latter technique is used in the sonometer shown in Figure 12-10.

Figure 12-10
A Type of Sonometer, Adjusted by Means of the Peg Technique

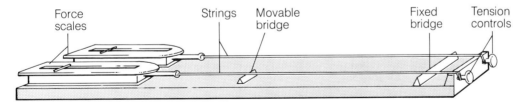

WAVES, SOUND, AND MUSIC

This experiment may be analysed either qualitatively or quantitatively. If you use qualitative analysis, you will have to listen very carefully to the frequencies produced, and hope that someone in your group or class can recognize frequency changes and relationships.

However, if you analyse the experiment quantitatively, you will have to use tuning forks and a frequency generator to determine the actual frequencies of the sounds. Then you will use a mathematical technique such as ratios or graphing to determine the mathematical relationships.

PURPOSE

To determine the relationship between the frequency of a vibrating string and its length, diameter, tension, and, if possible, density.

APPARATUS

sonometer with two or more strings of known diameter; metre stick; tuning forks or frequency generator (for quantitative analysis only)

PROCEDURE

1. Predict the relationship between the frequency of the string and each of the four variables named in the Introduction and the Purpose.
2. For a string of constant diameter, tension, and density, determine the relationship between the frequency of the string and its length. Figure 12-11 shows how to pluck the string and where to place the movable bridge to obtain lengths of the string in the ratio of 4:2:1.

Figure 12-11
Varying the Length of the String

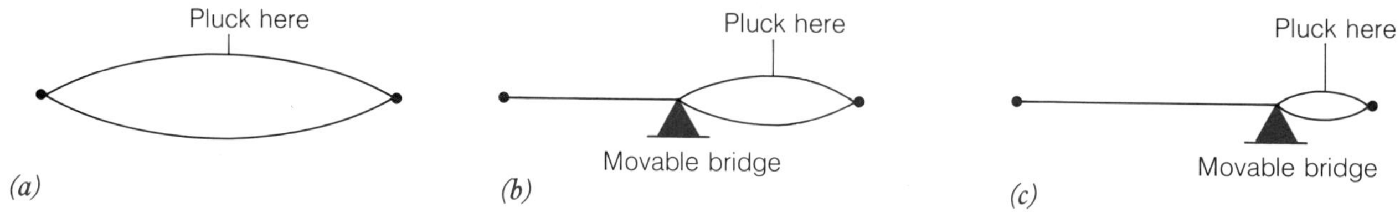

(a) (b) (c)

3. For strings of constant length, tension, and density, determine the relationship between the frequency of the string and its diameter.
4. For a string of constant length, diameter, and density, determine the relationship between the frequency of the string and its tension measured in newtons. (*Hint*: If you are doing this experiment quantitatively, notice that the relationship involves the square root of the tension.)
5. If the apparatus is available, determine the relationship between the frequency of the string and its density when the length, tension, and diameter are constant. (*Hint*: This relationship also involves a square root.)

CONCLUSIONS...

QUESTIONS

1. What happens to the frequency of a vibrating string if each of the following is increased?
   (a) length   (c) diameter
   (b) tension   (d) density

2. Repeat Question 1 for each variable increasing by a factor of 4.
3. A note sounded on a guitar string is "flat". What must be done to obtain the proper frequency?
4. Write an equation for each of the relationships discovered quantitatively in this experiment. For example, if $f \propto a$, then $f_1/f_2 = a_1/a_2$. Use f for frequency, L for length, d for diameter, F for force or tension, and D for density.
5. A 40 cm string on a violin vibrates at 440 Hz. What is the new frequency if the violinist changes the length to each of the following?
   (a) 30 cm   (b) 32 cm
6. A steel string of diameter 1.0 mm produces a fundamental frequency of 880 Hz. What is the fundamental frequency of a steel string with the same length and tension if its diameter is as follows?
   (a) 2.0 mm   (b) 0.40 mm
7. A 50 cm string under a tension of 25 N emits a fundamental note having a frequency of 400 Hz. What must be the new frequency if the tension is changed to 36 N?
8. An aluminum string, $D = 2700$ kg/m^3, has a frequency of 1000 Hz. Find the frequency of an iron string, $D = 7860$ kg/m^3, of the same length, diameter, and tension.

12.5 Stringed Instruments

Stringed instruments consist of two main parts – the vibrator and the resonator. The **vibrator** is the string, and the **resonator** is the case, box, or sounding board that the string is mounted on. A string by itself does not give a loud or, necessarily, even a pleasing sound. It must be attached to a resonator, through which forced vibrations help improve the loudness and quality of the sound. Even a tuning fork has a louder and better sound if its handle touches a desk, wall, or resonance box.

Stringed instruments can be played by plucking, striking, or bowing them. The quality of sound is different in each case. The quality also depends on what part of the string is plucked, struck, or bowed. For example, a string plucked gently in the middle produces a strong fundamental frequency or first harmonic. This is shown in Figure 12-12(a) on page 282. A string plucked at a position one-quarter of its length from one end vibrates in several modes, two of which are seen in Figure 12-12(b). In this case, the second harmonic is added to the first harmonic, changing the sound quality. Figure 12-12(c) shows the third harmonic superimposed on the first harmonic, again resulting in a different quality of sound.

Stringed instruments that are usually *plucked* include the banjo, guitar, mandolin, ukulele, and harp (Figure 12-13 on page 282). The harp, with 46 strings, is a complex instrument. It consists of a hollow soundboard (the resonator), a vertical pillar, and a curved neck. The strings are stretched between the pins on the soundboard and the pegs on the curved neck. A pedal mechanism enables the player to raise the frequency of individual strings. The other stringed instruments that are plucked resemble one another. They have from four to eight strings, as

WAVES, SOUND, AND MUSIC

Figure 12-12
Changing the Quality of Sound of a Vibrating String

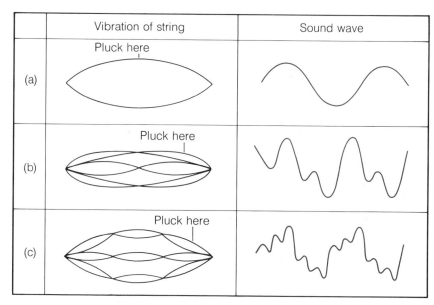

Figure 12-13
Harp

(a) Playing the harp

(b) Structural detail

Figure 12-14
Inside an Upright Piano

well as frets to guide the placement of the fingers. The thick wires under low tension create the lower notes, while the thin wires under greater tension create the higher notes.

The best-known stringed instrument that is *struck* is the piano. A piano key is connected by a system of levers to a hammer that strikes the string or strings to produce a certain note. See Figure 12-14. A modern piano has 88 notes, with a frequency range from 27.5 Hz to 4186 Hz. The short, high-tension wires produce high-pitched notes, and the long, thick wires produce low-pitched notes. The sounds from the strings are increased in loudness and quality by the wooden sounding board of the piano.

Stringed instruments that are usually *bowed* belong to the violin family. This family consists of the violin (Figure 12-15 on page 284), viola, cello, and bass. One side of each bow consists of dozens of fine fibres which are rubbed with rosin to increase the friction when stroked across a

Figure 12-15
Violin

string. Each instrument has four strings, and wooden sounding boards at the front and back of the case. The members of the violin family have no frets. Thus, the frequency can be changed gradually, not necessarily in steps as in the guitar family. (The only other instrument capable of changing frequency gradually is the trombone.) The smallest instrument, the violin, produces high-frequency sounds; the largest, the bass, low-frequency sounds.

Stringed instruments do not give out a great amount of power. The maximum acoustic power from a piano is 0.44 W, while that from a bass drum is 25 W! This explains why an orchestra needs many more violins than drums or trumpets.

PRACTICE

20. When the resonator of a stringed instrument vibrates, does it do so as a result of sympathetic vibrations? Explain your answer.
21. **Activity** Using a sonometer or a stringed instrument, pluck a string at various positions and compare the resulting frequencies and qualities. If possible, view the waveforms using an oscilloscope.
22. The fundamental frequency of a string is 220 Hz. What is the frequency of the following?
    (a) the second harmonic
    (b) the fourth harmonic
23. Explain how the production of beats can be used to tune a stringed instrument. Refer to Section 11.4.
24. A 440 Hz tuning fork is sounded with the A-string on a guitar, and a beat frequency of 3 Hz is heard. Then the tension in the string is decreased, and a new beat frequency of 4 Hz is heard.
    (a) What is the frequency of the guitar string when the beat frequency is 4 Hz?
    (b) What should be done to tune the string to 440 Hz?
25. **Activity** For a stringed instrument of your choice, research and report on the questions below. If experimentation is necessary to solve a problem, outline the procedure you will use.
    (a) *Societal Implications*
        • Who uses the instrument, and for what purposes is it used?
        • What is the cost of the instrument? What is the price range for similar instruments of the same type?

- What impact has the instrument had on music and society?
- What careers involve a knowledge of the instrument or the ability to play it?

(b) *Applications*
- What are the different types of the instrument?
- What other stringed instrument performs a similar function?
- How does this instrument compare with other stringed instruments?
- How is the instrument tuned?

(c) *Experiments*
- What factors affect the frequency of the strings, and in what way do they affect it?
- How does the body of the instrument affect the sound produced by the instrument?
- How can beats be used to tune the instrument?

12.6 Vibrating Columns of Air

Air can be made to vibrate to produce sound. Vibrating columns of air are the basis of wind instruments, including the most common instrument of all, the human voice.

Resonance occurs in an air column when a standing wave fits neatly into the column. You have seen that standing waves result from interference. A device that illustrates resonance of sound in an air column is shown in Figure 12-16. The flask can be raised or lowered to adjust

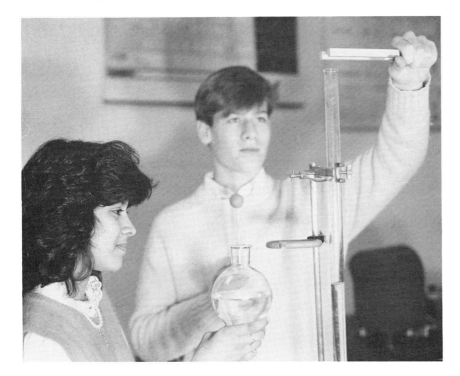

Figure 12-16
Resonance Apparatus

the level of water in the tube, which in turn changes the length of the air column. If a tuning fork having a certain frequency is sounded and held near the open end of the air column, resonant sounds will be heard only at certain lengths of the air column.

To understand how resonant sounds are caused by standing waves, consider Figure 12-17. It shows sound waves of constant wavelength in an air column open at one end. (In reality, the waves are longitudinal standing waves, but they are drawn here as transverse standing waves to make the concepts easier to visualize.) A sound wave which starts from the open end of the tube is reflected off the closed end where the water is located, and the incoming wave interferes with the reflected wave. Resonance occurs whenever the sound wave has a wavelength that allows a standing wave pattern to be set up. The air molecules at the closed end cannot vibrate easily. Thus, a minimum amplitude, called a **displacement node**, exists there. At the open end, vibration of the air molecules is easy. Thus, a maximum amplitude, called a **displacement antinode**, occurs there. The first resonance occurs when the length of the air column is one-quarter of the wavelength of the sound used.

Figure 12-17
Standing Wave Patterns of Constant Wavelength in Columns of Air Closed at One End

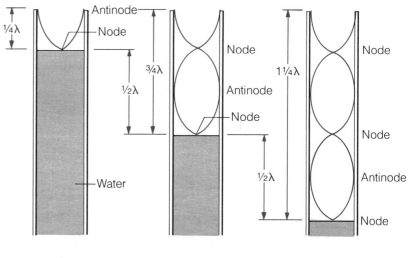

(a) First resonance   (b) Second resonance   (c) Third resonance

In Figure 12-17, the second column, (b), is half a wavelength longer than the first column. The third column is also half a wavelength longer than the second column. This means that the distance from one maximum sound to another in an air column is one-half the wavelength ($\frac{1}{2}\lambda$) of the sound, assuming that the wavelength remains constant. The distance from one node to another is also $\frac{1}{2}\lambda$. (Recall that this was also true for standing waves on ropes, studied in Chapter 10.)

Similar calculations can be made for columns of air that are open at both ends. Figure 12-18 shows that the first resonance occurs when the air column length is $\frac{1}{2}\lambda$ and that the distance from one maximum to the next is again $\frac{1}{2}\lambda$. (We assume that the wavelength is constant.) In these columns, maximum displacements or antinodes occur at both ends, because the air molecules vibrate easily in both locations.

These concepts will be applied to an experiment involving columns of air. Then they will be used to explain wind instruments, which create sound by means of vibrating air in columns.

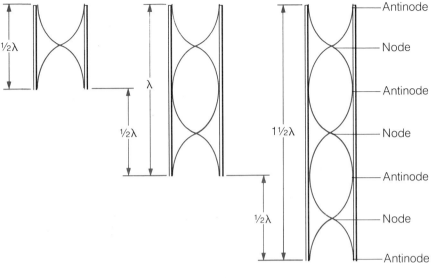

Figure 12-18
Standing Wave Patterns of Constant Wavelength in Air Columns Open at Both Ends

(a) First resonance   (b) Second resonance   (c) Third resonance

SAMPLE PROBLEM 1

A vibrating tuning fork is held near the mouth of a column filled with water. The water level is lowered, and the first loud sound is heard when the air column is 9.0 cm long. Calculate the following:
(a) the wavelength of the sound from the tuning fork
(b) the length of the air column for the second resonance

Solution

(a) $\frac{1}{4}\lambda = 9.0$ cm
∴ $\lambda = 36$ cm

Thus, the wavelength of the sound is 36 cm.

(b) The air column length is increased by $\frac{1}{2}\lambda$, or 18 cm, to obtain the second resonance. Thus, the total length is 9.0 cm + 18 cm = 27 cm, or $\frac{3}{4}\lambda$.

PRACTICE

26. What vibrates to create sound in a column of air?
27. In an air column, the distance from one resonance length to the next is 21.6 cm. What is the wavelength of the sound producing resonance if the column is (a) closed at one end and (b) open at both ends?
28. **Activity** Listen to the sound of water being added to a graduated cylinder. What happens to the pitch of the sound as the water is added? Explain why.

29. A 1000 Hz tuning fork is sounded and held near the mouth of an adjustable column of air open at both ends. If the air temperature is 20°C, calculate the following:
   (a) the speed of the sound in air
   (b) the wavelength of the sound
   (c) minimum length of the air column that produces resonance

---

Experiment 12B: Sound in a Column of Vibrating Air

INTRODUCTION
The distance between resonant or loud sounds in air columns can be used to find the wavelength of a sound, as described in Section 12.6. If the frequency of the sound is known, the universal wave equation ($v = f\lambda$) can be used to find the speed of the sound in air. If the air temperature (T) is known, the equation $v = [332 + 0.6\,(T)]$ m/s can be used to verify the speed found using the first equation. If the speeds are the same, the experiment is likely successful.

As you perform this experiment, notice that low-frequency sounds need long columns of air, and high-frequency sounds need short columns. This relates closely to what you will learn about wind instruments in the next section.

PURPOSE
To use the resonance of sound in air columns to find the wavelength and speed of sounds.

APPARATUS
3 or 4 tuning forks of known frequency (between 384 Hz and 2000 Hz); resonance apparatus (Figure 12-16); rubber hammer; thermometer; metre stick; two open-ended air columns, one of which fits inside the other (*i.e.*, an adjustable open-ended air column)

PROCEDURE

CAUTION:
Do not allow the vibrating metal tuning fork to touch the glass tube, because the tube will shatter easily.

1. Adjust the water level in the long glass tube by raising the supply flask until the water is near the top of the tube. Have one person strike the first tuning fork and hold it close to the mouth of the tube, keeping in mind the Caution in the margin. Have another person slowly lower the level of the water until the first resonance is heard. Check this level repeatedly until its location is certain. Then measure and record the length of the air column.
2. Repeat Procedure Step 1 for the second resonance and for a third, if it can be found. Set up a table of data that includes the tuning fork frequency, the resonance number, and the length of the air column.
3. Repeat the entire Procedure thus far, using tuning forks of different frequencies.

4. Repeat the Procedure thus far, using an adjustable air column open at both ends.
5. If time permits, perform at least one trial outside, where the air temperature is likely to be different from inside. Remember to measure the air temperature in the classroom.

ANALYSIS
1. Use the data from Procedure Steps 1 to 4 to determine an average wavelength of the sound from each tuning fork used in the experiment.
2. Use the wavelengths and frequencies to calculate the speed of sound in air in the classroom. Calculate the average of all the speeds.
3. Use the air temperature to find the speed of the sound. Compare this value to that found in Analysis Question 2.

CONCLUSIONS . . .

QUESTIONS
1. How would a higher air temperature affect the lengths of the resonating air columns found in this experiment? Why?
2. How would the results of the experiment differ if the air in the column were replaced with helium, in which the speed of sound is 927 m/s at 20°C?
3. A 500 Hz tuning fork is used to determine the resonances in an adjustable column of air open at both ends. The length of the air column changes by 33.4 cm between resonant sounds. Calculate the following:
   (a) the wavelength of the sound in the air column
   (b) the speed of the sound in air
   (c) the air temperature

12.7 Wind Instruments

All wind instruments contain columns of vibrating air molecules. The frequency of vibration of the air molecules, and thus the fundamental frequency of the sound produced, depend on whether the column is open or closed at the ends. As is the case with all vibrating objects, the large instruments create low-frequency sounds, and the small instruments create high-frequency sounds.

In some wind instruments like the pipe organ in Figure 12-19(a), the length of each air column is fixed. However, in most wind instruments, such as the trombone, the length of the air column can be changed.

To cause the air molecules to vibrate, something else must vibrate first. There are four general mechanisms for forcing air molecules to vibrate in wind instruments.

First, in **air reed instruments**, air is blown across or through an opening. The moving air sets up a turbulence inside the column of the instrument. Examples of such instruments are the pipe organ in Figure

Figure 12-19
Air Reed Instruments

(a) This beautiful pipe organ is located in a famous church in Czestochowa, Poland.

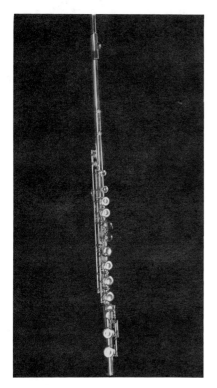

(b) Flute

Figure 12-20
Saxophone

12-19(a), flute in Figure 12-19(b), piccolo, recorder, and fife. The flute and piccolo have keys that are pressed to change the length of the air column. The recorder and fife have side holes that must be covered with fingers to change the length of the air column and thus control the pitch.

In **single-membrane reed instruments**, moving air sets a single reed vibrating, which in turn sets the air in the instrument vibrating. Examples of these instruments include the saxophone (Figure 12-20), clarinet, and bagpipe. In the bagpipe, the reeds are located in the four drone pipes attached to the bag, not in the mouth pipe. Again, the length of the air column is changed by holding down keys or covering side holes.

Next, in **double-mechanical reed instruments**, moving air forces a set of two reeds to vibrate against each other. This causes air in the instrument to vibrate. Examples include the oboe (Figure 12-21), English horn, and bassoon. Keys are pressed to alter the length of the air column.

The final mechanism is found in **lip reed instruments**, also called **brass instruments**. In this type of instrument, the player's lips func-

MUSIC, MUSICAL INSTRUMENTS, AND ACOUSTICS

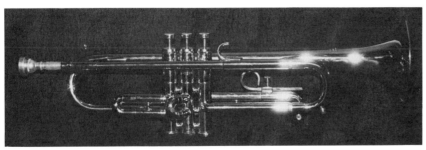

Figure 12-22 Trumpet

Figure 12-21
Oboe

tion as a double reed. They vibrate, causing the air in the instrument to vibrate. None of the air escapes through side holes, as in other wind instruments; rather, the sound waves must travel all the way through a brass instrument. Examples of such instruments are the bugle, trombone, trumpet (Figure 12-22), French horn, and tuba. The length of the air column is changed either by pressing valves or keys that add extra tubing to the instrument, or, in the case of the trombone, by sliding the U-tube.

The quality of sound from wind instruments is determined by such factors as the construction of the instrument and the experience of the player. However, just as with stringed instruments, it also depends on the harmonics produced by the instrument. In Section 12.6 and Experiment 12B you learned that a standing wave pattern is set up when resonance occurs in air columns. The standing waves create the fundamental frequencies or first harmonics. Let us now use an air column open at both ends to see how sound quality improves when other harmonics are present.

Figure 12-23(a) shows an air column open at both ends with a length one-half the wavelength of the resonating sound. This situation produces the first harmonic (f); the corresponding waveform is shown adjacent to the air column. Now, with the length of the column constant, as is the case at any given instant with a wind instrument, a second harmonic can be created by various means, such as blowing harder. Figure 12-23(b) shows that the wave must be half the length of the first wave in order to produce a standing wave pattern. Thus, the frequency here is double ($2f$), and we have the second harmonic. Figure 12-23(c) shows the first and second harmonics added together to give sound of higher quality. In air columns open at both ends, other harmonics may also be present, enhancing the quality further still.

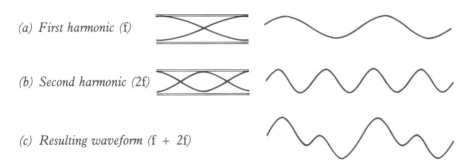

(a) First harmonic (f)

(b) Second harmonic (2f)

(c) Resulting waveform (f + 2f)

Figure 12-23
Improving the Quality of Sound in Wind Instruments

WAVES, SOUND, AND MUSIC

It is left to the Practice Questions to prove that an air column closed at one end has only odd numbers of harmonics (f, $3f$, $5f$, and so on). Thus, the quality of this sound differs from that from air columns open at both ends. In fact, every instrument, whether it is a wind instrument, a stringed instrument, or whatever, has its own harmonic structure and therefore its own unique quality. This fact is used to create "artificial" music on music synthesizers, which are described in Section 12.11.

PRACTICE

30. From the pairs of instruments listed, choose the one that has the higher range of pitches. (It will be helpful to discuss in class the size of the instruments.)

    (a) piccolo, flute (c) oboe, English horn
    (b) bassoon, English horn (d) tuba, trumpet

31. Use diagrams to help you prove that an air column closed at one end has only odd numbers of harmonics.

32. **Activity** Obtain a single reed from a clarinet and a double reed from an oboe and demonstrate whether or not sound can be produced by blowing across each one. The reeds or mouthpieces of other wind instruments could also be used.

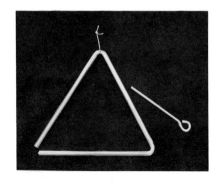

Figure 12-24
Triangle

12.8 Percussion Instruments

Percussion is the striking of one object against another. Percussion instruments are usually struck by a firm object such as a hammer, bar, or stick. These musical instruments were likely the first invented because they are relatively easy to make. (Doctors use percussion when they tap a patient's chest or back and listen for sounds that indicate either clear or congested lungs.) Percussion instruments can be divided into three categories.

Single indefinite pitch instruments are used for special effects or for keeping the beat of the music. Examples include the triangle (Figure 12-24), the bass drum, and castanets.

Multiple definite pitch instruments have bars or bells of different sizes that, when struck, produce their own resonant frequencies and harmonics. Examples are the tuning fork, orchestra bells, the marimba (Figure 12-25), the xylophone, and the carillon.

Variable pitch instruments have a device used to rapidly change the pitch to a limited choice of frequencies. An important example is the timpani, or kettle drum, which has a foot pedal for quick tuning.

Some instruments, such as the accordion and harmonica (Figure 12-26), are difficult to classify as a single type of instrument. The accordion and harmonica use moving air to set reeds vibrating. However, they do not have resonating air columns, so they are not usually called wind instruments. They are better classified as percussion instruments in which air knocks against reeds, causing them to vibrate.

MUSIC, MUSICAL INSTRUMENTS, AND ACOUSTICS

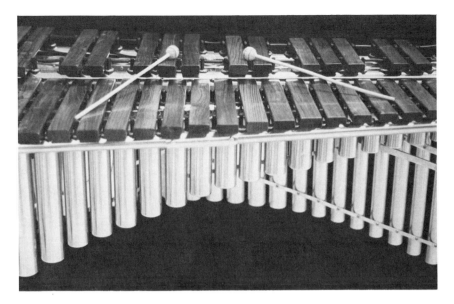

Figure 12-25
Marimba

Figure 12-26
Harmonica

PRACTICE

33. **Activity** Strike a tuning fork at various positions along one prong, listen carefully, then describe the sounds produced. Is the quality of the fundamental frequency enhanced or lessened by changing the striking position? Where, approximately, is the best position to strike the fork to eliminate the high-frequency sound?
34. Describe factors which you think affect the quality of sound from percussion instruments.

12.9 The Human Voice

The human voice is a fascinating instrument. The main parts of the body which help produce sound are shown in Figure 12-27(a). Figure 12-27(b) shows how the flow of air from the lungs causes sound.

The human voice consists of three main parts:

(a) the **source of air** (the lungs)
(b) the **vibrators** (the vocal folds or vocal cords)
(c) the **resonators** (the lower throat or pharynx, mouth, and nasal cavity)

293

WAVES, SOUND, AND MUSIC

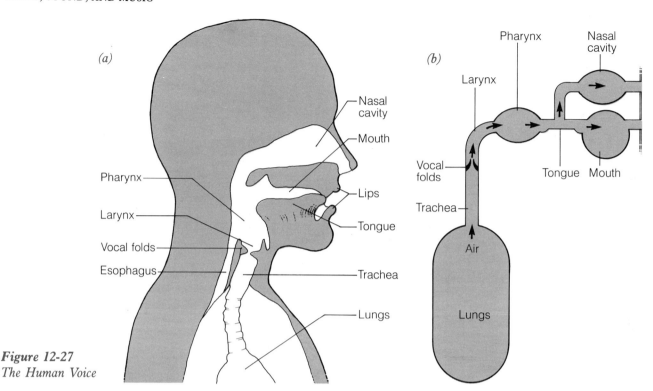

Figure 12-27
The Human Voice

Did You Know?
Human infants cannot produce most sounds we understand as articulate speech until after they are a year old when the larynx drops down in the throat.

To create most sounds, air from the lungs passes by the vocal folds, causing them to vibrate. The vocal folds are two bands of skin that act like a double reed. Loudness is controlled by the amount of air forced over the vocal folds. The pitch is controlled by muscular tension as well as by the size of the vibrating parts. Since, as you have seen, larger instruments have lower resonant frequencies, in general, male voices are lower in their frequency range than female voices. Refer to Table 12-5.

Table 12-5 Approximate Frequency Ranges of Singers

Type of Singer	Frequency Range (Hz)	Type of Singer	Frequency Range (Hz)
Bass	82–294	Alto	196–698
Baritone	110–392	Soprano	262–1047
Tenor	147–523		

The quality of sound from the voice is controlled by the resonating cavities and the parts in them, such as the tongue and lips. You can observe an interesting demonstration of this by holding a microphone to your throat and then to your mouth while making the same musical sound with your mouth open. If the signals are viewed on an oscilloscope screen, the benefit of having resonating cavities can be seen.

Of course, the quality of sound may also be improved by proper training. Good singers can control such effects as vibrato and tremolo. **Vibrato** is a slight, periodic changing of frequency (frequency modulation, or FM), and **tremolo** is a slight, periodic changing of amplitude (amplitude modulation, or AM).

Waveforms of certain sounds from the human voice were shown in the first section of this chapter. If you try to recreate those waveforms, you will likely be unsuccessful. Human voice waveforms are very complex to analyse or to reproduce accurately. However, the use of computers now makes it possible to recognize or reproduce many sounds. This has many useful applications, one of which is to manufacture robots that obey the oral commands of handicapped people who can speak but who cannot get around easily.

Did You Know?
Computer analysis is used to compare the writing styles of musical composers. In analysing piano sonatas, for example, it has been found that the composer Haydn used notes separated by an octave twice as often as Mozart or Beethoven did.

PRACTICE
35. Discuss whether the human voice should be classified as a stringed, wind, or percussion instrument.
36. **Activity** Determine the frequency range of your voice using either a musical instrument or an audio frequency generator, then classify your singing voice according to Table 12-5.

12.10 Electrical Instruments

Electrical instruments are made of three main parts—a **source** of sound, a **microphone**, and a **loudspeaker**. At hockey and football games, for instance, the announcer's voice directs sound energy into a microphone. The microphone changes sound energy into electrical energy which, after amplification, causes vibrations in a loudspeaker. These vibrations reproduce the original sound with an amplified loudness.

Many of the musical instruments discussed in the previous sections can be made into electrical instruments through the addition of a microphone and a loudspeaker. This is often done with stringed instruments that normally give out low amounts of power. A microphone is attached directly to the body of the instrument. In some cases, the design of the instrument is altered. An electric guitar, for example, may have a solid body rather than the hollow body of acoustic guitars. See Figure 12-28.

Loudspeakers are important in determining the quality of sound from an electrical instrument. A single loudspeaker does not have the same frequency range as our ears, so a set of two or three must be used to give both quality and frequency range. Table 12-6 lists details of the three common sizes of loudspeakers used in electrical sound systems. Figure 12-29 shows a typical set of loudspeakers.

Figure 12-28
Electric Guitar

Table 12-6 Details of Loudspeakers

Name	Approximate Size (cm)	Frequency Range (Hz)	Wavelength Range (cm)
Woofer (low-range)	25–40	25–1000	34–1400
Squawker (mid-range)	10–20	1000–10 000	3.4–34
Tweeter (high-range)	4–8	3000–20 000	1.7–11

The final column of Table 12-6 indicates that the sound waves from the tweeter have much shorter wavelengths than those from the woofer.

Figure 12-29
Typical Loudspeakers

Long wavelengths are diffracted easily through doorways and around furniture and people. However, the short waves from a tweeter are not diffracted around large objects, so their sound tends to be directional. As a result, a listener must be in front of the tweeter to get the full sensation of its sound, especially in the very high frequency range.

An interesting phenomenon occurs with the use of headphones used on portable radios and cassette players. Some of these headphones do not create the low frequencies usually associated with the woofer, yet the listener actually "hears" low-frequency sound. The higher harmonics produced by the headphones cause a sensation that makes us believe we are hearing the absent first and second harmonics.

PRACTICE

37. **Activity** If an electrical instrument such as an electric guitar is available, produce sounds with the electricity on and then off. Describe the differences between the sounds, especially in the quality.
38. Figure 12-30 shows a top view of a speaker system having a woofer and a tweeter facing to the right. Compare the sounds heard by the observers at A, B, and C.

Figure 12-30

12.11 Electronic Instruments

No doubt you have heard of music synthesizers or electronic synthesizers. These instruments, in contrast with stringed, wind, and percussion instruments, produce vibrations using electronic components such as resistors and transistors.

An electronic instrument consists of four main parts:

(a) The **oscillator** creates the vibrations.
(b) The **filter circuit** selects the frequencies that are sent to the mixing circuit.
(c) The **mixing circuit** adds various frequencies together to produce the final signal.
(d) The **amplifier and speaker system** makes the sound loud enough to be heard.

Synthesizers and electronic organs are common electronic instruments (Figure 12-31). The shape of the sound waves they produce can be con-

Figure 12-31
A Portable Electronic Synthesizer

PROFILE

Bruce W.J. Pennycook, D.M.A.
Associate Professor of Music
Associate Professor of Computing and Information Science
Director, Computer Music Faculty
Queen's University

When I was a student at Earl Haig Secondary School in Willowdale, Ontario, I majored in sciences in anticipation of studying dentistry at university. This was, in part, because I had been expelled from my first love, music. I had been totally bored with the conventional band repertoire, and was constantly trying to analyse the harmonic nature of various pieces by improvising. I suppose the teacher felt my improvisations were not suitable for a certain Bach chorale we were playing one day in class, and thus I found myself cut off permanently from any high school music courses.

Despite this academic setback, my musical interests continued to flourish. As a result, my first year at the University of Toronto as a Bachelor of Science student was a disaster – the result of too many hours spent performing coupled with too few hours in the science labs.

Bowing to the inevitable, I re-entered the University in 1969 as a Bachelor of Music student. My interests in music theory and electronic music composition soon developed into specialization in contemporary music composition and innovative music technology. Upon completing my fourth year, I received several awards and scholarships to pursue graduate studies in composition. That same year, 1972, I also had my first – and fateful – encounters with computers both in recording studios where I was performing and in large-scale concert set-ups. Soon I concluded that computers and music were going to meet in all forms of music making. So, with the help of a Doctoral Fellowship from the Canada Council, I went to the heart of computer music research – The Center for Computer Research in Music and Acoustics at Stanford University in what is now Silicon Valley.

Since that time, I have devoted my research and much of my compositional activities to problems relating to computer applications in music. In a very real sense, I function as an artist and a scientist combined. It is ironic, perhaps, that the career I abandoned for music has become very much the reality.

As a scientist, I work in two basic areas. The first of these involves processing musical sound by computer – either through straightforward recording techniques or through mixing, modifying, synthesizing, and playing back high fidelity sound. Secondly, I am involved in developing computer programs which are used by composers to help them as they create new pieces of music.

As an example, one major project I am currently involved in concerns the development of a special kind of "listening" computer program. To use this program, an individual composer will feed his or her compositions into a computer using an instrument such as an electronic keyboard or just his or her voice. The computer program will then analyse the music to determine what are the regular features of that composer's style. Later, when the composer decides to create a new musical work, the computer will be able to help by making suitable suggestions that are in keeping with the composer's individual style.

In one sense, at least, my methods differ radically from the traditional scientific approach. The development of new synthesis or compositional techniques is most certainly determined by considerations of taste and style.

On the other hand, I do approach the development of such instruments as the audio computer according to the basic scientific principles of testing and experimentation. Ultimately, I would like to be able to contribute to the development of computer programs that will be able to listen to music, speech, and other sounds such as the wind in the trees. In this way, we will be able to learn about the source and structure of sound. Several people at other centres are working toward this goal as well and I expect that moderately intelligent computer "listeners" will exist within the next ten years.

Like the research in robot vision and hand/eye coordination, audio research contributes to our understanding of how humans perceive, think, and move. That alone, in my opinion, is worth the effort.

WAVES, SOUND, AND MUSIC

Figure 12-32
The waveforms shown can be used to generate various other waves.

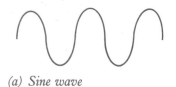

(a) Sine wave

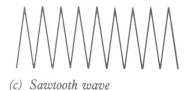

(b) Square wave

(c) Sawtooth wave

(d) Triangular wave

trolled; as a result, these instruments can emit sounds that resemble the sound of almost any musical instrument. The basic shapes of the waves used to create more complex waves are shown in Figure 12-32.

Electronic instruments can also control the attack and decay patterns of a sound. The **attack** occurs when the sound is first heard. It may be *sudden*, *delayed*, or *overshot*, as illustrated in Figure 12-33(a). The **decay** occurs when the sound comes to an end. It may be *slow*, *fast*, or *irregular*, as shown in Figure 12-33(b).

Figure 12-33
Control Patterns

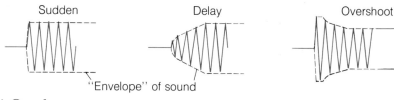

(a) Growth patterns

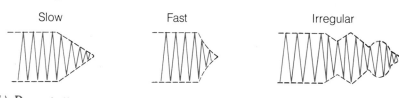

(b) Decay patterns

PRACTICE

39. **Activity** Listen to the sounds produced by an audio frequency generator that creates sine waves, square waves, and triangular waves, all at a constant frequency. Describe the differences in the sounds, and try to explain why the differences exist.
40. **Activity** If a musical synthesizer is available, have someone demonstrate its capabilities.
41. A synthesizer adds a square wave, S, ($\lambda = 4.0$ cm, $A = 2.0$ cm) and a triangular wave, T, ($\lambda = 4.0$ cm, $A = 1.0$ cm) to obtain a new sound. Draw two wavelengths of S and T, then use the principle of superposition to show their addition.

12.12 Acoustics

Some people claim that their singing voice is better in the shower than anywhere else. If this is true, it could be the result of the many sound reflections in a small room.

The qualities of a room or auditorium that determine how well sound is heard are called **acoustics**.

The acoustics of a room depend on the shape of the room, the contents of the room, and the composition of the walls, ceiling, and floor. Sounds in a large, empty room are subject to echo and are poor in quality. When rugs and furniture are put in the room, the acoustics improve.

In auditoriums and theatres, special design features are used for improving the acoustics, especially as they affect music. By trial and error, it has been discovered that the approximate "audience density" should be about one person for every 0.8 m² of floor area. Also, walls, ceilings, and any furnishings must be designed with a good balance of reflection and absorption of sound in mind. The photograph at the start of this chapter shows an auditorium with such features. Here, for example, the devices hung from the ceiling help control both the amount and direction of the sound reflected to the audience.

You have learned that harmonics structure helps determine the quality of musical sounds. Similarly, the physical quantity called "reverberation time" helps determine the acoustics of a chamber. **Reverberation time** is the time for the intensity level of a sound to drop to 10^{-6} (one millionth) of its original value. High-quality acoustics are possible with a reverberation time ranging from about 0.5 s in a small room used for speaking, to over 3 s in a large hall used for performances of orchestral music. These times vary for different frequencies of sounds, so you can likely appreciate the difficulty of designing an auditorium with excellent acoustics.

A typical graph illustrating the experimental determination of the reverberation time in a concert hall is shown in Figure 12-34. The sound intensity level is allowed to build up and, for the example shown, reaches a maximum value of 70 dB. This level is sustained briefly, then the sound is suddenly turned off (at $t = 6.4$ s). Within 1.8 s (at $t = 8.2$ s), the intensity level has dropped by 60 dB to 10 dB (*i.e.*, to 10^{-6} of the maximum value). Thus, the reverberation time is 1.8 s.

Did You Know?
There is a castle straight out of fantasy-land in the Bavarian Alps of West Germany whose original owner loved music. Neuschwanstein Castle, built in the later nineteenth century, has its own music hall with singular features for improving the acoustics. For example, the ceiling consists of a series of thin panels that vibrate with the music. The elaborate castle cost a fortune to build. For this and other excesses, the owner was sometimes called "Mad King Ludwig" of Bavaria.

Figure 12-34
Experimental Determination of Reverberation Time

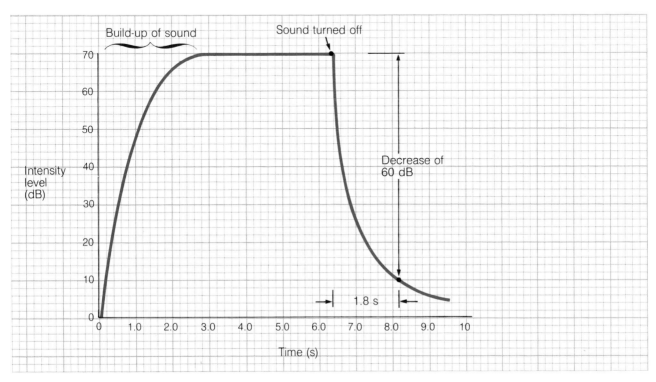

WAVES, SOUND, AND MUSIC

Figure 12-35
Four Structures Designed with Specific Acoustical Purposes in Mind

(a) Anechoic chamber

(b) The Hollywood Bowl in Hollywood, California, is designed to direct sound from the stage to the audience.

(c) This outdoor theatre is located in Tallinn, Estonia. The shell is designed to direct sound from a 30 000-member choir to the audience. In the background is the Baltic Sea.

(d) This 25 000-seat outdoor theatre (top right) was built in Ephesus, Turkey, by the Romans about 2000 years ago. The theatre has excellent acoustical properties.

Sometimes, rooms or buildings are needed for special purposes, four of which are shown in Figure 12-35. Figure 12-35(a) shows the design of a recording or sound-testing studio, also called an **anechoic chamber**. Such chambers have irregular shapes to prevent the setting up of standing waves, and the floors, walls, and ceilings are covered with a layer of sound-absorbing baffles. These baffles "deaden" the sound, allowing audio equipment to be tested without unwanted sound interference. (The reverberation time in an anechoic chamber is typically about 0.05 s.)

Figure 12-35(b) shows a band shell in a park directing sound to an outdoor audience. Notice the spheres, which are used to reflect the short-wavelength, high-frequency sounds in several directions. Figure 12-35(c) shows an open-air theatre designed especially for choirs; Figure 12-35(d) shows a 25 000-seat open-air theatre built by the Romans on the west coast of Turkey almost 2000 years ago. A person standing at the centre of the stage who speaks with ordinary loudness can be heard everywhere in the theatre. Perhaps the Romans could not explain acoustics scientifically the way we can today, but they certainly could design theatres with excellent sound characteristics.

This brings us to the end of the subject of sound in this text. However, we have barely scratched the surface of the fascinating topic of sound – an important type of energy that surrounds us at all times.

PRACTICE

42. Discuss the acoustics of both your physics classroom and the school auditorium. In each case, consider these questions:
    (a) What has been done to provide good acoustics?
    (b) What could be done to improve the acoustics?
43. By how many decibels does the intensity level of a sound decrease after the reverberation time in any room has elapsed?
44. Determine the intensity level of a sound after reverberation time if the original loudness is as follows:
    (a) 76 dB   (b) 67 dB   (c) 60 dB
45. The human ear can distinguish an echo from the sound that caused the echo if the two sounds are separated by an interval of about 0.10 s or more. At an air temperature of 22°C, what is the shortest length for an auditorium which could produce a distinguishable echo of a sound emanating from one end?

Review Assignment

1. The diagrams in Figure 12-36 show sound waveforms displayed on an oscilloscope. In each case, state a probable source of the sound and describe the sound heard.
2. Name the three characteristics of musical sounds, and state what each depends on.

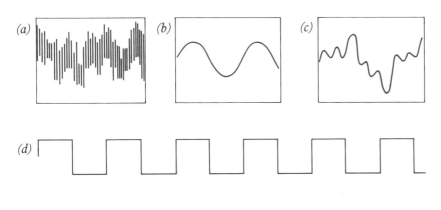

Figure 12-36

3. For a frequency of 900 Hz, state the frequency of the following notes:
   (a) three octaves higher   (b) two octaves lower
4. Arrange the following ratios of frequencies in decreasing order of dissonance.
   1:2; 3:2; 13:12; 5:4; 1:1
5. Use Table 12-2 and Figure 12-5 to calculate the frequency of a note one octave above each of the following:
   (a) F on the scientific scale
   (b) B on the scientific scale
   (c) C_1 on the musicians' scale
   (d) G_4 on the musicians' scale
6. A singer with an alto voice would like to sing a song using 400 Hz as the standard frequency. Calculate the frequencies of the notes of the octave below 400 Hz. (The scale has twelve intervals.)
7. One person has a threshold of hearing of 10 dB; another has a threshold of 30 dB at the same frequency. Which person has better hearing?
8. By what factor is a sound intensity level of 70 dB greater than one of 30 dB?
9. A 120 dB sound is reduced to 10^{-7} times the intensity level after 0.5 s. What is the new intensity level?
10. Refer to the graph in Figure 12-7. Determine the approximate frequency range of the sounds most easily heard by the human ear.
11. Describe practical ways to reduce noise pollution at a busy intersection in a city.
12. Assume that a 900 Hz note is the third harmonic of a sound. What are the first, second, and fourth harmonics?
13. Describe how the quality of a musical sound is affected by the harmonics structure of the sound.
14. What is the relationship between the frequency of a vibrating string and the string's length, diameter, tension, and density?
15. A string of length 80 cm and diameter 0.80 mm is under a tension of 64 N. When plucked, it emits a fundamental frequency of 660 Hz. What is the new frequency in each of these situations?
    (a) The length is decreased to 60 cm.
    (b) The diameter is decreased to 0.50 mm.
    (c) The tension is increased to 100 N.
    (d) All three factors in (a), (b), and (c) above are altered simultaneously.
16. Name the two main parts of a stringed instrument, and state the function of each part.
17. State examples of stringed instruments that
    (a) have frets
    (b) have no frets
    (c) are struck.
18. The note G_5 is played on a violin string. Describe ways in which the quality of the sound emitted can be altered.

19. For a column of air that is closed at one end, compare the amplitude of vibration of the air molecules at the two ends of the column.
20. The note B_4 ($f = 494$ Hz) is played at the open end of an air column that is closed at the opposite end. The air temperature is 22°C. Calculate the length of the air column for the first three resonant sounds.
21. **Activity** Determine the frequency of a tuning fork mounted on a wooden resonance box in your classroom. Then, using the air temperature in the room, calculate the length of the air column for the first resonant sound. Compare this value to the inside length of the wooden resonance box.
22. Say you are constructing a wooden resonance box open at both ends on which you want to mount a 320 Hz tuning fork. What is the minimum length of the box? (Assume a room temperature of 20°C.)
23. (a) List four methods of forcing air to vibrate in wind instruments.
    (b) Name an instrument for each method in (a) above.
24. An air column has a fundamental frequency of 330 Hz. What are the next two harmonics if the air column is (a) open at both ends and (b) closed at one end?
25. Name a percussion instrument that has
    (a) a single, indefinite pitch
    (b) multiple, definite pitches
    (c) a variable pitch.
26. In Section 12.5, a piano was described as a stringed instrument. Do you think it could also be described as a percussion instrument? Why or why not?
27. Describe methods by which the quality of a singer's voice can be controlled.
28. Electrical and electronic instruments differ in the way in which they create sound vibrations. Explain the difference.
29. For a person with equal sensitivity in both ears, would the direction from which a sound is coming be more easily identified if the sound has a high pitch or a low pitch? Explain your reasoning. (*Hint*: Consider the diffraction of sound waves.)
30. A synthesizer produces a constant-frequency sound with a delayed attack, then a sustained amplitude, and finally a slow decay. Draw a diagram of the waveform of the sound.
31. Sound tests are performed in a concert hall where echoes seem to be too powerful. The reverberation time for sounds in the frequency range of 2000 Hz to 4000 Hz is found to be 2.9 s, longer than the optimum value. What techniques could be used to decrease the reverberation time?
32. The impact sound from a drum registers 72 dB on a sound intensity level meter in an auditorium. After the auditorium's reverberation time has elapsed, what is the reading on the meter?
33. **Activity** Listen to the sound produced by a small loudspeaker connected to an audio frequency generator. Now cut a hole the size of the speaker in a large board or piece of cardboard. Hold the speaker at the hole and listen again. Use the concepts studied in this Unit on sound to explain what you hear.

Key Objectives

Having completed this chapter, you should now be able to do the following:

1. Describe scientifically the differences between noise and music.
2. Name the three characteristics of musical sounds, and describe what each depends on.
3. Describe the difference between consonance and dissonance.
4. State what consonance depends on.
5. Given the frequency of one musical note, calculate the frequency of a note an octave above or below it.
6. Recognize the scientific musical scale.
7. Recognize the musicians' scale, and state its standard frequency.
8. Given the nth root of the number two, calculate any n-interval equitempered musical scale based on any standard frequency.
9. State the units used to measure the intensity level of sounds.
10. Define the threshold of hearing and the threshold of pain.
11. Compare the intensities of any two sounds, given their intensity levels in decibels.
12. Describe the dangers of loud sounds.
13. Describe how changing the harmonic structure of a sound wave alters the quality of the musical sound.
14. Given the fundamental frequency of a vibrating object, calculate the frequencies of the higher harmonics.
15. State how the pitch of a vibrating string depends on the string's length, diameter, tension, and density.
16. Apply the equations that summarize the relationships stated in Key Objective 15.
17. Name the two main parts of every stringed instrument.
18. List various types of stringed instruments, and state how they are played.
19. Describe how resonance of sound waves is created in vibrating columns of air.
20. Given the wavelength of a sound resonating in an air column or the distance between resonant sounds, calculate the other quantity.
21. Determine the wavelength and speed of sound in air experimentally.
22. Describe four methods for making air in wind instruments vibrate.
23. List various types of wind instruments.
24. List various types of percussion instruments.
25. Describe the function of the body parts that produce and control the sound of the human voice.
26. Compare electrical and electronic instruments.
27. Describe methods of improving the quality of the sound of musical instruments.
28. Define acoustics, and describe factors affecting the acoustics of various chambers.
29. Define reverberation time, and state what factors affect it.
30. Given the original intensity level of a sound, calculate the intensity level after the reverberation time has elapsed.

UNIT V
Light and Colour

The information in this Unit will be useful if you plan a career in ophthalmology, optometry, physical medicine, pediatrics, ophthalmic technology, optics, photography, film-making, television, surveying, theatrics, fashion design, or interior decoration.

LIGHT AND COLOUR

CHAPTER 13
The Nature and Reflection of Light

What evidence indicates whether this photograph is upright or inverted? Would photography be possible if light were composed of particles emitted from the eyes?

What is light? Certain ancient Greeks, such as Plato (428 B.C.–348 B.C.), thought that light consisted of streams of particles emitted from the eye. If that is true, we might ask, why are we unable to "see" in the dark? And how could cameras take photographs of light, such as the one shown?

Other ancient Greeks, among them Aristotle, were of a different opinion. They thought that light consisted of tiny particles sent from objects to the eye. Both ideas were conjectures, and did not involve the close observation needed to understand why light acts the way it does.

In this chapter you will learn answers to the following questions: What is the general nature of light? How does light act when it is reflected off various objects? And how can we apply reflection of light in practical ways?

13.1 Sources of Light Energy

The ancient belief that light consists of streams of tiny particles was accepted for hundreds of years. However, we now know that **light** is a form of energy that is visible to the eye. It consists of tiny packages of energy which sometimes display properties of waves and sometimes properties of particles.

All objects that act as sources of light energy are called **luminous**. Luminous objects are classified according to the reason they emit light.

The most common luminous objects, called **incandescent sources**, emit light because they are at a high temperature. The sun, in which nuclear reactions produce extremely high temperatures, is our most important incandescent light source. It provides the energy needed for plants to grow, and it helps keep the temperature on the earth suitable for life. Other incandescent sources include fires, which gain a high temperature from chemical potential energy, and incandescent light bulbs, which gain a high temperature from electrical energy. See Figure 13-1.

Other types of luminous objects include those which are fluorescent, phosphorescent, chemiluminescent, and bioluminescent. **Fluorescent sources**, such as the gases used in neon and fluorescent lights, are luminous only when they are struck by high-energy waves or particles. **Phosphorescent materials** become luminous when struck by high-energy waves or particles, and remain luminous for a while. An oscilloscope screen gives off light for a short time after the oscilloscope is turned off; luminous dials on some clocks and watches emit light for several hours after they absorb energy. **Chemiluminescent materials** react chemically to produce light without a noticeable increase in temperature. Some safety lights emit light energy in this manner. **Bioluminescent animals**, such as fireflies and certain fish, can emit light energy because of a chemical reaction. See Figure 13-2.

Figure 13-1
The heating element on an electric stove emits light when it is at a high temperature.

Figure 13-2
Fireflies in the Waitomo Caves in New Zealand display bioluminescence.

LIGHT AND COLOUR

Certain light sources serve specific functions. For example, an electronic stroboscope, which can be used to study moving objects, emits short flashes of light at regular intervals. The photograph in Figure 13-3 shows a moving hand that is visible only when the strobe light in the otherwise dark room flashes on. The strobe light is luminous, but the hand is not. The hand becomes visible only when light energy bounces off it. Another special light source, the laser, will be discussed in Section 16.6.

Figure 13-3
A stroboscope is a special light source used to study moving objects such as the hand shown. In this case, the stroboscope flashed on every 0.10 s.

PRACTICE

1. Classify the luminous objects listed below as one of incandescent, fluorescent, phosphorescent, or chemiluminescent.
   (a) white-hot molten iron
   (b) material on a Halloween outfit that gives off light
   (c) television screen
   (d) liquid light (Two liquids, when combined, produce a faint glow in the dark.)
   (e) car headlight
   (f) lights in your classroom
2. List three reasons why light is important to us.
3. Electric lights in homes and schools are either incandescent or fluorescent. One type is much more energy efficient than the other. Explain which type is more efficient, and why. (*Hint:* Electrical energy changes to both heat and light energy. You might try touching a fluorescent light that has been on a long time.)
4. In the populated regions of Canada, the average power per unit area received from the sun during the year is 150 W/m². Given that rate, estimate the amount of energy (in megajoules) that falls *per annum* on a roof measuring 10 m by 20 m. Now calculate this amount of energy. (*Hint:* The equation for energy in terms of power and time was discussed in Chapter 6. In this case, area is also involved, so if you have difficulty with the problem, consider how area relates to the unit watts per square metre.)
5. **Activity** Set up a demonstration to view an example of **triboluminescence**, which is the emission of a faint glow of light from crystals being crushed. (The word stems from the Greek *tribein*, which means "to rub".) You will need a pair of pliers and some hard candy, such as a wintergreen Lifesaver. Darken the room completely, allow several minutes for your eyes to adapt to the dark, then crush the candy with the pliers. Describe what you observe. (Triboluminescence is neither common nor well understood. It occurs when hard materials containing certain sugars glow when crushed in the presence of gases such as nitrogen, a component of air.)
6. As you study the topic of light, you will learn more about the statement that light sometimes behaves like particles, sometimes like waves. Based on your understanding of the nature of particles and waves studied in previous chapters, predict some experiments that might be used to provide evidence for each type of behaviour. (*Hint:* Consider such concepts as mass, velocity, energy, reflection, diffraction, and interference.)

13.2 The Transmission of Light

Once light is emitted from an energy source, it travels through space or certain materials, displaying several properties or characteristics. Three of these properties are discussed in this section.

First consider the evidence that light travels in straight lines. A beam of light from a laser or projector, when shone in a dusty room, is seen to travel in a straight line, as Figure 13-4 shows.

Figure 13-4
This laser beam, which is made visible by being reflected off dust particles and the mirror, verifies that light travels in a straight line.

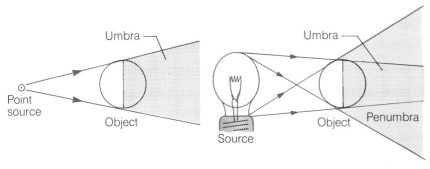

(a) Point source   (b) Large source

Figure 13-5
Shadows

Shadow formation also provides evidence that light travels in a straight line. If light from a point source is blocked by an object, a dark shadow called an **umbra** forms behind the object. See Figure 13-5(a). If light from a large source or from two separate sources is blocked by an object, the resulting shadow consists of two regions. The umbra is the darker region where no light falls, and the **penumbra** is the lighter part where some light falls. See Figure 13-5(b). Notice in Figure 13-5 that the lines drawn from each source represent paths of light called **light rays**. Light ray directions should always be shown in such diagrams.

Shadows in the solar system cause **eclipses**. An eclipse of the moon, called a **lunar eclipse**, occurs when the moon passes into the earth's shadow. An eclipse of the sun, or **solar eclipse**, occurs when the earth

LIGHT AND COLOUR

passes through the moon's shadow. Figure 13-6 illustrates a solar eclipse. Observers at position A on the earth see a **total eclipse** of the sun, while observers at B see a **partial eclipse**. Sometimes the moon is farther from the earth than the length of its shadow, so the umbra does not touch the earth, as shown in Figure 13-7. In this case, observers at A view an **annular eclipse**, in which the outer edge of the sun is still visible.

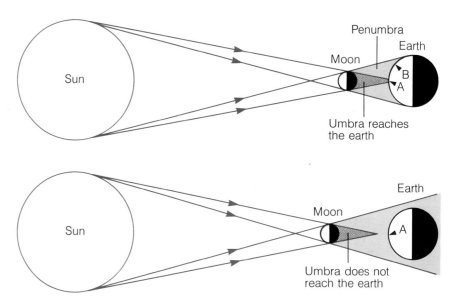

Figure 13-6
A Total Solar Eclipse

Figure 13-7
An Annular Eclipse

A particle model of light is used to explain properties of light acting like a particle; a wave model is used to explain properties of light acting like a wave. Does the straight-line propagation of light, for example, support either model? Based on our common experience with particles which can travel in straight lines, it is easy to imagine that if light were composed of particles, it could travel in a straight line. Based on our knowledge that sound, which is transmitted as a wave, can easily be diffracted around corners or obstacles, it is logical to assume that if light were composed of waves, it could do the same. Thus, at least when considering the formation of shadows, the wave model of light is not as adequate as the particle model.

Another property of light energy is that light does not need a material, or **medium**, in which to travel. Thus, light from the sun and stars can reach us through the vacuum of outer space. In fact, light energy travels fastest in a vacuum where there are no particles at all. This property also raises a difficulty for the wave model of light. Waves on ropes and water, as well as sound waves, require a medium through which to travel. The particle model, however, fits well, because there is no doubt that particles can travel easily through a vacuum. (In the nineteenth century, scientists who argued in favour of the wave theory hypothesized that there must be a material in outer space that transmits light. They named this unknown material "the ether".)

One other important property of light is its tremendously fast speed of propagation. Even today, this speed is difficult to measure with ordinary equipment. The first fairly successful measurement of the speed of light was made in 1676 by a Danish astronomer, Olaf Roemer (1644–1710). Roemer's unique method involved many careful observations through a telescope of the moons of Jupiter. Jupiter is a large planet located much farther from the sun than the earth, as depicted in Figure 13-8. The figure also shows the earth revolving around the sun, and Jupiter's innermost moon, Io, just emerging from Jupiter's umbra. With the earth in position A, Roemer made note of the time when Io emerged from the umbra. Then he observed that (about) 42.5 h later Io again emerged from the umbra. Roemer expected that, as the months passed and more measurements were taken, the period (42.5 h) would remain constant. To his surprise, and because of his careful observations, he discovered that as the earth progressed in its orbit to position B, the amount of time actually increased. Roemer reasoned, correctly, that the time difference resulted from the fact that the light travelling from Io to the earth had to cover an extra distance from A to B. Using his data, Roemer determined that the amount of time for light to travel a distance equal to the diameter of the earth's orbit (from C to D) was almost 22 min (or about 1300 s). The orbital diameter was believed at the time to be about 3×10^{11} m, so Roemer calculated the speed of light to be

$v = d/t$ (speed = distance/time)
  $= (3 \times 10^{11} \text{ m})/(1.3 \times 10^3 \text{ s})$
  $= 2.3 \times 10^8 \text{ m/s}.$

This high speed (230 million kilometres per second!) is actually lower than the currently accepted value.

Since Roemer's discovery, several other methods of determining the speed of light have been tried. A technique that provided an accurate result was carried out by an American scientist, Albert Michelson (1852–1931). He set up a light source on one mountain in California and a mirror on a second mountain 35.0 km away. He then pulsed a

Figure 13-8
Roemer's Method of Finding the Speed of Light

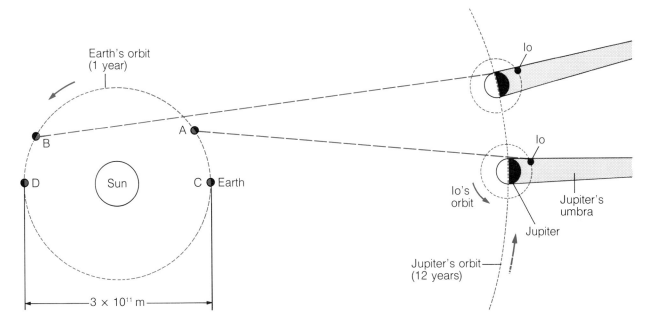

LIGHT AND COLOUR

Did You Know?
The speed of light in a vacuum, found using a laser, is now known to 10 significant digits. Its value is 299 792 456.2 m/s (± 0.1 m/s).

light beam at a rotating octagonal mirror, as shown in Figure 13-9, and timed how long it took the light to travel from A to M and back to mirror B, which had during that time moved into position C. Thus, the octagonal mirror had rotated 1/8 of a turn. Michelson discovered that in order for the experiment to work, the octagonal mirror had to rotate at a frequency of 535 Hz, in other words, with a period of rotation of 1.87×10^{-3} s. (Recall from earlier work that period is the reciprocal of frequency.) It is left as an exercise to prove that these values yield a speed of light close to today's accepted value of 3.0×10^8 m/s, somewhat higher than Roemer's value.

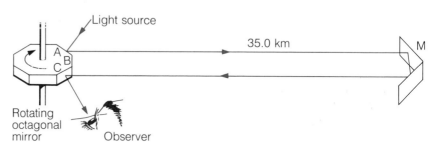

Figure 13-9
Michelson's Method of Finding the Speed of Light

Figure 13-10
A Ray Box

PRACTICE

(*Note*: Unless otherwise stated, assume that the speed of light in a vacuum, to three significant digits, is 3.00×10^8 m/s.)

7. **Activity** Design and carry out an experiment to illustrate the formation of shadows behind a small object such as a rubber stopper. A convenient light source for this type of experiment is called a ray box, shown in Figure 13-10. Aim light from the ray box or boxes from two or three directions toward the object. Draw the pattern observed, including the umbra and any penumbras.
8. Draw a diagram illustrating a lunar eclipse.
9. Calculate the percentage error of Roemer's method of obtaining the speed of light, assuming that the accepted value is 3.0×10^8 m/s. (Refer to Section 2.4.)

10. If light from the sun takes 5.0×10^2 s to reach us, what is the radius of the earth's orbit? (Assume that the orbit is circular.)
11. The average distance from the earth to the moon is 3.84×10^8 m. How long does it take a beam of laser light to travel from the earth to the moon and back again?
12. Use the data given in the explanation of Michelson's experiment to calculate his determination of the speed of light.
13. What would Michelson have observed in his experiment if he had doubled the frequency of rotation of the octagonal mirror? Would his calculations have yielded the same result?
14. **Activity** Between the time of Roemer and Michelson, other scientists developed their own ways of determining the speed of light. Research and report on the methods used by these French scientists:
    (a) Armand Fizeau, 1819–1896
    (b) Jean Foucault, 1819–1868
15. The distance light travels through a vacuum in one year is called a **light year**.
    (a) Calculate this distance.
    (b) The distance from the earth to the Andromeda Galaxy is about 1.9×10^{22} m. Express this distance in light years.
16. Estimate the number of times you think light could travel around the earth in one second, assuming that the light could somehow be bent to do so. Now calculate the value, given that the earth's radius is 6.38×10^6 m.
17. **Activity** If a light intensity meter is available, use it to perform an experiment to verify that the intensity of light from a point source varies inversely as the square of the distance from the source. Try using a projector beam to approximate a point source. Use a graph to analyse the results, and describe sources of error in the experiment. This relationship represents yet another property of light energy.

13.3 The Interaction of Light with Matter

Materials may be classified according to how they treat light that strikes them. **Transparent materials**, such as clear glass and shallow water, allow light to be transmitted easily. A clear image can be seen through them. **Translucent materials**, such as waxed paper and stained glass, allow the transmission of some light, but no clear image can be seen through them. **Opaque materials**, such as concrete and wood, allow no light to pass through; all the light is either absorbed or reflected, and no image at all is seen through them.

Light energy is easily absorbed by dark, dull surfaces, and easily reflected by shiny, smooth surfaces. A device for demonstrating this fact is *Crookes' radiometer*, named after Sir William Crookes, an English scientist who lived from 1832–1919. The radiometer, shown in Figure 13-11, has a set of four vanes balanced on a pivot. One side of

Figure 13-11
A Radiometer

Figure 13-12
Three Types of Reflection

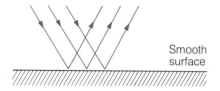

(a) Regular reflection

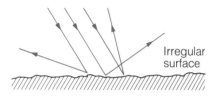

(b) Diffuse reflection

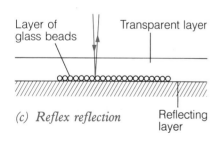

(c) Reflex reflection

each vane is black; the other is white or silver. Most, but not all of the air particles have been evacuated from the device. When bright light is aimed toward the radiometer, the vanes spin rapidly. Why does this happen? The shiny surfaces reflect light, and the dark surfaces absorb light energy which changes to thermal energy. This thermal energy warms nearby air molecules, increasing their kinetic energy. The fast-moving air molecules then strike the dark surface, forcing the vanes to spin in the appropriate direction.

When light is reflected off a surface, the type of reflection depends on the smoothness of the surface. **Regular** or **specular reflection** occurs when light strikes a smooth, shiny surface such as a mirror. See Figure 13-12(a). **Irregular** or **diffuse reflection** occurs when light strikes an irregular surface such as a painted ceiling; see Figure 13-12(b). **Retro-** or **reflex reflection** occurs when light from a source strikes a specially manufactured surface and is reflected back to the source. Traffic signs and safety garments are made with retro-reflective materials that bounce light back toward vehicle headlights. Such materials consist of three layers: an inner reflecting layer, a thin layer of tiny glass beads, and an outer transparent layer. See Figure 13-12(c).

When light is transmitted through a substance, it sometimes undergoes other, less common phenomena. For example, light that enters water droplets or particles of flesh can be scattered or diffracted. Medical

Figure 13-13
In this time-exposure photograph, the diffraction of light is most evident wherever the light sources are brightest.

researchers use this fact to discover such problems as a collapsed lung in newborn infants. When light passes through thin cloth or through light filters with closely spaced lines, patterns resulting from the diffraction of light appear. The photograph in Figure 13-13 was taken with a diffraction filter. (Diffraction of water waves and sound waves was discussed in Chapter 10.) Here we have evidence supporting the wave model of light. Light can be diffracted through openings if the openings are spaced very closely. This means that the light waves must have a very small wavelength.

Scientists continue to study the interaction of light energy with matter in order to learn more about the dual wave/particle nature of light.

PRACTICE
18. Name three examples other than those in this section of each of the following types of materials:
    (a) transparent
    (b) translucent
    (c) opaque
19. What type of clothing would be most appropriate on a hot, sunny day? Why?
20. Name at least one example other than those in this section, of (a) specular reflection and (b) diffuse reflection.
21. Does the action of a radiometer tend to support a particle model or a wave model of light? (It may help you to know that before William Crookes tried his own device, he thought the vanes would spin in a direction opposite to what was later observed.)
22. **Activity** View and describe the diffraction pattern of light produced when an intense, monochromatic (single-colour) source of light passes by a very sharp edge, as shown in Figure 13-14, or through a very narrow slit. Does this activity provide evidence in favour of the wave or the particle model of light? Explain.

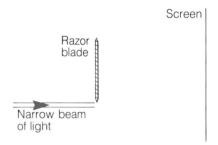

Figure 13-14

Experiment 13A: Plane Mirrors

INTRODUCTION
The following definitions, illustrated in Figure 13-15, are important in this experiment. An **incident ray** is a ray of light travelling from a source to some object, such as a mirror. A **reflected ray** is a ray of light that has bounced off an object. A **normal** (N) is a line perpendicular to the surface where the incident ray strikes; it is not a light ray. The **angle of incidence** ($\angle i$) is the angle between the incident ray and the normal. The **angle of reflection** ($\angle r$) is the angle between the reflected ray and the normal.

PURPOSE
(a) To study the law of reflection for plane mirrors.
(b) To learn how to locate an image in a plane mirror.

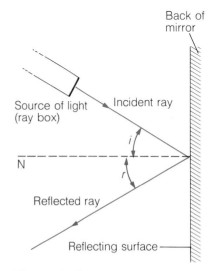

Figure 13-15
Plane Mirror Definitions

LIGHT AND COLOUR

APPARATUS

ray box with a single-slit window; plane mirror; protractor

Figure 13-16
Set-ups for Experiment 13A

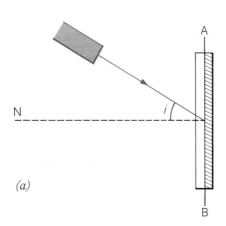

(a)

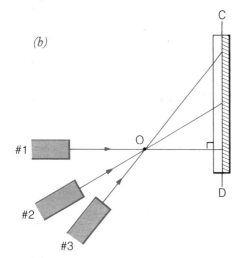

(b)

PROCEDURE

1. Draw a line AB on a piece of paper and place the mirror so that its reflecting surface, which is probably at the back of the glass, lies along AB, as in Figure 13-16(a). Aim an incident ray from the ray box at the mirror and use small dots to mark the incident and reflected rays. Remove the ray box and mirror and use a straight-edge to draw the rays.
2. Use a protractor to draw a normal from the point where the rays meet at the mirror. Label and measure the angles of incidence and reflection.
3. Repeat Procedure Steps 1 and 2, using two new diagrams and distinctly different angles.
4. Draw a line CD and a point object O on a piece of paper, as shown in Figure 13-16(b). Locate the image (O') of the object (O) in the mirror in the following way.
   (a) Aim a single ray [ray #1 in Figure 13-16(b)] through O, so that it strikes the mirror and reflects back onto itself. Draw the ray(s). Now move the ray box twice more (rays #2 and #3) and draw the incident and reflected rays.
   (b) Remove the mirror and extend the reflected rays straight back behind the mirror, because that is where they appear to come from.
   (c) Find the point at which the extended lines intersect. This is the location of the image (O'). Label and measure the distance from O to the mirror and the distance from O' to the mirror.
5. Look into a plane mirror and describe the image. Is it upright or inverted? Is it laterally inverted; *i.e.*, if you wink your right eye, which eye, from the point of view of the image, winks? How does the image size compare to the object size?

ANALYSIS

1. How does the angle of incidence compare to the angle of reflection? (Your answer should lead to a conclusion about a law of reflection for plane mirrors.)
2. State the angle of reflection if the angle of incidence is (a) 23°; (b) 78°; (c) 0°.
3. In a plane mirror, how does the distance from the object to the reflecting surface compare with the distance from the image to the reflecting surface?
4. A person is moving toward a plane mirror at a speed of 20 cm/s.
   (a) At what speed is the person's image approaching the mirror?
   (b) At what speed are the person and the image approaching each other?

CONCLUSIONS . . .

THE NATURE AND REFLECTION OF LIGHT

13.4 Ray Diagrams for Plane Mirrors

The purpose of a ray diagram for a plane mirror is to determine the location and characteristics of the image of an object seen in the mirror. The rays in the diagram are drawn so that they obey the following two **laws of reflection for plane mirrors**:

(1) **The angle of incidence equals the angle of reflection.**
(2) **The incident ray, the normal, and the reflected ray all lie in the same plane.**

Consider object OB in Figure 13-17(a). From point O two separate incident rays are drawn, then the corresponding normals and reflected rays are drawn. The location of the image (O′) of point O is found where the reflected rays intersect. However, the reflected rays are diverging (spreading apart), so they cannot meet in front of the mirror. Thus, they are extended straight back behind the mirror (using broken lines) until they intersect at point O′. To an observer in front of the mirror, this is simply the position from which the light appears to be coming. A similar procedure is used to find the image (B′) of point B. Then the total image, O′B′, is drawn.

In Figure 13-17(a), the distance from any point on the object to the mirror equals the distance from the corresponding point in the image to the mirror. This provides a shortcut for finding an image in a plane mirror, shown in Figure 13-17(b).

Figure 13-17
Locating an Image in a Plane Mirror

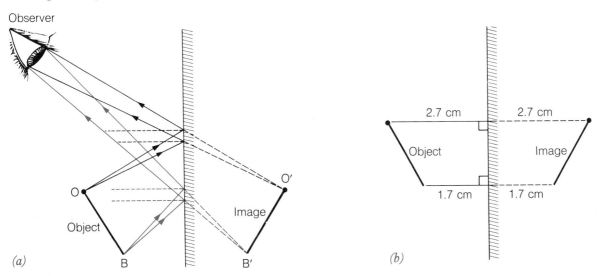

Once an image has been located in a ray diagram, we can describe it. Four **characteristics** are generally needed to describe each image.

(a) **Attitude** The attitude of the image is either upright or inverted relative to the object.

(b) **Magnification** The magnification (M) of the mirror is found by calculating the ratio of the image height (h_i) to the object height (h_o). Thus,

$$M = \frac{h_i}{h_o}.$$

If $M > 1.0$, the image is larger than the object. If $M < 1.0$, the image is smaller, and if $M = 1.0$ the image and object are the same size. Notice that magnification has no units because it is a ratio of heights.

(c) **Type** An image is either real or virtual. A **real image** can be placed onto a screen, as you will observe later in the chapter. The light rays that produce a real image actually meet each other. A **virtual image**, also called an *imaginary image*, cannot be placed onto a screen. The light rays which create such an image never actually intersect; they only appear to intersect behind the mirror. To see a virtual image in a mirror, you must look into the mirror. In a single mirror a virtual image is always upright.

(d) **Location** The image is located either in front of or behind the optical device. The image distance can be compared to the object distance.

The above list of four characteristics of an image does not include lateral (left-to-right) inversion. This type of inversion, common in all single plane mirrors, is not revealed in a ray diagram. From the point of view of an independent observer, what is on the right side of the object ends up on the right side of the image; thus, lateral inversion is not a true inversion.

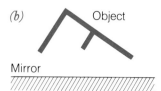

Figure 13-18

PRACTICE

23. Describe the image in a plane mirror by stating its
    (a) attitude; (b) magnification; (c) type; and (d) location.

24. Copy each of the diagrams in Figure 13-18 into your notebook and locate the image of the object.

25. A 160 cm girl is standing 50 cm from a long plane mirror. What is
    (a) the distance from the image to the mirror?
    (b) the distance from the girl to her image?

25. **Activity** View and describe the formation of a real image by placing a converging lens (a simple magnifying glass) between a source of light and a screen. (Adjust the position of the lens until you obtain a clear image of the light source on the screen.)

27. Figure 13-19 shows a point object (O) in front of two mirrors perpendicular to each other.
    (a) Draw a detailed ray diagram to locate all the images produced.
    (b) **Activity** Place two small mirrors onto your diagram in (a) to check your diagram.

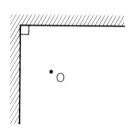

Figure 13-19

13.5 Applications of Plane Mirrors

Plane mirrors are commonly used as looking glasses, but they also have other interesting applications. For example, some machines in an arcade shooting gallery utilize a plane mirror. Figure 13-20 shows the type of machine in which a rifle shoots light that is reflected off a mirror at a target below. A direct hit is recorded if the light strikes a sensitive part of the target. To the observer, the target image appears straight ahead and at the distance shown in the figure.

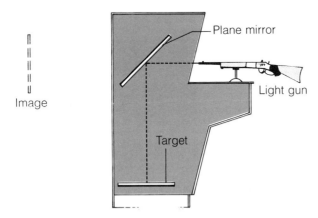

Figure 13-20
Changing the Direction of Light in a Shooting Gallery

"One-way mirrors", special plane mirrors, also have their uses. For example, they are used by law enforcement officers in interrogation rooms. The suspect stands in a brightly lit room looking toward a glass wall that reflects much of the light back into the room. The witness or any other observer watches from an adjacent room which is dimly lit. Light from the bright room can enter the dark room, but very little light travels the other way. Often the glass wall has a thin layer of a reflective material on the side adjacent to the dark room.

Plane mirrors are used in the field of medicine as well. Dentists use front-surfaced mirrors to observe hard-to-see regions of the mouth. Eye doctors view the inside of a patient's eye by using an ophthalmoscope. This device, illustrated in Figure 13-21, has a small but bright light source that directs light to a small mirror. The light travels from the mirror to the eye, and the physician views the illuminated eye through a small opening.

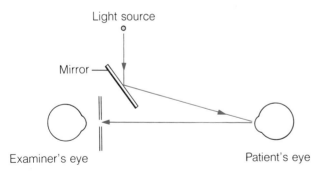

Figure 13-21
The Ophthalmoscope

LIGHT AND COLOUR

Figure 13-22
A plane mirror is used on a dial of an electric meter.

Meters, among them voltmeters and electric meters, often have plane mirrors behind the needle to prevent parallax (Figure 13-22).

Another application of plane mirrors is designed on the principle that when two mirrors are placed at an angle to each other, multiple images appear. The smaller the angle, the greater is the number of images. At 90° between the mirrors, three images are formed (see Practice Question 27); at 60°, five images appear. A kaleidoscope, shown in Figure 13-23, has two plane mirrors at an angle of 60° to one other. Coloured glass crystals are placed between the mirrors. Five identical images of the crystal pattern, in addition to the crystal pattern itself, can be seen at a given time. As the kaleidoscope is turned, new patterns form.

Figure 13-23
The Kaleidoscope

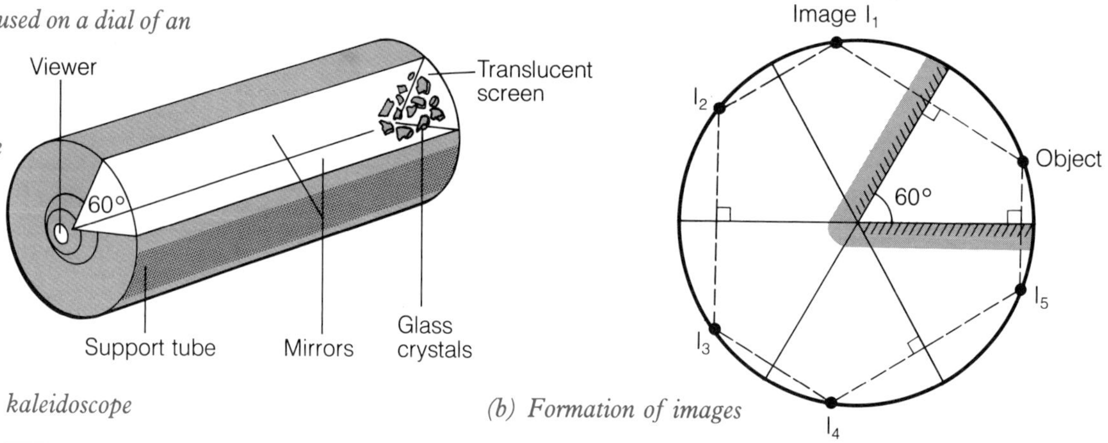

(a) Design of the kaleidoscope

(b) Formation of images

Did You Know?
As engineers race to reduce the size and aerodynamic resistance of automobile components, many new designs are being developed. One alternative to conventional headlamps uses a plane mirror like that in Figure 13-24. The face of this headlamp is only 8 cm wide and 4 cm high.

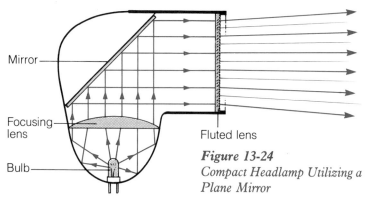

Figure 13-24
Compact Headlamp Utilizing a Plane Mirror

PRACTICE

28. Describe the circumstances required for a window in your home to act as a one-way mirror in the following situations:
    (a) viewed from the inside
    (b) viewed from the outside
29. Describe some applications of plane mirrors other than those already mentioned.
30. **Activity** An equation that can be used to determine the number of images formed by two mirrors at specific angles to each other is $n = (360°/a) - 1$, where n is the number of images and a is the angle between the plane mirrors. Design and perform an experiment to determine whether or not this equation works for these angles: 180°, 90°, 60°, and 45°.

13.6 Curved Mirrors

Although plane mirrors are very common, they are not the only type of mirror used. Curved mirrors also have many applications.

Curved mirrors are classified according to how they reflect light rays from a distant source. A **converging mirror** causes such light rays to converge or come together. Its reflecting surface is concave (curved inward), as shown in Figure 13-25(a). A **diverging mirror** causes the light rays to diverge or spread apart. Its reflecting surface is convex (curved outward), as in Figure 13-25(b). Notice also in Figure 13-25 that point C, called the **centre of curvature**, is the imaginary centre of the circle of which the mirror is a section. Distance r, the **radius of curvature**, is the distance from C to the reflecting surface.

Other definitions for curved mirrors are illustrated in Figure 13-26. The **principal axis** of a curved mirror is a line drawn through C which passes through the middle, or **vertex** (V), of the mirror. The **focal point** (F) is the position where incident rays parallel and close to the principal axis meet, or appear to meet, after they are reflected. The **focal length** (f) is the distance from the focal point to the vertex of the mirror; i.e., the distance VF.

Figure 13-25
Curved Mirrors

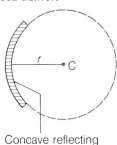

Concave reflecting surface

(a) Converging

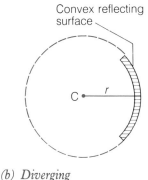

Convex reflecting surface

(b) Diverging

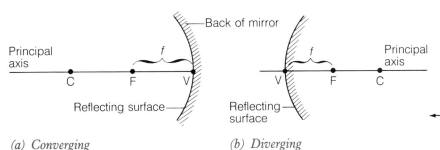

(a) Converging   (b) Diverging

← *Figure 13-26*
Curved Mirror Definitions

Curved mirrors that are shaped as if they had been cut out of a regular cyclinder are called **cylindrical** or **circular mirrors**. Such mirrors were shown in Figure 13-25. Curved mirrors which are designed as part of a sphere are called **spherical mirrors**. Curved mirrors are also designed in yet another way – in the shape of a curve called a parabola, illustrated in Figure 13-27. Such mirrors, called **parabolic mirrors**, have special functions that you will discover in the rest of the chapter.

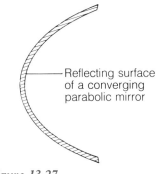

Reflecting surface of a converging parabolic mirror

Figure 13-27
A Parabolic Mirror

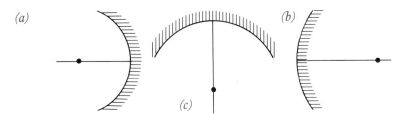

Figure 13-28

PRACTICE
31. For each circular mirror shown in Figure 13-28,
    (a) state whether it is converging or diverging
    (b) measure its radius of curvature.

Experiment 13B: Converging Mirrors

INTRODUCTION
For certain parts of this experiment you will need to draw a normal to a circular curved mirror. To do so, you can use the fact that, for all circles, the normals intersect at the centre of curvature.

In the Purpose of this experiment, part (d) is optional, but it reveals an important problem of circular converging mirrors called **circular aberration**. The problem can be corrected using a parabolic mirror. (The expression "spherical aberration" applies to spherical mirrors.)

PURPOSE
To do the following for a converging mirror:
(a) verify the laws of reflection
(b) find the ratio of the focal length to the radius of curvature
(c) find three convenient rules for light rays striking the mirror
(d) find the cause of and correction for circular aberration
(e) determine the characteristics of images seen in the mirror

APPARATUS
ray box with single-slit and triple-slit windows; three converging mirrors (circular, parabolic, and spherical)

PROCEDURE
1. Set the circular mirror flat on a piece of paper and **draw** its shape. Aim a single light ray so that it is reflected onto itself, and draw the ray. Move the ray box around to find a second single ray that is reflected onto itself, as shown in Figure 13-29(a). Repeat this procedure for a third ray. Then locate the intersection of the reflected rays. Because the rays are normal, the point of intersection is the centre of curvature. Measure the radius of curvature and label the diagram.
2. Use the diagram from Procedure Step 1 to determine the laws of reflection for a converging mirror. Aim a single ray so that it strikes the mirror where one of the normals does, but not along the normal. Draw the reflected ray and measure the angles of incidence and reflection. Repeat this procedure at least once more, using a distinctly different angle of incidence. Label your diagram.
3. Start a new diagram for this step. Aim three parallel rays of light toward the mirror so that the middle ray strikes the vertex and is reflected onto itself, as in Figure 13-29(b). Draw all the rays and label the point of intersection of the reflected rays, F. Measure and label the focal length, f.
4. Start a new diagram again. Aim a single ray toward the vertex of the mirror so that it is reflected onto itself, as in Figure 13-29(c). Draw in this line, and label it as the principal axis. Use the distances you discovered in Steps 1 and 3 above to label points C and F. Then, using a single ray for each, determine three important rules for con-

Figure 13-29
Procedures for Experiment 13B

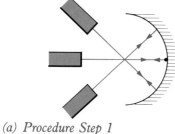

(a) Procedure Step 1

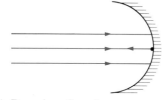

(b) Procedure Step 3

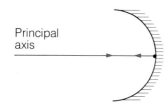

(c) Procedure Step 4

verging mirror reflection. Draw all the rays on your diagram. The rays used to determine the rules are as follows:
   (a) an incident ray parallel to the principal axis
   (b) an incident ray through the focal point
   (c) an incident ray through the centre of curvature at a small angle to the principal axis
5. Using the diagram from Step 3, aim another single ray toward the mirror parallel to the original rays but quite far away from the principal axis. Draw the reflected ray. You will likely notice a problem called **aberration**. (*Hint*: Does the reflected ray pass through the focal point?) Now use three or four parallel rays aimed toward a converging parabolic mirror, as shown in Figure 13-30, to determine how it corrects for aberration. Draw a ray diagram of the parabolic reflection.
6. With a spherical converging mirror, determine the attitude, type, and approximate magnification of the image seen when the mirror is held at these locations:
   (a) at a large distance from you
   (b) fairly close to your face

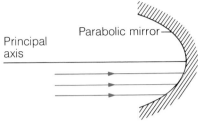

Figure 13-30
Correcting for Circular Aberration

ANALYSIS
1. What are the laws of reflection for converging mirrors? How do they compare to the laws of reflection for plane mirrors?
2. What is the ratio of the focal length to the radius of curvature for a circular converging mirror?
3. What is the focal length of a circular converging mirror whose radius of curvature is as follows?
   (a) 12 cm   (b) 60 mm   (c) 4.6 cm
4. What is the radius of curvature of a circular converging mirror whose focal length is as follows?
   (a) 2.1 cm   (b) 25 mm   (c) 5.8 cm
5. Describe where each incident ray will be reflected in Figure 13-31.
6. Describe the characteristics of each converging mirror image shown in Figure 13-32.

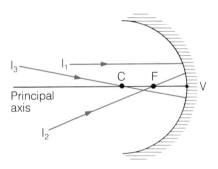

Figure 13-31

Figure 13-32
Images in Converging Mirrors

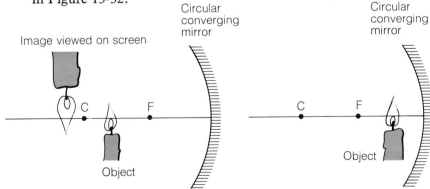

Image seen when looking into the mirror

(a) Object located between the centre of curvature and the primary focal point

(b) Object located inside the primary focal point

CONCLUSIONS . . .

LIGHT AND COLOUR

Figure 13-33
Set-ups for Experiment 13C

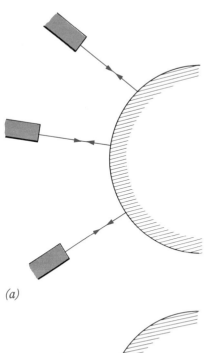

(a)

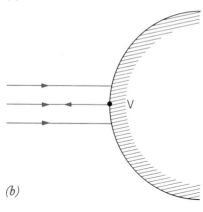

(b)

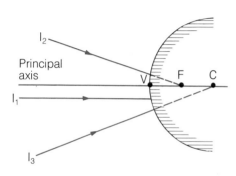

Figure 13-34

Experiment 13C: Diverging Mirrors

INTRODUCTION
In the previous experiment you learned how light behaves when it is reflected from a converging mirror. The Procedure for the present experiment is almost identical to that for Experiment 13B, with the exceptions that a diverging mirror is used here, and the problem of aberration is not investigated. As you will learn in this experiment, light rays that are reflected off a diverging mirror spread apart; as a result, they do not meet. In order to make such rays meet, or at least appear to meet, you must extend the reflected rays behind the mirror, just as you did for plane mirrors.

PURPOSE
To do the following for a diverging mirror:
(a) find the ratio of the focal length to the radius of curvature
(b) verify the laws of reflection
(c) find three convenient rules for light rays striking the mirror
(d) determine the characteristics of images seen in the mirror

APPARATUS
ray box with single-slit and triple-slit windows; two diverging mirrors (one circular and one spherical)

PROCEDURE
1. Carry out a procedure to determine Purpose (a). Use as references Figures 13-33(a) and (b) as well as Procedure Steps 1 and 3 in the previous experiment.
2. Carry out a procedure to determine Purpose (b), using as reference Procedure 2 in the previous experiment.
3. Determine three important rules for diverging mirror reflection and illustrate them on a separate diagram. To determine these rules, use the following rays:
   (a) an incident ray parallel to the principal axis
   (b) an incident ray aimed toward F
   (c) an incident ray aimed toward C
4. Examine the image of an object in a spherical diverging mirror. Describe the image, stating its attitude, type, and approximate magnification.

ANALYSIS
1. How does the ratio of the focal length to the radius of curvature for diverging mirrors compare to the same ratio for converging mirrors?
2. How do the laws of reflection for diverging mirrors compare to the laws of reflection for other mirrors?
3. Describe where each ray in Figure 13-34 will be reflected.
4. The focal point of a diverging mirror can be called a virtual focal point. Explain why.

CONCLUSIONS . . .

THE NATURE AND REFLECTION OF LIGHT

13.7 Ray Diagrams for Curved Mirrors

A ray diagram can be used to locate the image of an object situated in front of a curved mirror. The method you will use here applies the rules for curved mirror reflection that you learned from doing the previous two experiments. (The rules are actually based on the laws of reflection for all mirrors.)

In the Sample Problems below, note that the reflected rays must meet in order for a real image to be formed. In the second one, the reflected rays do not meet, but their extensions do, behind the mirror. Therefore, this image is virtual.

SAMPLE PROBLEM 1
Use a ray diagram to find the image of a 1.5 cm high object located 7.4 cm from a circular converging mirror with a focal length of 1.8 cm. State the attitude, type, location, and magnification of the image.

Solution
If $f = 1.8$ cm, then $r = 3.6$ cm. Use a compass to draw a curve with $r = 3.6$ cm, then locate C and F in the diagram. Draw the object, in this case an arrow, resting on the principal axis 7.4 cm from the vertex of the mirror. Use three incident and reflected rays, according to the rules, to locate the image. (At least two rays are required to find an image; the third ray can act as a check.)

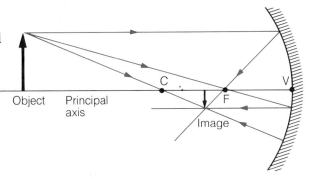

The image is inverted, real, and located between the focal point and the centre of curvature. Its magnification is

$$M = \frac{h_i}{h_o} = \frac{0.50 \text{ cm}}{1.5 \text{ cm}} = 0.33.$$

SAMPLE PROBLEM 2
Repeat Sample Problem 1, using a circular diverging mirror and an object distance of 4.5 cm and a focal length of 2.3 cm.

Solution
The image is upright, virtual, and located behind the mirror, between the vertex and the focal point. Its magnification is

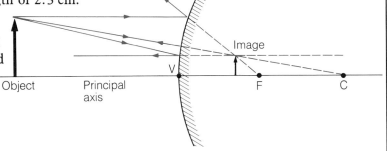

$$M = \frac{h_i}{h_o} = \frac{0.50 \text{ cm}}{1.5 \text{ cm}} = 0.33.$$

LIGHT AND COLOUR

PRACTICE

32. Draw a fully-labelled ray diagram to locate the image of the object for each situation described in Table 13-1. In each diagram use two rules. For each image found, state its attitude, type, and location, and calculate its magnification.

Table 13-1

Type of Mirror	Focal length (cm)	Height of Object (cm)	Distance from Object to Mirror (cm)
(a) converging	3.0	1.5	12.0
(b) converging	4.0	2.0	8.0
(c) converging	4.0	1.5	6.0
(d) converging	4.0	1.0	4.0
(e) converging	4.0	1.5	2.0
(f) diverging	3.0	2.0	10.0
(g) diverging	3.0	1.5	3.0

33. Based on your ray diagrams in Practice Question 32, state conclusions about drawing ray diagrams to locate images in curved mirrors.
34. Use a ray diagram similar to the one shown in Sample Problem 1 to verify that $\frac{1}{f} = \frac{1}{d_o} + \frac{1}{d_i}$, where d_o is the distance from the object to the vertex and d_i is the distance from the image to the vertex. (*Hint*: Omit the ray that passes through C, and use the ratios of h_i/h_o for two sets of similar triangles in the ray diagram. Use a small object to reduce possible error.)

13.8 Applications of Curved Mirrors

Converging Mirrors

Figure 13-35 shows four uses of converging mirrors. Such applications require a parabolic rather than a cylindrical or spherical shape.

A common type of headlight, shown in Figure 13-35(a), has its source of light near the focal point of the mirror. Light is reflected off the mirror to form a directed beam.

Converging mirrors are also used to concentrate sunlight. A solar cooker, illustrated in Figure 13-35(b), has a cooking pot located at the focal point of the mirror. Figure 13-35(c) demonstrates the way in which a converging mirror is used in a reflecting telescope. Huge telescopes up to 6.0 m in diameter collect much light, to allow astronomers to view distant stars which are not visible through smaller telescopes. Figure 13-35(d) shows the largest solar collector in the world, located in the Pyrenees Mountains in Southern France. Several large plane mirrors on a nearby hillside (not shown in the photograph) can be adjusted to reflect the sun's light to the eight-storey converging mirror. The mirror directs the light to a concentrated area in the smaller building, where scientific research is conducted.

Did You Know?
Researchers at Québec's Laval University plan to build a reflecting telescope using a liquid mirror. The mirror would consist of a layer of liquid mercury about 30 m in diameter. It would be placed horizontally to prevent spillage, and would be rotated slowly to produce a concave shape. Since the telescope would be aimed vertically, only a brief view of any one part of the sky would be possible. Thus, images photographed on successive nights would be added by computer to obtain a usable image.

Figure 13-35
Applications of Converging Mirrors

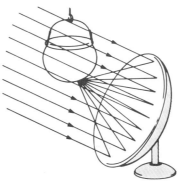

(a) Headlamp of a car   (b) Solar cooker

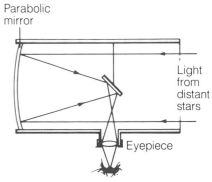

(c) Reflecting telescope

(d) Solar reflector

Diverging Mirrors

Images in diverging mirrors are always upright and smaller than the object. Therefore, diverging mirrors allow a viewer to see over a wide angle. Such mirrors are used in stores to discourage shoplifting, and as rear-view mirrors on trucks, buses, and motorcycles (Figure 13-36). When using a diverging mirror, a driver must be careful because a vehicle is much closer than it appears in the mirror.

LIGHT AND COLOUR

Figure 13-36
A diverging mirror gives a wide-angle view.

PRACTICE

35. List uses of converging mirrors other than those mentioned in this section.
36. A statement printed on the diverging mirror attached to the front right-hand door of a car reads "CAUTION: Objects in mirror are closer than they appear."
    (a) What does the statement mean?
    (b) How could the statement be improved in order to state what it really means?

Review Assignment

1. List the properties of light examined in this chapter.
2. State four examples of incandescent objects, and in each case name the type of energy that is being converted into light energy.
3. What evidence do we have that light travels in straight lines?
4. What conditions are necessary in order for each of the following types of eclipses to occur?
   (a) lunar (b) solar (c) annular
5. (a) What is the speed of light in a vacuum?
   (b) At that speed, how far can light travel in one millisecond?
6. How long does it take light to travel across Canada, a distance of 6.0×10^3 km?
7. Our galaxy, the Milky Way Galaxy, is approximately 1×10^5 light years across. Express this enormous distance in metres.

8. Approximately 400 years ago, Galileo Galilei tried to measure the speed of light in air using two lanterns, which we will name A and B, separated by a distance of several kilometres. Galileo uncovered A for an instant; when his partner saw the light from A, he uncovered B for an instant. Galileo measured the time between his initial sending of the light and the receiving of the light from B. From his observations, he concluded that light travels either instantaneously or extremely fast. What were some sources of error or uncertainty in Galileo's experiment?
9. Compare the effect of light on each pair of materials listed.
   (a) a black surface; a white surface
   (b) a transparent material; an opaque material
   (c) a smooth surface; a rough surface
   (d) a dull surface; a shiny surface
10. State the characteristics of a substance which prevents glare from reflected light.
11. State the laws of reflection for plane mirrors.
12. If you stand 60 cm in front of a plane mirror, how far are you from your image in the mirror?

Figure 13-37

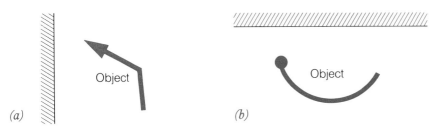

Figure 13-38

13. Each diagram in Figure 13-37 shows an object in front of a plane mirror. Draw the diagrams in your notebook and locate the images.
14. For the diagram shown in Figure 13-38, predict the image that you will observe when you place a plane mirror along edge AB, then AC, and finally BD. Verify your predictions.
15. Describe the differences between a real image and a virtual image.
16. Figure 13-39 shows the image of an ordinary clock as seen in a mirror.
    (a) Which way are the hands turning?
    (b) What is the correct time?
17. A box is painted black on the inside, and a jar and a piece of glass are placed as in Figure 13-40.
    (a) Copy the figure into your notebook, and show where a candle should be placed to give an observer the illusion that it is in the jar.
    (b) Complete a ray diagram to explain your answer in (a).
18. A patient must be 6.0 m from a certain eye chart in order for the eye test to be valid. The chart is available with either normal or backwards printing. How can a doctor whose office is only 4.0 m long arrange the situation using a plane mirror? Draw a diagram to ilustrate your reasoning.

Figure 13-39

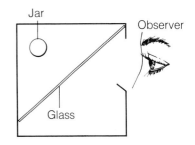

Figure 13-40

LIGHT AND COLOUR

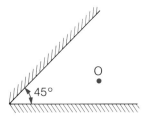

Figure 13-41

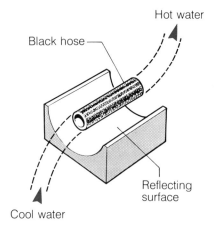

Figure 13-42

19. Two plane mirrors are positioned at an angle of 45° to each other, as in Figure 13-41.
    (a) Calculate the number of images of object O that should appear.
    (b) Draw a diagram to show the locations of the images.
20. For circular curved mirrors, state the following ratios:
    (a) focal length to radius of curvature
    (b) radius of curvature to focal length
21. Draw a fully-labelled ray diagram to locate the image of the object in each case below. State the characteristics of the image, including the magnification. If you have to draw a diagram to scale, indicate the scale used.
    (a) A person whose face is 24 cm long is looking into a cosmetic mirror from a distance of 30 cm. This type of mirror is a converging mirror, and it has a focal length of 75 cm.
    (b) The mirror described in (a) above is used to place the image of a glowing light bulb onto a screen. The bulb is 12 cm high and is located 110 cm from the mirror.
    (c) The person in (a) above is looking at a diverging mirror of focal length 55 cm from a distance of 40 cm.
22. Assume that you have a shiny tablespoon. How would you hold the spoon to obtain the following images of yourself?
    (a) small and virtual
    (b) small and real
    (c) large and virtual
23. Figure 13-42 shows one way of heating water using solar energy.
    (a) Describe ways in which this is an application of topics in this chapter.
    (b) How would you determine the position of the black hose for the best heating effects?
24. What causes the image near the outside edge of a spherical converging mirror to appear distorted? What type of mirror can be used to correct this problem?
25. Use the equation given in Practice Question 34 to check the value of the image distance in the ray diagram in Review Assignment 21(b).
26. Describe evidence related to this chapter supporting (a) the particle nature of light and (b) the wave nature of light.

Key Objectives

Having completed this chapter, you should now be able to do the following:
1. State reasons for the importance of light.
2. List sources of light energy and explain which are incandescent, fluorescent, phosphorescent, chemiluminescent, or bioluminescent.
3. Cite evidence that light travels in straight lines.
4. Name the parts of a shadow.
5. Draw diagrams of lunar, solar, and annular eclipses.

6. Explain how light differs from sound in its transmission.
7. State the value of the speed of light in a vacuum and understand techniques used to measure that speed.
8. Define a light year, and calculate its value.
9. Define and give examples of materials that are transparent, translucent, and opaque.
10. Distinguish among specular, diffuse, and reflex reflection, and give examples of each.
11. Recognize examples of the scattering and diffraction of light.
12. State the laws of reflection for mirrors.
13. Draw a ray diagram to locate the image of an object in a plane mirror.
14. For each image seen in a mirror, describe its attitude, type, location, and magnification.
15. State the meaning of each of the following terms for curved mirrors: converging, diverging, concave, convex, centre of curvature, radius of curvature, principal axis, vertex, focal point, focal length, circular aberration, parabolic.
16. Know and apply the ratio of the focal length to the radius of curvature for circular curved mirrors.
17. State the three rules used to draw ray diagrams for (a) converging mirrors and (b) diverging mirrors.
18. Draw a ray diagram to find the image of an object at various distances from a curved mirror.
19. Describe applications of all types of mirrors.
20. Use a ray diagram for a circular converging mirror to prove that
$$\frac{1}{f} = \frac{1}{d_o} + \frac{1}{d_i}.$$
21. Cite evidence supporting both the particle nature and the wave nature of light.

LIGHT AND COLOUR

CHAPTER 14
Refraction and Lenses

The sparkle of a diamond can be attributed to the way light travels into and out of the crystal.

Have you ever used a magnifying glass to start a campfire on a sunny day in the summer? Or, looking down, have you ever noticed a distorted view of your legs as you walk in clear, waist-deep water? Such effects occur because of the refraction, or bending, of light. In this chapter you will learn what causes this phenomenon, as well as how to measure amounts of refraction. Then you will apply the knowledge to answer such questions as these: Why do real diamonds, such as the one in the photograph, sparkle more than counterfeit ones? How is it possible to trap and use light energy in solid plastic fibres? How does light behave when it is refracted upon entering various substances and lenses?

14.1 Refraction and the Speed of Light

In the previous chapter, you studied evidence that light travels in straight lines. That principle is true as long as the light is travelling in a uniform substance. However, when light travels from one transparent substance into another, it can bend. This bending is called the **refraction of light**.

Refraction occurs, for example, when light travels from air into a block of glass at an angle other than 90° to the surface, as shown in Figure 14-1(a). (Some of the light is reflected off the surface of the block, but we are not concerned with that here.) The reason the light is refracted as shown is that its speed decreases when it enters the glass from the air. Another example of how a change in speed causes a change in direction is illustrated in Figure 14-1(b). One wheel of the set reaches the sand first and slows down. The other wheel momentarily continues travelling at a greater speed on the pavement, causing the set of wheels to change its direction of motion.

Light travels at different speeds in different transparent materials. Thus, when light is refracted upon entering one medium from another, the amount of refraction depends on the relative speeds of light in the two materials. For example, light is refracted more when entering glass from air at a given angle than when entering water from air at the same angle. We therefore say that glass has a greater **optical density** than water; light travels more slowly in the glass. (Optical density has no relationship to physical density, mass, or volume.)

For light that is refracted as it travels from material 1 into material 2, we can calculate a quantity called the **index of refraction** (symbol n) which indicates the ratio of the speeds of light in the two materials:

$$n_{1\to 2} = \frac{v_1}{v_2}.$$

Here, $n_{1\to 2}$ is the index of refraction for light travelling from material 1 into material 2, v_1 is the speed of light in material 1, and v_2 is the speed of light in material 2. The units for the ratio of the speeds divide out, so index of refraction has no units.

Often, one of the "materials" involved in calculating the index of refraction is a vacuum, which is actually no material at all. An even more common medium is air. The speed of light in a vacuum is 2.99792×10^8 m/s, and the speed of light in air (at 0°C and atmospheric pressure) is 2.99705×10^8 m/s. These speeds are so close that, to two or three significant digits, they are equal. Thus,

$$v_{\text{vacuum}} = v_{\text{air}} = 3.00 \times 10^8 \text{ m/s (to three significant digits)}.$$

Notice as you read Sample Problem 1 on page 334 that the index of refraction for light travelling from air into alcohol is the reciprocal of the index of refraction for light travelling from alcohol into air. This can be verified by finding the product of the two indexes, *i.e.*, $1.36 \times 0.737 = 1.00$.

Figure 14-1
Refraction

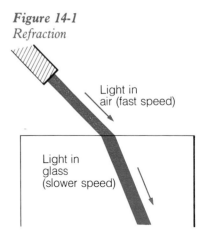

(a) *Refraction of light*

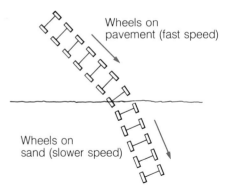

(b) *Refraction of a set of wheels*

LIGHT AND COLOUR

> **SAMPLE PROBLEM 1**
> Light travels in ethyl alcohol at a speed of 2.21×10^8 m/s. Calculate the index of refraction for light travelling from (a) air into ethyl alcohol and (b) ethyl alcohol into air.
>
> **Solution**
>
> (a) $n_{\text{air} \to \text{alc}} = \dfrac{v_{\text{air}}}{v_{\text{alc}}}$
>
> $= \dfrac{3.00 \times 10^8 \text{ m/s}}{2.21 \times 10^8 \text{ m/s}}$
>
> $= 1.36$
>
> ∴ The index of refraction is 1.36.
>
> (b) $n_{\text{alc} \to \text{air}} = \dfrac{v_{\text{alc}}}{v_{\text{air}}}$
>
> $= \dfrac{2.21 \times 10^8 \text{ m/s}}{3.00 \times 10^8 \text{ m/s}}$
>
> $= 0.737$
>
> ∴ The index of refraction is 0.737.

Table 14-1 lists the indexes of refraction for light travelling from a vacuum (or from air) into various materials and *vice versa*. The speed of light in each medium is also given. Notice that as the speed of light in materials decreases, the index of refraction (relative to air or a vacuum) increases. You will discover how this fact relates to angles of incidence and refraction in the next section.

Did You Know?
Certain fish have adapted so well to overcoming the effects of the refraction of light when looking from the water toward air that scientists are puzzled by their skills. This type of fish can shoot a high-pressure jet of water from its mouth out into the air toward a stationary prey, such as an insect, located up to 3 m above the surface of the water. The fish takes aim from beneath the surface of the water, where refraction of light must be taken into consideration in order to record a "hit". The startled prey falls to the water where, in a sort of reversed fisherman's role, the fish scoops up its next meal.

Table 14-1 The Speed of Light and Indexes of Refraction

Medium	Speed of Light in Medium (m/s)	Index of Refraction (Air → Medium)	Index of Refraction (Medium → Air)
air	3.00×10^8	1.00	1.00
ice	2.31×10^8	1.30	0.769
ethyl alcohol	2.21×10^8	1.36	0.735
benzene	2.00×10^8	1.50	0.667
glass			
fused quartz	2.05×10^8	1.46	0.685
crown glass	1.97×10^8	1.52	0.658
light flint	1.90×10^8	1.58	0.633
heavy flint	1.82×10^8	1.65	0.606
zircon (a gemstone)	1.58×10^8	1.90	0.526
diamond	1.24×10^8	2.42	0.413

Scientists have known about the refraction of light since about the second century A.D., when Ptolemy, a Greco-Egyptian scientist, thought he knew how to predict the amount of refraction that occurs when light travels from one material to another. His calculations had weaknesses that you will discover in the next experiment. More than eight centuries after Ptolemy, when the Arabs were leaders in world science, the Arabian mathematician Alhazen (965–1038) studied refraction in detail but failed to find an acceptable mathematical analysis of it. In the seventeenth century, Willebrord Snell, a Dutch mathematician (1591–1626), discovered how to calculate the index of refraction using the angles of incidence and refraction for light travelling from one medium into another (Figure 14-2). Snell was not able to explain why light acts as it does, but his discovery helped later scientists explain the nature of light.

Figure 14-2
Willebrord Snell (1591–1626)

PRACTICE

(*Note*: The usual practice in this book is to use two significant digits. However, for calculations involving the index of refraction, three significant digits will be used.)

1. Calculate the index of refraction for light going from air into a material in which the speed of light is (a) 2.40×10^8 m/s and (b) 1.80×10^8 m/s.

2. Calculate the index of refraction for light going from each material in Practice Question 1 into air.

3. What is the relationship between $n_{\text{air} \to \text{material}}$ and $n_{\text{material} \to \text{air}}$?

4. The index of refraction for light travelling from air into material M is 1.60. What is the index of refraction for light travelling from M into air?

5. Calculate the speed of light in three materials with the following indexes of refraction for light entering from air:
   (a) 2.00   (b) 2.50   (c) 1.10

6. Light beams in air are aimed at an angle of 45° to the surfaces of two different substances, ice and zircon, both of which are listed in Table 14-1. In which substance do you think the amount of refraction will be greater? Explain your answer.

7. **Activity** Place a solid object such as a stick, pencil, or ruler into water in a beaker or other container, as shown in Figure 14-3. Observe the object from various directions, and try to explain what you observe.

8. **Activity** For this activity you will need a coin, an opaque evaporating dish, a small beaker, and some water. Place the coin in the middle of the base of the evaporating dish, and position your eyes at a level where you just miss seeing the coin. Refer to Figure 14-4. Slowly add water to the dish without moving the coin, and observe the results. Explain your observations, given that your eyes "believe" that light travels in straight lines.

Figure 14-3
An Example of the Refraction of Light

Figure 14-4
Set-up for an Activity Illustrating Refraction

Experiment 14A: Refraction and the Index of Refraction

INTRODUCTION
Ray diagrams for the refraction of light resemble those for reflection of light studied in the previous chapter. Figure 14.5 show a typical labelled ray diagram of light being refracted twice: first as it leaves air and enters a rectangular prism, then as it leaves the prism and returns to the air. The normals are drawn perpendicular to the surfaces where the rays enter and leave the prism. The angles of incidence ($\angle i$), refraction ($\angle R$), and emergence ($\angle e$) are measured from the normal to the ray.

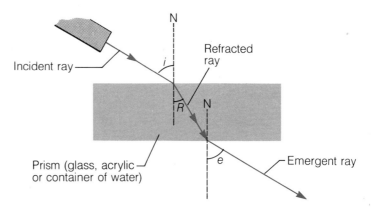

Figure 14-5
A Labelled Ray Diagram of the Refraction of Light

The analysis of data in this experiment is important. In the second century A.D., Ptolemy believed that, for any two transparent materials, the ratio of the angle of incidence to the angle of refraction remains constant. In 1621, Willebrord Snell used the branch of mathematics called trigonometry to calculate the ratio of the sine of the angle of incidence to the sine of the angle of refraction. You can perform both calculations in the analysis of this experiment, although the first one is optional. (If you have not studied trigonometry, refer to Appendix E at the back of the text for an explanation of sines of angles.)

PURPOSE
To study the refraction of light in various materials and to determine the index of refraction of those materials.

APPARATUS
ray box with single-slit window; rectangular solid prism (made of glass, acrylic, or other plastic substance); thin, transparent rectangular dish to hold liquids; protractor; water; a second transparent liquid such as glycerin or mineral oil (optional) (*Note*: The instructions are written for rectangular prisms, but semicircular ones may be used for most of the experimental procedures. With semicircular prisms and dishes, the incident ray should be aimed toward the middle of the flat edge.)

PROCEDURE
1. Place the solid prism on a piece of paper and draw its outline. Remove

the prism so that you can draw a normal (broken line) and an incident ray (solid line) with ∡i = 60°, as shown in Figure 14-6(a). Label the angle and rays. Repeat this procedure using three more diagrams and angles of incidence of 40°, 20°, and 0°. (Remember that these angles are measured from the normal.)
2. Place the prism back onto the first diagram and aim a single ray from the ray box along the incident ray. Draw the ray that emerges on the opposite side of the prism, then remove the prism and draw the entire path of light. Draw a second normal at the surface where the light emerges from the prism. Measure and label the angle of refraction (∡R) in the prism and the angle of emergence back into the air (∡e).
3. Repeat Procedure Step 2 using the other angles of incidence.
4. Repeat Steps 1 to 3 using water in a transparent container. If the container is large you can draw all the incident rays on one diagram if you place them near one corner, as in Figure 14-6(b). If a liquid other than water is available, perform this step once more.
5. Tabulate your measurements in a table similar to the one shown below.

Figure 14-6
Set-ups for Experiment 14A

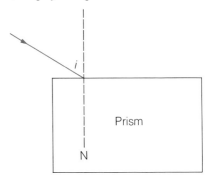

(a) Procedure Step 1

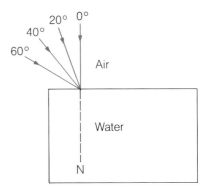

(b) Procedure Step 4

Measurements and Calculations	Acrylic				Water
Angle of incidence (∡i) in air	60°	40°	20°	0°	60°...
Angle of refraction (∡R) in prism					
Angle of emergence (∡e) in air					
Ratio of ∡i/∡R (optional)					
sin ∡i					
sin ∡R					
Ratio of sin ∡i/sin ∡R					

ANALYSIS

1. For each ray of light drawn, calculate the ratio of ∡i to ∡R to three significant digits and enter the data in your table. For light entering one material from another, does the ratio ∡i/∡R remain constant as Ptolemy thought it should?
2. Use either a calculator or a table of trigonometric values (Appendix F) to calculate the sine of each angle of incidence and refraction. Enter the data in your table. Then calculate the ratio of sin ∡i/sin ∡R for each ray and enter those data as decimal numbers. For light entering one material from another, does this ratio remain constant as Snell thought it should? For each material named in your table, calculate an average value of the ratio of sin ∡i/sin ∡R. This ratio is equal to the index of refraction of the material.

CONCLUSIONS ...

QUESTIONS

1. When light travels at an angle from a medium of low optical density (such as air) to one of higher optical density (such as water), is it refracted away from or toward the normal?

2. How does the direction of the incident ray toward a rectangular prism compare to the direction of the emergent ray? (*Hint*: Compare the angle of incidence in air to the angle of emergence in air for each set of rays.)
3. Explain the statement, "A rectangular prism causes a light ray to undergo **lateral displacement** for angles of incidence different from 0°."
4. Under certain circumstances in this experiment, light is reflected off the *inside* surface of the prism. This is called total internal reflection, which is applied in the use of devices such as binoculars and prism periscopes. Research to determine: What are these devices used for? Who uses them? What do they cost? In what careers are they used? Of what advantage or disadvantage are they to society? What are other applications of the same phenomenon?

14.2 Snell's Law of Refraction

Snell discovered that for light travelling from one material into another, the ratio of the sine of the angle of incidence to the sine of the angle of refraction remains constant. This ratio, the index of refraction, is the same as the ratio of speeds (Section 14.1). His finding is summarized in a statement called **Snell's law of refraction**:

> For light travelling from material 1 into material 2, the ratio of the sine of the angle in material 1 to the sine of the angle in material 2 is constant, and is called the index of refraction.

Snell's law can be expressed by this equation:

$$n_{1 \to 2} = \frac{\sin \angle 1}{\sin \angle 2}$$

Notice in the diagram in Sample Problem 2 that the angle of emergence (60°) is equal to the angle of incidence in the air. This means that for a prism with light entering and leaving two parallel sides, the emergent ray is parallel to the initial incident ray. The dotted line in the diagram in this Sample Problem is used to indicate that the path of the emergent ray has not deviated from the path of the incident ray. It has, however, been *laterally displaced*.

It would naturally be difficult to find the speed of light in a diamond crystal or plastic prism directly. However, Snell's law of refraction can be applied to allow us to do so, because the index of refraction equals not only the ratio of the speeds but also the ratio of the sines of the angles:

$$n_{1 \to 2} = \frac{\sin \angle 1}{\sin \angle 2} = \frac{v_1}{v_2}.$$

In applying this equation, it is useful to remember that the speed of light in air or a vacuum, to three significant digits, is 3.00×10^8 m/s.

REFRACTION AND LENSES

SAMPLE PROBLEM 2
A ray of light enters a glass prism at an angle of incidence of 60° and follows the path shown in the diagram. Determine the index of refraction for light travelling from (a) air into the prism and (b) the prism back into air.

Solution

(a) $n_{A \to G} = \dfrac{\sin \angle 1}{\sin \angle 2}$

$= \dfrac{\sin 60°}{\sin 35°}$

$= \dfrac{0.866}{0.574}$

$= 1.51$

∴ The index of refraction for light travelling from air into the prism is 1.51.

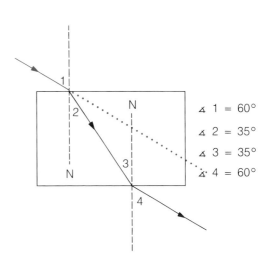

$\angle 1 = 60°$
$\angle 2 = 35°$
$\angle 3 = 35°$
$\angle 4 = 60°$

(b) $n_{G \to A} = \dfrac{\sin \angle 3}{\sin \angle 4}$

$= \dfrac{\sin 35°}{\sin 60°}$

$= \dfrac{0.574}{0.866}$

$= 0.663$

∴ The index of refraction for light travelling from the prism into the air is 0.663. (This answer can also be found by determining the reciprocal of the answer in (a) above.)

SAMPLE PROBLEM 3
In the illustration, material A is air and material B is some liquid. Determine the speed of light in B.

Solution

Since $\dfrac{\sin \angle A}{\sin \angle B} = \dfrac{v_A}{v_B}$

$v_B = \dfrac{v_A \sin \angle B}{\sin \angle A}$

$= \dfrac{3.00 \times 10^8 \text{ m/s} \times \sin 42°}{\sin 70°}$

$= \dfrac{3.00 \times 10^8 \text{ m/s} \times 0.669}{0.940}$

$= 2.14 \times 10^8 \text{ m/s}$

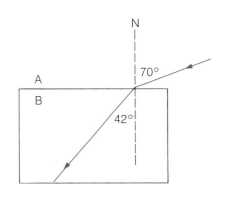

Thus, the speed of light in B is 2.14×10^8 m/s.

LIGHT AND COLOUR

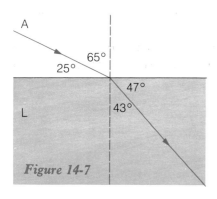

Figure 14-7

PRACTICE

9. A beam of light is aimed from air into three different materials, X, Y, and Z, such that the angle of incidence in each case is 56°.
   (a) Determine the index of refraction of X, Y, and Z, given that the angles of refraction are 26°, 40°, and 30°, respectively.
   (b) Use Table 14-1 to determine the identity of materials X, Y, and Z.
10. Determine the speed of light in each medium you used in Experiment 14A.
11. Figure 14-7 shows a ray of light travelling from air (A) into a liquid (L). Determine the speed of light in L.

Figure 14-8
Explaining Total Internal Reflection

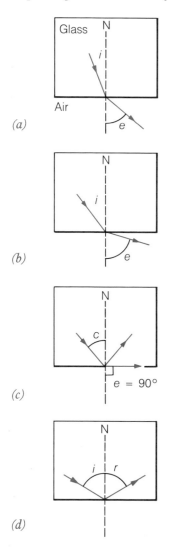

(e) *Total internal reflection of laser light in water*

Experiment 14B: Total Internal Reflection

INTRODUCTION

You have learned that as light travels from a material of high optical density (such as plastic) into one of lower optical density (such as air), it is refracted away from the normal. This means that the angle of emergence in the less dense material is greater than the angle of incidence in the more dense material. This fact is shown in Figure 14-8(a) for light travelling from glass into air.

In Figure 14-8(b), the angle of incidence in the glass has increased, and the angle of emergence in the air is almost 90°. When this happens, the white light splits up into the colours of the rainbow. Also, some light is internally reflected; in other words, some light is reflected off the inside surface of the glass.

When the emerging light just disappears along the surface of the glass, as in Figure 14-8(c), the angle in the glass is called the **critical angle**, $\angle c$. Any angle of incidence greater than the critical angle results in **total internal reflection** in the more dense material. See Figure 14-8(d). Total internal reflection can occur in any transparent material that is adjacent to an optically less dense material. Figure 14-8(e) shows laser light being internally reflected in water.

REFRACTION AND LENSES

The fact that the critical angle in a material results in the light's disappearance along the edge of the material allows us to use Snell's law to calculate unknown quantities. For these calculations, we assume that the theoretical value of the angle of emergence in the less dense material is 90°. For example, when light travels from glass into air,

$$n_{\text{glass}\to\text{air}} = \frac{\sin(\angle \text{ glass})}{\sin(\angle \text{ air})}$$

$$= \frac{\sin \angle i}{\sin \angle e}$$

$$= \frac{\sin \angle c}{\sin 90°}$$

$$= \sin \angle c \quad (\sin 90° = 1).$$

In general, the sine of the critical angle in material 1 equals the index of refraction for light travelling from material 1 toward material 2. Thus, if either the critical angle or the index of refraction is known, the other quantity can be found.

PURPOSE
(a) To determine the critical angle in various materials surrounded by air.
(b) To study an application of total internal reflection.

APPARATUS
ray box with single-slit and double-slit windows; semicircular solid prism (made of glass or plastic); semicircular plastic dish for liquids; water; 2 triangular solid prisms having angles of 45° and 90°; a second liquid such as glycerin or mineral oil (optional); polar graph paper (optional)

PROCEDURE
1. Place the semicircular solid prism on a piece of paper and draw its outline. (If polar graph paper is available for this procedure, your teacher will explain its use.) Aim a single light ray ($\angle i = 30°$) from the curved side of the prism directly toward the *middle* of the flat edge, as shown in Figure 14-9(a). Complete your drawing, including the normal and all rays and angles.
2. With the set-up from Procedure Step 1, slowly move the ray box to increase the angle of incidence until rainbow colours occur in the air. Then move the box slightly further, until the light in the air just disappears, as in Figure 14-9(b). Mark the rays, remove the prism, and measure the critical angle, $\angle c$, for the solid. Have your teacher check your value.
3. Start a new diagram with the prism used in Steps 1 and 2. Aim a single ray from the curved side so that the angle of incidence in the prism is greater than $\angle c$. Complete your diagram and determine whether the first law of reflection ($\angle i = \angle r$) is valid.
4. Repeat Steps 1 to 3, using water in a semicircular plastic dish. If another liquid is available, perform this step once again.

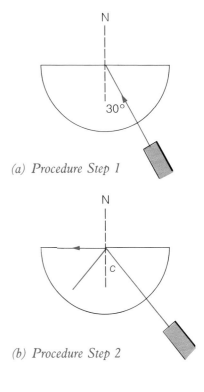

Figure 14-9
Set-ups for Experiment 14B

(a) Procedure Step 1

(b) Procedure Step 2

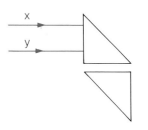

(c) Procedure Step 5

5. A periscope is an application of total internal reflection. To observe how light travels in a periscope, set up the two triangular prisms as in Figure 14-9(c). Aim two rays, X and Y, as shown, and draw the paths they follow. Determine whether the final emergent rays are upright or inverted when compared to the incident rays.

CONCLUSIONS...

QUESTIONS
1. The critical angle in benzene for light travelling toward air is 42°. Which of the following angles of incidence of rays in benzene would result in total internal reflection?
   (a) 35°   (b) 50°   (c) 43°   (d) 3°
2. In Experiment 14A you determined the index of refraction for certain materials. Use each index of refraction and the Snell's law equation to calculate the critical angle for each medium. (Refer to the example shown in the Introduction to Experiment 14B.) Compare the calculated values to your experimental values.
3. Calculate the critical angle in a material whose index of refraction for light travelling toward air is
   (a) 0.600 and (b) 0.850.
4. A student who is asked to determine the identity of materials A and B discovers that the critical angle in A is 47°, and in B, 32°. Both angles result when light is aimed toward air. Help the student identify each material. (Use Table 14-1 as a reference.)

14.3 Applications of Total Internal Reflection

Did You Know?
If you look closely at the front fenders of some cars, you will see an application of total internal reflection. Bundles of optical fibres take light from each signal light and internally reflect it around toward the driver. A similar design at the rear of the car allows the driver, by looking in the rear-view mirror, to judge whether the signal lights have been left on unintentionally or whether the brake lights are functioning.

Plane mirrors reflect light, as do solid transparent prisms that allow total internal reflection. However, plane mirrors have disadvantages in that they tarnish easily and so do not last as long as prisms; as well, they absorb more light energy than prisms. Thus, the capacity of prisms to reflect internally is useful when mirrors are either inconvenient or unsatisfactory for reflection.

Figure 14-10 illustrates four applications of total internal reflection. Figure 14-10(a) shows a periscope, studied in Experiment 14B. This device is used on submarines to view the seascape above the surface while the submarine remains hidden underwater. The periscope has two prisms which internally reflect light. Bicycle reflectors, shown in Figure 14-10(b), also use total internal reflection. Light from a vehicle behind the bicycle strikes the reflector and bounces back to the driver of the vehicle. Internally reflecting prisms are also used in binoculars, illustrated in Figure 14-10(c). Without prisms, the binoculars would have to be made longer to give the same magnification of images. Finally, Figure 14-10(d) shows solid plastic tubing in which a light beam is internally reflected every time it strikes an inside surface. The study of this concept, called **fibre optics**, is being applied to the transmission of tele-

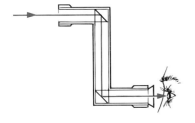
(a) Periscope

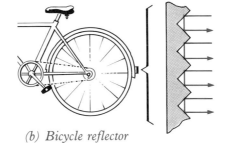
(b) Bicycle reflector

Figure 14-10
Applications of Total Internal Reflection

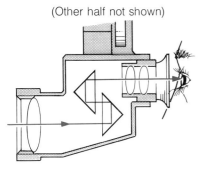
(c) Prism binoculars

(d) Fibre optics

Communication systems of the future will use thin fibres (top of photograph) rather than bulky cables (bottom of photograph).

phone and television messages on laser beams in thin fibres. Fibre optics is also important in industry and medicine. For example, thin, flexible tubes are used to view internal regions of machines or human bodies without the need for cutting or major surgery.

As another example of total internal reflection, consider a question posed in the chapter introduction: Why do real diamonds sparkle more than counterfeit ones? To answer this question, we shall begin with some calculations. According to Table 14-1, the index of refraction of diamond in air is 2.42. Thus, the index of refraction for light travelling from diamond toward air is $1/2.42 = 0.413$. The critical angle for diamond is therefore the angle whose sine is 0.413, or $\angle c = 24.4°$. Light rays that enter a real diamond will be totally internally reflected if they strike a surface within the diamond at an angle greater than 24.4°. Keep in mind that a diamond is cut in such a way that when its motion relative to an observer is even very slight, the light that enters and exits will do so from a different surface. Thus, the chances are very good that light entering a diamond will be totally internally reflected many times before it exits. The result is a sparkling effect as the diamond is moved. Fake diamonds, often made of the gemstone zircon, have a critical angle that exceeds 30°, so the sparkling effect is not as noticeable. (The quality of a fine diamond is determined not only by its purity or lack of flaws, but also by the craftsmanship of its polished faces.)

PRACTICE
12. Use the information in Table 14-1 to calculate the critical angles of crown glass and heavy flint glass. Which substance would sparkle more if used as a gem? Explain why.

13. The fibres used in fibre optics are surrounded by a thin, transparent film.
    (a) Should the film have a lower or higher optical density than the central fibre? Explain your answer.
    (b) What do you think is the function of the film? (*Hint*: Hundreds of fibres are bundled together to form a fibre tube.)
14. Canada is a major exporter of copper, an important metal in the manufacture of cables used for telephone lines. Canada also finds itself in a world of rapidly increasing communication technology, where a laser pulse along a single fibre of glass can carry about five million times as many phone calls as a single copper wire. As the use of glass fibres in fibre optics increases, the need for copper decreases. Should governments in Canada support the development of fibre optics or the maintenance of an economically important copper industry? Give reasons for your answer.
15. A **mirage** is an optical illusion that commonly occurs on hot, summer days. It is often seen as a pool of water some distance ahead of an observer on a highway or in a desert. It occurs as a result of internal reflection of light in a cool layer of air which lies above a warmer layer near the surface of the road or desert. Figure 14-11(a) shows a mirage photographed on a large, dried-up water bed called the Etosha Pan in Namibia, Africa. (A pan is a shallow depression that generally contains water only during the rainy season.) Figure 14-11(b) shows the start of a ray diagram to explain the formation of a mirage. Copy the diagram into your notebook, then complete it and explain what is happening.

Figure 14-11
Mirages

(a) A mirage

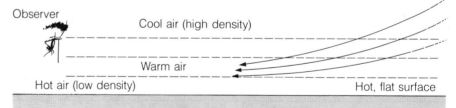

(b) Start of a mirage ray diagram

REFRACTION AND LENSES

14.4 Lenses

Anyone who has used a magnifying glass to light a campfire or used a camera to take a photograph has made use of the refraction of light in lenses, as has anyone who wears eyeglasses.

A **lens** is a transparent device with at least one curved edge. Light passing through a lens obeys Snell's law of refraction ($n_{A \to B} = \sin \angle A / \sin \angle B$). One example of the behaviour of light as it passes through a lens is shown in Figure 14-12.

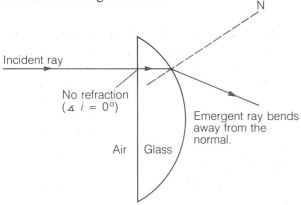

Figure 14-12
A lens alters the direction of a light ray.

Lenses may be either converging or diverging. A **converging lens** causes light rays from a distant source to converge to a focal point. Such a lens is thicker in the middle than at the outside edge. A **diverging lens** causes light rays from a distant source to spread out as if it originated from a virtual focal point. It is thicker at the outside edge than in the middle. Figure 14-13 shows the general shapes of convex and concave lenses.

Figure 14-13
The Design of Lenses

 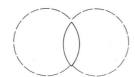 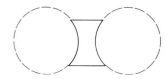

(a) Single convex lens   (b) Double convex lens   (c) Single concave lens   (d) Double concave lens

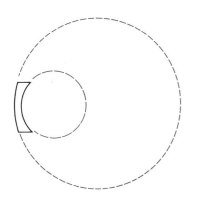 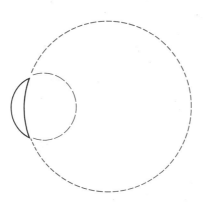

(e) Convexo-concave lens   (f) Concavo-convex lens

LIGHT AND COLOUR

Figure 14-14

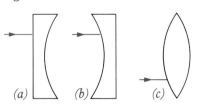

PRACTICE
16. The diagrams in Figure 14-14 show light rays in air approaching glass lenses. Copy the diagrams into your notebook, then draw the approximate direction of each light ray as it travels into the glass and back into the air. (*Hint*: Draw the appropriate normals wherever a ray strikes a surface.)

Experiment 14C: Converging Lenses

INTRODUCTION
A lens has a principal axis, just as a curved mirror does. However, a lens has two focal points rather than the single one of curved mirrors. For a converging lens, the **primary focal point** (**PF**) is located on the side of the lens opposite to the source of light. The **secondary focal point** (**SF**) is located on the same side as the source. Refer to Figure 14-15. For a thin double convex lens, the focal length (f) is measured from either one of these focal points to the middle of the lens, as indicated in Figure 14-15. (Notice that we have not defined the centre of curvature or the radius of curvature for the curved part of a lens. A lens maker is concerned with them, but we do not need them for our experiments or ray diagrams.)

Figure 14-15
A Converging Lens

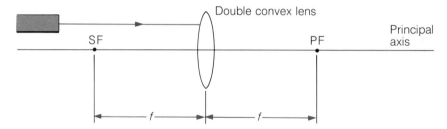

In the Purpose of this experiment, Part (c) is optional, but it reveals an important problem of converging lenses called **chromatic aberration**. The Greek prefix *khroma* means "colour", so you can expect to see colours produced as a result of this aberration. (A converging lens also has spherical aberration for light of a single colour. Spherical aberration, a problem of circular converging mirrors also, was described in Chapter 13.)

To describe the characteristics of an image in a lens you should state the image's attitude (upright or inverted), type (real or virtual), magnification ($M = h_i/h_o$), and location. These characteristics of images are described in the previous chapter.

PURPOSE
To do the following for a converging lens:
(a) find the focal length
(b) find three convenient rules for light rays striking the lens
(c) find the cause of chromatic aberration
(d) determine the characteristics of images seen in the lens

REFRACTION AND LENSES

APPARATUS
ray box with single-slit and triple-slit windows; 2 converging lenses (one cylindrical, for use with the ray box; the other spherical, for use as a viewing lens); if Procedure Step 8 is to be performed, an optical bench or similar apparatus

PROCEDURE
1. Place the cylindrical lens at the middle of a piece of paper and draw its outline. Remove the lens and draw a line joining the top and bottom of the lens, shown as line AB in Figure 14-16. Then draw the perpendicular bisector of line AB, which is line CD in the diagram. Line CD is the principal axis of the lens.

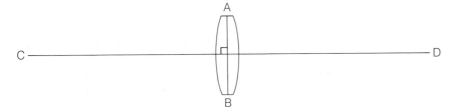

Figure 14-16
Set-up for Experiment 14C, Procedure Step 1

2. Replace the lens on your diagram, then aim three parallel rays toward the lens so that the central ray falls along its principal axis. Draw the rays. The point where the emergent rays meet is the primary focal point (PF). Label it and measure the focal length, f, to the middle of the lens.
3. Repeat Procedure Step 2 with the light coming from the other side of the lens. Label the point of intersection the secondary focal point (SF).
4. With the ray box back on the original side of the lens, use a single ray for each rule to determine three rules for converging lenses. Draw all the rays on your diagram. The rays used to determine the rules are as follows:
   (a) an incident ray parallel to the principal axis (but not far from it)
   (b) an incident ray through the secondary focal point
   (c) an incident ray through the middle of the lens at a small angle to the principal axis
5. Using the same ray diagram, aim another single ray toward the lens parallel to the principal axis but striking the lens near the outside edge. Draw the resulting ray, and take note of the problem of chromatic aberration.
6. Devise a technique for using the ray box to determine the focal length of the spherical or "viewing" lens. Now hold the lens at a distance equal to its focal length from an object and describe the observed image. (You may have to move the lens backward or forward slightly to view the effect.)
7. Hold the same spherical lens at arm's length and view objects around you. Describe three of the four characteristics of the image (attitude, type, and approximate magnification) when the object viewed is
   (a) closer to the lens than the focal point
   (b) far beyond the focal point.

LIGHT AND COLOUR

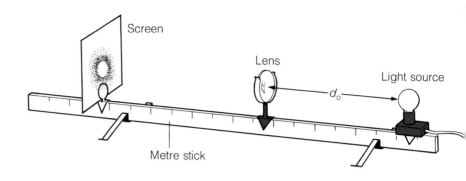

Figure 14-17
Set-up for Experiment 14C, Procedure Step 8

8. To study Steps 6 and 7 above more quantitatively, you can arrange a set-up so that light coming from a bright source converges through a spherical lens to produce a real image on a screen. Refer to Figure 14-17. Determine the focal length of the lens. Then alter the distance of the object to the lens (d_o) and describe both quantitatively and qualitatively what occurs when the object position changes as indicated below. (The screen position can be adjusted.)
   (a) $d_o > 2f$
   (b) $d_o = 2f$
   (c) $2f < d_o < f$
   (d) $d_o = f$
   (e) $d_o < f$

CONCLUSIONS . . .

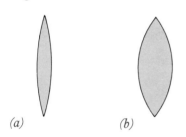

Figure 14-18

QUESTIONS

1. Which shape of lens shown in Figure 14-18 do you expect to have the greater focal length? Explain your answer. (The lenses are made of the same material.)
2. Describe where each ray in Figure 14-19 will be refracted.
3. Compare chromatic aberration with spherical (or circular) aberration. Where have you seen examples of chromatic aberration occurring in an everyday situation?
4. When a converging lens is used as an ordinary magnifying glass, is the image it gives real or virtual?
5. Assume that you are given a magnifying glass. Describe two methods you could use to find its focal length.

Figure 14-19

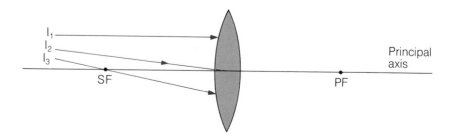

Experiment 14D: Diverging Lenses

INTRODUCTION
In the previous experiment you learned how light behaves when it is refracted through a converging lens. The present experiment is almost identical to the previous one in its Procedure, with the exceptions that a diverging lens is used here, and the problem of aberration is not considered.

Figure 14-20 shows that for a diverging lens the primary focal point (PF) is located on the same side of the lens as the source of light. (This is the opposite of the situation for a converging lens.) The light rays which are parallel to the principal axis determine where the primary focal point is located. Those light rays diverge, or spread apart, after they pass through a diverging lens, so the refracted rays must be extended backward to find where they meet or, rather, appear to meet. The secondary focal point (SF) is located on the opposite side of the lens. For a double concave lens, the focal length, f, is measured from either the PF or the SF to the middle of the lens.

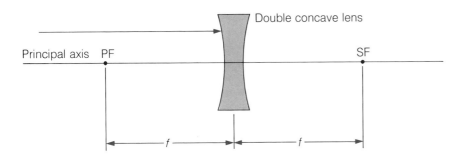

Figure 14-20
A Diverging Lens

PURPOSE
To do the following for a diverging lens:
(a) find the focal length
(b) find three convenient rules for light rays striking the lens
(c) determine the characteristics of images seen in the lens

APPARATUS
ray box with single-slit and triple-slit windows; 2 diverging lenses (one cylindrical, for use with the ray box; the other spherical, for use as a viewing lens)

PROCEDURE
1. Carry out a procedure to determine Purpose (a). Use as references Figure 14-20 as well as Procedure Steps 1 and 2 in Experiment 14C. In the diagram you obtain, include both focal points.
2. Use the diagram you drew for Step 1 above to determine three rules for diverging lenses. Draw all the rays on your diagram. To determine these rules, use the following:
   (a) an incident ray parallel to the principal axis

LIGHT AND COLOUR

(b) an incident ray toward the secondary focal point (*Hint*: Remove the lens and aim an incident ray *toward* the SF, then replace the lens.)

(c) an incident ray toward the middle of the lens at a small angle to the principal axis

3. Use the diverging spherical lens to view objects around you. Describe three of the four image characteristics (attitude, type, and approximate magnification). Do the image characteristics depend on the distance between the lens and the object?

Figure 14-21

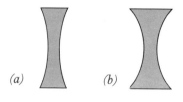

CONCLUSIONS . . .

QUESTIONS

1. The lenses in the diagrams in Figure 14-21 are made of the same material. Which lens has the greater focal length? Explain your answer.

Figure 14-22

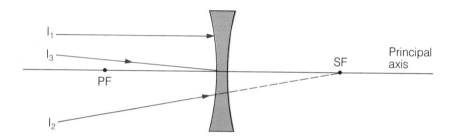

2. Describe where each ray in Figure 14-22 will be refracted.
3. Can an image in a diverging lens be placed onto a screen? Why or why not?

14.5 Ray Diagrams for Lenses

A ray diagram can be used to find the image of an object located near a lens. Ray diagrams for lenses resemble those for mirrors. Any two of the three rules learned in the lens experiments may be used to locate an image. After an image is located, its characteristics can be stated.

Light passing through a lens can be refracted twice, first when entering the lens and again when leaving. This double refraction is not easy to draw accurately, so a shortcut can be used. A straight line is drawn to represent the lens, and the type of lens is indicated in the centre of the diagram. The light rays in the diagram are shown as being refracted only once, but because they obey the rules, the results obtained are adequate.

In the Sample Problems that follow, note that the refracted rays must meet in order to produce a real image. In the second one, the refracted rays do not meet unless they are extended behind the lens. Therefore, this image is virtual.

SAMPLE PROBLEM 4

A 1.5 cm high object is located 8.0 cm from a converging double convex lens of focal length 2.5 cm. Draw a ray diagram to locate the image of the object and state its characteristics.

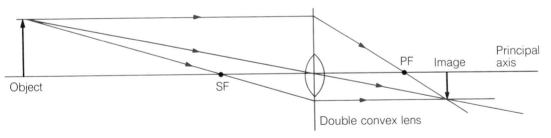

Solution
The diagram is drawn according to the instructions, with the object, in this case an arrow, resting on the principal axis. Three incident and refracted rays are drawn, according to the rules for locating the image. (Two rules help locate the image; the third acts as a check.) The resulting image, seen in the ray diagram, is inverted and real; its magnification is

$M = h_i/h_o$
$ = 0.70 \text{ cm}/1.5 \text{ cm}$
$ = 0.47.$

The image is located on the side of the lens opposite to the object and beyond the primary focal point.

SAMPLE PROBLEM 5

Repeat Sample Problem 4 using a diverging double concave lens.

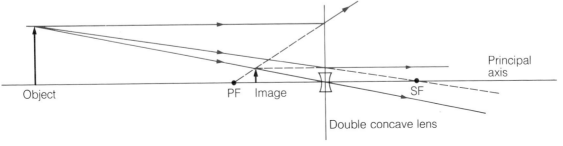

Solution
Again, three incident and refracted rays are drawn using the rules you found experimentally. The refracted rays must be extended straight back to where they meet on the side of the lens where the object is located. The resulting image, shown in the ray diagram, is upright and virtual; its magnification is

$M = h_i/h_o$
$ = 0.40 \text{ cm}/ 1.5 \text{ cm}$
$ = 0.27.$

The image is located between the primary focal point and the lens, on the same side of the lens as the object.

Several optical instruments, including the camera, human eye, microscope, and refracting telescope, use lenses. These applications are discussed in the next chapter.

LIGHT AND COLOUR

PRACTICE

17. Draw a fully-labelled ray diagram to locate the image of the object for each situation described in the table below. State the characteristics of each resultant image.

Type of Lens	Focal Length (cm)	Height of Object (cm)	Distance of Object to Lens (cm)
(a) converging	3.0	2.0	6.0
(b) converging	3.0	1.5	4.5
(c) converging	4.0	1.5	4.0
(d) converging	4.0	1.5	2.0
(e) diverging	4.0	2.0	4.0

18. Analyse the results of your ray diagrams in Practice Question 17, above, and Practice Question 32 in Chapter 13. Then describe how mirrors compare with lenses.

19. Use a ray diagram similar to the one in Sample Problem 4 to verify that $\frac{1}{f} = \frac{1}{d_o} + \frac{1}{d_i}$ where d_o is the distance from the object to the middle of the lens and d_i is the distance from the image to the centre of the lens. (*Hint*: Omit the ray that passes through the middle of the lens, and use the ratios of h_i/h_o for two sets of similar triangles in the ray diagram.) Also prove that $h_i/h_o = d_i/d_o$.

20. Use the first equation in Practice Question 19 to calculate the value of d_i in Practice Question 17(a). Check whether your diagram verifies the calculation.

21. Verify that the second equation in Practice Question 19 applies to the ray diagrams in Practice Question 17(a) and (b).

Review Assignment

1. When light travels from air into water at an angle of incidence other than 0°, why is it refracted?
2. In which substance shown in Figure 14-23, A or B, is light travelling more slowly? Which substance has the lower optical density?
3. Light travels in ocean water at a speed of 2.17×10^8 m/s. Calculate the index of refraction for light travelling from
   (a) air into ocean water
   (b) ocean water into air
   (c) ocean water into benzene.
   (The speed of light in various materials is given in Table 14-1.)
4. A person is spear-fishing from the edge of a pond. Draw a diagram to show where the person should aim the spear to stand a chance of striking a fish beneath the surface.
5. The index of refraction for light travelling from air into ruby is 1.54. What is the index of refraction for light travelling in the opposite direction?
6. Calculate the speed of light in materials having the following indexes of refraction for light travelling toward air:
   (a) 0.55   (b) 0.92

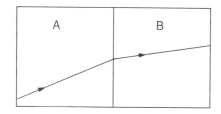

Figure 14-23

7. Draw a ray diagram to explain the Activity described in Practice Question 8 in this chapter.
8. **Activity** Place a rectangular glass or plastic prism onto a printed page of this text and view the print from various directions. Explain what you observe.
9. The angle of incidence of a ray into a rectangular prism is 0°. What is the size of the angle of (a) refraction and (b) emergence?
10. Write the equation for Snell's law of refraction for light travelling from substance X into substance Y.
11. Assume that you are given a sample of an unknown transparent liquid.
    (a) How would you find the speed of light in the liquid?
    (b) How would you determine the identity of the liquid using your calculated value for the speed?
12. Draw a labelled diagram of a 3.0 cm by 5.0 cm solid rectangular prism which has a ray of light from air striking its long edge such that $\angle i = 50°$ and $\angle R = 30°$.
    (a) What is the angle of emergence? Draw the emergent ray in your diagram.
    (b) Calculate the index of refraction of the material for light entering from air.
    (c) What is the speed of light in the prism?
    (d) Use Table 14-1 to determine a possible identity of the substance.
    (e) What is the critical angle of light in the prism when it is surrounded by air?
    (f) Measure the lateral displacement of the emergent ray in millimetres. (*Hint*: Extend the incident ray.)
13. What is meant by the term "critical angle"? What is the relationship between the optical density of a substance and the size of the critical angle within it for light travelling toward air?
14. State two conditions necessary for total internal reflection.
15. Copy Figure 14-24 into your notebook and determine the direction of the ray (or rays) of light.
16. Use the data in Table 14-1 to find the critical angle in (a) zircon and (b) ice.
17. Calculate the index of refraction of a substance surrounded by air if the critical angle in the substance is (a) 35° and (b) 50°.
18. A right-angled periscope with a single prism can be used for seeing around corners. Draw a diagram showing how you would design such a periscope while applying the principle of internal reflection.
19. **Activity** Place a glass or plastic prism into a beaker and slowly add glycerin until the prism is covered. View the prism from various directions and explain what you observe.
20. What is a mirage? Why does it occur?
21. A certain type of lens can be used to start a fire in bright sunlight. Draw a diagram of such a lens to explain why it can start a fire.
22. Describe how you would determine the focal length of the following:
    (a) a solid glass tube
    (b) a corrective eyeglass lens which is actually a diverging lens

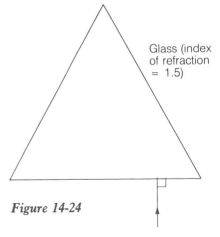

Figure 14-24

23. Draw a fully-labelled ray diagram to locate the image of the object in each situation described below. State the characteristics of each resulting image.
    (a) A detective is inspecting part of a broken toothpick with a magnifying glass (converging lens) of focal length 65 mm. The piece of toothpick is 15 mm long and is held 30 mm from the middle of the magnifying glass.
    (b) A demonstration of a real image is set up using a candle flame (2.0 cm high) as the object. The flame is located 8.0 cm from the appropriate type of lens of focal length 5.5 cm. (In this diagram, show where the screen should be placed to view the real image.)
    (c) A 30 cm high object is located 90 cm from a diverging lens of focal length 40 cm.
24. For each optical device described blow, state the type of lens which could be used, and the approximate location of the object.
    (a) photographic enlarger (image real and enlarged)
    (b) human eye (image real and small)
    (c) projection lamp or spotlight (light is bright, but no image exists)
    (d) copy camera (image real and not magnified)
    (e) magnifying glass (image virtual and enlarged)
    (f) camera lens (image real and small)
    (g) fish-eye lens in a doorway (image virtual and small)
25. Use the first equation given in Practice Question 19 to check the value of the image distance in your ray diagram in Review Assignment 23(b) above.
26. Predict a possible technique for reducing the problem of chromatic aberration in converging lenses.

Key Objectives

Having completed this chapter, you should now be able to do the following:
1. Define refraction of light, and explain why it occurs.
2. Recognize the relationship between a material's optical density and the speed of light in the material.
3. Know the speed of light in air or a vacuum to three significant digits.
4. Given any two of the three variables, index of refraction and the speed of light in two adjacent materials, calculate the third variable for light travelling from one material into the other.
5. Given the index of refraction for light travelling from material A into material B, determine the index of refraction for light travelling from B into A.
6. Explain demonstrations of refraction.
7. Measure the angles of incidence, refraction, and emergence in a diagram of the refraction of light in a prism.

8. Verify Snell's law of refraction of light experimentally.
9. Given any two of the index of refraction, angle of incidence, and angle of refraction, determine the third quantity.
10. Recognize that the index of refraction found using Snell's law is equivalent to the index of refraction found using the ratio of speeds of light in two adjacent substances.
11. Explain the term "lateral displacement" for light passing through a rectangular prism.
12. Describe the conditions required for total internal reflection.
13. Define critical angle, and be able to find it experimentally.
14. Given either of the index of refraction of a substance surrounded by air or the critical angle within it, find the other quantity.
15. Explain applications of total internal reflection.
16. Name and draw the shapes of converging and diverging lenses.
17. Recognize the location of and method for finding the primary and secondary focal points of converging and diverging lenses.
18. State three rules for drawing ray diagrams for both converging and diverging lenses.
19. Determine the characteristics of images seen in a lens.
20. Draw a ray diagram to find the image of an object located at various distances from a lens.
21. Use a ray diagram for a converging lens to prove that $\frac{1}{f} = \frac{1}{d_o} + \frac{1}{d_i}$ and that $M = h_i/h_o = d_i/d_o$.
22. Define chromatic aberration in lenses and compare it with spherical (or circular) aberration in curved mirrors.

LIGHT AND COLOUR

CHAPTER 15
Optical Instruments

A Surveyor's Transit

An optical instrument is a device which produces an image of an object. The most important optical instrument is the eye, which allows us to view images of our surroundings. If we use our imagination, a word that stems from "image", we can describe in words the images created by other optical instruments. A camera, for instance, produces permanent images that resemble the real world we view with our eyes. A microscope helps us view images of tiny cells and other objects invisible to the unaided eye. A refracting telescope, in the form of the transit seen in the photograph, allows a surveyor to determine the angle between a reference point and a distant object. How do these and other optical instruments apply the principles studied in the previous two chapters to produce images? And what are some of the limitations associated with their use?

OPTICAL INSTRUMENTS

15.1 The Functions of Optical Instruments

Every optical instrument fulfills one or more of the four important functions listed below. For each function named, at least one example is given.

(a) *Recording images.* Eyes record images temporarily, while cameras and photocopiers record images permanently.
(b) *Improving impaired vision.* Eyeglasses and contact lenses help people with impaired vision see clearly.
(c) *Producing enlarged images.* A microscope can make an image appear hundreds of times larger than the original object; a projector places a large image onto a screen from a small photographic film or slide.
(d) *Viewing distant objects.* Binoculars and telescopes make distant objects appear much closer and clearer than does the unaided eye. Figure 15-1 shows a photograph of two nebulae (singular, nebula)

Figure 15-1
The Horsehead nebula (dark foreground) in the constellation Orion absorbs light from the more-distant stars as well as from the emission nebula seen in the background.

357

LIGHT AND COLOUR

in outer space taken with the use of a large telescope. The bright nebula in the background consists of a huge cloud of particles which emits its own light. It is possible that portions of the cloud will eventually evolve to form a star. The dark nebula in the foreground consists of particles that absorb light. Such phenomena would be impossible to observe and record without the use of both a telescope and a camera using a long exposure time.

The experiments and concepts in this chapter will give you a basic understanding of a limited number of optical instruments. Each instrument you will study applies knowledge gained from either or both of the previous two chapters on the reflection and refraction of light.

PRACTICE
1. Which function (or functions) of optical instruments is served by each of the following?
   (a) an overhead projector used in school classrooms
   (b) a pair of opera glasses
   (c) a television camera
2. Which of the two types of images does a projector project onto a screen? (If you cannot recall the difference between a real image and a virtual image, refer to Section 13.4.)

Experiment 15A: The Pinhole Camera

INTRODUCTION
The first photograph ever taken with a camera was produced in 1826 in France. The type of camera used at that time was similar to the type you will use in this experiment.

The image produced by a pinhole camera results from the fact that light in a single material (in this case, air) travels in a straight line. This fact will be applied when you draw ray diagrams to explain your observations in the first part of this experiment.

The instructions for this experiment are based on the design of a simple pinhole camera like the one shown in Figure 15-2. If a commercially made camera is used, it may not be possible to follow all the steps of the procedure.

Figure 15-2
A Simple Pinhole Camera

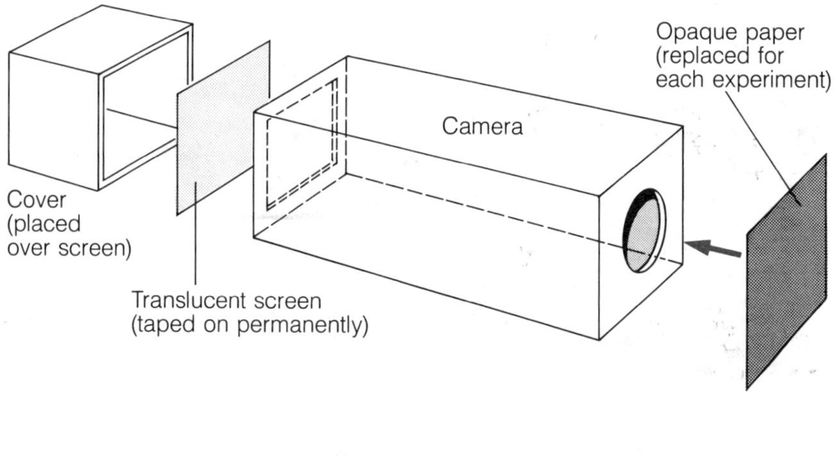

OPTICAL INSTRUMENTS

The second part of this experiment, which is optional but interesting, requires that you develop your own black-and-white photographs taken with the pinhole camera. If a special darkroom is available in your school for developing photographs, it may be more convenient to use than your science classroom.

PURPOSE
(a) To study the images formed by a pinhole camera.
(b) To develop photographs taken with a pinhole camera and determine how to control the camera to obtain successful photographs.

APPARATUS

Part A:
pinhole camera; light source; pin; piece of opaque paper; converging lens having a focal length greater than the length of the camera; tape

Part B:
pinhole camera; piece of thin sheet metal; piece of opaque paper; pin; fine sandpaper; double-sided tape; photographic paper; black-and-white photograph developing materials (developer, stop bath, fixer, thermometer, and all related containers); darkroom safety light; source of water

CAUTION:
It is recommended that the light source used in Part A be a small electric lightbulb. However, a candle may be used as a source if the following precautions are observed: The camera should never be brought close to the flame, and a fire extinguisher and fire blanket should be readily available.

PROCEDURE

Part A: Images in a Pinhole Camera
1. Tape the opaque paper over the open end of the camera and poke a pinhole in the middle of it. With the room lights off, aim the pinhole toward the light source from a distance of about 30 cm. Describe the image seen on the translucent screen by stating its attitude, type, location, and approximate magnification.
2. Determine what happens to the image on the screen when the camera is moved closer to and farther from the light source.
3. Add two more pinholes, one about 6 mm below the first hole and the other about 3 mm to the right of the first hole. View the light source and describe what you observe.
4. Add several more pinholes approximately 3 mm apart in the shape of the letter "R" and describe the effect. (A diagram of the image would help your description.)
5. Stand about 1 m from the light source and hold the converging lens between the source and the camera. Adjust the position of the lens until you obtain a single bright image of the light source on the screen.
6. Remove the dark paper and try to produce a clear image of the room or a scene outside using the lens in front of the camera. (In this case, the room lights may have to be on.)

Part B: Using the Pinhole Camera
(*Note*: Read the entire procedure before you begin the experiment in order to be adequately prepared to carry out all the steps.)

1. Seal the edges of the camera so that when the photographic paper is later installed no light will leak in. Place a piece of double-sided tape inside the camera at the end opposite the pinhole. Refer to Figure 15-3.

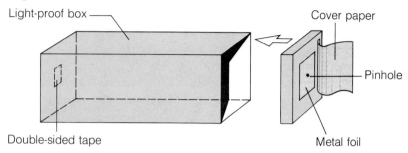

Figure 15-3
Pinhole Camera Set-up for Photographing

2. Poke a small pinhole (about 0.50 mm in diameter) in the metal foil and use a fine sandpaper to smooth the edges. Attach the metal foil to the inside of the camera lid, then attach a cover paper on the outside of the lid, as shown in Figure 15-3.
3. Mix the developing chemicals (the developer, stop bath, and fixer) according to the manufacturer's instructions, and pour them into the appropriate trays for use later.
4. In a darkened room with only the safety light on, cut the photographic paper into pieces about 4 cm by 4 cm. Attach one piece to the tape on the inside of the camera. Place the lid on the camera and be sure all openings are sealed. (Store all excess pieces of photographic paper in a light-proof container.)
5. Find an appropriate subject to photograph (a light-coloured car or a scene with several light and dark spaces) and set up the camera so that it can be held steady as it faces the subject. Open the cover flap for a known time (try 4 s, 5 s, or 6 s, depending on the lighting conditions).
6. Back in the darkened room, develop the photograph according to the manufacturer's instructions, and rinse the photograph in water for several minutes.
7. Compare your photograph with those of other students and predict what you should do to try to obtain the best possible photograph. Consider such variables as the size of the pinhole, the exposure time, the lighting conditions, and the contrast of the subject photographed. If possible, verify your predictions.

ANALYSIS
1. When an image is formed in a pinhole camera, what property of light is illustrated? Explain your answer.
2. Draw a ray diagram to show each of the following:
   (a) An object is located close to a pinhole camera with a single pinhole.
   (b) An object is located far from the camera in (a).
   (c) An object is located far from a pinhole camera with two pinholes.
3. In Part A, Step 6 you obtained an image similar to one in Section 14.5, Practice Question 17. Which diagram in Practice Question 17 relates to Step 6?

4. The photograph obtained with a pinhole camera has the dark and light spaces reversed with respect to the original subject. Explain why this occurs.

CONCLUSIONS . . .

15.2 Lens Cameras and Photography

A lens camera is a light-proof box in which a converging lens forms a real image on a film. The lens gathers light from the scene to be photographed so that, under normal conditions, the film is exposed to the light for only a fraction of a second. This process is, of course, different from the method for producing a pinhole camera photograph. The latter requires a long exposure time because of the small opening in the pinhole camera, as you learned if you performed Experiment 15A, Part B.

The principle of the camera was described as early as the tenth century by the Arabian scholar Alhazen, who used a pinhole camera to view a solar eclipse safely. The great Italian scientist and artist Leonardo da Vinci (1452–1519) described the *camera obscura*, a device used to project an image onto a screen from which an artist could copy a scene. ("*Camera obscura*" is the Latin for "dark room".) In da Vinci's era, the *camera obscura* produced a faint image because only a small amount of light passed through the pinhole. By the middle of the sixteenth century, however, lenses were being used to help produce a brighter and sharper image. See Figure 15-4.

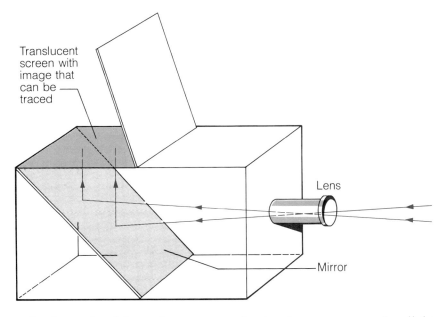

Figure 15-4
A Camera Obscura *in Use about 300 Years Ago*

In the early eighteenth century, scientists became aware that light causes silver salts to turn dark. Within the next hundred years, this discovery made possible the invention of a photographic film that was sensitive to light. Thus, true photography began in the nineteenth century.

During much of the nineteenth century, lens cameras used a single photographic plate that was both bulky and fragile. In the 1880s George Eastman, an American inventor (1854–1932), introduced roll film, and soon thereafter daylight-loading, roll-film cameras came into use. In 1935, colour film which required only one exposure and could be used in ordinary cameras was introduced. Since then, great improvements in convenience and specialized uses have made photography a major industry and important hobby in much of the world.

Today, the field of photography is open to almost anyone who is interested. For the hobbyist, inexpensive, easy-to-use equipment is available in the form of instant cameras, movie cameras, or, most commonly, roll-film and disc-film cameras. Professional photographers use more sophisticated equipment in studios, or for press and industrial photography. Specialized equipment is available for the movie and television industries as well as for such applications as high-speed, underwater, aerial, microscope, or infrared photography.

All modern lens cameras, whether simple or sophisticated, contain the following main parts:

(a) the **light-proof box**, which supports the entire apparatus
(b) the **converging lens**, or system of lenses, which gathers light and focuses it onto a film
(c) the **diaphragm**, which controls the amount of light passing through the lens when the picture is taken
(d) the **film**, which records the image on a light-sensitive mixture of gelatin and silver bromide which is later developed to give a permanent record of the scene photographed
(e) the **viewfinder**, which allows the photographer to see what he or she is photographing

One type of modern camera, the **single-lens reflex (SLR) camera**, is described here because it is simple enough to be understandable yet complex enough to illustrate several principles of physics. The SLR camera is a kind of 35 mm camera, so called because 35 mm of film is

Figure 15-5
A Single-Lens Reflex Camera

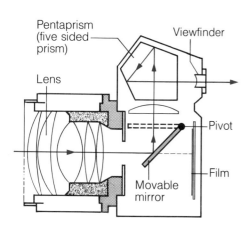

(a) Simplified diagram showing path of light to the viewfinder

(b) Cutaway view of a single-lens reflex camera

exposed each time a picture is taken. Light enters the camera's lens system and strikes a plane mirror, as shown in Figure 15-5(a). The light is reflected up to the *pentaprism* or five-sided prism, where total internal reflection guides the light through the viewfinder. When the photograph is snapped, the plane mirror lifts up, allowing the light to strike the film. The advantage of this system is that the photographer is able to see an upright image that corresponds exactly to the scene that will be exposed on the film, although the image on the film is inverted. Figure 15-5(b) shows a cutaway view of an SLR camera.

Using SLR cameras, and certain other types of cameras, a photographer is able to control four important variables:

(a) The **exposure time**, also called the shutter speed. This is the amount of time the shutter remains open when a photograph is taken. It is stated in seconds or fractions of a second, for instance, 1/60 and 1/1000.

(b) The **focus**, the adjustment of the distance from the lens (actually, the lens system) to the film. Adjusting the focus allows objects at various distances from the camera to be seen clearly (in focus). SLR cameras have a focus control that is operated by turning the lens, which is mounted on a spiral screw thread. Simpler cameras have a fixed focus in which everything beyond about 1.5 m is reasonably clear. Some cameras adjust the focus automatically by using sonar, a process for determining distance which was detailed in Chapter 11.

(c) The **focal length** of the lens system, which can be controlled by either interchanging lenses or adjusting a zoom lens. SLR camera lenses can easily be interchanged because they have bayonet (push and twist) fittings. A zoom lens allows the photographer to adjust the focal length by means of a process illustrated in Figure 15-6.

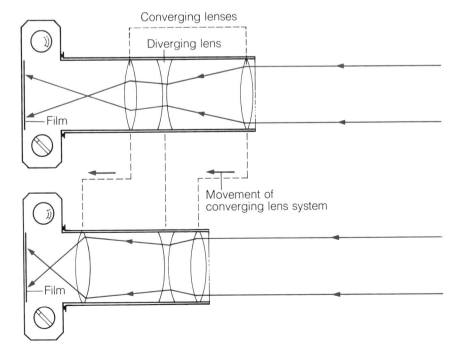

Figure 15-6
The Zoom Lens, with Sample Rays

(a) *Telephoto setting: longer focal length, larger image, smaller angle of view*

(b) *Wide-angle setting: shorter focal length, smaller image, larger angle of view*

LIGHT AND COLOUR

Figure 15-7
The Effect of Varying the Focal Length

(a) This scene was photographed with a zoom lens of focal length 85 mm.

(b) The same scene, photographed with the zoom lens set at 200 mm.

Lenses with a short focal length (about 25 mm to 35 mm) see a wide-angle view, while those with a long focal length (about 200 mm) see a small angle but an enlarged image. Figure 15-7 shows photographs taken with lenses of long and short focal length.

(d) The **aperture** (d) is the diameter to which the shutter opens. It is controlled by the ***f*-stop number**, which is the ratio of the focal length of the lens to the diameter of the opening. The available f-stop numbers range from 1.4 to 32, depending on the lens, and have these typical values: 32, 22, 16, 11, 8, 5.6, 4, 2.8, 2, 1.4. When the f-stop is changed from one number to the next, larger or smaller, the amount of light that enters the lens changes by a factor of very nearly 2. Thus, at $f/8$ there is approximately twice as much light as at $f/11$, so obtaining the same exposure requires about half the exposure time. Refer to Table 15-1.

Table 15-1 The Range and Explanation of *f*-Stop Numbers

f-Stop Number	32	22	16	11	8	5.6	4	2.8	2	1.4
Meaning	$f/32$	$f/22$	$f/16$	...						
Diameter when $f = 40$ mm (in millimetres)	1.25	1.8	2.5	3.6	5	7.1	10	14.3	20	28.6
Diagram of some examples					○	○				
Sample explanation	Diameter increases by a factor of $\sqrt{2}$ or 1.4, so the area increases by a factor of $\sqrt{2}^2$ or 2.									

OPTICAL INSTRUMENTS

SAMPLE PROBLEM 1

When a lens of focal length 50 mm is set at $f/16$, the correct exposure time for a particular scene is 1/100 s. Then the lens aperture is adjusted to $f/8$.
(a) Calculate the aperture (d) at $f/16$ and at $f/8$.
(b) What is the new exposure time at $f/8$?

Solution

(a) At $f/16$, $d = \dfrac{f}{16} = \dfrac{50 \text{ mm}}{16}$

$= 3.125$ mm

At $f/8$, $d = \dfrac{f}{8} = \dfrac{50 \text{ mm}}{8}$

$= 6.25$ mm

∴ The aperture is about 3.1 mm at $f/16$ and about 6.2 mm at $f/8$.

(b) At $f/8$ the diameter of the lens is 2 times as large as at $f/16$, so the surface area is 4 times as large. Thus, the exposure time should be 1/4 of the original value, or 1/400 s.

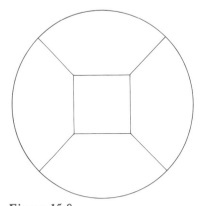

Figure 15-8
Five-segment Imaging

Did You Know?

The use of tiny computer systems in new SLR cameras can make photography easier for professional and amateur photographers alike. One use of such a computer is providing different exposure times for different segments of the film. The lens views the scene in individual segments in a pattern like that in Figure 15-8. The computer compares the brightness of the segments with numerous pre-programmed examples, and adjusts the camera to give the best possible exposure in each of the segments. This type of lens is not easily "fooled" by bright regions surrounding dark subjects, or *vice versa*.

Controlling the aperture makes possible special effects. At a small aperture ($f/22$), only the central portion of the lens is used, and much of the scene photographed is in focus. Such a photograph has a large **depth of field**, the range of object distances at which the image is clear.

Figure 15-9
This photograph has a relatively small depth of field. Only the first two geese are in clear focus. Beyond them the image becomes increasingly blurry.

At a large aperture (*f*/2), the outside portion of the lens is used. The result is a photograph with subjects in focus within a small range of distances from the lens. A photograph with a small depth of field can be seen in Figure 15-9.

This summary of cameras and photography merely skims the surface. Numerous reference books and magazines are available for interested learners.

PRACTICE

3. Describe the differences between a pinhole camera and a lens camera by comparing
   (a) their methods of exposure
   (b) the image obtained with each.
4. Describe how you would design a *camera obscura* to help you draw a portrait of a person's face.
5. State the function of each of the following:
   (a) shutter   (c) silver bromide
   (b) viewfinder   (d) focus
6. State some possible exposure times needed to obtain a clear photograph of the following:
   (a) a fast-moving object
   (b) a dimly lit room in a building
7. **Activity** If a SLR camera is available, use it to view the effect of adjusting the focus, aperture, and focal length as you look through the viewfinder.
8. Assume that a lens is in focus when you are viewing a distant object. You then aim the camera at a nearby object, which appears out of focus. Should the lens be moved closer to or farther from the film to obtain proper focus? Relate your answer to experimentation with lenses in the previous chapter. (*Hint*: Draw two ray diagrams using a lens of constant focal length. Use two distinctly different object distances, both of which must be greater than the focal length.)
9. Compare the light-gathering ability of a lens that has its aperture changed in the following ways:
   (a) from *f*/2 to *f*/4
   (b) from *f*/5.6 to *f*/1.4
   (c) from *f*/2.8 to *f*/22
10. A lens of 200 mm focal length is set at 1/60 s and *f*/11. What is the new exposure time if the aperture is changed to
    (a) *f*/16   and   (b) *f*/4?
11. A lens of 40 mm focal length is set at 1/250 s and *f*/5.6.
    (a) What is the diameter of the aperture?
    (b) What is the new aperture setting if the time is changed to 1/1000 s?
12. Assume that you are taking a photograph of a flower, with a sparkling brook in the background. Under what conditions would you obtain a low depth of field? Describe the effect of the low depth of field.

15.3 The Human Eye

The human eye is a fascinating optical instrument. The details of human vision are complex and not fully understood, but a brief explanation here will help you appreciate just how amazing your eyes are.

The human eye has many features that aid our vision. Figure 15-10 shows the basic structure of the eye.

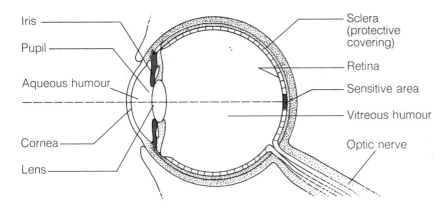

Figure 15-10
The Structure of the Right Eye (as viewed from above)

The average diameter of the eye ranges from 20 mm to 25 mm. This valuable instrument is protected by the tough, light-proof **sclera**, which covers the entire eye except the cornea. Light rays that enter the eye are focused by both the cornea and the lens. The **cornea** is a fixed transparent layer at the front of the eye; it has an index of refraction close to that of water. The cornea does about two-thirds of the focusing of the light rays. The converging **lens**, which is composed of a fibrous substance, then refracts the rays further, because it has a slightly higher index of refraction than the cornea. An inverted real image is focused on the back of the eye, just as the image in a camera is focused on the film.

The lens is somewhat flexible, so its shape can be controlled by the **ciliary muscles**. When you view a distant object, the ciliary muscles are relaxed, and the lens has a normal shape. When you view a nearby object, the muscles force the lens to become thicker so the image can remain in focus. This process of changing the focal length of the eye is called **accommodation**. It allows the human eye to have a large range over which it can focus on objects, although if you view nearby objects for a long time, the ciliary muscles remain under tension; the result is eyestrain. This problem is alleviated by looking at distant objects from time to time.

The **pupil** of the eye is the "window" through which light enters the lens. It appears black because most of the light that enters a human eye is absorbed inside. The pupil is surrounded by the **iris**, the coloured portion of the eye, which controls the size of the pupil. In bright light the pupil becomes small, and in dim light the pupil enlarges to let in more light. You have likely experienced walking into a dark room or theatre where you are unable to see until your pupils have enlarged.

One way to overcome this problem is to cover or close one eye for a while before you enter.

Two important liquids, the **aqueous humour** and the **vitreous humour**, help the eyeball maintain its shape. The aqueous humour, which is a watery liquid, has the added function of supplying cells to repair damage to the cornea or lens. The index of refraction of both these liquids is approximately equal to that of the cornea, but less than that of the lens.

At the rear of the eye is the light-sensitive **retina**, which consists of numerous blood vessels, receptors, and nerves. There are two types of receptors, rods and cones. The approximately 120 million **rods** in each eye are sensitive to black and white, while the more than 6 million **cones** are sensitive to colours. (Colour vision is described in greater detail in the next chapter.) When you look straight at an object, the clearest part of the image is located in the region marked "sensitive area" in Figure 15-10. See also Figure 16-12.

The light energy received by the rods and cones is transformed into electrical energy, which is transmitted via the retinal nerves to the larger **optic nerve**. This nerve sends the signals to the brain, which in turn interprets what you see. At the location in each eye where the retinal nerves join the optic nerve, a **blind spot** occurs. The blind spot is noticeable only when one eye is closed. Figure 15-11 illustrates how you can find your own blind spot.

Figure 15-12
Optical Illusions

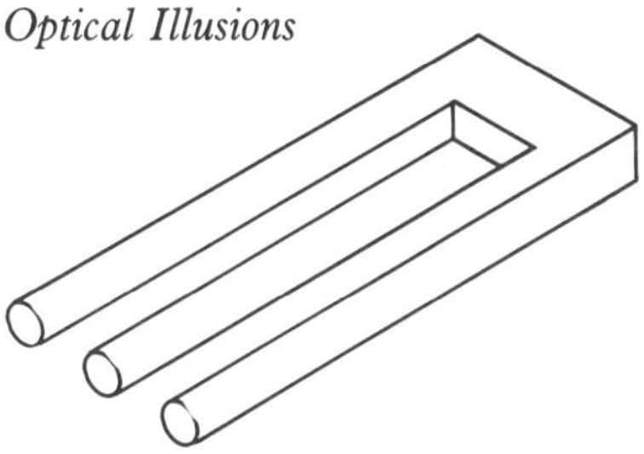

Figure 15-11
Determining Your Blind Spot

Hold the book at arm's length. Cover your left eye and stare at the number 1 with your right eye. Move the book toward you until the 2 disappears. This shows the location of your blind spot. If you move the book closer to or farther from you, the 2 will reappear.

(a) Our experience informs us that the figure shown has three prongs. However, on close inspection you would agree that the object would be impossible to build.

(b) Does the image alternate between a rabbit and a duck?

Our vision is also enhanced by other mechanisms. For example, having two eyes (binocular vision) rather than just one allows us to judge distances more accurately. Viewing a nearby object forces our two eyes to aim closer to each other and, with experience, we can then judge the approximate distance to the object. (If you studied Chapter 2, you will recognize that this situation relates to parallax.) Yet another example is the ability of our eyes to retain an image for about 1/25 s after the object is removed. If a new image replaces the first one after a short period (1/30 s, for example), the image appears to be in constant motion. This capacity of the eye allows the operation of motion pictures. Each movie frame is shown for a fraction of a second, but its image remains in the eye until the image from the next frame arrives.

The images we see may also be influenced by our experience and by the surroundings of what we are viewing. Figures 15-12(a) and (b) show two examples of optical illusions that trick the eye.

OPTICAL INSTRUMENTS

PRACTICE

13. Compare the action of the human eye with that of a lens camera. Include a description of the parts of the eye that correspond to the aperture, lens, diaphragm, and film of the camera.
14. What is the function of each of the parts of the eye listed below?
    (a) the cornea
    (b) the sclera
    (c) the ciliary muscles
    (d) the aqueous humour
    (e) the rods
    (f) the optic nerve
15. **Activity** Roll a sheet of paper into a tube and hold it in your right hand up to your right eye, as shown in Figure 15-13. Place your open left hand halfway along the tube and look through both eyes. Can you see through your left hand? Explain the illusion.
16. **Activity** Design and construct a device to illustrate the retention of images by the human eye. If you place one half of a diagram on one side of a card and the other half on the opposite side, the two diagrams will appear as one when the card is spun fairly quickly on a stick or tube.

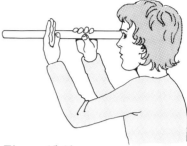

Figure 15-13

15.4 Vision Defects and Their Corrections

Normal, healthy eyes allow the formation of a clear image on the retina of the eye for objects at distances from about 25 cm (the **near point**) to infinity (the **far point**). A person with normal eyes is said to have 6/6 vision, which means that the eye sees clearly at a distance of 6.0 m from the object viewed. Thus, someone with 6/12 vision must be 6.0 m away from an object in order to see it as clearly as someone with normal eyes would at a distance of 12.0 m.

More than half the population of North America suffers from vision defects with the result the image received on the retina is not in focus. The four general types of focusing problems are discussed below.

In **myopia**, or **near-sightedness**, the image in the eye comes to a focus in front of the retina. This problem is usually caused by an eyeball that is too long or a cornea that is too sharply curved. To a person with myopic vision, distant objects appear blurry – out of focus. A diverging lens corrects this fault, as Figure 15-14 demonstrates.

Did You Know?
Since 6 m is approximately equal to 20 ft (feet) in the Imperial system, the former expression "20/20 vision" corresponds to the metric expression "6/6 vision", both of which indicate normal vision.

Figure 15-14
Myopia or Near-sightedness

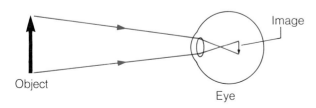

(a) The image comes to a focus in front of the retina.

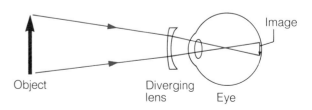

(b) A diverging lens corrects the defect.

In **hyperopia**, or **far-sightedness**, the image comes to an imaginary focus behind the retina. The usual cause is a shortened eyeball. To a person with hyperopic vision, it is objects close to the eye that appear blurred. Figure 15-15(b) shows that this defect can be corrected by using a converging lens. (Hyperopia is also called *hypermetropia*.)

Figure 15-15
Hyperopia or Far-sightedness

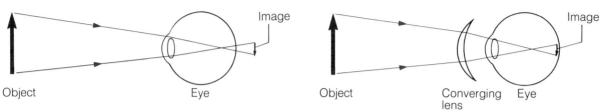

(a) *The image comes to a focus behind the retina.*

(b) *A converging lens corrects the defect.*

Astigmatism, or asymmetrical focusing, results when the cornea has an uneven surface which causes better focusing in one plane than another. To a person with astigmatic vision, equal-sized lines in different directions appear to have different thicknesses. Figure 15-16 gives a simple test for astigmatism. This defect can be corrected by using a cylindrical lens with a different focal length in one plane than in the other. It can also be corrected by wearing contact lenses, which cause a layer of tears to form between the contact lens and the cornea. This layer helps correct the astigmatism.

Figure 15-16
A Test for Astigmatism

Hold the book at arm's length. View the diagram with one eye at a time. If all the lines appear equally bright, you do not have astigmatism.

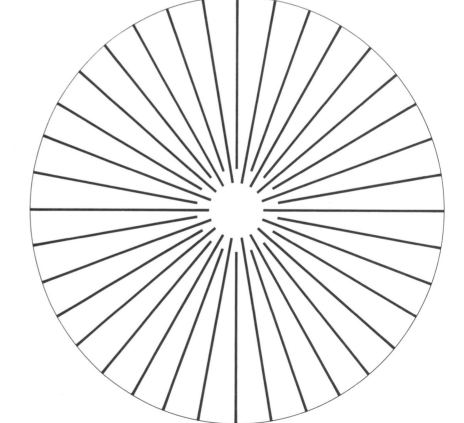

Presbyopia is a focusing problem that occurs when a person's eyes can no longer accommodate to view nearby objects. It is normally associated with advancing age. The lens becomes less pliable with continued use and the ciliary muscles have ever-greater difficulty in focusing the lens. To a person with presbyopic vision, nearby objects, especially printed words, appear out of focus. The problem can be corrected by using reading glasses with converging lenses. If a person who has presbyopia also has another vision defect, such as myopia, bifocal lenses (Figure 15-17) must be worn. The lower lenses are for reading.

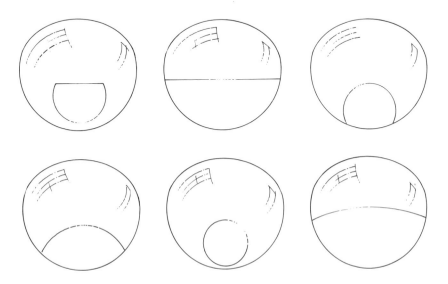

Figure 15-17
Typical Designs of Bifocal Lenses

A popular alternative to eyeglasses are contact lenses. Athletes and workers in some hazardous occupations who require corrective lenses often choose contact lenses for safety reasons, while many other people who wear them do so for cosmetic reasons. Contact lenses, which have been in use since the 1950s, are made of either hard or soft plastic. The lens rests on a film of tears on the cornea. Hard contact lenses require a long break-in period (perhaps up to one year), while soft lenses require a short break-in period because they are much more flexible and comfortable. However, soft contact lenses are more expensive than hard ones, and are not as effective in correcting serious astigmatism problems.

Colour blindness is a defect in which certain shades of colour are not clear to the sufferer. Its cause is that some cones in the retina do not respond to the light energy received. About 8% of all males and 0.5% of all females have some form of colour blindness. (Colour vision is discussed in greater detail in Chapter 16.)

The eye is a very sensitive optical instrument that can suffer from defects other than the common ones described above. We should take proper care of our eyes by reading in appropriate lighting conditions, wearing eye protection during sports and work activities, refraining from getting an excess of ultraviolet radiation, and in general, not straining our eyes.

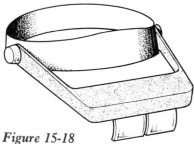

Figure 15-18
This "optivisor", which provides a magnification up to 3 times, can be tilted out of the way when not in use.

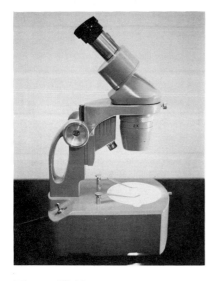

Figure 15-19
A Student Microscope

Figure 15-20
A Simple Refracting Telescope

PRACTICE

17. Which eye has better vision, one with 6/8 vision, or one with 6/16 vision? Explain your answer.
18. A person has a near point of 13 cm and a far point of 65 cm.
    (a) What type of defect does the person have?
    (b) Can the person see an object clearly at a distance of 5 cm? 25 cm? 80 cm?
    (c) What type of lens will allow the person to see distant objects clearly?
19. State the type of lens that is used to correct each of the following:
    (a) myopia   (c) hyperopia
    (b) astigmatism (d) presbyopia
20. **Activity** Analyse the vision problems of a friend or family member who wears framed corrective lenses by holding the lenses close to a printed page and using the following criteria:

    • A diverging lens held close to an object makes the object appear smaller.
    • A converging lens held close to an object makes the object appear larger.
    • A cylindrical lens, when rotated, causes the object viewed to appear to change shape.

21. Explain how each of the criteria listed in Practice Question 20 can be used to analyse a person's vision.

15.5 The Microscope and the Refracting Telescope

Lenses have four general functions, described in Section 15.1. We have discussed how lenses record images and improve impaired vision. Now we will discuss how they produce enlarged images and help us view distant objects.

A converging lens can be used as a magnifier in which the image appears two or three times larger than the object. Figure 15-18 shows a convenient device which can be worn by people, such as jewellers, who require a magnified view.

A magnification no greater than two or three times is insufficient for viewing small objects such as skin cells. To obtain larger images, two or more lenses are combined to make an instrument called a **microscope**, one type of which is shown in Figure 15-19.

Similarly, a single lens does not help the normal eye view distant objects. Two or more lenses must be used simultaneously to enlarge the image of distant objects such as stars, planets, and the moon. One optical instrument that uses lenses to make distant objects appear larger and closer is called a **refracting telescope**. It is the basis of operation of the surveyor's transit shown at the start of this chapter. See also Figure 15-20. (A reflecting telescope, which uses a converging mirror, was described in Chapter 13.)

OPTICAL INSTRUMENTS

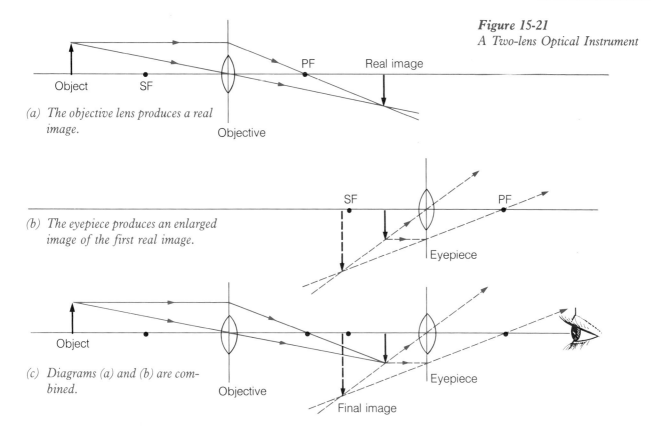

Figure 15-21
A Two-lens Optical Instrument

(a) The objective lens produces a real image.

(b) The eyepiece produces an enlarged image of the first real image.

(c) Diagrams (a) and (b) are combined.

It is possible for you to make a basic microscope or refracting telescope by using two converging lenses. The lens closest to the object is called the **objective lens**, or simply the **objective**, and the lens the eye looks into is called the **eyepiece lens** or **eyepiece**.

To see how an image is produced in a two-lens instrument, refer to the three ray diagrams in Figure 15-21. Figure 15-21(a) shows an object located beyond the focal point of the objective lens. The real image formed by the objective lens then becomes the "object" for the eyepiece in Figure 15-21(b). This object is located between the eyepiece and its secondary focal point. The final image, found by using the usual rules for ray diagrams, is virtual, inverted relative to the original object, larger than the original object, and located beyond the secondary focal point of the eyepiece. Figures 15-21(a) and (b) are combined to give (c), a complete ray diagram of a two-lens system.

Many microscopes and refracting telescopes have three or more lenses. Extra lenses are included to produce an even larger image or an upright image.

PRACTICE

22. Determine the magnification of the final image compared to the original object in Figure 15-21.
23. Does Figure 15-21 represent a microscope or a refracting telescope? In what way(s) would the diagrams differ if they were drawn for the other instrument?

Experiment 15B: The Microscope and the Refracting Telescope

INTRODUCTION
In any multi-lens optical instrument, the objective lens and all other lenses positioned before the eyepiece must produce a real image. Then the eyepiece will enlarge that real image. In order for this to happen, the eyepiece should be fairly close to the image being magnified.

In this experiment, both the microscope and the two-lens refracting telescope use two converging lenses and create an inverted image. A telescope that produces an inverted image is called an **astronomical telescope**; astronomers do not care if the stars they are viewing are inverted. To produce an upright image, a telescope requires a third lens. The third lens, called the **erector lens**, is located between the objective and the eyepiece. Such a telescope, generally used for viewing objects on the earth, is called a **terrestrial telescope**.

Converging lenses of different focal lengths are needed in this experiment. If the focal lengths of the lenses are not labelled, they can be found by aiming parallel light rays (from a ray box or a distant light source) toward the lens, finding the focal point on a screen, and measuring the focal length.

PURPOSE
To make and compare microscopes and telescopes with two or three lenses.

APPARATUS
converging lenses of various focal lengths; optical bench apparatus with corresponding supports

PROCEDURE

Part A: Comparing Lenses
1. Obtain two or three lenses of different focal lengths, and determine the focal lengths.
2. Determine how the magnification of a distant object viewed through a lens depends on the focal length of the lens. (Hold the lens about 30 cm from your eye.)
3. Determine how the magnification of an object near a lens depends on the focal length of the lens. (Again hold the lens about 30 cm from your eye.)

Part B: The Astronomical Telescope
1. Obtain two lenses of different focal lengths and determine the focal lengths. Choose which lens will be the objective and which will be the eyepiece. (*Hint*: Apply the results from Procedure Part A.)
2. Place the eyepiece in a holder at one end of the optical bench. Place the objective at a distance equal to the sum of the focal lengths

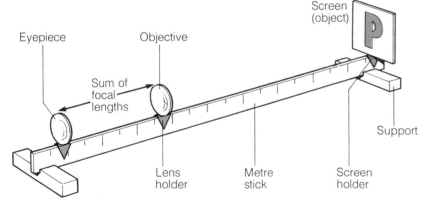

Figure 15-22
An Optical Bench Set-up

away from the eyepiece, as shown in Figure 15-22. Look through the eyepiece at a distant object and move the lenses and your eye until you obtain the clearest and largest image of the object. Describe that image.
3. Determine the magnification of the telescope by drawing two sets of equally-spaced lines on a piece of paper. View the lines from a distance through the telescope as well as with the unaided eye.
4. Try various combinations of lenses to discover which will give the largest image of the same object. For instance, try objective $f = 5$ cm and eyepiece $f = 10$ cm; objective $f = 5$ cm and eyepiece $f = 5$ cm. Describe your observations.

Part C: The Microscope
1. Obtain two lenses of relatively low focal lengths and determine the focal lengths.
2. Place the lenses in the supports on the optical bench so that they are separated by the sum of their focal lengths. Place the screen holder slightly beyond the focal point of the objective. Insert a piece of paper with small print into the screen holder.
3. Look through the eyepiece and move the lenses and your eye back and forth until you find the clearest and largest image. The image should be inverted and larger than the object if the system is acting like a microscope. Devise a technique for determining the magnification of the microscope.
4. Repeat Part C, Steps 1 to 3 using various combinations of lenses; for example, objective $f = 5$ cm and eyepiece $f = 5$ cm; objective $f = 5$ cm and eyepiece $f = 10$ cm. Try to discover which combination gives the largest and clearest image.

Part D: The Terrestrial Telescope
(*Note*: Attempt this part of the experiment only if you fully understand the previous ones.)
1. Choose three lenses, two of which worked well in Part B. Determine the focal lengths of all the lenses.
2. On the optical bench, arrange the three lenses so that the erector lens is located between the objective and the eyepiece. Your experience in the previous parts of this experiment should help you decide where to place the lenses.

3. Discover how to obtain an enlarged, clear, upright image of a distant object. Describe what you discover.

ANALYSIS
1. Three lenses, $f_A = 5.0$ cm, $f_B = 10$ cm, and $f_C = 20$ cm, are available for use in an optical instrument. Which one should you choose to be the following?
   (a) the objective of an astronomical telescope
   (b) the eyepiece of an astronomical telescope
   (c) the objective of a microscope
2. For the astronomical telescope, calculate the ratio of the focal length of the objective to the focal length of the eyepiece. How does this compare to the magnification found experimentally?
3. Describe the advantages and disadvantages of astronomical and terrestrial telescopes.
4. Draw ray diagrams to illustrate the formation of images in the instruments used in this experiment.

CONCLUSIONS . . .

15.6 Ray Diagrams for Other Optical Instruments

Ray diagrams provide a convenient means of determining the characteristics of images seen in optical instruments. If you have studied all of Chapters 13, 14, and 15 you should be able to draw ray diagrams for these optical devices: plane and curved mirrors; converging and diverging lenses; reflecting telescope; periscope; prism binoculars; pinhole camera; lens camera; human eye; correcting lenses for eyes; simple magnifier; microscope; refracting telescope. Other optical instruments are described below.

Opera Glasses

Opera glasses are conveniently small optical instruments that provide a low magnification. They are manufactured in a wide variety of forms, from simple toys to expensive status symbols. Their operation is based on the two-lens system invented by Galileo Galilei in 1609. That system,

Figure 15-23
The Galilean Telescope

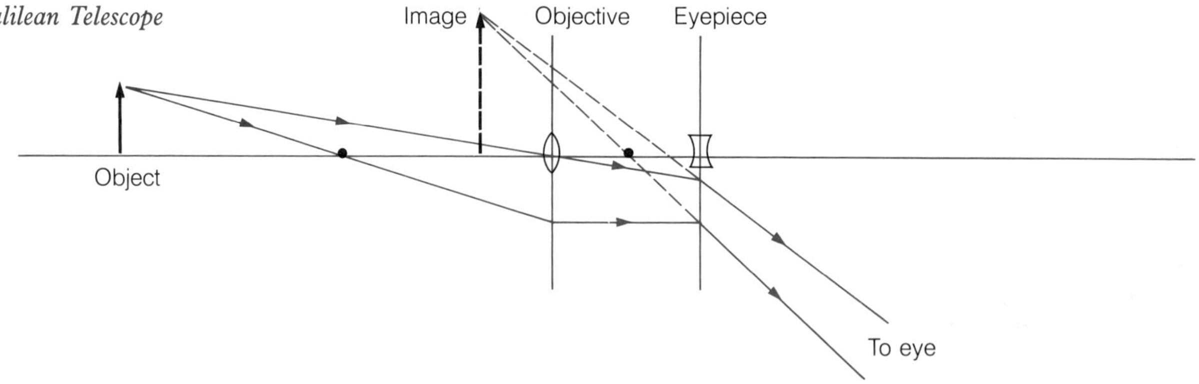

now called the "Galilean telescope", consists of a converging lens as the objective and a diverging lens as the eyepiece. Refer to Figure 15-23. In order to produce an upright image, the diverging lens must intercept the light rays from the objective before a real image is formed. Opera glasses made in this fashion have a typical magnification of about three times.

The Projector and Photographic Enlarger

The film or slide projector uses a converging lens (the objective) to place a real, enlarged image of a transparent photograph onto a screen. Figure 15-24 shows the main parts of a projector. The photograph, in the form of a film or a single slide, is located just beyond the focal point of the objective lens. A high-powered lamp is located at the focal point of a converging lens system called the **condenser**. Light rays emerging from the condenser are parallel to each other, thereby providing uniform illumination to all parts of the photograph.

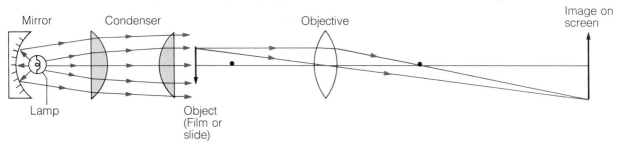

Figure 15-24
The Projector

The photographic enlarger operates on the same principles as the projector. However, the enlarger uses a piece of photographic paper in place of the screen.

The Photocopier

The photocopier, like the projector, utilizes several optical devices. Figure 15-25 shows the basic operation of the photocopier from the point of view of light. The subject to be copied is placed so that the bright light which scans it from one side to the other will be reflected to the first mirror, m_1, a front-surfaced mirror. The light then travels to the second front-surfaced mirror, m_2, where it reflects toward a converging lens (L). This lens produces a real image of the subject on a

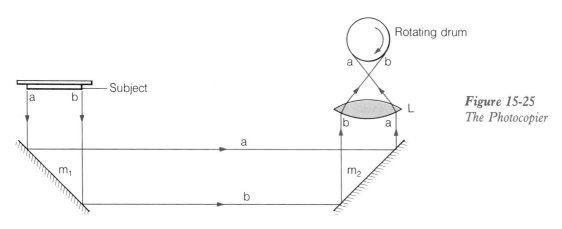

Figure 15-25
The Photocopier

piece of paper held onto the rotating drum. As the original light scans the subject from "a" to "b", the drum rotates from "a" to "b" as shown. A process using static electricity causes ink particles to adhere to the areas on the drum where no light arrived. The timing mechanism for the entire process is complex, and involves a thorough understanding of mechanics and electricity as well as light.

Of course, there are many optical instruments which have not been described in this book. Numerous references are available for interested students. One last important optical instrument, the laser, will be described in the next chapter.

PRACTICE

24. **Activity** Predict focal lengths of lenses that would provide both a clear image and ample magnification for a Galilean telescope (or a pair of opera glasses). Verify your prediction experimentally.
25. In the diagram of the projector, Figure 15-24, a converging mirror is shown as part of the lamp. What is the function of that mirror?
26. In the photocopier shown in Figure 15-25, both mirrors, m_1 and m_2, are front-surfaced mirrors. For what reasons are common mirrors with the silvered surface on the back of a piece of glass not used?

Review Assignment

1. Name two optical instruments that are used to do each of the following:
   (a) record images
   (b) improve impaired vision
   (c) produce large images
   (d) help view distant objects
2. What is the relationship between the magnification of the image in a pinhole camera and the distance between the camera and the object?
3. Pinholes are poked in the front of a pinhole camera in the shape shown in Figure 15-26. Draw a diagram of the image seen on the camera's screen when the camera is aimed toward a flame.
4. Describe the steps used in developing black-and-white photographic paper. What advice would you give to a person attempting the developing process for the first time?
5. Under what conditions is a photograph taken with a pinhole camera likely to turn out well?
6. How does a lens camera differ from a pinhole camera?
7. Place the advancements in photography listed below in chronological order.

   (a) introduction of colour film
   (b) use of lenses
   (c) invention of roll film

Figure 15-26

(d) first photograph ever taken
(e) use of the *camera obscura*
(f) discovery of the sensitivity of silver salts to light

8. State the main function of these parts of a lens camera:
   (a) diaphragm (b) zoom lens (c) aperture control
9. Figure 15-27(a) shows a lens of fixed vocal length focused on an object about 4 m from the camera; Figure 15-27(b) shows the same lens focused on the object from a different distance. Is that distance greater or less than 4 m? Explain your answer.
10. What happens to the light-gathering ability of a lens when its aperture changes in the following ways?
    (a) from $f/16$ to $f/2.8$ (b) from $f/4$ to $f/11$
11. A lens of focal length 75 mm is set for proper exposure at $f/11$ and 1/15 s.
    (a) What is the disadvantage of using an exposure time of 1/15 s?
    (b) If the exposure time is adjusted to 1/60 s, what is the correct aperture setting?
12. A lens of focal length 40 mm is set at 1/30 s and $f/22$ for the correct exposure of a scene.
    (a) Is the depth of field large or small?
    (b) If the aperture is now adjusted to $f/4$, what is the new exposure time?
    (c) At $f/4$, how does the depth of field compare to that at $f/22$?
13. State whether the image viewed in each of the following instruments is real or virtual:
    (a) pinhole camera   (e) simple magnifier
    (b) human eye   (f) microscope
    (c) television camera   (g) opera glasses
    (d) projector   (h) refracting telescope
14. State the main function of these parts of the human eye:
    (a) pupil (c) lens
    (b) iris   (d) retina
15. What is meant by the term "accommodation of the eye"? How is it controlled?
16. If you wish to see an object clearly, why is it best to look straight at the object?
17. **Activity** Set up an experiment to determine the angle from your line of vision ($\angle a$ in Figure 15-28) at which your blind spot occurs. Compare your value with that of other students.
18. (a) What feature of the eye allows us to enjoy movie cartoons?
    (b) Describe how you would make such a cartoon.
19. What changes occur in the normal human eye when it goes from looking at a brightly lit book to viewing a dark, distant object?
20. **Activity** Determine the approximate near point and far point of each of your own eyes. If you wear corrective lenses, perform this activity both with and without the lenses.
21. **Activity** Discover the "floating forefinger trick" in the following manner. At arms' length, position your outstretched index fingers horizontally in line with each other and about 6 mm to 8 mm apart.

Figure 15-27

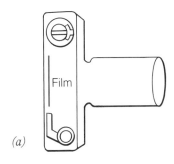

(a)

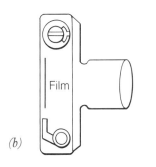

(b)

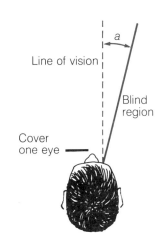

Figure 15-28

Aim your eyes just above your fingers at a distant object. Let your eyes relax. If you can see the "trick", explain it. If you cannot see it, try again.

22. In each case below, the near point and far point, respectively, of a person's vision are given. State the type of vision (or defect) for each person, and name the type of lens (if any) needed to correct the defect.
    (a) 15 cm; 12 m
    (b) 25 cm; infinity
    (c) 70 cm; infinity
    (d) 60 cm; 8.0 m

23. What is astigmatism, and how can it be corrected?

24. Draw a ray diagram to locate the final image in a two-lens instrument, using Figure 15-29 as a guide. Label your diagram and state the characteristics of the final image.

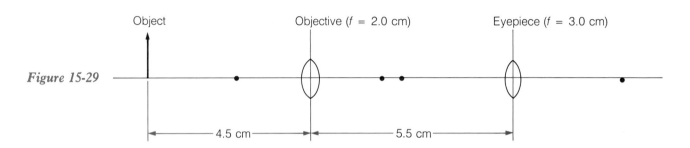

Figure 15-29

25. Describe the similarities and differences between a microscope and an astronomical refracting telescope.

26. What is the function of the erector lens in a terrestrial telescope? Is the image it produces real or virtual? Explain your answer.

27. The table below shows the lenses available in a certain physics lab for experimenting with optical instruments.

Type of Lens	Focal Length (cm)	Number Available
converging	50	2
converging	20	2
converging	5	2
diverging	20	2
diverging	5	2

State which lenses you would use to make each of the instruments listed below. Explain your choice.
(a) a compound microscope
(b) a refracting telescope
(c) a Galilean telescope
(d) a terrestrial telescope
(e) a projector (only the objective lens is required)

OPTICAL INSTRUMENTS

Key Objectives

Having completed this chapter, you should now be able to do the following:
1. List four functions of optical instruments, and describe examples of instruments that perform each function.
2. Draw a ray diagram to locate the image in a pinhole camera, and describe that image.
3. Develop photographs taken with a pinhole camera and state how to control the camera to obtain successful photographs.
4. Recognize the historical developments and current importance of photography.
5. Describe the functions of the main parts of a lens camera.
6. List the four variables that can be controlled using a single-lens reflex camera, and describe the effect of altering each variable.
7. State the meaning of f-stop, and describe its relationship to exposure time.
8. Given the f-stop and the focal length of a lens, calculate the aperture.
9. Given the correct f-stop and exposure time for a particular scene, calculate other combinations of f-stop and exposure time that will yield the proper exposure.
10. Define the depth of field of a lens, and describe how to vary it.
11. State the names and functions of the main parts of the human eye.
12. Compare and contrast the use of the pinhole camera, the lens camera, and the human eye.
13. State what advantages we gain because our eyes have accommodation power, binocular vision, and image retention.
14. Explain various optical illusions.
15. State the meaning of 6/6 vision and 6/x vision, where $x \neq 6$.
16. Recognize the meaning of the near point and far point of the human eye.
17. Describe defects of the human eye and corrections for those defects.
18. Describe advantages and disadvantages of wearing contact lenses.
19. Given a lens, determine its focal length experimentally.
20. Draw a ray diagram to show how an enlarged image is produced by optical instruments such as a simple magnifier, a refracting telescope, and a microscope.
21. Given a choice of lenses, make each of the following optical instruments: an astronomical telescope; a compound (two-lens) microscope; a terrestrial telescope (optional); and a Galilean telescope.
22. Explain how to determine the magnification of a two-lens system.
23. Describe the basic operation of a pair of opera glasses, a projector, and a photocopier.

CHAPTER 16
Colour and Light Theory

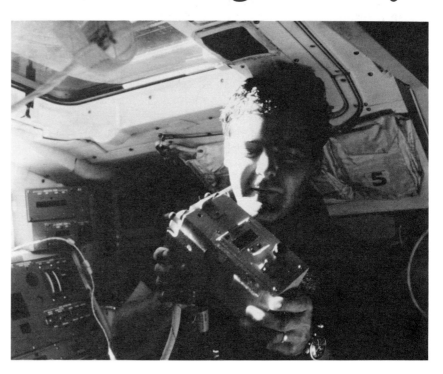

Light has been studied experimentally for hundreds of years. Here, measurements of light from the sun form part of a space shuttle mission.

An important characteristic that humans have in common with bees, apes, and goldfish is the ability to see colours. Most animals' eyes are sensitive to black and white only. To us, their world would seem very dull.

The human eye is sensitive to a range of colours which form part of a larger range of energies. Although much is known about light and our sense of sight, much remains to be learned. An example of modern experimentation, performed by a Canadian astronaut aboard a space shuttle, is shown in the photograph. The hand-held instrument aimed at the sun is used to obtain data which will increase our knowledge of light.

In this chapter, you will study the properties of light that allow us to see different colours, as well as theories used to explain light. The experience should help you answer such questions as these:

- What is white light?
- Why does a red rose appear red in white light?
- Why does a red rose appear black in blue light?
- How are the colours on a colour television screen produced?
- What causes rainbows?
- How does a laser operate, and why are its applications important?

16.1 The Dispersion and Recombination of White Light

For thousands of years it was known that precious jewels such as diamonds cause sparkling colours to appear when viewed in white light. Philosophers and scientists alike thought that the colours were created by something within the jewels. Then, in 1666, when he was just 23 years old, Sir Isaac Newton performed an important experiment verifying his hypothesis that colours are a property of light.

In the first part of his famous experiment, Newton passed a beam of sunlight through a triangular glass prism and discovered that white light can be separated into many colours, a process called **dispersion**. See Figure 16-1. The resulting band of colours, called the **visible spectrum**, is made up of colours that blend into one another and range from red to violet. See also Colour Plate 1.

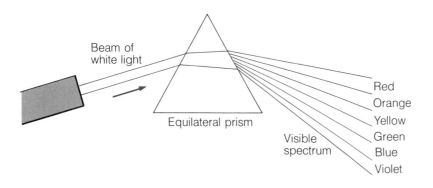

Figure 16-1
Demonstrating the Dispersion of White Light

(Before you read on, it would be wise to perform the activities in Practice Questions 1 and 2. They will help you better understand the remainder of this section.)

Dispersion of white light occurs because the different colours of light making up the visible spectrum are refracted at different angles upon travelling from one material into another. The amount of refraction of each colour depends on the change in the speed of the light upon entering the second material. (This change, in turn, depends on the wavelength of the light, a variable that will be discussed later in the chapter.)

Since the speed of light in a material is related to the material's index of refraction, we can use the index to indicate the order in which the spectral colours occur. Table 16-1 on page 384 lists the index of refraction for the six basic colours of the spectrum travelling from air into crown glass, a very clear type of glass used in making optical instruments. Each index listed is an average of the range of indexes of refraction for the colour named.

Notice in Table 16-1 that the lowest index of refraction is that of red light. This indicates that red is the colour which slows down the least and thus is refracted the least.

LIGHT AND COLOUR

Table 16-1 The Index of Refraction for the Spectral Colours Travelling from Air into Crown Glass

Colour	Index of Refraction
red	1.514
orange	1.516
yellow	1.517
green	1.520
blue	1.524
violet	1.531

Newton reasoned that, if white light is composed of several colours, he should be able to combine them to produce white light. In the second part of his famous light experiment, he performed the **recombination** of the spectral colours to obtain white light. This provided further evidence that colours are a property of light. Figure 16-2 shows three techniques for recombining the spectral colours. The first two use refraction of light to bring the colours together. The third technique, using a quickly rotating set of colours called "Newton's disk", takes advantage of the image retention of the eye.

Figure 16-2
Recombination of Spectral Colours

(a) Using a triangular glass prism

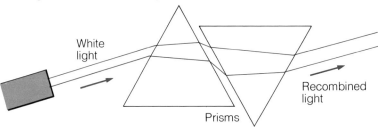

(b) Using a converging lens

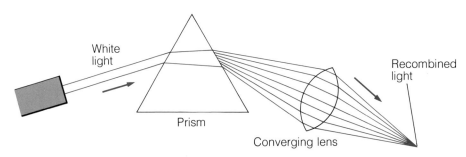

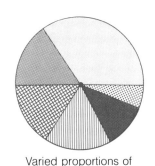

Varied proportions of spectral colours

(c) Using a Newton's disk (The disk must be spun rapidly in bright white light so that all the colours strike the eye before the previous images leave.)

Thus, to understand why a diamond sparkles with brilliant colours, we must realize that white light disperses greatly upon entering the diamond from air. (Recall from Chapter 14 that diamond has an index of refraction of 2.42.) Then each colour undergoes numerous total internal reflections, causing the colours to be well separated when they emerge from the diamond.

PRACTICE

1. **Activity** Set up an experiment to demonstrate the formation of the visible spectrum using a white beam of light from a ray box or projector. (Refer to Figure 16-1.) Draw a diagram of the spectrum produced, and compare the angle of deviation for red and violet.

The **angle of deviation** is the angle between the direction of the emergent ray and the direction of the initial incident ray in the air.

2. **Activity** Perform an experiment using at least one of the techniques shown in Figure 16-2 for recombining the spectral colours. Is the resulting white as pure as the original white? Explain.
3. Sunlight consists of more radiant energies than are visible to our eyes. The sun also emits invisible radiations called **infrared** and **ultraviolet**. Include these radiations in the positions where you think they belong in your diagram for Practice Question 1. How do you think the prefixes "infra" and "ultra" relate to the energies of the radiations?
4. A certain converging lens has a focal length of 10 cm for yellow light. Does the focal length increase or decrease for the following colours of light?
   (a) red   (b) green   (c) violet
5. White light is aimed toward crown glass at an angle of incidence of 40° C. Calculate the angle of refraction for the following components of the light:
   (a) red   (b) blue
6. Why did you not observe dispersion when you viewed the emergent ray from the rectangular prism in Experiment 14A?
7. What is the critical angle for orange light in crown glass?
8. Calculate the speeds of red light and violet light in crown glass.

Experiment 16A: Adding Light Colours and Viewing Colour Shadows

INTRODUCTION
In Section 16.1 you learned that white light can be separated into its spectral colours using a single prism, and that those colours can be recombined to give white light. However, not all the colours of the spectrum are needed to produce white light. Only three of them, known as the **primary light colours**, are required.

Two colours observed in this experiment may be new to you, so they are defined here. *Cyan* is a greenish-blue colour, and *magenta* is pinkish-purple.

When observing colours in this experiment you may notice that what should be white appears to be yellowish. Better results can only be obtained with high-quality equipment not always available in schools.

PURPOSE
(a) To study the addition of light colours.
(b) To observe and describe colour shadows.

APPARATUS
ray box with 2 mirrors at one end (or 3 ray boxes); set of 6 plastic or glass filters (red, green, blue, yellow, cyan, and magenta); white screen

LIGHT AND COLOUR

PROCEDURE

1. Place a green filter and a red filter in the ray box, as shown in Figure 16-3(a). Adjust the mirror so that the two colours of light overlap on a nearby white screen. Record the colour you observe at the overlap.

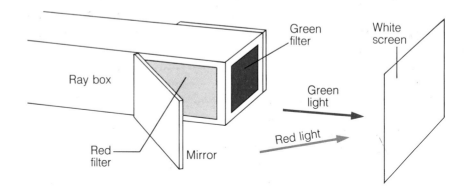

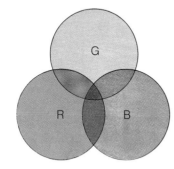

Figure 16-3
Adding Light Colours

(a) Adding red light and green light

(b) Colour chart

2. Repeat Procedure Step 1 for the following colour combinations:
   (a) green and blue
   (b) blue and red
   (c) red, green, and blue
   (d) blue and yellow
   (e) red and cyan
   (f) green and magenta

   (*Note:* Try interchanging the positions of the filters in each case to determine whether you can improve your results.)

3. Draw a colour chart like the one in Figure 16-3(b). Use the first letter of each colour (R, G, B, Y, C, M, and W) to complete it. Ask your teacher to check your chart before you continue.

4. With the red, green, and blue filters placed in separate positions at the end of the ray box, aim the colours so that they overlap on the screen. Place a pen or pencil between the ray box and the screen to obtain three colour penumbras and a black umbra. Explain how the shadows occur. (Umbras and penumbras were described in Section 13.2.)

ANALYSIS

1. List the three primary colours of light.
2. List the three **secondary light colours**, which result from the mixing of pairs of primary colours.
3. List three sets of **complementary light colours**, which are two colours, one primary and one secondary, that when added together produce white light.
4. In Figure 16-4, R is a source of red light and G is a source of green light. Both are aimed in an otherwise dark room toward an opaque object. Redraw the diagram in your notebook and label all the colours (as well as black) within the rectangle. (Two parts of the diagram are already labelled.)
5. Repeat Analysis Question 4, using a third light source (blue) located between the other two sources.

CONCLUSIONS . . .

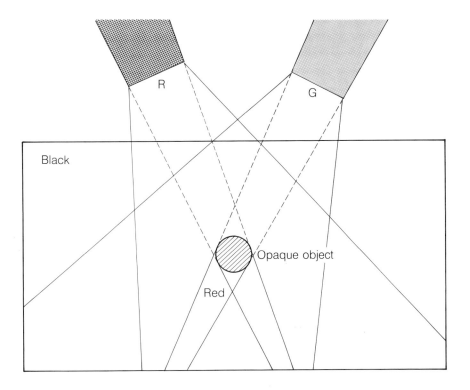

Figure 16-4
Colour Shadows

16.2 Additive Colour Mixing

Additive colour mixing is the process of adding light to light, observed in the previous experiment. When separate colours such as red, green, and blue are added together, the light becomes lighter and brighter. See Colour Plate 2.

An interesting application of additive colour mixing is in theatrical and musical productions. A bank of spotlights with primary colour filters can be used to produce numerous shades of colour falling on the stage. For example, if green and blue lights of equal intensity strike the stage, the resulting colour of a white object is cyan. If a more greenish colour is required, the intensity of the blue is reduced and that of the green is increased.

It is important to distinguish between adding and subtracting light colours. Colours can be subtracted by being absorbed when the light strikes an opaque object or a coloured transparent filter. The next experiment deals with the subtraction of light. Later in the chapter you will study applications that combine both additive and subtractive colour mixing.

PRACTICE
9. The lighting crew in a live theatre uses lights with red, green, and blue filters. What filters should the lighting crew use if the director wants a white object to appear the following colours?
   (a) yellow   (b) magenta   (c) white

Experiment 16B: Subtracting Light Colours

Figure 16-5
Predicting the Colour after Reflection and Transmission

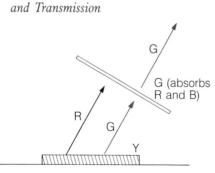

(a) A yellow object viewed through a green filter appears green.

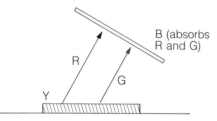

(b) A yellow object viewed through a blue filter appears black.

INTRODUCTION

In the previous experiment, colours of light were added together to yield lighter colours. In this experiment, colour filters will be used to absorb or subtract certain colours. For instance, a blue filter is a piece of plastic or glass that transmits blue light but absorbs red and green light. (Most filters do not have pure colours, so the absorption is not total.)

If your colour chart based on Figure 16-3(b) is correct, it can be used to predict which light energies are transmitted by colour filters or reflected by coloured opaque objects. An object with a primary light colour reflects only its own colour. An object that has a secondary light colour reflects the two primary colours of which it is composed. Thus, a yellow rose is yellow in white light because it reflects red and green light. Of course, it also reflects yellow light.

Figure 16-5 illustrates how to predict the colour viewed when light that is reflected off an opaque object is transmitted through a filter. The prediction is not always accurate, however, because the colours of opaque objects and filters are not pure.

You may wish to design an alternate procedure for this experiment. One alternative is to mix small quantities of poster paints and view the results in white light. Another is to look at an opaque object through two colour filters rather than one for each observation. Yet another alternative is to use coloured lights while viewing coloured objects.

PURPOSE
To study the subtraction of light colours.

APPARATUS
set of colour filters (or coloured lights); set of coloured opaque objects (Both sets should include these colours: red, green, blue, yellow, cyan, and magenta. Note that the colours shown in Colour Plate 2 could be used as the coloured "objects".)

PROCEDURE

1. In your notebook, set up a table of observations similar to Table 16-2. Use your colour chart to predict the resulting colours and enter the colours in the appropriate spaces.

2. View a red object through each of the colour filters, one at a time. Record the colours seen in the appropriate spaces in the table of observations.

3. Repeat Procedure Step 2 for a green object, then a blue one, and so on. Complete the table.

COLOUR AND LIGHT THEORY

Table 16-2 Observations for Experiment 16B

Colour of Filter or Light	Colour of Opaque Object in White Light					
	red	green	blue	yellow	cyan	magenta
red						
green						
blue			SAMPLE ONLY			
yellow						
cyan						
magenta	• •					

Predicted Colour   Observed Colour

ANALYSIS

1. State the primary light colours reflected by an opaque object of each of these colours:
   (a) red   (b) yellow   (c) cyan
2. What primary light colours are transmitted by these light filters?
   (a) green   (b) magenta
3. White light strikes a red rose. Why does the rose appear red?
4. Explain why a red rose viewed through a green filter or in green light should appear black.
5. Explain why a magenta sweater viewed in red light or through a red filter should appear red.

CONCLUSIONS . . .

16.3 Subtractive Colour Mixing

You have likely had the experience of mixing two paints or colour pigments together to obtain a third colour. For example, if you mix yellow and cyan pigments, you obtain green pigment. The process of mixing colour pigments together to obtain new colours is called **subtractive colour mixing**.

To understand why colours are subtracted when pigments are mixed together, we will summarize the facts based on the previous experiment. Filters and opaque objects are similar to each other in that they absorb (subtract) certain colours. Figures 16-6 and 16-7 on page 390 show what happens theoretically when white light, which can be considered to consist of the three primary light colours, strikes filters and opaque objects. Study these figures, and work through Sample Problems 1 and 2.

LIGHT AND COLOUR

Figure 16-6
These diagrams illustrate the colours that can be transmitted by colour filters struck by white light.

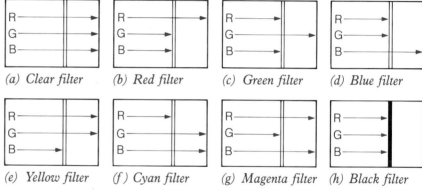

(a) Clear filter (b) Red filter (c) Green filter (d) Blue filter

(e) Yellow filter (f) Cyan filter (g) Magenta filter (h) Black filter

Figure 16-7
These diagrams illustrate the colours that can be reflected by opaque objects struck by white light.

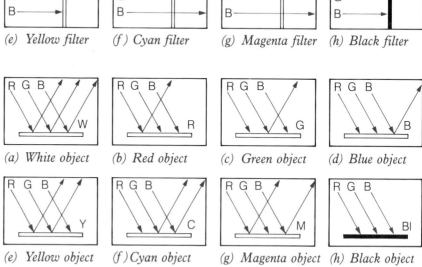

(a) White object (b) Red object (c) Green object (d) Blue object

(e) Yellow object (f) Cyan object (g) Magenta object (h) Black object

---

SAMPLE PROBLEM 1
What colour is transmitted when cyan light (B,G) strikes the following colour filters?
(a) blue   (b) red   (c) yellow

Solution
(a) blue (A blue filter transmits blue. Cyan is made up of blue and green.)
(b) no colour, or black (A red filter absorbs both green and blue.)
(c) green (A yellow filter absorbs blue.)

SAMPLE PROBLEM 2
What colours are reflected when magenta light strikes a yellow object?

Solution
Magenta light is made up of red and blue. A yellow object absorbs blue, so only red can be reflected. The object therefore appears red.

Colour pigments absorb light energies just as filters and opaque objects do. Colour pigments, however, can be mixed together to absorb other light energies. Figure 16-8 gives two examples of mixing colour pigments.

COLOUR AND LIGHT THEORY

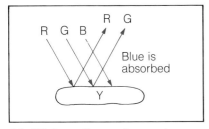
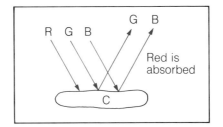
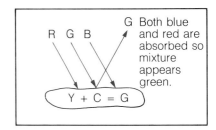

(a) Mixing yellow and cyan pigments

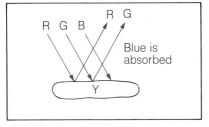
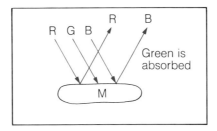

(b) Mixing yellow and magenta pigments

Figure 16-8
Mixing Colour Pigments in White Light

Three colour pigments can be mixed to yield black, or almost black, pigment. Those three pigments, yellow, cyan, and magenta, are called the **primary pigment colours**. When two pure primary pigments are mixed, a **secondary pigment colour** results. The secondary pigment colours are red, green, and blue. Try not to confuse light colours with pigment colours. The primary light colours (R, G, and B) are the secondary pigment colours, and the secondary light colours (Y, C, and M) are the primary pigment colours. See Colour Plates 2 and 3.

The predicted results of either adding or subtracting light colours are theoretical and cannot be exact. The scientific explanation of colour, especially colour vision, is not fully understood, and the process of mixing light or pigments is much more complex than this book indicates. Thus, we must accept that some of our observed results may not coincide with current theoretical predictions.

PRACTICE

10. State the colour(s) transmitted in each case:
    (a) White light (R, G, B) strikes a red filter.
    (b) White light (R, G, B) strikes a cyan filter.
    (c) Red light strikes a green filter.
    (d) Yellow light (R, G) strikes a red filter.

11. State the colour(s) reflected off each opaque object:
    (a) White light (R, G, B) strikes a magenta object.
    (b) Blue light strikes a green object.
    (c) Blue light strikes a magenta object.
    (d) Cyan light strikes a yellow object.

12. What colour results when the following pigments are mixed?
    (a) Y + M   (b) Y + C   (c) C + M   (d) Y + C + M

LIGHT AND COLOUR

Figure 16-9
Observer Location for Viewing a Rainbow

Figure 16-10
Formation of a Primary Rainbow

16.4 Applications of Colour

The Rainbow

To see a rainbow in the sky, an observer must be located between the raindrops and the sun, as Figure 16-9 indicates. The sun's rays travel to the water droplets, where some of them are internally reflected. Then they emerge from the droplets and travel to the observer. The rainbow that results from single internal reflections is called the **primary rainbow**. Other orders of rainbows, which are much more difficult to observe than a primary rainbow, occur when the light is internally reflected more than once. For instance, a secondary rainbow, which is located higher in the sky than the primary, occurs after two internal reflections. See Colour Plate 4.

To understand how the spectral colours are produced in a rainbow, consider Figure 16-10(a). It shows a beam of light entering a single drop of water, where it is refracted and disperses into its spectral colours. The various colours are reflected off the inner surface of the droplet (total internal reflection). As the light leaves the droplet, it is refracted again, but remains split into the same colours as those observed in the dispersion experiment. In Figure 16-10(a) violet light is entering the observer's eye from a region of the sky approximately 40° to the line of sight. The observer would have to look slightly higher in the sky to see the colours blue through red from the other droplets. The resulting arrangement of colours of the primary rainbow is shown in Figure 16-10(b).

(a) *White light is dispersed in each raindrop and undergoes a single internal reflection.*

(b) *Arrangement of primary rainbow colours*

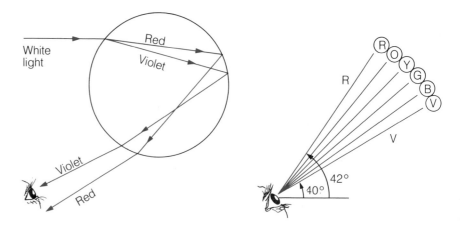

Figure 16-11 shows that a secondary rainbow occurs in an arc at an angle of about 52° to the observer's line of sight. See also Colour Plate 4.

COLOUR AND LIGHT THEORY

Figure 16-11
Formation of a Secondary Rainbow

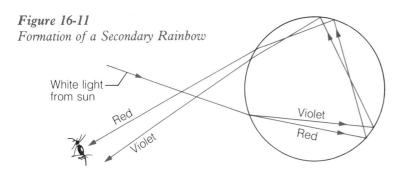

(a) White light is dispersed in each raindrop and undergoes two internal reflections.

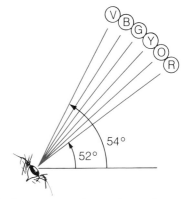

(b) Arrangement of secondary rainbow colours

Colour Vision

The normal human eye is well adapted to see colours of light. The retina at the back of the eye has two types of receptors, rods and cones, as we discussed in Chapter 15. The rods are sensitive to black and white, while the cones, which are what we are concerned with here, are sensitive to colours. See Figure 16-12.

It is generally thought that there are three types of cones, each type responding to one primary light colour (red, green, or blue). The cones can respond in various combinations to give all the shades of colour we see. For instance, if the cones sensitive to green and blue are stimulated equally, we see cyan.

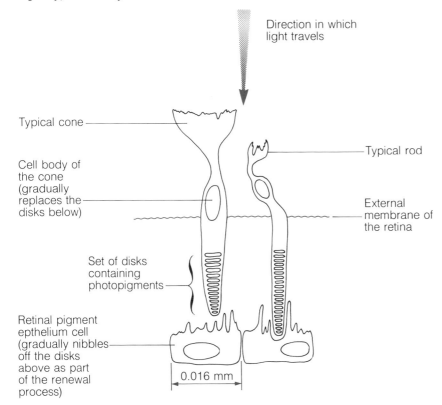

Figure 16-12
The Rod and Cone Structure of the Eye

LIGHT AND COLOUR

An interesting demonstration of evidence supporting the three-cone theory involves staring at a coloured object until the cones in your eyes become fatigued. To observe **retinal fatigue**, place a small, bright red object on a white background. Stare at the centre of the object for about 40 s. Then stare intently at one spot on a piece of white paper. Can you explain what you observe?

When you stare at a red object, the cones sensitive to red in your retina become tired. Therefore, when you stare at a white surface, which reflects red, green, and blue light, only the cones sensitive to green and blue respond normally. Thus, you observe the colour cyan in the shape of the original red object.

The defect of colour blindness, discussed briefly in Section 15.4, occurs in many different forms, some more serious than others. In one form of colour blindness, the red cones do not exist and the number of colours seen is much fewer than normal. In another type of colour blindness, the red and green cones respond simultaneously to all colours of the red-to-green end of the spectrum, so most reds and oranges appear to be yellow. Other forms of colour blindness are usually less severe.

Colour vision is very complex, and the theory that three types of cones are responsible for colour vision does not adequately explain all observations about the eye. Scientists are still researching the question of how we see colour to discover what other factors are involved.

Colour Television

The operation of a colour television is based on additive colour mixing. A colour television screen is coated with an orderly pattern of thousands of sets of bars (or sometimes dots), each set arranged in groups of three. Every bar consists of a phosphorescent material which, when struck by high-speed electrons from the television tube, emits one of the three primary light colours. The bars are called red, green, and blue, depending on the colour they emit.

Three electron guns, one for each bar colour, are located at the rear of the television. At selected instants they send high-speed electrons toward the bars on the screen. Each time a bar is struck by electrons it gains energy, then emits much of that energy in the form of light. To produce blue light, only blue bars are struck. To produce yellow light, both green and red bars are struck, while black results when no bars are struck. A great range of colours can be produced by proper control of the bars to be struck by electrons. See Figure 16-13.

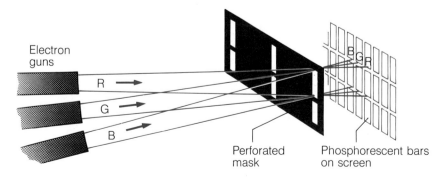

Figure 16-13
The Basic Operation of a Colour Television

PRACTICE

13. In which direction, east or west, should an observer look to try to see a rainbow in the early (a) morning and (b) evening?
14. **Activity** Design and carry out an experiment to view a primary rainbow (and a secondary one, if possible) using a fine spray from a garden hose. The sun should be no higher in the sky than 42° from the horizontal. Determine the angle between the horizontal and the primary (and/or secondary) rainbow.
15. Which cones in the human eye must be activated in order to see (a) magenta and (b) cyan?
16. If you stare at a bright yellow circle and then at a white piece of paper, you see a blue circle. Explain why this happens.
17. In a colour television, which phosphorescent bars must be struck by high-speed electrons in order to produce the following?
    (a) yellow   (b) white   (c) black

16.5 Light Theories and the Electromagnetic Spectrum

In Chapter 13 you were asked to consider whether the particle model or the wave model was preferable for explaining the observed behaviour of light. This question is not easy to answer, and has perplexed scientists for centuries. Now that you have studied several concepts related to light, you will be able to appreciate the development of light theories from the seventeenth century to the present.

During Sir Isaac Newton's lifetime, both a particle model and a wave model of light were proposed. The effects of coloured fringes as light was diffracted through small openings had been observed, and some scientists tried to explain these effects with a wave theory. Newton, however, favoured the particle theory, and proposed that diffraction could occur as a result of the interaction of tiny light particles with the edges of an opening.

There were many other observations providing evidence in favour of one theory or the other. However, because Newton leaned toward the particle theory, and because his works on mechanics were so well established, the particle theory tended to predominate the issue until the end of the 1700s.

In the early 1800s, physicists verified through experimentation that light displays certain properties of waves. Using the interference effects that occur when light is diffracted through two very narrow, closely spaced slits, the approximate wavelengths of the colours of visible light were found. Thus, the particle theory faded from importance and the wave theory tended to predominate. (The interference of pulses and waves on ropes and springs was described in Chapter 10. The concepts there related to interference in one dimension only. The interference pattern that occurs when light passes through a double slit is an easily observed two-dimensional phenomenon. However, the mathematical

LIGHT AND COLOUR

Figure 16-14
Heinrich Hertz (1857–1894)

Figure 16-15
Max Planck (1858–1947)

analysis of two-dimensional interference was not considered in this text because of space limitations.)

Science in the 1800s, like science today, continued to advance. Physicists did research to discover more about the nature of light. By the mid-nineteenth century, the speed of light in water had been determined, offering further evidence supporting the wave theory of light.

Then, in the late 1800s, new discoveries showed that it was necessary to rethink the wave theory once again. The German physicist Heinrich Hertz (1857–1894) (Figure 16-14) discovered that, under certain conditions, light energy can cause electrons (negatively charged particles) on the surface of a metal to escape. This phenomenon, called the **photoelectric effect**, can be explained only if light is assumed to have some properties of particles. In 1900, another German scientist, Max Planck (1858–1947) (Figure 16-15), developed a theory of light combining the particle and wave theories. He proposed that light consists of tiny bundles of energy called **photons** or **quanta** of energy. ("Quanta" is the plural of the Latin *quantum*, which means "distinct amount".) According to Planck, each photon has its own unique energy (a quantum of energy). This theory differs from the wave theory, which proposes a continuous set of energies. Thus, the **photon theory of light** states that light consists of distinct packages of energy (called photons) that display both wave-like and particle-like properties.

Today, the photon theory of light is widely accepted because experimental evidence continues to support it. Photons of visible light are created when electrons within an atom drop from a higher energy level to a lower one. The photon energies that result from these transitions are transmitted by waves called **electromagnetic waves** (e.m. waves). (Electric and magnetic forces are discussed in Unit VI.)

Electromagnetic waves have properties similar to those of other waves, which were detailed in Chapter 10. A brief outline of the important definitions and equations is repeated here for convenience. If the source of e.m. waves is periodic, then the source is said to have a frequency of vibration. Frequency (f) is the number of vibrations per second, measured in hertz (Hz). After a wave is produced by an acceleration it moves outward from its source, transferring energy elsewhere. The wave travels at a speed (v) which depends on the medium through which the wave is travelling. In air, for example, light travels at a speed of 3.00×10^8 m/s. The travelling wave has a wavelength (λ), which is simply the length of one wave. The universal wave equation ($v = f\lambda$) relates the speed of a periodic wave to its frequency and wavelength. From this equation it follows that as the frequency of the source increases, the wavelength of the photons decreases.

Each colour of the visible spectrum has its own range of frequencies and wavelengths. The energy transitions that produce red light have a lower frequency than the transitions that produce the other colours. Thus, red light has the longest wavelengths and violet the shortest in the visible spectrum. Figure 16-16 shows the approximate ranges of frequencies and wavelengths of the colours of the visible spectrum. Notice that as the frequency increases, the energy associated with the waves also increases. See also Colour Plate 1.

COLOUR AND LIGHT THEORY

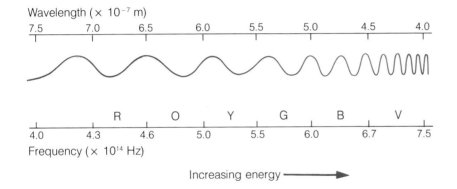

Figure 16-16
The Visible Spectrum

SAMPLE PROBLEM 3
A sodium lamp emits a bright yellow light with a wavelength in air of 5.89×10^{-7} m. Calculate the frequency of this light.

Solution

$f = \dfrac{v}{\lambda}$ (from $v = f\lambda$)

$= \dfrac{3.00 \times 10^8 \text{ m/s}}{5.89 \times 10^{-7} \text{ m}}$

$= 5.09 \times 10^{14}$ s^{-1} or Hz

∴ The frequency is 5.09×10^{14} Hz.

Visible light is only a small portion of the entire set of electromagnetic waves. The entire set, called the **electromagnetic spectrum**, is illustrated in Figure 16-17. Once again, notice that high-frequency waves such as gamma rays and X rays have high energies. Also note that infrared light, which is often used for heat treatment, has lower frequencies and energies than visible light. Ultraviolet light, the type of light that causes sunburns and can also lead to skin cancer, has higher frequencies and energies than visible light.

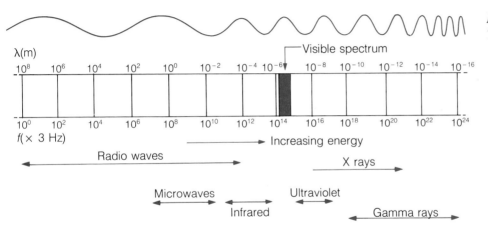

Figure 16-17
The Electromagnetic Spectrum

LIGHT AND COLOUR

Where does the scientific world stand now in its quest for an understanding of the nature of light? Though we have advanced tremendously during the twentieth century toward a better understanding of light, as our use of lasers and fibre optics amply demonstrates, we must realize that the current theories are still limited. We must therefore strive to learn even more. As the twenty-first century approaches, we can look forward to many varied and exciting developments in this area of science.

PRACTICE

18. One theory of light states that, as light approaches a more dense medium from air, it is attracted to the particles of the more dense medium. This attraction causes the light to speed up and thus bend toward the normal as it enters the medium.
    (a) Which theory of light explains refraction in this manner?
    (b) What is the weakness of the above explanation?
19. **Activity** Many scientists have influenced the development of the wave theory of light. Research and report on the contributions made to this theory by one or more of the following scientists:
    (a) Francesco Grimaldi (1618–1663)
    (b) Christian Huygens (1627–1695)
    (c) Thomas Young (1773–1829)
20. One of the bright lines emitted from a hydrogen gas source of light has a wavelength of 4.10×10^{-7} m.
    (a) Calculate the light's frequency.
    (b) What is the colour of the bright line?
21. To which colour of the visible spectrum do you think the human eye is most sensitive? Give a reason supporting your choice of colour. (*Hint*: Consider which energies of the electromagnetic spectrum are most difficult to see, in other words, are invisible.)
22. Calculate the wavelength of the following:
    (a) an X ray with a frequency of 2.0×10^{18} Hz
    (b) a radio wave with a frequency of 5.0×10^3 kHz
    (c) an electric wave with a frequency of 60 Hz

16.6 The Laser: An Incredible Application of Light

Lasers provide an appropriate conclusion to the topic of light for a variety of reasons. First, they operate as a result of the physical principles of light you have studied. In addition, they have countless exciting and important applications.

Characteristics of Lasers

Most sources of light energy, such as an incandescent lightbulb, emit light because of spontaneous accelerations of particles. The result is that the photon energies have many different values, so the emitted light has many different wavelengths. See Figure 16-18(a). A laser is different – it emits only one wavelength (or a controlled set of discrete wavelengths) of light, so all the waves are in phase (in step) with one another, as Figure 16-18(b) demonstrates.

Because of the single-wavelength, in-phase nature of laser light, lasers have other unique characteristics. Light from lasers spreads very little as it travels. Lasers can be controlled to emit either continuous beams or pulses as short in duration as a few billionths of a second. They can also be controlled to emit light with very high energy. These features make possible numerous applications of lasers, some of which will be described after you read about how a laser operates.

Figure 16-18
Comparing Incandescent Light with Laser Light

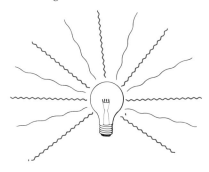

(a) *White light from an incandescent source consists of waves of many wavelengths.*

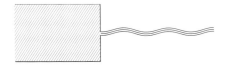

(b) *Light from a laser consists of waves which are in phase and have a single wavelength.*

Laser Operation

A clue about how a laser produces its nearly "perfect" light is found in the word itself. **LASER** is an acronym for **L**ight **A**mplification by the **S**timulated **E**mission of **R**adiation. Since the invention of the first laser in 1960 by the American Theodore Maiman, many types of lasers have been developed. They all operate on the principle of the stimulated emission of radiation. Only certain substances have atoms that are capable of "lasing". These substances include chromium, carbon dioxide gas, helium-neon gas mixtures, and such unusual materials as food dye and liquid bleach.

To understand the principle of stimulated emission, consider the example of one type of laser, the ruby laser (Figure 16-19). It is made of a long, thin rod of ruby, with two precisely aligned converging mirrors at the ends. The ruby rod contains chromium atoms, which are ultimately responsible for producing the stimulated emission. Electrical energy is given to the flashlamp surrounding the ruby rod. The flashlamp in turn pumps light energy into the ruby, and this light energy "excites" the atoms of chromium. Let us look closely at what happens with the chromium atoms.

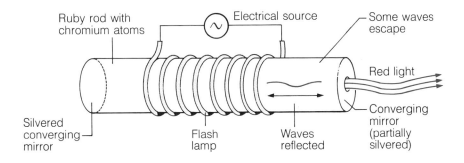

Figure 16-19
A Basic Ruby Laser

Initially, the electrons in the chromium atoms are in a normal, low-energy state, E_{low}. When the flashlamp pumps light energy into the atoms, the electrons in many of the atoms jump to a higher energy level, E_{high}. This is depicted in Figure 16-20(a), which shows an electron jumping from E_{low} to E_{high}. The increase in the electron's energy, ΔE, is simply $E_{high} - E_{low}$. Thus, the electron is now in the excited state.

Ordinarily, an electron in the excited state will spontaneously lose its energy, dropping back down to the lower energy level, E_{low}, in one,

LIGHT AND COLOUR

two, or more steps. This process results in photons of energy being emitted at random, in an uncontrolled way, as shown in Figure 16-20(b). But stimulated emission is different. In the case of chromium, for example, if a photon having an energy equal to $\Delta E = E_{high} - E_{low}$ strikes a chromium atom with an excited electron, the electron will be stimulated to drop down to E_{low}. A photon will be emitted with the same energy as the first photon. Furthermore, the second photon will jump out *with* the first one and will travel in the same direction. Thus, the two photons, having the same energy, frequency, wavelength, and speed, will travel together and in phase. This is shown in Figure 16-20(c).

Figure 16-20
Explaining Stimulated Emission

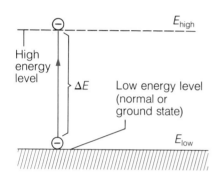

(a) *Energy from an external source causes an electron to jump from its normal low-energy state to a high-energy state.*

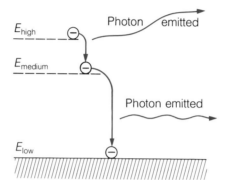

(b) *Uncontrolled emission of energy occurs if the electron loses its energy in two or more steps.*

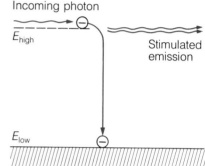

(c) *Stimulated emission of energy: An incoming photon strikes the excited atom and stimulates the electrons to drop to E_{low} and emit a photon in phase with the initial photon.*

Did You Know?
The cover of this book shows a pattern of light as it is reflected off an optical disk. Such a disk resembles an ordinary phonograph in appearance, but it can hold thousands of times as much information as a magnetic floppy disk of the same size. Tiny crystal lasers are used to write information onto the disk or to read the information stored there.

Once the stimulated emission has begun, one main condition is required to build up the emissions and keep them coming. The chromium atoms must be constantly pumped with light energy so that a large population of the atoms has electrons in the excited state, E_{high}. Soon more and more photons are travelling back and forth in the ruby rod and, in a sort of chain reaction, the light energy is amplified. The laser's front mirror reflects a high percentage of the light that strikes it. The remainder of the laser light emerges from the small hole in the front mirror and travels as a fine, monochromatic (single-coloured) beam. (Refer to Figure 16-18.) Other lasers operate on similar principles of pumping energy into atoms and then stimulating the emission of in-phase photons, all having the same energy.

Uses of Lasers

Lasers, though a recent invention, have already been put to an extraordinary number of uses in diverse fields. The fine straight beams of laser light provide accurate alignment during the construction of bridges, roads, tunnels, and skyscrapers. Lasers help survey rugged terrain that is difficult or impossible to approach. They are used in conjunction

COLOUR AND LIGHT THEORY

with prisms, lenses, and music to entertain people at concerts and spectacular indoor and outdoor laser shows. Lasers are improving the field of communications in other ways, too, as telephone and other messages are transmitted along optical fibres using the total internal reflection of laser light. In industry, intense beams of laser light drill fine holes in rubber, aluminum, and diamond, or cut steel saw blades with great precision. In police work, lasers can help "see" fingerprints previously impossible to detect. See Figure 16-21 as well as Colour Plate 8.

Did You Know?
Canada is a leader in the field of optoelectronics, the science that links lasers and fibre optics with electronics. One of the numerous successful developments is called **lidar**, which stands for **l**ight **d**etection **a**nd **r**anging. It is the laser (or light) equivalent of radar. Applications of lidar include surveying and monitoring air pollution.

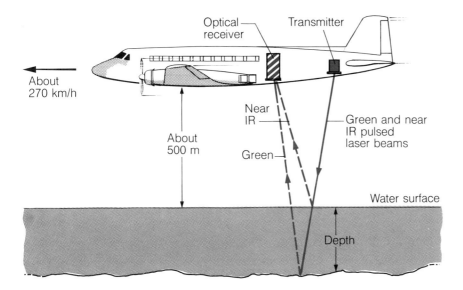

Figure 16-21
An airborne laser system is used to survey inaccessible coastlines, such as those in Canada's north. The near infrared beam reflects off the surface of the water and the green beam reflects off the bottom. The speed of light in water and the time delay between the two pulses received back at the plane are used to find the water depth.

In the field of measurement, lasers provide valuable assistance. Laser reflectors, which use total internal reflection in a manner similar to a bicycle reflector, have been placed on the moon and on satellites in orbit around the earth. Laser light pulsed from the earth strikes a reflector, and part of it returns to the source. Precise measurements of the timing of the laser signals have allowed scientists to measure the earth-moon distance to within 15 cm as well as observe shifts in the earth's crust. The latter measurements help scientists predict earthquakes and monitor continental drift.

In medicine the intense heat of a pulsed laser can be used to cut like an extremely sharp knife and simultaneously to heal the wound by means of the localized high temperature. Even the colour of the laser beam used is critical in operations. A blood clot, which consists of hardened red blood cells, can easily be vaporized by using a blue or green laser. (Remember from subtractive colour theory that red objects absorb blue or green light.) The surrounding cells are left untouched. Certain cancer diagnosis and cancer treatment procedures are also more easily performed using lasers.

The electronics and computer industries are changing rapidly, in part thanks to the use of lasers. Tiny semiconductor lasers, no larger than a single grain of salt, are combined with other optical devices such as lenses and prisms to produce an optical computer. Such computers operate on tiny pulses of light rather than electricity, and perform operations thousands of times faster than electronic digital computers.

Did You Know?
Pulsed lasers are useful for delivering information very quickly. The lower the period of duration of the pulse, the greater is the amount of information which can be transmitted in a given amount of time. Pulse periods are in the order of 10^{-10} s and lower.

PROFILE

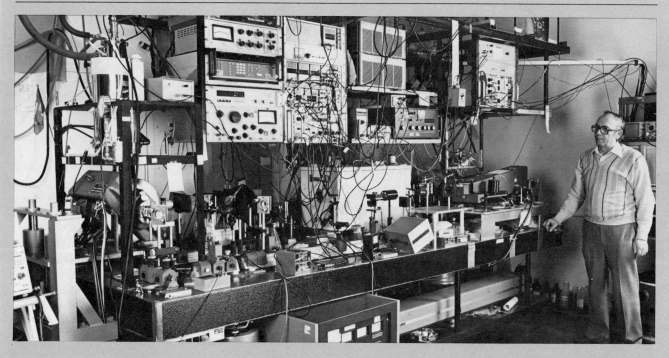

**Alexander Szabo, Ph.D.
Senior Research Officer
National Research Council of Canada**

I do not recall any particular incident which steered me into science. It's something I've always been interested in. However, I do recall reading a lot of science fiction as a boy and being an avid fan of *Buck Rogers* movie serials. I was also later captivated by the television series, *Star Trek*.

After graduating with degrees in science from Queen's and McGill Universities in the 1950s, I began working on microwave devices (masers). Masers work on a principle of stimulated emission. For example, if a substance is exposed to radiation (or "excited by it", as we say) some of the electrons in the atoms of the substance will absorb the extra energy. The electrons in an atom normally circle the nucleus at a certain orbit, but when they absorb the extra energy from the radiation, they will move to a higher orbit. The electrons are very unstable at this new orbit, so they soon fall right back to their old orbits; and as they fall back, the electrons release the excess energy – usually in the form of light or invisible radiation.

Lasers work on the same principle of stimulated emission. When you shine a light through a crystal, you can see it break up into a rainbow of colours. These colours are actually different frequencies of light, and lasers are essentially a single frequency of high-intensity light. When the laser was first

Physicists are performing sophisticated and expensive research to try to harness the greatest available source of energy – nuclear fusion. This type of nuclear reaction, which occurs in the sun and other stars, involves the joining of the dense nuclei of elements such as hydrogen, with a subsequent release of a relatively large amount of energy. Extremely high temperatures are required for a fusion reaction, and lasers are considered one possible means of providing such conditions.

demonstrated in 1960, it seemed a natural step for me to move into this new and exciting field. Soon I was designing a maser that would be stimulated by a laser.

If you ask me what my "scientific method" is, I would have to answer that there is no method as such. In my case, I use the trial and error method. Eventually, if I'm lucky, I manage to ask nature the right question and get something more illuminating than a silly answer (the standard reward for a silly question).

In advanced research, it has been my experience that the unexpected is always lurking in the background. It is always present, in the shadow of every idea, of every measurement; and when it emerges, it can turn a research project upside down. So it was in the case of my laser-excited maser. I would never have predicted it, but my early work with lasers has led to a revolutionary method for storing information in a computer.

Essentially, my invention is the process I call "hole-burning". During the 60s, scientists at Bell Labs in the United States were studying how lasers interact with gases. By directing different lasers at bottles of pure gas, they found that certain molecules in the gas would fluoresce; that is, they would emit a narrow frequency of light.

At the time, scientists did not believe that solids would react in the same way as gases when excited by lasers. After considering the matter, I decided that they could be wrong. So I created an experiment using a ruby laser directed onto a piece of ruby. The whole operation was very difficult because it had to be done in the single millisecond that the ruby laser flashed. To my delight and surprise and awe – this isn't usually the way things happen – the thing worked on the first shot. I was kind of delirious because it worked just the way I thought it would.

Essentially, what had happened was that instead of absorbing the light, the ruby had actually transmitted it. It was similar to burning a hole in a piece of paper using a magnifying glass and then shining the light through the hole. I realized that this had profound implications for computers. By shining different frequency lasers onto a tiny slab of crystal, I could burn holes similar to those used in computer punch cards but infinitely smaller. Theoretically, we would be able to store many trillions of bits of information in a single square centimetre of crystal.

In 1972, I applied for a patent on the hole-burning process for computer memory. It was granted in both Canada and the United States. However, some major problems remained to be solved before this technique could be put to practical use. Since then, scientists working at industrial bases such as IBM and Bell Labs have solved most of these difficulties.

From my thirty years of research, I conclude that the future of science is *completely* unpredictable! That is why we do basic research. Certainly in my side interest of optical computing, the fun is just beginning and some awesome possibilities for computers lie ahead. I think that the current generation of young scientists is fortunate to be presented with such interesting and worthwhile challenges.

For them, as for scientists in the past, *motivation* is the real key to solving the problems of science. For without motivation, nothing happens. What basic research really comes down to is (1) curiosity, and (2) the thrill of discovery. When you think about it, these probably apply to other creative endeavours as well, such as music and art. As to how to foster motivation, I have no suggestions. My observation of students and research assistants who have worked with me would indicate that it comes with the genes.

The only dismal note is that the amount of basic research in Canada is too small. The National Research Council is one of the very few places, outside of universities, where basic research is done. We need much more support from private industry for basic research.

Laser Holography

Holography is yet another exciting application of the laser with countless potential uses. **Holography** is the process of making a three-dimensional photograph on a two-dimensional surface without the use of lenses. It was first conceived in 1947 by Denis Gábor (1900–1979), a Hungarian physicist, who could not perfect the process until the invention of the laser several years later.

In one technique used to make a hologram, a laser beam is split into two parts. One of them goes directly to the photographic film. The other part strikes the object to be photographed, then is reflected to the photographic film, where it interferes with the first part of the beam. (Refer to Figure 16-22.) After development, the resulting pattern can be viewed using laser light; the image seen is a strikingly faithful reproduction of the original subject. (The word "hologram" is derived from the Greek *holos*, meaning "whole", and *gramma*, meaning "something written or recorded".) Techniques are also available to produce holograms which can be viewed in white light.

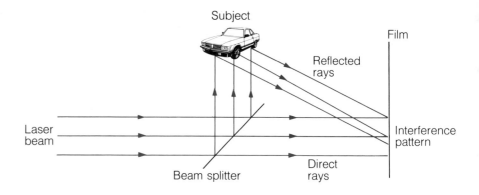

Figure 16-22
The Making of a Hologram

Figure 16-23
An Example of the Universal Product Code

Numerous uses of holography exist or are being developed. A common use may be found at your local supermarket checkout, where a laser scanner reflects light from the Universal Product Code, the set of bars found on most packaged products (Figure 16-23). The light is reflected off the coding, through a holograph scanner, to a detector. The product's name and price appear on your bill and, at the same instant, the store's inventory is brought up-to-date.

The Future of the Laser

Shortly after the invention of the laser in 1960, scientists began calling it "a solution in search of a problem". That description is no longer valid. With all its current applications, the laser certainly has solved its share of problems. New types of lasers and methods of using them will continue to be developed.

The laser and the theory of light provide a link between the topics of light and electricity, which is the next Unit in this text. Studying electricity will help you understand more completely certain concepts described in the present Unit.

PRACTICE
23. List differences between laser light and ordinary white light.
24. What are the main components of a laser?
25. A double balloon consists of a small green balloon inside a colourless, transparent balloon. Both are fully blown up. A laser beam aimed

toward the green balloon causes it to burst inside the transparent balloon.
  (a) What are two possible colours of the laser beam? Explain your choices.
  (b) Why does the outer balloon not break?
26. List uses of lasers of which you are aware. (Do not list uses already mentioned in the text.)
27. **Activity** Use recent periodicals to determine some other modern uses of lasers. Report on your findings; include a list of the references you used.

Review Assignment

1. Explain how a prism can be used to cause the dispersion of white light.
2. List the colours of the visible spectrum in the order of lowest to highest energy.
3. Explain why green light is refracted more than orange light when the two colours enter a prism from the same direction. (*Hint*: Consider the changes of speed as the colours travel from one medium into the other.) Is your explanation supported better by the particle theory or the wave theory of light?
4. When a Newton's disk is spun rapidly in white light, the reflected light appears to recombine.
   (a) On what property of the eye does the recombination rely?
   (b) Predict what would be observed if the disk were made using only the three primary light colours.
5. A diverging lens has a focal length of 8.0 cm for white light. Does the focal length increase or decrease for light that is
   (a) red and (b) violet?
6. A beam of white light is incident upon a block of crown glass and produces a spectrum of colours within the glass. For the green component of the spectrum, calculate
   (a) the critical angle in the glass and
   (b) the speed in the glass.
7. When white light passes through a clear glass window, why does it not disperse into its spectral colours?
8. In an equilateral prism, the dispersion of light is most easily seen when the light in the middle of the visible spectrum (yellow light) is parallel to one edge of the prism. Calculate the angle of incidence of yellow light in air that creates this situation in a prism made of crown glass.
9. What colour results when the following pairs of light colours are added together?
   (a) red and green
   (b) magenta and green
   (c) blue and yellow

10. Bright white fabrics gradually become yellowish. Their whiteness can be restored by using a material that adds a certain colour to the yellowed fabrics.
    (a) What primary light colours are reflected by yellow?
    (b) What other primary light colour should be added to yellow to produce white?
    (c) Why do detergents have "bluing agents"?
11. Three stage lights, blue, red, and green, are used to illuminate an actress's yellow dress. State the colour of the dress when
    (a) only the blue light is on;
    (b) both the blue and red lights are on; and
    (c) all three lights are on.
12. Repeat Analysis Question 4 in Experiment 16A, using three light sources—red in the middle, and blue and green on either side of red.
13. An art student, needs blue paint, but the only pigments available are R, G, C, M, and Y. How can the student obtain blue using some of those pigments?
14. A stained-glass window and a clear window are exposed to direct sunlight for an equal amount of time.
    (a) Predict which window, if either, would feel warmer.
    (b) Explain your answer in terms of the law of conservation of energy.
15. Photographers rely on colour filters to improve the colours of subjects such as portraits under certain conditions. State the colour filter you would recommend in each case described below, and explain your choice.
    (a) At sunrise and sunset, the sky appears reddish.
    (b) At noon on a sunny day, the sky appears unnaturally whitish.
    (c) On a cloudy day, the subjects in a photograph appear abnormally bluish.
16. Fluorescent lighting in stores has a predominance of blue light and little red light. Explain how this affects the buying of such commodities as clothing, cosmetics, and household decorating supplies.
17. Compare the colours resulting from the dispersion of white light in a triangular prism with the colours of (a) a primary rainbow and (b) a secondary rainbow.
18. When seen from the ground, a rainbow has the shape of an arc. What shape do you think it would have when viewed from an airplane high above the raindrops?
19. Describe the similarities between colour vision and colour television.
20. Yellow is an important colour in highway safety. Warning signs, maintenance equipment, and fog lights are often predominantly yellow. Explain why this colour is a wise choice.
21. Do not let the following question destroy your faith in magic. . . . Certain marked cards used by magicians can be distinguished only if the magician wears lenses of the appropriate colour. For example, assume that the Jack of Hearts has a faint green dot on the back of the card, not visible to the unaided eye. What colour filter would you suggest the magician should wear to see the dot? Elaborate on your answer.

COLOUR PLATES

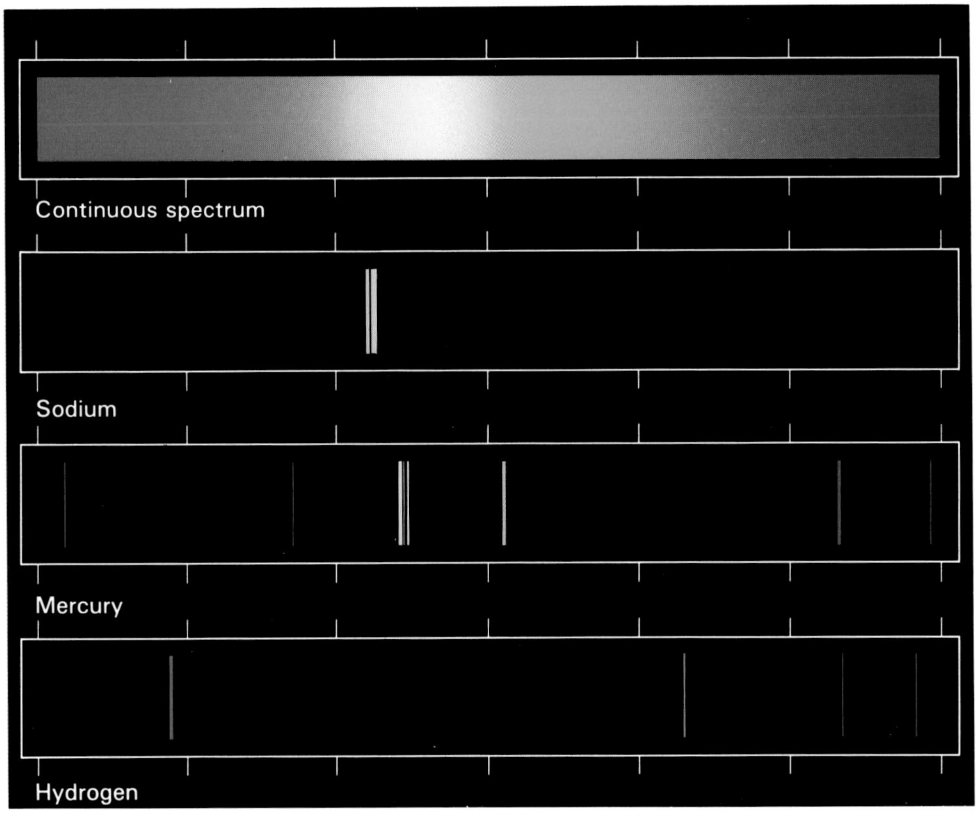

Colour Plates

Plate 1
A continuous white light spectrum and spectra of excited atoms of the elements sodium, mercury, and hydrogen. Wavelengths of the light are given in nanometres (nm). (See page 383.)

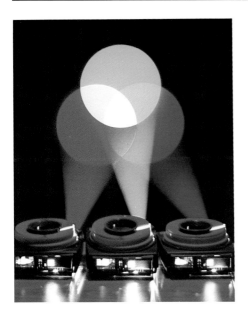

Plate 2
The primary light colours, red, green, and blue, overlap to produce the secondary light colours, yellow, cyan, and magenta, as well as white. (See page 387.)

Plate 3
The primary pigment colours, yellow, magenta, and cyan, absorb the white light from the background, resulting in the secondary pigment colours, red, blue, and green, as well as black. (See page 391.)

COLOUR PLATES

Plate 4
This double rainbow was seen over Red Deer, Alberta. Notice that the colours of the secondary rainbow are in the reverse order of the colours of the much brighter primary rainbow. (See page 392.)

Plate 5
The Aurora Borealis, or Northern Lights, often illuminate the night sky in regions near the magnetic north pole. This example is seen near Fairbanks, Alaska. (See page 495.)

COLOUR PLATES

Plate 6
The beam from an argon ion laser, seen at the top of the photograph, provides the pumping energy to the dye laser at the bottom. In this case, light from the dye laser is used to conduct research in a laboratory.

Plate 7
Scientists use high energy lasers and sensitive imaging equipment to study soot formation in the harsh environment of turbulent flames. This research is directed toward improving the efficiency and emissions of future automobile engines.

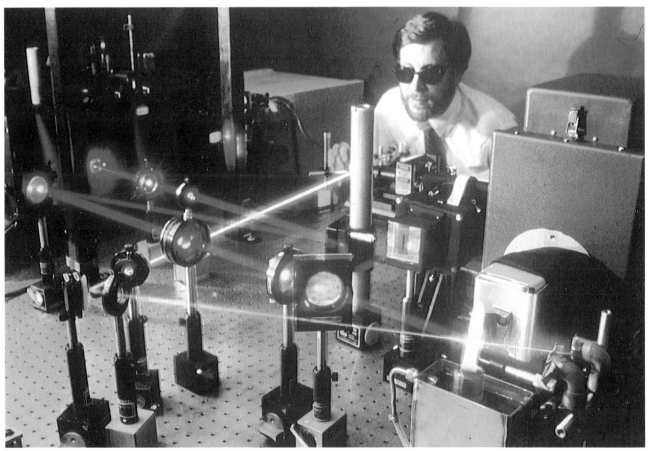

COLOUR PLATES

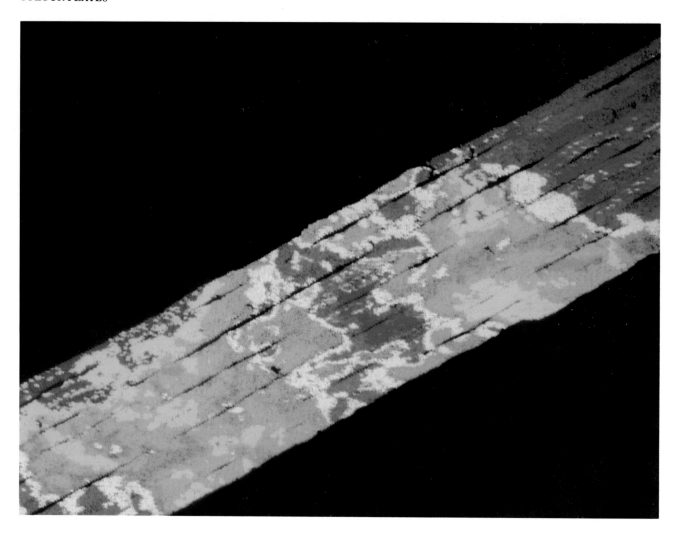

Plate 8
Airborne Laser Surveying
The large figure illustrates the water depth data produced by the laser system shown in Figure 16-21. The data consists of 11 parallel flights of the airplane, with each flight covering a path 260 m wide. The smaller figures show the data at magnifications of 50 times and 2500 times respectively. Depths are indicated to the closest tenth of a metre.

COLOUR AND LIGHT THEORY

22. **Activity** This activity, which involves a fascinating phenomenon called the *Pulfrich effect*, requires a pendulum about 1 m in length and at least one colour filter. One person should operate the pendulum, while the viewers stand about 5 m or more from the pendulum.
    (a) Hold a red filter over your left eye. Keeping *both* eyes open, watch the pendulum as it swings perpendicularly to your line of sight. Describe the motion observed.
    (b) Repeat (a) with the filter over your right eye.
    (c) Repeat the experiment, using filters of different colours. If observations differ for different colours, try to explain why.
23. Name three theories used throughout the last several hundred years to explain the nature of light. At present, which theory is most satisfactory? Why?
24. What causes electromagnetic waves?
25. Describe the similarities and differences between visible light and the other parts of the electromagnetic spectrum.
26. The frequency of one of the bright colours emitted by a hydrogen discharge tube is 4.57×10^{14} Hz.
    (a) Calculate the wavelength of this light.
    (b) What colour is the light?
27. Choose your favorite AM and FM radio stations. For each station, name or calculate the following:
    (a) its call letters
    (b) the frequency (Express your answer in hertz.)
    (c) the wavelength of the emitted waves
28. If you studied wave diffraction in Experiment 10C, explain why it is possible for a radio to receive signals from the radio station emitter even though there are obstacles between the emitter and the receiver.
29. **Activity** Research and explain the meaning of AM and FM, which are the two types of modulation of the signals emitted from a radio station. On which type of modulation is the use of laser communication based? Explain.
30. In Section 2.1, it was stated that the metre was formerly defined as 1 650 763.73 wavelengths of a certain light emitted by krypton-86 atoms in a vacuum. Find the wavelength, frequency, and colour of this krypton light.
31. A certain laser produces electromagnetic waves having a wavelength of 2.0×10^{-10} m. Are the waves visible?
32. Explain what is unique about laser light and why it is important.

Key Objectives

Having completed this chapter, you should now be able to do the following:
1. Define dispersion of white light, and explain how and why it occurs.
2. List the colours of the visible spectrum in the order in which they occur.

LIGHT AND COLOUR

3. Relate the speed of light and the index of refraction to dispersion.
4. Define recombination of light, and describe methods used to demonstrate it.
5. Know the position of infrared and ultraviolet light relative to the visible spectrum.
6. Define primary, secondary, and complementary light colours.
7. Draw a diagram to illustrate the formation of colour shadows when the sources of light are the primary light colours.
8. Distinguish between additive colour mixing and subtractive colour mixing.
9. Predict the colour of an opaque object seen through a colour filter.
10. Define primary and secondary pigment colours.
11. Predict the colour of a pigment composed of a mixture of two or three primary pigments.
12. Explain how a primary rainbow is formed.
13. Compare colour vision with colour television.
14. Describe how retinal fatigue can be demonstrated, and explain its cause.
15. Explain why retinal fatigue supports the three-cone model of colour vision.
16. Contrast and compare the theories of light.
17. Given any two of a wave's speed, frequency, and wavelength, calculate the third quantity.
18. Relate the colour of light to its wavelength, frequency, and energy.
19. Define electromagnetic spectrum, and recognize where the visible spectrum fits into it.
20. Describe the characteristics, operation, and various uses of a laser.
21. Define holography, and describe how a hologram is made.

UNIT VI
Electricity and Electromagnetism

The information in this Unit will be useful if you plan a career in nursing, medicine, cardiology, neurology, radiology, psychiatry, audio/video systems production, electrical engineering, aeronautical engineering, electrical contracting, electric power distribution, microelectronics, optoelectronics, computer design, robotics, appliance manufacturing or repair, auto mechanics, instrumentation technology, pollution control, laboratory technology, telecommunications, astronomy, or atmospheric science.

ELECTRICITY AND ELECTROMAGNETISM

CHAPTER 17
Static Electricity

Today, as in the past, the tallest structure in many European towns is the bell tower.

During the Middle Ages, the highest point in most European towns was the bell tower of the local church or cathedral, as the photograph shows. Many people believed that in a thunderstorm, the lightning could be driven away by ringing the church bells. Unfortunately, lightning often struck the bell tower, killing the bell ringer who was trying to save others from the lightning!

In this chapter you will learn what static electricity is and why lightning is included in the topic. You will discover answers to such questions as these: How could the bell ringers' lives have been spared? What does the study of static electricity reveal about the structure of matter? How can electricity be controlled for useful applications?

17.1 The Force of Electricity

No doubt you have experienced a clinging force when you stroked a comb through dry hair or removed synthetic clothing from your body or a clothes dryer. The clinging force is caused by **static electricity**, the build-up of electric charges on an object. The word "static" means "at rest", so static electricity is different from the type of electricity that flows through a wire (current electricity).

After an electric charge is built up on an object, it is possible for the charge to be transferred off the object. This transfer of charge is called a **discharge**. A spark is actually a rapid discharge. When you shuffle across a rug on a dry day in the winter, the shock you feel as you touch a metal doorknob or another person is an example of a discharge. A lightning bolt is nature's most spectacular example of a static electricity discharge (Figure 17-1).

Figure 17-1
Lightning over Kamloops, B.C.

The force exerted on an object with an electric charge is called the **force of electricity**, or **electric force**. It is a component of one of the fundamental forces of nature, **electromagnetic force**, which is the interaction of the force of electricity and the force of magnetism (to be studied later). The force of electricity is responsible for other forces such as friction, mechanical pushes and pulls, cohesion, and adhesion. Recall from Chapter 5 that scientists now consider that there are four fundamental forces: gravitational, electromagnetic, strong nuclear, and weak nuclear. The Activity in Practice Question 2 involves a comparison of the force of gravity and the force of electricity.

The force of electricity was studied and described in Greece as early as the seventh century B.C. A man named Thales (636 B.C.–546 B.C.) discovered that when amber, a hard, yellowish resin from dead trees, was rubbed with fur, it attracted small objects such as feathers, hair, and pieces of straw. Because the Greek word for "amber" is *elektron*,

ELECTRICITY AND ELECTROMAGNETISM

modern words related to electricity stem from Thales's discovery.

Little more was learned about the force of electricity between the time of Thales and the Renaissance. Then, in 1600, the English scientist William Gilbert (1540–1603), who was court physician to Queen Elizabeth I, published an important book entitled *De Magnete*. In it he described controlled experiments in which he studied properties of magnetism and static electricity. He noted that amber was not the only substance that could accept an electric charge, although he did not discover how to charge metal objects. He is also credited with naming electricity and with being one of the most important influences in the development of controlled experimentation.

Following the publication of Gilbert's work, many static electricity experiments were carried out. Perhaps the most famous one was performed by Benjamin Franklin (1706–1790), an American statesman, inventor, and scientist (Figure 17-2). Predicting that lightning might be a form of static electricity, Franklin wanted to test the clouds during a thunderstorm. His experimental apparatus consisted of a long hempen cord with a kite at one end and a metal key at the other end, a static electricity testing apparatus, and a dry silk cloth which he used to hold onto the cord. As a lightning storm approached, he flew the kite while keeping dry under a shelter. (He must have realized that his experiment would be even more dangerous if he got wet!) He observed that the kite did indeed draw electric charges from the clouds and that the static electricity acted the same way as that produced in the laboratory. (Franklin was lucky he was not killed. Some scientists who have tried to repeat his experiment were not so fortunate. So do not try it!)

A mathematical analysis of the force of electricity is found in the Supplementary Topic at the end of this chapter.

Figure 17-2
Benjamin Franklin (1706–1790)

PRACTICE

1. List examples of static electricity other than those described in this section.
2. **Activity** Perform an investigation to learn how the forces of electricity and gravity compare. You will need some small bits of paper and a chargeable object such as a plastic comb stroked through your hair or a blown-up balloon rubbed on a piece of clothing.
   (a) As the charged object approaches the bits of paper, how does the force exerted depend on the distance between the object and the bits of paper? Compare this to the relationship between gravitational force and distance. (*Hint*: The force of gravity is inversely proportional to the square of the distance between two objects.)
   (b) Does the charged object have to touch the bits of paper in order for the force to be observed? How does electric force compare to the force of gravity, which acts at a distance?
   (c) Draw a diagram of a single bit of paper, showing the forces acting on it when it is accelerating toward the charged object. Which is the maximum force? In general, how does the force of electricity compare in size to the force of gravity per unit mass of the object causing the force?

STATIC ELECTRICITY

Experiment 17A: The Laws of Electric Charges

INTRODUCTION
The **laws of electric charges** may be called the **law of repulsion** (pushing apart) and the **law of attraction** (pulling together). Both repulsion and attraction result from the force of electricity.

In this experiment, two types of plastic are suggested. One type, cellulose acetate or simply *acetate*, is clear. The other type, *polyethylene*, is white. When polyethylene is rubbed with wool or cat's fur, it takes on a charge that is defined as *negative*. This fact is used to determine the types of charges on other objects. (In some laboratories, vinylite or ebonite is used instead of polyethylene, and glass is used instead of acetate. Glass should be rubbed with silk.)

If a charge is opposite to the type of charge on polyethylene, it is called *positive*. An object with no charge is called *neutral*.

You will use a piece of plastic as an electroscope in this experiment. An **electroscope** is a device which can be electrified with a known charge used to find the type and size of a charge on another object.

PURPOSE
To investigate the repulsion and attraction of electric charges.

APPARATUS
2 retort stands with clamps and supports; 2 polyethylene and 2 acetate strips (or their alternatives); woollen cloth; cotton cloth; plastic comb; plastic pen

PROCEDURE
(*Note*: Before you begin each step, read the instructions and predict what you will observe. Then verify your predictions.)
1. Suspend the polyethylene strip as shown in Figure 17-3. Hold the suspended strip in the middle and rub both ends with wool. Rub one end of the other polyethylene strip with wool and hold it close to one end of the first strip. Repeat this a few times and describe your observations.
2. Again rub both ends of the suspended polyethylene with wool. Now hold the part of the wool you used for rubbing close to the polyethylene. Describe what happens. Then try holding your finger close to the polyethylene; describe what you observe.
3. Rub an acetate strip with cotton (or wool) and hold the strip close to the suspended, charged polyethylene. Describe the effect. What type of charge is on the acetate?
4. Suspend an acetate strip from the retort stand and repeat Procedure Steps 1 and 2 using two acetate strips and cotton cloth.
5. Rub a polyethylene strip with wool and hold the strip close to the suspended, charged acetate. Describe what happens.

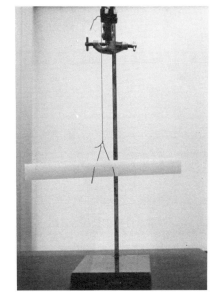

Figure 17-3
Set-up for Experiment 17A

413

ELECTRICITY AND ELECTROMAGNETISM

6. Set up two electroscopes, one with a charged polyethylene strip and one with a charged acetate strip. Use them to determine the size (small, medium, or large) and type (positive or negative) of charge on the following:
   (a) a plastic comb rubbed through your hair
   (b) a plastic pen rubbed on a piece of clothing
   (c) other appropriate objects

ANALYSIS

1. State which objects used in this experiment
   (a) became negatively charged
   (b) became positively charged
   (c) remained neutral.
2. What effect do like charges have on each other?
3. What effect do unlike charges have on each other?
4. Why was the repulsion of a suspended strip the only sure way to determine an unknown charge in this experiment?
5. The force of gravity is an attraction force only. How does this compare with the force of electricity?

CONCLUSIONS . . .

17.2 The Atomic Theory of Matter

Based on experimental observations, scientists have developed theories to explain matter and the way it behaves. To try to explain observations related to electricity, you will use the theory of the structure of matter. In this chapter you will learn about the concepts needed to explain the action of electric charges. Then in Chapter 22 you will discover the developments leading to the present-day theories.

Structure of the Atom

All matter is composed of tiny particles called atoms. An **atom** is the smallest part of an element which can take part in a chemical reaction.

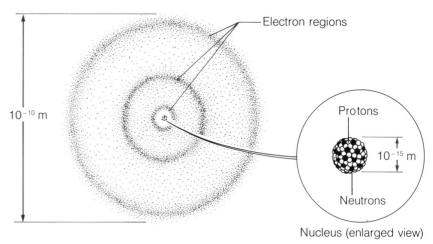

Figure 17-4
The Representation of an Atom. (The sizes given are order-of-magnitude values.)

The word stems from the Greek word *atomos* which means "indivisible", although we now know that the atom *is* divisible. Each atom consists of a dense, central **nucleus** surrounded by fast-moving **electrons**. In the nucleus are found two types of particles, **protons** and **neutrons**. Figure 17-4 shows what an atom might look like if we were able to see it. The nucleus is so tiny compared to the atom that its components are redrawn in the corner of the diagram. The ratio of the diameter of the nucleus to the diameter of the atom is approximately equal to the ratio of the length of a kernel of corn to the length of a football field. Thus, an atom consists mainly of empty space.

Charges on Particles

All particles can have one of three possible charges – negative, positive, or neutral. The charge on a piece of amber rubbed with fur was defined as negative by Benjamin Franklin, long before anyone knew about electrons and protons. That definition is still used, and since we now know that charged amber has an excess of electrons, we say that electrons have a **negative** charge. Protons have the opposite charge, **positive**. The magnitude of the charge on one electron equals the magnitude of the charge on one proton – only the sign differs. An atom, or any object, having an equal number of electrons and protons has a **neutral** charge. (A neutron has no charge.)

The Periodic Table of the Elements

An **element** is a pure substance that cannot be broken down into anything simpler by ordinary chemical means. Hydrogen (symbol H) and oxygen (symbol O) are two common elements. When they combine chemically in a ratio of 2:1 by volume, they form the substance water, H_2O.

There are 92 naturally occurring elements and approximately 14 artificial ones. They form groups or families having common characteristics. Thus, they can be arranged in a chart called the **periodic table of the elements**. The entire table is shown in Appendix B, but for now we will consider the first 20 elements only, shown in Table 17-1.

Each element consists of atoms that have the same number of protons. For example, every hydrogen atom has one proton and every oxygen atom has 8 protons. The number of protons in one atom is called the element's **atomic number**. The elements in Table 17-1 are listed in the order of increasing atomic number, from 1 to 20.

An element's **mass number** is the sum of the number of protons and neutrons in an atom of the element. For instance, an atom of oxygen with 8 protons and 8 neutrons has a mass number of 16. (The mass of a proton or neutron is almost 2000 times as great as the mass of an electron, so electrons are not considered in the mass number.)

Electrons travel in regions surrounding the nucleus. Each region can hold a certain maximum number of electrons. For example, the region closest to the nucleus can hold up to two electrons and the next region up to eight. The electron arrangement for each element is shown in

Table 17-1 The First 20 Elements of the Periodic Table

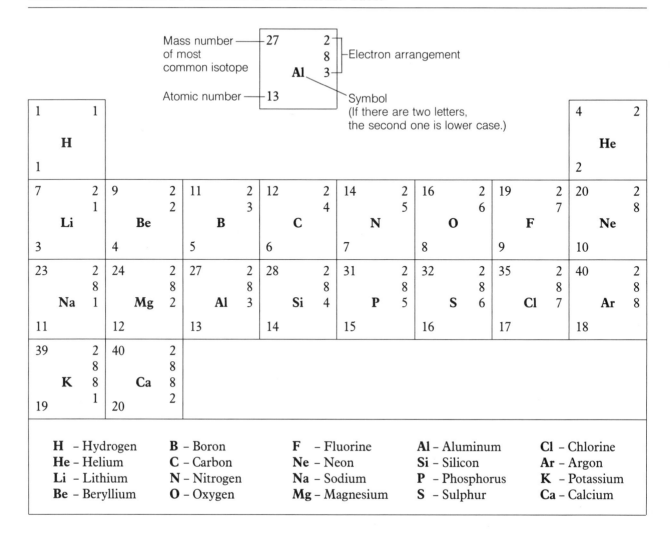

Table 17-1. Electron arrangements are important in this unit on electricity and electromagnetism.

An atom with an equal number of electrons and protons is neutral. However, if an atom loses or gains electrons, it becomes charged because the number of electrons differs from the number of protons. Such a charged particle is called an **ion**.

Models of Atoms

The atomic theory of matter and the information in Table 17-1 can be used in drawing models of atoms. Although the models are not drawn to an exact scale, they at least help visualize the concepts that are important in the study of static electricity. Figure 17-5(a) shows a model of an atom of aluminum; Figure 17-5(b) shows a simpler model.

These basic ideas will be used to explain many events observed in the topic of electricity. Table 17-1 will serve many times as a reference.

PRACTICE

3. State the type of charge on each of the following:
   (a) a proton
   (b) a neutron
   (c) an electron
   (d) a nucleus
   (e) an ion having an excess of electrons
   (f) an ion having a deficiency of electrons
4. For each of the objects described below, state whether it has an excess or a deficiency of electrons.
   (a) a polyethylene strip which has been rubbed with wool
   (b) the wool used in (a)
   (c) an acetate strip which has been rubbed with cotton
5. Assume that for a certain atom the nuclear diameter is 1/50 000 the atomic diameter. If your head represents the nucleus, what size is the atom? (Express your answer in kilometres.)
6. State the number of neutrons in an atom of each of the following elements:
   (a) nitrogen   (b) hydrogen   (c) phosphorus
7. Draw a simple model of an atom of each of these elements:
   (a) beryllium
   (b) helium
   (c) phosphorus.

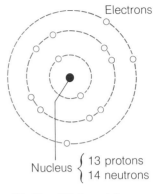

Figure 17-5
Models of the Aluminum Atom
(a) Planetary model

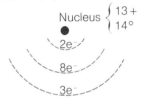

(b) Simplified model

17.3 Transferring Electric Charge by Friction

The atomic theory of matter can be used to explain the effects observed in electricity experiments. For example, in Experiment 17A you obtained a negative charge on a strip of polyethylene (or a similar material) by rubbing it with wool. Let us see why.

Before rubbing occurs, the polyethylene has an equal number of protons and electrons and so is neutrally charged. The wool is also neutrally charged. This is shown in Figure 17-6(a). The polyethylene atoms have a stronger attraction for electrons than the wool atoms do. When the wool and polyethylene are rubbed together, the rubbing makes possible greater contact between wool and polyethylene atoms. The polyethylene attracts some electrons and becomes negatively charged, with an excess of electrons. The wool becomes positively charged because it now has a deficiency of electrons. See Figure 17-6(b).

When the charged polyethylene and wool are held close together, they attract one another. This occurs because opposite charges attract, as Figure 17-6(c) demonstrates.

Every material has atoms with their own characteristic attraction for electrons. For instance, rubber atoms attract electrons more readily than wool or fur atoms do. Table 17-2 lists several materials in order of increasing attraction for electrons. The list is called the **static electricity series**. A material will lose electrons when rubbed with a material *lower* in the list.

Figure 17-6
Charging Polyethylene with Wool

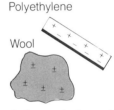

(a) Neutral charges before rubbing

(b) Rubbing causes electrons to go from the wool to the polyethylene.

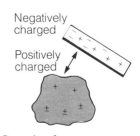

(c) Opposite charges attract.

ELECTRICITY AND ELECTROMAGNETISM

Table 17-2 The Static Electricity Series

+ Atoms have a poor attraction for electrons.
cat's fur
acetate
glass
wool
lead
silk
paraffin wax
ebonite
copper
rubber
amber
sulphur
gold
− Atoms have a large attraction for electrons.

SAMPLE PROBLEM 1
What types of charges result on rubber and silk when they are rubbed together?

Solution
According to Table 17-2, rubber atoms have a greater attraction for electrons than silk atoms do. Thus, the rubber will become negatively charged and the silk positively charged.

PRACTICE
8. Use the atomic theory of matter to explain the build-up of charge when glass and silk are rubbed together.
9. State the types of charges that result on each material when each pair is rubbed together:
   (a) lead, wool   (c) glass, paraffin wax
   (b) acetate, fur   (d) sulphur, silk
10. Predict the location of polyethylene and cotton in the static electricity series. Explain the procedure you would use to verify your predictions. (After you learn more about charging objects, you may be able to carry out an experiment to test your procedure.)

17.4 Electric Conductors and Insulators

The atomic theory states that all matter is composed of atoms which have electrons. Some materials have electrons that are bound tightly to the nucleus and are not free to travel to a neighbouring atom. Such materials are called **electric insulators**. The opposite type of material, an **electric conductor**, has electrons, found in the outermost regions of

STATIC ELECTRICITY

the atoms, which are free to travel. If a charged object is held near a conductor, the forces observed can be determined by applying the laws of electric charges, namely that like charges repel and unlike charges attract. Therefore, if a negative charge is brought near a conductor, the electrons in the conductor will be repelled, and if a positive charge is brought near a conductor, the electrons will be attracted. Table 17-3 lists several good electric conductors and several non-conductors or insulators. Notice that metals are good conductors.

Table 17-3   Electric Conductors and Insulators (in alphabetical order)

Conductors	Insulators
aluminum	air
copper	amber
gold	glass
iron	paper
nickel	rubber
silver	silk
	wool

To transfer a charge to a metal conductor, the conductor must be insulated from its surroundings. The metal sphere in Figure 17-7(a), which is supported on an insulating stand, can be charged negatively or positively by contact with a charged object. In Figure 17-7(b) it has a negative charge. Once the conductor is charged, you can discharge it by touching it. In this case the excess electrons travel from the conductor through your body to the earth. The process of conducting a charge to or from the earth is called **grounding** (symbol ⏚), shown in Figure 17-7(c). Grounding a small metal conductor may be compared to pouring a cup of water into an ocean. It makes a lot of difference to the cup but no difference to the ocean.

Figure 17-7
Charging and Discharging an Insulated Conductor

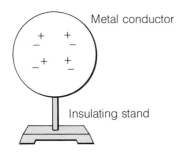

(a) An insulated conductor with a neutral charge

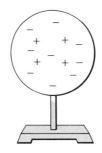

(b) The conductor is charged negatively.

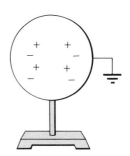

(c) The conductor is grounded and becomes neutral again.

An important application of grounding is used during aircraft refuelling. There is a danger that a static electricity discharge, from either a charge on the airplane or the friction between the fuel and the nozzle, may cause the fuel to explode. To prevent this danger, the aircraft and

ELECTRICITY AND ELECTROMAGNETISM

Figure 17-8
An aircraft is grounded during refuelling to prevent a static electricity discharge.

the fuel-hose nozzle are grounded before refuelling begins (Figure 17-8).

Although most materials are either conductors or nonconductors, some materials are classified as **semiconductors**. One way in which they differ from other conductors is that, as their temperature rises, they become better conductors. The reverse is true for ordinary conductors. The conductivity of semiconductors can be increased tremendously by the addition of small amounts of other elements. Semiconductors have many uses in controlling electronic circuits in such devices as calculators and computers.

PRACTICE

11. William Gilbert, the English scientist mentioned earlier who studied static electricity, discovered several insulators that would accept an electric charge, but no conductors. Give possible reasons why he did not make this discovery.
12. What kind of charge results on a grounded insulated conductor?
13. How many conductors listed in Table 17-3 are elements? (If necessary, refer to Appendix B.) How many insulators are elements? Speculate on a possible reason for the difference.
14. While metal car bodies are being spray-painted at the factory, they are grounded. Explain the advantage(s) of this. (*Hint*: As the spray leaves the nozzle at high speeds, it becomes charged.)
15. **Activity** Research and report on the use of semiconductors in electronic devices.

17.5 Mapping Electric Fields

An **electric field** exists in the space surrounding a charged object in which the electric force acts. If another electric charge enters the field of the first charge, it will experience a force of attraction or repulsion.

In order to identify the characteristics of the electric field around a charge, a series of electric field lines can be drawn or "mapped" on a diagram.

Mapping electric fields can be compared to the process of mapping other, more common quantities. Figure 17-9(a) shows the mapping of contour lines of equal elevation on a field of land. Each solid line represents a constant elevation ranging from 200 m above sea level to 240 m above sea level. From the map you should be able to judge the location of the hill and which side of the hill is steepest. Figure 17-9(b) shows the mapping of isobars (regions of equal air pressure) on a weather map. You should be able to see where the pressure is changing most rapidly, indicating a weather change. Figure 17-9(c) shows the mapping of an electric field around a single positive charge. In this map, the field lines are closer to each other near the charge. This means that the electric field is strongest near the charge, and diminishes as the distance from the charge increases.

Notice in Figure 17-9(c) that the field lines around the charge have been given a direction. This direction is determined by the direction of

Did You Know?
Sharks are sensitive to the small electric field surrounding a possible prey, such as a fish. Experiments have shown that a shark will attack an artificial electric field and ignore a nearby piece of food.

STATIC ELECTRICITY

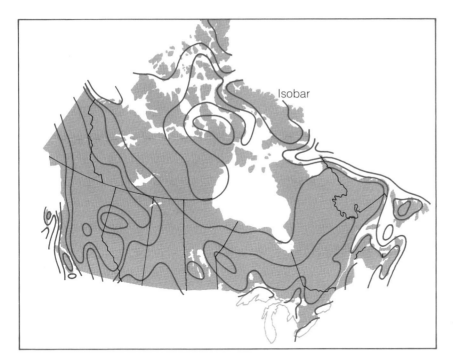

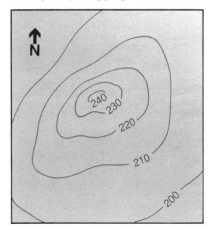

Figure 17-9
Examples of Mapping

(a) Contour map showing lines of equal elevation

◂ (b) Weather map showing lines of equal air pressure

the force experienced by a small *positive test charge* placed in the field of the charge. Figure 17-10 shows the direction of a single field line for a negatively charged object which attracts the positive test charge. Thus, electric field lines are directed toward a negative charge (opposite charges attract) and away from a positive charge (like charges repel).

Figure 17-10
The direction of electric field lines can be found by using a small positive test charge. Since opposite charges attract, the direction is toward a negative charge.

Mapping electric fields becomes more complex when more than one electric charge is present. Figure 17-11 shows the electric fields surrounding pairs of charges.

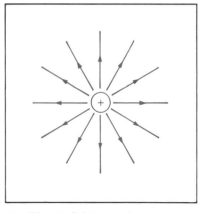

(c) Electric field around a single positive charge

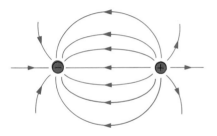

(a) Opposite charges   (b) Like charges

Figure 17-11
Electric Field Lines for Pairs of Charges

421

ELECTRICITY AND ELECTROMAGNETISM

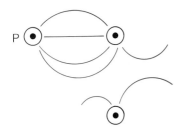

Figure 17-12

PRACTICE

16. Assuming that point P in Figure 17-12 is a proton, what conclusions can you draw?
17. Draw a straight horizontal line about 5 cm long in your notebook to represent a positively charged wire. Map the electric field above and below the wire.
18. Draw two straight horizontal lines about 5 cm long and parallel to each other, separated by about 2 cm. If the top line represents a positively charged plate and the bottom one a negatively charged plate, map the electric field between the plates. (This field is uniform except near the ends of the plates.)

17.6 Distribution of Charges on Insulators and Conductors

How are charges distributed when they are deposited on either an insulator or a conductor? We can predict that the charges do not move on an insulator and that they spread out on a conductor. To verify these predictions, we can test the charge distribution with a **proof plane**, a conducting metal disk attached to an insulating handle. When touched to a charge, it takes on part of that charge, which can then be tested using a known charge on an electroscope. (A simple electroscope was introduced in Experiment 17A.)

First, let us consider insulators. If one end of an object made of an insulating material is given a charge by means of friction, only that end becomes charged. For example, when polyethylene is rubbed with wool, the electrons pile up at the location of the rubbing, as shown in Figure 17-13(a). Figure 17-13(b) shows the use of the proof plane to test the charge at one location on the polyethylene.

Figure 17-13
Testing the Charge Distribution on an Insulator

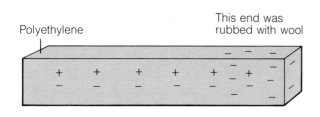

(a) Part of the insulator is given a charge.

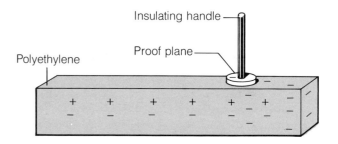

(b) A neutralized proof plane is touched momentarily to one part of the object to be tested. If a charge exists, the proof plane will acquire the same type of charge, which can then be tested with an electroscope (not shown).

Next consider a conductor in the shape of a hollow metal sphere, open at the top and mounted on an insulating stand. A charge is deposited onto the conductor using, for example, contact with a charged polyethylene strip. A proof plane is used to test the charge both inside and outside the sphere. As Figure 17-14 demonstrates, there is no charge on the inside of the sphere; it is entirely on the outside. This occurs for two reasons: first, charges move readily in a conductor; second, the charges repel one another so that they move as far as possible from each other, which on a sphere is to the outer surface.

On a spherical conductor the charge spreads out evenly. On other shapes, however, the charges tend to repel one another toward the more pointed surfaces. Figure 17-15 illustrates why this occurs. In Figure 17-15(a), two electrons are shown at the surface of a conductor with a rounded surface. The repulsive force of one electron on the other is in line with their centres, so it is very nearly parallel to the surface, as shown with the coloured arrows. Thus, the electrons repel each other as far as possible on the conducting surface. (Of course, there are numerous other electrons in the vicinity, not shown in the diagram.) In Figure 17-15(b), two electrons are shown at the surface of a conductor with a more pointed surface. In this case, the repulsive force, shown in colour, is not parallel to the surface. The repulsive force can be considered to have two components, one parallel to the surface, the other perpendicular to the surface. Since the force parallel to the surface is fairly small, as Figure 17-15(b) shows, the electrons are not repelled as far along the surface as they were on the rounded surface. The force pushing away from the conductor, perpendicular to the surface, is fairly large, which leads to the conclusion that charges on a pointed conductor tend to be repelled away from the point. The resulting charge distribution and electric field lines are shown in Figure 17-16.

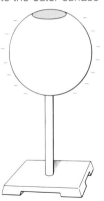

Figure 17-14
Charge Distribution on a Hollow Metal Conductor

Figure 17-15
Forces on Rounded and Pointed Conductors

(a) *Like charges on a rounded conductor repel one another along the surface.*

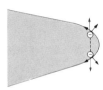

(b) *Like charges on a pointed conductor repel one another away from the surface.*

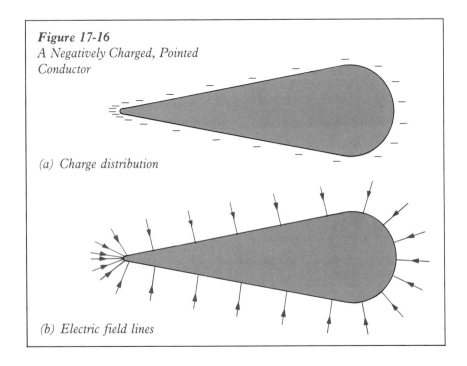

Figure 17-16
A Negatively Charged, Pointed Conductor

(a) *Charge distribution*

(b) *Electric field lines*

A pointed conductor discharges quickly, especially if the surrounding air is moist. Air contains many ions, both positive and negative. A negatively charged conductor attracts the positive ions in the air, and neutralization occurs. A positively charged conductor attracts the negative ions. See Figure 17-17.

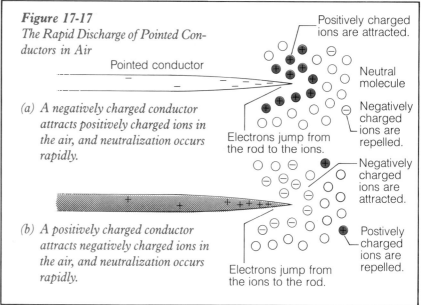

Figure 17-17
The Rapid Discharge of Pointed Conductors in Air

(a) A negatively charged conductor attracts positively charged ions in the air, and neutralization occurs rapidly.

(b) A positively charged conductor attracts negatively charged ions in the air, and neutralization occurs rapidly.

Figure 17-18
The lightning rod is higher than any other part of the building. In this case, the metal conductor also serves as a decorative feature.

An important application of pointed conductors is the **lightning rod**. This device, invented by Benjamin Franklin, helps prevent lightning from striking a tall structure such as a cathedral's bell tower, or an isolated building such as a barn or a farmhouse. The rod is a pointed metal conductor, placed higher than any other part of the building (Figure 17-18). Whether the charge travels from the clouds to the ground or from the ground to the clouds, it can drain quickly through the conductor. This prevents the lightning from starting a fire.

PRACTICE
19. Figure 17-19 shows a negatively charged object, A, brought close to a neutrally charged conductor on an insulating stand. A proof plane is touched to the conductor and then brought close to the negatively charged electroscope, D. Describe what happens, and why, when the proof plane tests first point B, then point C.

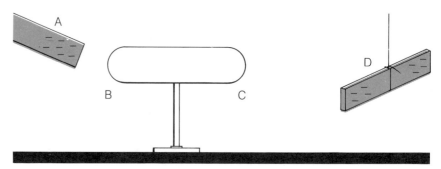

Figure 17-19

20. If you drag your feet across a rug on a dry winter day, an electric charge builds up on your body. In order to feel the shock of discharge, you would touch a metal doorknob, rather than a wooden door.
    (a) Explain why the effect is greater on a dry day.
    (b) Why is it greater when you touch a metal doorknob?
21. At the start of the chapter, the question "How could the bell ringers' lives have been spared?" was posed. Can you suggest an answer?

17.7 Charging Electroscopes by Conduction and Induction

In Experiment 17A you used a charged plastic strip as an electroscope. Its purpose was to find the type and size of an unknown charge. As an electroscope, such a large object is crude, reacting well only to large charges. For measuring small charges, a more sensitive instrument is needed.

Various types of sensitive electroscopes are used in science laboratories. One common type is called a **leaf electroscope**. It consists of one or two thin metal leaves attached to the bottom of a conducting strip that is supported in an insulating stand. Atop the conducting strip is a conducting metal cap (Figure 17-20). (Although the diagrams and descriptions in this section relate to a double-leaf electroscope, the use of a single-leaf electroscope, or other types, is similar.)

An electroscope must have a known charge in order for us to be able to judge other, unknown charges. One way to charge an electroscope or other conductor is by **conduction**. (This is also called *charging by contact*, a process described earlier in the chapter.) Figure 17-21 shows a series of diagrams explaining how a leaf electroscope becomes positively charged by means of a charged acetate strip. Notice that the representative number of protons remains constant in each step. Only the electron motion is important in considering the transfer of charge in metal conductors. (Each " + " or " − " represents billions of protons and electrons respectively.)

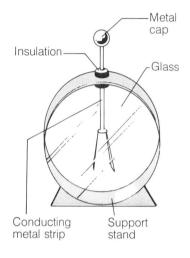

Figure 17-20
Double-leaf Electroscope

Figure 17-21
Charging an Electroscope by Conduction. (The insulating stand is not shown.)

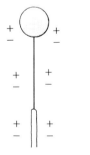

(a) Neutral electroscope

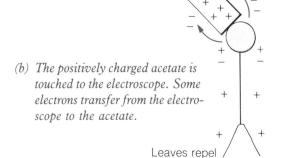

(b) The positively charged acetate is touched to the electroscope. Some electrons transfer from the electroscope to the acetate.

Leaves repel

(c) The acetate is removed, and the electroscope is left with a positive charge.

ELECTRICITY AND ELECTROMAGNETISM

Figure 17-22
Steps in Charging a Leaf Electroscope by Induction, Using a Negative Strip

(a) Neutral electroscope

(b) The negatively charged polyethylene is brought close to the electroscope. Electrons on the electroscope are repelled to the bottom, and the leaves repel one another because they have like charges.

(c) The electroscope is grounded (by being touched), and some electrons are repelled to the ground.

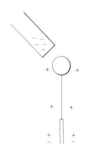

(d) The grounding is removed, but some electrons near the bottom of the electroscope cause the leaves to remain flat.

The second way to charge an electroscope or other conductor is by a process called **induction**. A known charge brought near the electroscope causes a separation of the charges on the electroscope. Then the electroscope is grounded for an instant, and charges are either repelled to or attracted from the earth. Figure 17-22 shows the steps in charging a leaf electroscope by induction using a negatively charged strip. In this case, the negative charges from the electroscope are repelled to the earth, leaving the electroscope positively charged.

A modern and much more sensitive version of the leaf electroscope is used in a portable device for measuring radiation from X ray machines or radioactive materials. The device is called a *direct-reading dosimeter*. It consists of a very thin U-shaped conducting fibre attached to another conductor, as shown in Figure 17-23. It is at least a million times more sensitive than a leaf electroscope used in a classroom.

PRACTICE

22. When a leaf electroscope is not charged, state what happens to the leaf (or leaves) when the following charged strips approach:
    (a) negatively charged   (b) positively charged
    (You may want to try this.)
23. Use your answer to Practice Question 22 to explain why an electroscope is more useful if it possesses a known charge.
24. Draw a series of diagrams to show the charge distributions on a leaf electroscope being charged by conduction by means of a negatively charged plastic strip.
25. When an electroscope is charged by conduction or contact, how does its final charge compare with the original charge on the charging device?
26. Draw a series of diagrams to show the charge distributions on a leaf electroscope being charged by induction by means of a positively charged plastic strip.
27. When you are charging an electroscope by induction, how does the final charge compare with the original charge on the charging device?

(e) The polyethylene is removed and the leaves, now with a positive charge, repel one another.

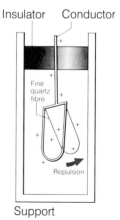

Figure 17-23
An Ultra-sensitive Electroscope: The Direct-reading Dosimeter

Experiment 17B: Induction of Electric Charges

INTRODUCTION
As you are performing this experiment, remember that induction refers to charging from a distance. The original charged object that causes the induction never touches the object on which the charge is induced.

One of the devices used in this experiment is called an **electrophorus**. It consists of a smooth metal plate and an insulating handle, with a separate base made of an insulating material. See Figure 17-24. The electrophorus can be used to store an electric charge.

PURPOSE
To study the effects of inducing electric charges.

APPARATUS
polyethylene strip; acetate strip; woollen cloth; cotton cloth; stream of water; leaf or needle electroscope; 2 metal spheres mounted on insulating stands; electrophorus set

PROCEDURE
1. Predict what will happen when a negatively charged polyethylene strip is brought close to a smooth stream of water from a tap. Check your prediction experimentally, and describe your observations.
2. Predict the result of bringing a positively charged acetate strip close to the stream of water. Again check your prediction and describe what you observed.
3. Use a negatively charged polyethylene strip to charge the electroscope by induction. (If you cannot recall how to do this, refer to Figure 17-22.) What type of charge is on the electroscope? How can you prove it? (This step is important. If you are unsure of what is happening, stop and ask for assistance now!)
4. Charge the polyethylene strip negatively and bring it close to a neutralized mounted sphere. Ground the sphere as shown in Figure 17-25(a). Predict the type of charge remaining on the sphere. Verify your prediction using the charged electroscope. Describe what you did and what you observed.
5. Place two metal spheres together and neutralize them by grounding them. Bring a negatively charged polyethylene strip close to one end of the pair, as in Figure 17-25(b). Then remove the sphere that is farther from the polyethylene strip. Predict the type of charge on each sphere. Check your prediction using the charged electroscope. Describe your observations.
6. Rub the insulating base of the electrophorus with wool. Place the metal base on the charged base. Ground the metal for a brief instant. Predict the type of charge on the metal plate and verify your prediction using the charged electroscope. If you now ground the metal plate, a small spark may result. (What else can you discover by experimenting with this device?)

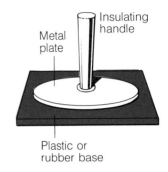

Figure 17-24
The Electrophorus

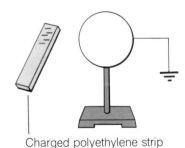

Figure 17-25
Charging by Induction

(a) By grounding

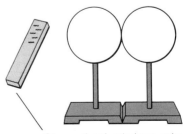

(b) By separation

ELECTRICITY AND ELECTROMAGNETISM

Figure 17-26

ANALYSIS

1. In Procedure Steps 1 and 2, did the water become charged? If so, what type of charge was on the water in each case? If not, why did it react in the manner you observed? (This problem might be more easily understood if you are aware of the shape of a water molecule, H_2O, which is shown in Figure 17-26.)
2. What are two methods of charging a conductor by induction?
3. Draw a series of diagrams to explain how charging occurred in each of Steps 4, 5, and 6. In each case, start with a neutral charge on the conductor, then show the charging process and the final result. Remember that only the electrons move in conductors.

CONCLUSIONS . . .

17.8 Static Electricity Generators

Anyone who has enjoyed the sensation shown in Figure 17-27 has helped serve one of the functions of static electricity generators – demonstrating principles of electric charge. The photograph shows the girl's hair displaying the force of repulsion. The hair acts as a pointed object, discharging the generator the girl is touching. Not shown is the insulating stool she is standing on for safety reasons.

A **static electricity generator** is a device that separates large quantities of charge. It is used to research and demonstrate principles of static electricity (Figure 17-28). Large ones are used to accelerate charged particles to high speeds and then smash them into other particles. Scientists study the collisions in these "atom smashers" and learn more about the structure of matter.

Figure 17-27
Demonstrating Electric Discharge through Hair

STATIC ELECTRICITY

Figure 17-28
This 1901 photograph shows a physicist named Nicola Tesla sitting calmly amidst numerous discharging sparks. Tesla separated the charges by using high-frequency alternating currents.

A common type of static electricity generator in science laboratories is the **Van de Graaff generator**. It was invented in 1931 by Robert Van de Graaff (1901–1967), an American physicist. Its basic design is shown in Figure 17-29. The source at the bottom of the machine separates charges rapidly, and the rubber belt carries them up to the charge collector. For our discussion we will assume that the charges are electrons, which are negative. (When used to accelerate positive charges, however, the generator is designed to send positive charges up the belt.) The electrons leave the belt at the pointed collector, which conducts them to the large hollow metal sphere. There, the electrons are spread rapidly and evenly over the outside surface of the sphere.

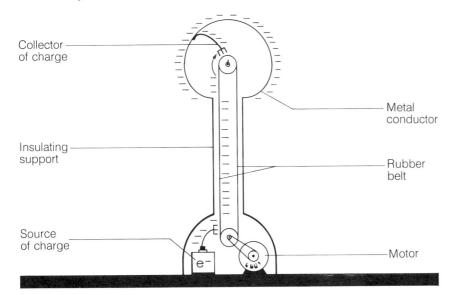

Figure 17-29
The Design of the Van de Graaff Generator

429

ELECTRICITY AND ELECTROMAGNETISM

Figure 17-30
The Smoke Condenser

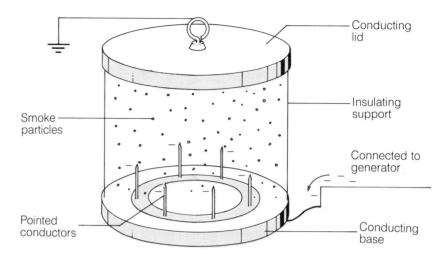

Figure 17-31
Static electricity discharge occurs between the Van de Graaff generator on the left and the grounding sphere on the right.

If a sharp, pointed conductor, such as a pin, is attached to the outer surface of a Van de Graaff generator, the charges leaving through that conductor can be used to demonstrate several phenomena. The action of a smoke condenser is a good example. Smoke is introduced into a chamber made of two conducting plates separated by a hollow insulating cylinder. On the bottom plate are pointed conductors, as shown in Figure 17-30. With the set-up shown, as soon as the generator is turned on, the smoke particles become ionized (charged) and travel to one of the metal plates. This device has important applications in the design of ways of reducing pollution from industrial smoke stacks.

(a) Charging the Leyden jar

Figure 17-32
The Leyden Jar

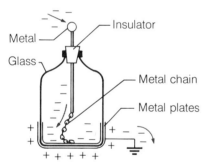

(b) Distribution of charges on the Leyden jar

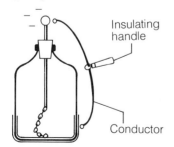

(c) Discharging the jar

430

Without no pin connected to the Van de Graaff generator, the charge that builds up on the metal sphere can be used directly for demonstrations. One example is shown in the photograph of a discharge in Figure 17-31. Another example involves a device called a **Leyden jar**. The jar is named after the Dutch town of Leyden, where it was invented in 1745. It consists of two conducting sheets separated by an insulating wall, and a conducting rod that is connected to the inside sheet. A large electric charge can be placed onto the jar. Because the jar has the capacity to store charge, it is called a **capacitor**. (Capacitors are used extensively in the electric circuits of radios, televisions, and so on.) Figure 17-32(a) shows the act of charging the device, and Figures 17-32(b) and (c) show how the Leyden jar has been charged and discharged.

CAUTION:
A spark from a Leyden jar can be extremely dangerous.

PRACTICE
28. A piece of fur is placed on top of the sphere of a neutral Van de Graaff generator. Then the generator is turned on. Predict what will happen, and explain why. (Diagrams might help.)
29. Figure 17-33 shows an arrangement for demonstrating one of Newton's laws of motion. The three-armed device is called a *pinwheel*.
    (a) Decide the direction in which the pinwheel will rotate when the generator is turned on, and explain why.
    (b) Which one of Newton's laws does this action illustrate?
30. Describe a safe technique that might be used to determine which type of charge is on a specific Van de Graaff generator.

Connected to the generator

Figure 17-33

17.9 More Applications of Static Electricity

Static electricity is often harmful, but sometimes beneficial. Several of its applications have already been described in this chapter. This section presents more applications, both unwanted and desirable.

Static electricity can be a nuisance, especially in a cold, dry climate. It causes clothes to cling to the body and lint to stick to clothing. It also causes dust particles to adhere to phonograph records and newly polished tables. One way to overcome these problems is to add moisture to the air with a humidifier. Moist air molecules become ionized easily, and thus help drain away excess charge. Another solution is to use antistatic spray.

Of course, lightning is a form of static electricity which presents a distinct hazard. During a thunderstorm you should not stand near a tall tree or any other solitary object. If you are standing up or lying down and lightning strikes, the hazard is far greater than if you are squatting. The best choice is to seek shelter, for example in a car. If someone has been struck by lightning, it is worthwhile to administer proper first-aid revival techniques, even if the person appears lifeless. Reported cases verify that revival is possible. See Figure 17-34 on page 432.

Did You Know?
At any time of year, up to 2000 thunderstorms may be moving through the earth's atmosphere, generating up to 100 lightning strokes per second. Every few million strokes, there is a giant discharge that can travel up to 6000 km/h from the clouds toward the ground, and up to 500 000 km/h in the opposite direction. Temperatures near a lightning bolt can reach as high as about 30 000°C.

ELECTRICITY AND ELECTROMAGNETISM

Figure 17-34
Laboratory tests reveal that a car is a relatively safe place to be during a lightning storm. In the test shown, the car's electronic systems, including the computer controls, were unharmed by the artificial lightning.

Smaller static electricity discharges, too, may cause fires and explosions. In hospital operating rooms, anaesthetics, which are explosive gases stored under pressure, are sometimes used. A spark could have lethal consequences in this situation, especially for the patient. To prevent spark discharges, the floor and the shoes and hose worn by the doctors and nurses must be good electric conductors.

Shocks received on the outside of the body must be relatively large to cause permanent damage. However, even tiny shocks received directly by the heart can result in death. Therefore, hospital personnel must again be careful not to cause even the smallest spark near a patient who has conducting tubes leading to the heart.

Despite this catalogue of dangers, static electricity does have its benefits. One simple use is in the wrapping of food. Static electricity forces help plastic wrap cling to food or containers, thus sealing the food from the surrounding air and maintaining its freshness.

An electronic air cleaner uses electric charging to purify the air circulating in a home with a central forced-air system. Millions of dust and pollen particles pass through the home's air ducts. The particles that are small enough to pass through a filter screen reach a cell where a large positive charge ionizes them positively. They are then attracted to the negatively charged collectors, where they accumulate. The collectors must be washed periodically to maintain the air cleaner's efficiency. Figure 17-35 shows the structure of this type of air cleaner. (The source of electric charge is not shown.)

Certain high-speed printers also apply the principles of static electricity. An example is shown in Figure 17-36. Ink particles pumped out of a nozzle acquire a charge as they pass by the charging plates. Once the droplets have a known charge, their direction can be controlled by another set of charged plates. The process is computer controlled, so it is very fast – up to 1000 lines per second! Several styles of print are available on most ink-jet printers.

Figure 17-35
The photograph shows an electronic air cleaner, with the collectors in the background.

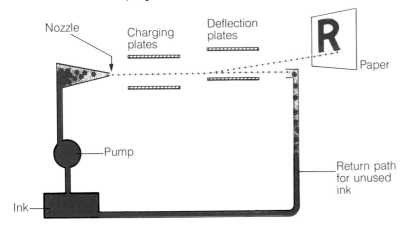

Figure 17-36
A High-speed Printer

A popular application of electric charging is in the use of negative-ion generators, which induce negative charges on air particles. Experiments seem to show that negatively ionized particles help make people feel energetic and healthy. The generators are now used in automobiles, homes, and offices.

There are many other applications of static electricity, too numerous to describe here. You will be asked to think of some of them in the Practice Questions that follow.

PRACTICE

31. Regional variations make it impossible for people to have the same experiences with static electricity. Compare your own experiences with what you expect people to experience in the following locales:
    (a) the desert regions of Arizona
    (b) the coastal regions of England
32. Several dangerous explosions have occurred when powerful jets of water were used to wash ocean-going oil tankers. What was the likely cause of these explosions? What could be done to prevent them?
33. Comment on the fact that many people say they feel better after having taken a shower. (In your answer consider charges, not cleanliness! Perhaps you could also compare showers and baths.)

34. Design a simple circuit that could be used as a smoke detector or smoke alarm. (*Hint*: Smoke particles can be ionized readily, and will conduct electricity across a narrow gap. You might try proving this by forcing smoke near a charged leaf electroscope.)
35. **Activity** Research and report on applications of static electricity not mentioned in this chapter. Some possibilities include arc welding, photocopying, making microphones with high-frequency response, coating short fibres on the insides of musical instrument cases, separating shells from nuts, and separating minerals from ores.

Review Assignment

1. Define the force of electricity and state two examples that illustrate it.
2. What is a static electricity discharge? Why does it occur?
3. When Benjamin Franklin did his kite experiment, why did he hold onto the rope with a silk cloth?
4. Explain why, on a cool dry day, a balloon rubbed on a piece of cloth will cling to a wall.
5. What is the purpose of an electroscope? What advantage does a leaf electroscope have over a suspended polyethylene strip used as an electroscope?
6. Describe how you would prove experimentally the following laws:
   (a) the attraction of electric charges
   (b) the repulsion of electric charges
7. A suspended polyethylene strip is given a negative charge. What can you conclude for certain if you bring another object close to the polyethylene and the two objects (a) repel one another and (b) attract one another?
8. Two objects, A and B, are originally neutral. When rubbed together, A loses electrons to B.
   (a) What type of charge is on A and on B?
   (b) What happens if A is brought close to a negatively charged leaf electroscope? Explain.
9. State the number of protons in an atom of (a) boron and (b) chlorine.
10. How many neutrons are there in an atom of (a) oxygen and (b) argon?
11. How many electrons are there in a neutral atom of (a) helium and (b) magnesium?
12. Draw a model of an atom of (a) neon and (b) sodium.
13. What particles are responsible for the build-up and transfer of a charge in conductors?
14. Why must a conductor be placed on an insulating stand in order to become charged?
15. Describe the differences between charging a conductor by conduction and induction.
16. State the type of charge on each material when the following pairs of materials are rubbed together:
    (a) rubber, paraffin wax   (b) lead, silk

17. **Activity** Perform the following test on several materials to determine whether or not they are good conductors. Charge a leaf electroscope by induction. Hold the (neutral) object to be tested in your hand, and touch a corner of the other end of the object to the top of the electroscope. (The sample should be at least 30 cm long.) Describe what you observe and conclude.
18. What shape of conductor is best suited to holding an electric charge? Why? Draw a diagram of the electric field surrounding the conductor to support your answer.
19. The diagrams in Figure 17-37 show negatively charged metal conductors on insulating stands. Redraw the diagrams in your notebook and indicate the approximate charge distribution and the electric field for each.
20. Explain how a pointed conductor ionizes nearby air particles when it is (a) positively charged and (b) negatively charged.
21. The diagrams in Figure 17-38 show a procedure for charging two metal conductors by induction. Redraw the diagrams in your notebook, indicating the charge distribution in each case.

Figure 17-37

(a)

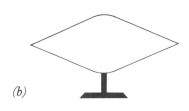
(b)

Figure 17-38

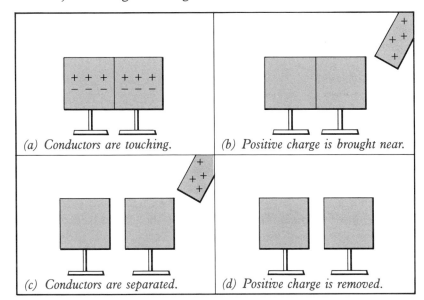

22. A double-leaf electroscope has been charged with an unknown charge, and its leaves are repelling one another. A polyethylene strip is rubbed with wool and brought toward the electroscope from a distance. At first, the leaves fall until they are vertical; then they begin to rise again.
    (a) Determine the type of charge on the electroscope.
    (b) Explain the action of the leaves. (Diagrams may help.)
23. The CN Tower in Toronto, Ontario, is the world's tallest free-standing structure (as of the time of writing). It is struck frequently by lightning, yet it suffers no severe damage. Explain how this is possible.
24. Describe how you would design a static electricity precipitator to capture the polluting smoke particles emitted from a coal-burning factory.

ELECTRICITY AND ELECTROMAGNETISM

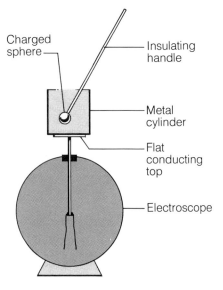

Figure 17-39

25. **Activity** Before you perform this Activity, read the instructions and predict the result. Set up two flat, parallel conductors vertically on insulating stands. Suspend a light-weight, metal-covered ball on thread between the conductors. Use induction to deposit opposite charges on the conductors, then move the conductors close enough for something to happen. Describe and explain your observations.

26. **Activity** For this Activity you will need the usual plastic strip and woollen cloth; two leaf electroscopes (one with a neutral charge and a flat top, one charged with a known charge); a tall, hollow can made of metal; and a metal sphere or proof plane with an insulating handle. Before each step be sure the can is neutral, and predict what will occur. After each step, explain what you observe.
    (a) Set up the apparatus as in Figure 17-39. Use induction to place a negative charge on the sphere, then hold the sphere inside the can without touching the sides. Try moving the sphere around.
    (b) Repeat Step (a) on the outside of the can.
    (c) Touch the sphere to the inside bottom of the can. After you observe what happens to the can, determine the sphere's charge using the second electroscope.
    (d) Neutralize the can, recharge the sphere, and repeat Step (c) on the outside of the can.

Supplementary Topic: Coulomb's Law and the Elementary Charge

You have learned that an electric charge can exert a force on another electric charge. The force can vary in magnitude, and can be either an attraction or a repulsion. A quantitative analysis of factors affecting electric force was first performed by Charles Coulomb (1736–1806), a French physicist. He devised an experiment to measure the force of repulsion between two charged objects, one of them suspended by a bar to a vertical wire, as illustrated in Figure 17-40. When charged object B was brought close to charged object A, it repelled A and caused the wire to twist a measurable amount. (Note the similarities between this experiment and the one performed by Cavendish about ten years later to find the universal gravitation constant, as discussed in Chapter 5, Supplementary Topic 1.)

The relationship discovered by Coulomb can be stated using these symbols:

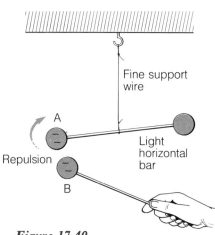

Figure 17-40
A Simplified Diagram of Coulomb's Investigation

F_E is the electric force between two charged objects.
Q_1 is the charge on one object.
Q_2 is the charge on the second object.
d is the distance between the centres of the two objects.

If Q_2 and d are constant, $F_E \propto Q_1$ (direct variation).
If Q_1 and d are constant, $F_E \propto Q_2$ (direct variation).

If Q_1 and Q_2 are constant, $F_E \propto \frac{1}{d^2}$ (inverse square variation).

Thus, $F_E \propto \frac{Q_1 Q_2}{d^2}$ (joint variation)

$\therefore \quad F_E = \frac{kQ_1 Q_2}{d^2}$, where k ≠ 1.

Thus, **Coulomb's law of electric charges** states:

> The electric force between two charged objects is directly proportional to the product of the charges on the objects and inversely proportional to the square of the distance between them. (The direction of the force extends along an imaginary straight line joining the centres of the two objects and is called positive for a repulsion and negative for an attraction.)

See Figure 17-41.

As you know, the SI units of force and distance are the newton and the metre, respectively. The unit of electric charge is the **coulomb** (C), named after Charles Coulomb. Using these units, the value of the constant in Coulomb's equation is 9.0×10^9 N·m²/C².

Thus, Coulomb's law of electric charges in equation form is

$$F_E = \frac{kQ_1 Q_2}{d^2}.$$

Here, $k = 9.0 \times 10^9$ N·m²/C². Notice the symmetry between this equation and the one for Newton's law of universal gravitation.

Figure 17-41
The Direction of the Electric Force between Charged Objects

(a) When charges are alike, the force is one of repulsion and is called positive.

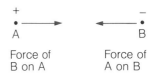

(b) When charges are opposite, the force is one of attraction and is called negative.

SAMPLE PROBLEM 2
Calculate the force of attraction between two charged spheres, each having a charge of $+2.0 \times 10^{-6}$ C, whose centres are separated by a distance of 1.0×10^{-1} m.

Solution

$F_E = \frac{kQ_1 Q_2}{d^2}$

$= \dfrac{9.0 \times 10^9 \frac{N \cdot m^2}{C^2} \times 2.0 \times 10^{-6} \, C \times 2.0 \times 10^{-6} \, C}{(1.0 \times 10^{-1} \, m)^2}$

$= 3.6$ N (Notice the cancellation of units.)

$\therefore$ The force is one of repulsion, with a magnitude of 3.6 N. (The magnitudes of the charges in this problem are typical of those experienced when performing static electricity experiments in a science classroom.)

Using Coulomb's equation, if the charges have opposite signs, the calculated force will have a negative sign. This means that the force is one of attraction, which is to be expected.

At the beginning of the twentieth century, an American physicist, Robert Millikan (1868–1953), performed an experiment to determine the quantity of charge on the particles making up the atom. Through careful measurements, he discovered that electric charges were multiples of a certain small charge called the **elementary charge** (e). In other words, the smallest charge he found was e, and all other charges were either e, or $2e$, or $3e$, and so on. The value of the elementary charge (e) in coulombs was found to be 1.6×10^{-19} C.

Since an electron is a negative elementary charge and a proton is a positive elementary charge, we can conclude that the charge on one electron is $-e = -1.6 \times 10^{-19}$ C, and the charge on one proton is $+e = +1.6 \times 10^{-19}$ C.

Thus, an object with an excess or deficiency of N electrons has a total charge, Q, given by the equation

$$Q = Ne$$

Here, $e = -1.6 \times 10^{-19}$ C for an *excess* of electrons, and $e = +1.6 \times 10^{-19}$ C for a *deficiency* of electrons.

PRACTICE

36. Calculate the electric force between two charged spheres, each with a charge of -3.0×10^{-7} C, whose centres are separated by a distance of 2.0×10^{-2} m.
37. Determine the electric force between two point charges, $Q_1 = +4.0 \times 10^{-8}$ C and $Q_2 = -8.0 \times 10^{-7}$ C, separated by distances of (a) 1.0 mm and (b) 2.0 mm.
38. Find the electric force between two electrons that are 3.0×10^{-5} m apart.
39. What is the charge in coulombs on an object that has
    (a) an excess of 5 electrons;
    (b) a deficiency of 10^3 electrons;
    (c) an excess of 6.25×10^{19} electrons.
40. Determine the number of elementary charges required for a charge of 1.0 C. (*Hint*: The equation $Q = Ne$ can be rearranged.)
41. A polyethylene strip has a charge of -4.8×10^{-7} C. What is the excess number of electrons on the strip?
42. **Activity** Research and report on the experiment performed by Millikan to measure the elementary charge, commonly called the Millikan oil-drop experiment. One source of information might be a computer program that simulates the experiment.

Key Objectives

Having completed this chapter you should now be able to do the following:
1. Define and give examples of static electricity charging and discharging.
2. Define and state examples of the force of electricity.

3. Recognize the contributions of Thales, William Gilbert, and Benjamin Franklin to the study of electricity.
4. List properties of electric force.
5. State and experimentally verify the laws of electric charges.
6. Use an electroscope of known charge to determine the unknown charge on another object.
7. Name the main components of an atom and the type of charge on each.
8. Recognize the reasons for the arrangement of the first 20 elements in the periodic table.
9. Define atomic number, mass number, and electron arrangement of atoms.
10. Define ion, and describe how a particle becomes ionized.
11. Draw a simple model of an atom of each of the first 20 elements.
12. Use the atomic theory of matter to explain events observed in static electricity experiments.
13. State the type of charge that results when any two materials in the static electricity series are rubbed together.
14. Describe the difference between electric insulators and conductors, and list examples of both types of materials.
15. Explain why grounding an object neutralizes it.
16. Map electric field lines around a single charge or a pair of charges.
17. Predict and test the distribution of electric charges on insulators and on conductors of various shapes.
18. Explain why pointed conductors allow rapid discharging.
19. Use words and diagrams to describe how to charge a leaf electroscope by conduction and induction.
20. Describe ways of charging conductors by induction.
21. Explain the basic function and operation of a static electricity generator.
22. Describe both harmful effects and beneficial applications of static electricity.

Supplementary

23. State how the electric force between two objects depends on the charges on the objects and their distance of separation.
24. Apply Coulomb's law of electric charges to solve problems involving electric force.
25. Define elementary charge.
26. Given either of the total charge on an object or the number of excess or deficient electrons on it, determine the other quantity.

ELECTRICITY AND ELECTROMAGNETISM

CHAPTER 18
Current Electricity

Miniature electric circuits are used in such devices as computers.

For millions of years, the human race got along without current electricity. Then in 1780, an Italian doctor named Luigi Galvani (1737–1798) discovered something that changed the course of history (Figure 18-1). While dissecting a dead frog, he noticed that its legs jerked violently if his knife was touching a nerve centre when a nearby static electricity generator produced a spark. Later he discovered that touching two dissimilar metals to the frog's leg caused the same effect, without the generator. Other scientists carried his experimentation further, and by 1800 current electricity was being produced.

Since that humble start just over 200 years ago, many developments related to electricity have changed our lives – machines, computers, space techology, lasers, miniature circuits (shown in the photograph) which can process up to thousands of pages of written data per second, the information explosion, and problems with pollution and energy resources. In this chapter, you will read about experiments that led to these developments, and learn how to determine and control the fundamental variables related to current electricity.

18.1 Electric Charges in Motion

When Luigi Galvani first discovered that a frog's legs reacted to being touched by metal, he believed that the electricity came from the frog. Another Italian scientist, Count Alessandro Volta (1745–1827) (Figure 18.2) read of Galvani's findings and carried out related experiments.

Figure 18-1
Luigi Galvani (1737–1798) is seen experimenting with frog's legs.

Figure 18-2
Alessandro Volta (1745–1827)

He proved that the electricity came from the two dissimilar metals, not the frog's legs. The frog was only a sensitive detector of electricity.

Volta applied his knowledge to the design of a source of current electricity, now called a **voltaic cell** in his honour. The voltaic cell consists of two dissimilar metals, called **electrodes**, placed in a liquid called an **electrolyte**. An example of a simple voltaic cell, using copper and zinc as the electrodes and a lemon as the electrolyte, is shown in Figure 18-3. Outside the cell is the load that uses the energy from the electricity, and the connecting wires. In this case, the load is an instrument which indicates a minute flow of electric charge; it is called a **galvanometer**, after Luigi Galvani.

The type of electricity produced by a voltaic cell is called current electricity. **Current electricity** is the movement of charged particles along a path. It clearly differs from static electricity (studied in Chapter 17), in which charges that move very little pile up on an object.

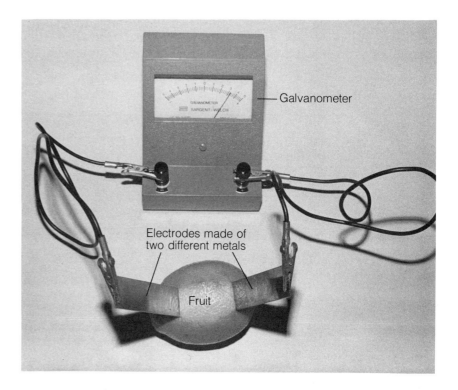

Figure 18-3
A Simple Voltaic Cell

It is possible for both positive and negative charges to move in current electricity. However, in this text we are concerned only with the motion of electrons free to travel in electric conductors. The expressions "electron flow" and "electron current" will appear frequently, and have the same meaning. (Some texts speak of "conventional current", which is the flow of positive charges and is opposite in direction to electron flow or electron current. That expression began long before it was discovered that it is the electrons which flow in metal conductors. Both ways of labelling current have advantages. Electron current is easier for beginning students to understand and apply, probably because it is true for the circumstances studied. Conventional current is useful for certain chemical reactions, electronics applications, and some applications of high-energy particle accelerators. Later in this unit you will encounter left-hand rules which apply to electron current. The corresponding rules for conventional current are right-hand rules.)

PRACTICE

1. Describe the difference between static electricity and current electricity.

2. **Activity** Set up a simple voltaic cell by placing two dissimilar metal electrodes (*e.g.*, copper and zinc) into a beaker partly filled with an electrolyte (*e.g.*, dilute sulphuric acid). Connect the cell to a galvanometer. Experimentally, discover as much as you can about the cell. Of course, further research and reading will add to your understanding. In your report, include any questions you may have. You may find the answers as you progress through this topic.

18.2 Sources of Electrical Energy

Scientists have learned how to change several types of energy into electrical energy. The energy source chosen depends greatly on the intended use of the electricity.

Most of the electrical energy used in our homes, schools, offices, and industries comes from generating stations, where huge rotating turbines produce electricity. Chapter 21 will describe more extensively the way in which rotating turbines produce current electricity. The method by which the turbines are made to rotate varies with the source of energy. Recall from Chapter 7 that non-renewable energy sources include fossil fuels (coal, oil, and natural gas) and fuel for nuclear fission (uranium); renewable sources include falling water (at hydraulic generating stations), geothermal energy, tides, wind, waves, and biomass.

Certain specific applications, such as the electricity requirements for automobiles, submarines, satellites, and portable devices, require other sources of energy. We begin with the source used by Alessandro Volta to produce the first electric current.

Chemical Energy

Chemical potential energy can be changed into electrical energy by means of a chemical reaction in a cell. If the chemical reaction can proceed only one way, the chemicals gradually become used up, and the cell is called a **primary cell** or **dry cell**. Such a cell cannot be recharged. If the chemical reaction is reversible (that is, it can be made to go both ways by using an outside source of energy), the cell can be recharged; it is called a **secondary cell** or **storage cell**. Two or more cells connected together form a battery of cells, or simply a **battery**.

THE VOLTAIC CELL

The principle of the operation of chemical cells is shown in Figure 18-4. The atoms of one electrode, often zinc, are dissolved into the liquid electrolyte, leaving behind electrons. The electrode therefore becomes negatively charged. The second electrode, composed of a different substance, becomes positively charged. When the two electrodes are connected to a conductor outside the cell, the electrons flow from the negative electrode through the conductor to the positive electrode. The energy possessed by the electrons is given to whatever is in their path. The reaction ends when one substance is used up; it is not reversible.

The voltaic cell is excellent for student experimentation, but it is not practical for ordinary use because it is wet, cumbersome, and easily spilled. Thus, other cells have been developed for everyday applications.

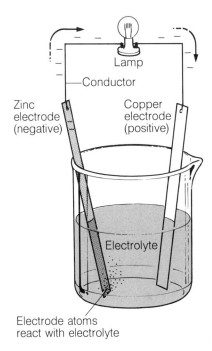

Figure 18-4
Operating a Voltaic Cell

PRIMARY CELLS

Modern primary cells, commonly called dry cells, are composed of two dissimilar metals separated by an electrolytic paste. In the versatile

carbon-zinc cell, often used in flashlights, fire alarms, and portable radios, zinc is the negative electrode and carbon is the positive. The electrolyte is a solution of salts, as shown in Figure 18-5.

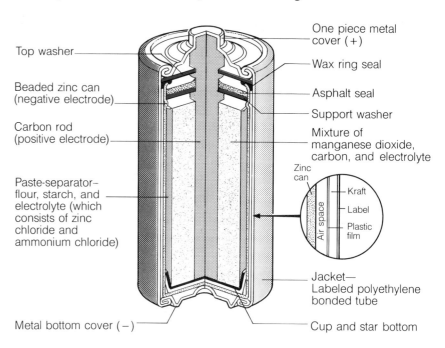

Figure 18-5
The Carbon-Zinc Cell

The **zinc-alkaline cell**, which uses the alkali potassium hydroxide as its electrolyte, has a high efficiency and functions well at low temperatures. Its uses are similar to those of the carbon-zinc cell. See Figure 18-6.

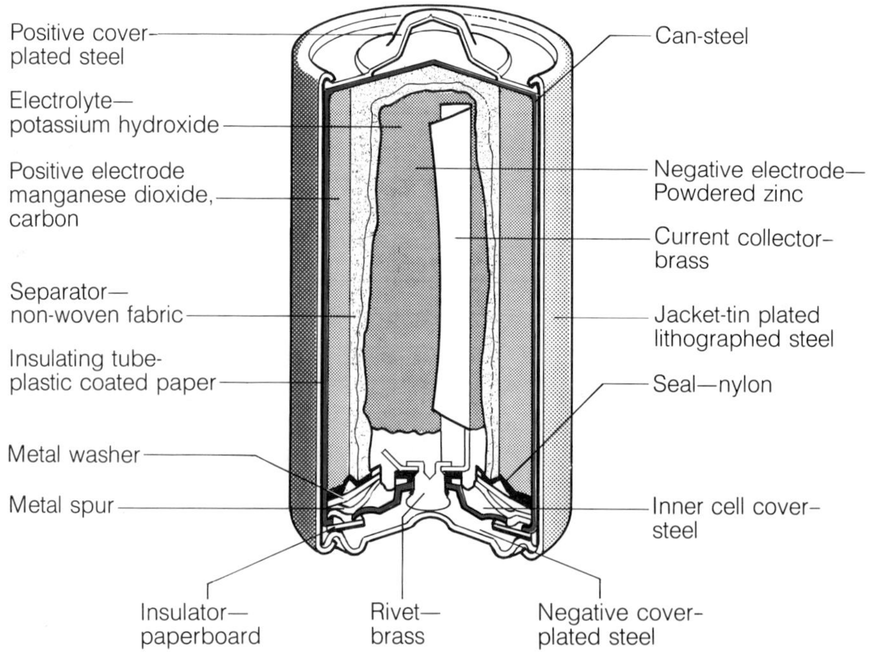

Figure 18-6
The Zinc-Alkaline Cell

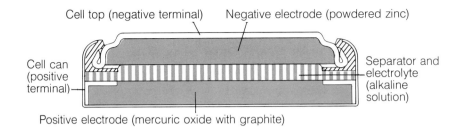

Figure 18-7
The Mercury Cell

When a continuous supply of electrical energy is required, a **mercury cell** is used. It uses zinc as the negative electrode, mercuric oxide as the positive electrode, and an alkali as the electrolyte. See Figure 18-7. The mercury cell is found in devices needing only a small flow of electric charge, such as electric watches, hearing aids, and heart pacemakers.

Other less common cells have other specific applications. Researchers are continuously trying to find ways of improving the efficiency of cells.

SECONDARY CELLS AND BATTERIES

Secondary cells, also called storage cells because they can be recharged, have the same fundamental parts as primary cells – two dissimilar metals and an electrolyte. The **lead-acid cell**, used to make typical automobile batteries, has lead as the negative electrode, lead-dioxide as the positive electrode, and a sulphuric acid solution as the electrolyte. See Figure 18-8. The chemical reaction that converts chemical energy into electrical energy is reversible. When the cell becomes weak, it can be recharged by having another source of electricity connected to it. In other words, recharging involves the conversion of electrical energy into chemical energy. Other secondary cells include **silver-cadmium**, **nickel-cadmium**, and **nickel-iron cells**. Although these cells are relatively expensive, their convenience is obvious in such devices as rechargeable flash units for cameras, and portable vacuum cleaners.

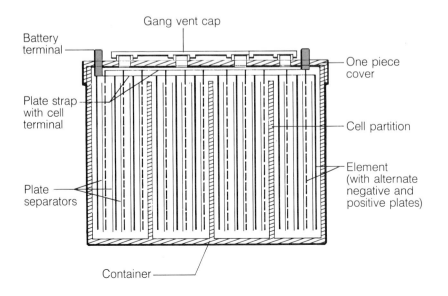

Figure 18-8
The Lead-Acid Secondary Battery

Thermal Energy

Thermal energy (or heat) can be changed into electrical energy by means of a device called a **thermocouple**. It consists of two dissimilar metal conductors joined at the ends, which are kept at different temperatures. A set-up demonstrating the operation of a thermocouple is shown in Figure 18-9. The electrical energy, in this case measured by the galvanometer, is used to determine extremes of temperature. With an iron-copper thermocouple, temperatures between 0°C and 275°C can be measured. A platinum-rubidium thermocouple has a range from 0°C to 1700°C, and a copper-constantin alloy thermocouple has a range from −180°C to 370°C. Two applications of thermocouples are detecting flames and finding the temperature of molten metal.

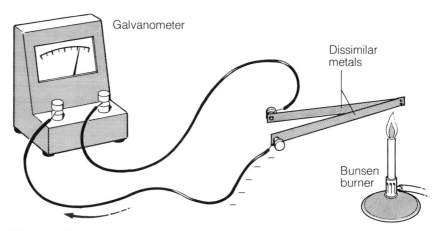

Figure 18-9
Demonstrating the Thermocouple

Did You Know?
Researchers who study the mechanics of the human body in sports apply the principle of piezoelectricity. They use an instrument called a force plate containing piezoelectric crystals which respond to variations in force. The force plate is positioned on a track or a gymnasium floor, and an athlete performs a typical manoeuvre on it. The resulting pattern of forces causes electron flow variations in the crystals. The data from these variations are analysed by computer and used to help athletes improve their techniques.

Piezoelectricity

There are many ways of changing mechanical energy into electrical energy. One is the process called **piezoelectricity**. In this technique, a mechanical force is exerted on a crystal cell composed of crystals of quartz or Rochelle salt. Doing this causes two surfaces on opposite sides of the cell to become charged negatively and positively, respectively. If a conducting wire is connected to the two sides, an electron current flows. Some cigarette lighters use this technique for igniting the flame. Piezoelectricity is also used in crystal microphones, some phonograph cartridges, and certain radiation detectors, such as the ultra-sensitive electroscope shown in Figure 17-23. The crystals last indefinitely.

Light Energy

When light energy strikes the surface of certain metals, electrons may be freed from the atoms near the surface. The freed electrons are emitted from the surface and conducted through an external circuit. To allow this action to continue, the metal is placed in an evacuated tube. The resulting device is known as a **photoelectric cell**. (The prefix "photo" means "light".) Two metals commonly used with visible light are potassium and cesium. Photoelectric cells are used in satellites and certain portable devices such as calculators. See Figure 18-10.

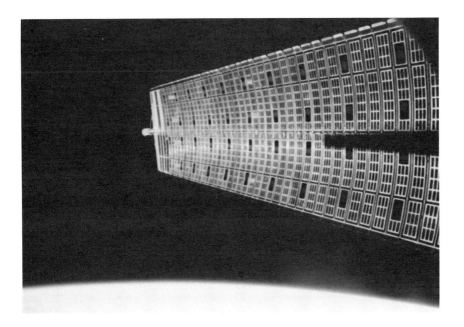

Figure 18-10
The cells in this solar energy panel attached to a space shuttle convert light energy into electrical energy. The panel shown was used for experimental purposes.

PRACTICE
3. State the most likely source of electrical energy needed to operate each of the following:
   (a) a home computer
   (b) a car radio
   (c) an electric watch
   (d) a submarine communication system
   (e) a miner's safety lamp
   (f) a telephone
   (g) a disposable thermometer at an iron-ore smelting factory
4. What are the main components of a chemical cell?
5. State the difference between each of the following pairs of devices:
   (a) a primary cell and a secondary cell
   (b) a cell and a battery
6. In one experiment, Luigi Galvani poked a brass rod through the upper part of a dead frog's leg and connected the rod to an iron support, as shown in Figure 18-11. When the leg relaxed, it touched the base of the support and immediately contracted in a violent fashion. Then the leg relaxed and the cycle began again.
   (a) How does Galvani's experiment relate to the operation of a voltaic cell?
   (b) What does the experiment indicate about the frog's leg?
7. **Activity** In the previous chapter, you learned about the static electricity series. Now you can try to develop a corresponding series for current electricity – the **electromotive series**. Design and perform an experiment to classify metals in the order of their tendency to lose electrons and become the negative electrode in a chemical cell. Some metals to try are lead, aluminum, zinc, nickel, iron, and copper. (*Hint*: When zinc and copper are used together in a voltaic cell, the zinc becomes the negative electrode. Thus, it precedes copper in the electromotive series.)

Figure 18-11

ELECTRICITY AND ELECTROMAGNETISM

18.3 Electric Current

Figure 18-12
Water Current and Electric Current

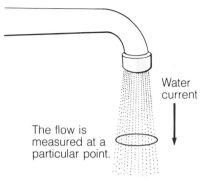

(a) Water current is a measure of the volume of water flowing past a given point per unit time. It can be measured in litres per second.

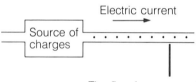

(b) Electric current is a measure of the number of electric charges flowing past a given point per unit time. It can be measured in coulombs per second.

Assume that you are asked to measure the flow of water coming from an open tap. You might try to count the number of water molecules that come out of the tap in a certain amount of time, and express the flow in molecules per second. However, the billions of molecules would be difficult to count, so you would likely measure the flow in a more convenient unit, for instance, litres per second. See Figure 18-12(a).

Electricity may be compared to flowing water. **Electric current** (I) can be defined as a measure of the total charge that passes by a particular point in a circuit each second. See Figure 18-12(b). (This is an *operational* definition, one that defines a quantity in terms of variables which can be readily determined. The *formal* definition of electric current is found in Appendix A.) Electric current can thus be measured as the number of electric charges per second. However, that number would be in the billions, so a more convenient unit, the coulomb (C), is used to measure electric current. (One coulomb of charge is equivalent to 6.24×10^{18} elementary charges. Refer to the Supplementary Topic in Chapter 17.)

We can now define electric current (I) in equation form, in terms of quantity of charge (Q) and time (t).

$$\text{current} = \frac{\text{charge}}{\text{time}} \quad \text{or} \quad I = \frac{Q}{t}$$

Thus, current is measured in coulombs per second, a unit that is given the name **ampere** or **amp** (A) in honour of André-Marie Ampère (1775–1836), a French physicist (Figure 18-13). Therefore, 1.0 A = 1.0 C/s.

SAMPLE PROBLEM 1

A battery delivers 5.0 C of charge in 20 s. What is the current from the battery, expressed in amps and milliamps?

Solution

$I = \dfrac{Q}{t}$

$= \dfrac{5.0 \text{ C}}{20 \text{ s}}$

$= 0.25 \text{ A}$

$= 2.5 \times 10^2 \text{ mA} \quad (1.0 \text{ A} = 1000 \text{ mA})$

∴ The current is 0.25 A or 2.5×10^2 mA.

Figure 18-13
André-Marie Ampère (1775–1836)

The equation $I = Q/t$ can be rearranged to determine any one of the variables, given the other two.

CURRENT ELECTRICITY

The instrument used to measure electric current is called an **ammeter**. It is connected directly into the path of the moving charges in what is called a **series connection**. Figure 18-14(a) shows the correct way to do this; Figure 18-14(b) shows the same set-up, but using symbols instead of drawings. Notice that the electrons leave the negative terminal of the cell and enter the negative terminal of the ammeter, causing the needle to swing to the right. After the electrons have given their energy to the lightbulb, they are attracted back to the source, where they gain more energy.

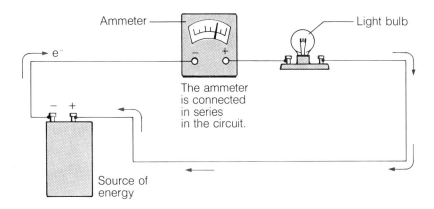

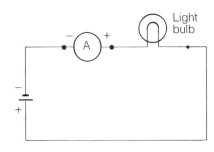

Figure 18-14
Measuring Electric Current

(a) *The electrical set-up*

(b) *The same set-up drawn with symbols*

An ammeter has from one to several scales. On a multi-scale meter, the scale read must correspond to the terminal used. For example, if the needle swings halfway across the 50 mA scale, the current is 25 mA. The questions that follow will give you practice in reading scales before you perform any experiment.

PRACTICE

8. Find the current in each case.
   (a) $Q = 4.2$ C, $t = 0.30$ s
   (b) A safety fuse in an electric circuit burns out when more than 60 C pass by in 8.0 s.
   (c) In one minute, 51 C of charge pass through a 100 W lightbulb.
9. Calculate the total charge in each situation.
   (a) $I = 15$ A, $t = 12$ s
   (b) A cell provides 0.15 A of current to a radio for 30 min.
   (c) A calculator draws 20 mA of current from a solar cell for 26 s.
10. Determine the time in each case.
    (a) $Q = 8.6$ C, $I = 2.0$ A
    (b) A current of 22 A delivers 1.1 C of charge.
    (c) A charge of 75 mC is delivered by a current of 150 mA.
11. State the current in amps for the ammeter shown in Figure 18-15 when the wire is connected to the following scales:
    (a) the 0–500 mA scale
    (b) the 0–5 A scale
    (c) the 0–25 A scale

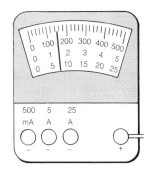

Figure 18-15

449

ELECTRICITY AND ELECTROMAGNETISM

Figure 18-16
Potential Energies

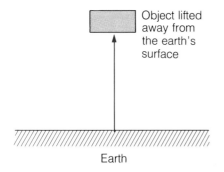

(a) Work is done on an object to move it farther from the earth. At the higher level, the object has increased gravitational potential energy.

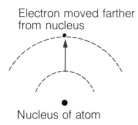

(b) Work is done on an electron to move it farther from the nucleus of an atom. At the new separation, the electron has increased electrical potential energy.

Did You Know?
An important part of the body's nervous system is the **neuron**, a nerve cell which can receive, interpret, or transmit electrical messages. An electric potential difference exists across the surface of every neuron because of a greater number of negative charges on the inside than on the outside. The potential difference is typically about 60 mV to 90 mV.

18.4 Electric Potential Difference

Recall from your study of gravitational potential energy (Chapter 6) that work must be done against gravity to raise an object to a higher level. There is a difference in the object's gravitational potential energy between the lower and higher positions. That difference means that the raised object is capable of doing work. See Figure 18-16(a).

In a similar fashion, an electrical energy source does work in separating electric charges, giving them electric potential energy. That potential energy is capable of doing work in an electric circuit. See Figure 18-16(b).

Electric potential is a measure of the electric potential energy per unit charge. (Try to remember that electric potential is not simply potential energy; it is potential energy per charge.) In an electric circuit, it is the **potential difference** that is important. This difference can be either a rise or a drop. The energy source produces a **potential rise**. The load, which transforms electrical energy into some other form of energy, causes a **potential drop**. The equation for electric potential difference, sometimes called **voltage**, is

$$\text{potential difference} = \frac{\text{change of energy}}{\text{charge}} \quad \text{or} \quad V = \frac{\Delta E}{Q}.$$

Here, ΔE is the change of energy measured in joules (J), Q is the charge measured in coulombs (C), and V is the potential difference or voltage measured in **volts** (V). The volt is named after Alessandro Volta.

Thus, for example, a potential rise of 1.0 V means that a source gives 1.0 J of energy to each coulomb of charge that passes through it.

SAMPLE PROBLEM 2
What is the potential rise of a cell that gives 0.90 J of energy to every 0.60 C of charge passing through it?

Solution

$V = \dfrac{\Delta E}{Q}$

$= \dfrac{0.90 \text{ J}}{0.60 \text{ C}}$

$= 1.5 \text{ V}$

∴ The potential rise is 1.5 V.

The equation $V = \Delta E/Q$ can be altered to determine any one of the quantities, given the other two. In the calculations involving potential difference in this text, we will assume that all the energy given by the source is consumed by the load. In actual circuits, a small amount of energy is consumed in the source itself and in the connecting conductors.

By combining the equation $V = \Delta E/Q$ with the equation $I = Q/t$, various expressions for variables can be found. It is left to the Practice Questions at the end of this section to derive those expressions.

The instrument used to measure potential difference is called a **voltmeter**. To measure potential rise, the voltmeter is connected across the source. To measure potential drop, the voltmeter is connected across the load. This type of connection, in which the electrons have at least two path choices, is called a **parallel connection**. It is illustrated in Figures 18-17(a) and (b). Compare this set-up with the series connection for ammeters, shown in Figure 18-14.

Notice in Figure 18-17 that the electrons leave the negative terminal of the source and enter the negative terminal of each voltmeter. This causes the needle in each voltmeter to swing correctly to the right. Reading a voltmeter is similar to reading an ammeter, as you will find in Practice Question 16.

Figure 18-17
Measuring Potential Differences

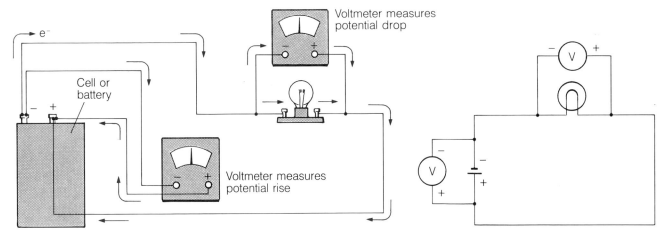

(a) The electrical set-up

(b) The same set-up drawn with symbols

PRACTICE

12. Copy and complete the table below in your notebook. Watch the units carefully.

	Potential Difference	Energy	Charge
(a)	? V	42 J	3.5 C
(b)	? V	6.0 J	50 mC
(c)	6.0 V	? J	0.25 C
(d)	9.0 V	? J	400 mC
(e)	2.2 V	1.1 J	? C
(f)	220 V	4.0 kJ	? C

13. Find the potential drop across a lightbulb if 7.2×10^3 J of work are required to move 60 C of charge across the bulb.

ELECTRICITY AND ELECTROMAGNETISM

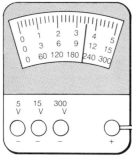

Figure 18-18

14. How much work is done on 4.6 μC of charge to increase its potential by 12 V?
15. A 6.0 V source does 3.9×10^{-4} J of work on a quantity of charge. Determine the quantity of charge, in coulombs and microcoulombs.
16. What is the reading on the voltmeter shown in Figure 18-18 when the wire is connected to the following scales?
    (a) the 0–5 V scale
    (b) the 0–15 V scale
    (c) the 0–300 V scale
17. Starting with the equations for current and potential difference ($I = Q/t$ and $V = \Delta E/Q$), prove that $\Delta E = VIt$. Is this equation more or less useful than $\Delta E = QV$ in a laboratory situation? Explain your answer.
18. Prove that the unit of the product VIt is identical to the unit of energy.
19. How much energy is consumed by a hair dryer that draws 14 A of current from a 120 V source for 12 min? Express your answer in joules and megajoules.

18.5 Electric Resistance

Electric conductors and insulators have already been discussed in Chapter 17. Conductors allow electric charges, namely, free electrons in metals, to move more easily than insulators do. However, even conductors resist current flow to some degree. The extent to which a conductor or insulator resists the flow of current is called **resistance**.

The human body has an electric resistance, which varies greatly with moisture content. Dry skin has a much higher resistance than wet skin. This principle is applied in the use of a **polygraph** or lie-detector. A stressful situation (such as telling lies) causes sweating, which lowers the electric resistance of the skin by an amount easily detected by the polygraph.

A more common use of electric resistance is found in appliances that produce heat and light. Consider, for example, the broken lightbulb shown in Figure 18-19. The thicker outside conductors have low resistance and gain little energy from electrons passing through them. The thin, coiled wire strung across the top has a high resistance to electron flow, so it gains much energy from the electrons, heats up, and emits light.

Figure 18-19
Resistance in an Electric Lightbulb

The resistance of an electric conductor depends on four factors:

(a) **Length** The resistance varies directly as the length of the conductor; doubling the length will double the resistance.
(b) **Cross-sectional area** The resistance varies inversely as the cross-sectional area. If the area doubles, the resistance becomes half as great because the electrons have more space in which to travel.
(c) **Type of material** A material that is an excellent conductor has many free electrons and a low resistance. Less efficient conductors have a higher resistance to electron flow. Silver, copper, and

gold have a very low resistance; aluminum and tungsten have a slightly higher resistance; and iron has an even higher resistance.

(d) **Temperature** The resistance of most materials, metal conductors in particular, becomes lower as the temperature drops. At extremely low temperatures (in the region of −270°C), the resistance of some materials drops to zero. These materials become superconductors. Using superconductivity would be an efficient way of transporting energy, and research continues in this field. However, there are some substances, called semiconductors, whose resistance becomes higher as the temperature drops. (Semiconductors were mentioned in Chapter 17.) These substances, among them carbon, silicon, and germanium, are used in manufacturing electronic parts.

The symbol for resistance is R. Its unit of measurement is the **ohm** (symbol Ω, the Greek letter *omega*), in honour of Georg Simon Ohm (1787–1854), a German physicist (Figure 18-20). Although resistance can be measured with an ohmmeter, we will use calculations instead and find it indirectly.

Resistors are devices manufactured with a known resistance. They are intended to be used in electronic devices and science laboratories. Two common materials used to make resistors are wire and granulated carbon. Figure 18-21 shows three resistors – a wire-wound resistor of fixed value, a colour-coded carbon resistor of fixed value, and a wire-wound variable resistor called a **rheostat**. The colour coding for resistors is described in Practice Question 22. Rheostats are used to control such devices as electric motors and dimmer switches.

Figure 18-20
Georg Simon Ohm (1787–1854)

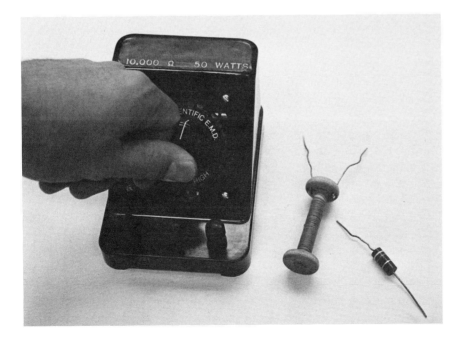

Figure 18-21
Three Types of Resistors

PRACTICE
20. If you were designing an electric toaster, would you use conducting wire of high resistance or low resistance? Why?

Figure 18-22
A Colour-coded Resistor

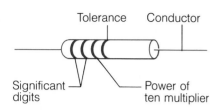

Example: If the colours, read from left to right, are brown, red, orange, and silver, the resistance is $12 \times 10^3 \, \Omega \pm 10\%$, or $1.2 \times 10^4 \, \Omega \pm 10\%$

21. Assume that the resistance of a certain iron conductor, A, is $0.22 \, \Omega$. Compare the resistance of a conductor B to A in each of the cases described below. If possible, give a numerical answer.
    (a) A is 2.4 m long and B is 1.2 m long.
    (b) The cross-sectional area of A is 1.1 mm^2 and of B is 2.2 mm^2.
    (c) The temperature of A is 20°C and of B is 150°C.
    (d) A is made of iron; B of gold.

22. **Activity** Obtain various colour-coded resistors and read each resistance value using the following information. The first two rings of colour are the significant digits, the third ring is the multiplier, and the fourth ring is the tolerance (percent possible error in the value). Refer also to Figure 18-22 and to Table 18-1.

Table 18-1 The Resistor Colour Code

Colour	Digit	Multiplier	Tolerance
black	0	10^0 or 1	
brown	1	10^1	
red	2	10^2	
orange	3	10^3	
yellow	4	10^4	
green	5	10^5	
blue	6	10^6	
violet	7	10^7	
gray	8	10^8	
white	9	10^9	
gold		10^{-1}	$\pm 5\%$
silver		10^{-2}	$\pm 10\%$
no colour			$\pm 20\%$

18.6 Electric Circuits

The main components of a simple electric circuit are the energy **source**, the **load**, and the **conductor** that transfers energy from the source to the load. However, most practical circuits are not so simple. Often they involve more than one load, a switch, as well as instruments for measuring unknown quantities.

Circuit diagrams are representations of the actual components, but they are much easier to draw. You have already seen a few of the symbols used in circuit diagrams; a more complete set of symbols appears in Figure 18-23.

Notice in Figure 18-23 that there are two ways of connecting cells to make a battery. When cells are connected in series, the negative terminal of one cell is joined to the positive terminal of the next one. Electrons in the circuit gain more energy as more cells are added. When cells are arranged in parallel, all the negative terminals are joined; similarly, all the positive terminals are joined. As more cells are added, the energy of the electrons does not change, but the current delivered

CURRENT ELECTRICITY

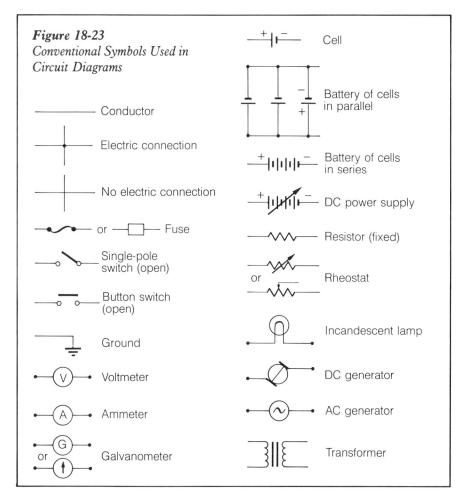

Figure 18-23
Conventional Symbols Used in Circuit Diagrams

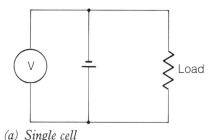

Figure 18-24

(a) Single cell

by each cell becomes less. Cells last longer using a parallel connection. (More about cells in series and parallel is found in the Practice Questions that follow.)

Electric circuits may be classified as closed or open. In a **closed circuit**, the switch is closed and the electrons have a path to follow. In an **open circuit**, the switch is open or the connections are severed, so the electrons cannot travel.

PRACTICE

23. Use the proper symbols to draw a diagram of the following circuit. Two batteries, each consisting of four cells, are connected in parallel. A switch controls the total electron current delivered to three lamps in parallel with each other. A fuse protects the circuit where the current is largest.

24. **Activity** Determine the potential rise of series and parallel cells in a circuit. You will need a voltmeter, 2 or 3 cells, connecting wires, and a load (for example, a 25 Ω resistor). Measure the potential rise of each cell connected to the load. Then arrange the cells in series and parallel and find the potential rise in each case. See Figure 18-24.

(b) Cells in series

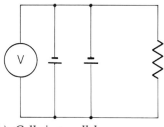

(c) Cells in parallel

455

ELECTRICITY AND ELECTROMAGNETISM

Experiment 18A: Electric Current, Potential Difference, and Resistance

The three variables, current (I), potential difference (V), and resistance (R), were defined earlier in the chapter, but not in terms of one another. In this experiment you will determine the relationship among these variables.

When performing an experiment involving three variables, you must keep one variable constant while you are finding the relationship between the others. In this case, the resistance is kept constant, electric potential is the independent variable (controlled by you), and current is the dependent variable.

PURPOSE
To determine the relationship between the current through a resistor and the potential drop across it.

CAUTION:
Never leave an electric circuit connected longer than necessary because overheating may result.

APPARATUS
at least 3 fixed resistors of different values; variable power supply (or a constant DC source connected to a rheostat); voltmeter; ammeter; switch; connecting wires

PROCEDURE
1. Set up an observation table based on Table 18-2. There will be at least three resistors (R_1, R_2, R_3, . . .), each with a minimum of three sets of data; therefore, nine or more rows of data are required in the final two columns.

Table 18-2 Sample Chart of Observations for Experiment 18A

Resistor	Labelled Resistance (ohms, Ω)	Potential (volts, V)	Current (amps, A)
R_1			
		SAMPLE ONLY	
R_2			

2. Set up the circuit as shown in Figure 18-25. Adjust the power supply so that the potential is an appropriate low value. Then measure the current through the resistor and record the data in your table.
3. Adjust the power supply to give a second, higher value of the potential. Again find the current through the resistor.
4. Continue increasing the potential and finding the corresponding current as advised by your teacher.
5. Repeat Procedure Steps 2 to 4 using the second resistor, R_2, then the third one, R_3, and so on.

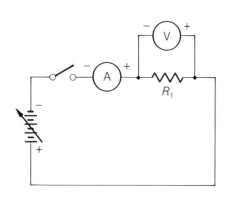

Figure 18-25
Circuit for Experiment 18A

ANALYSIS

1. On a single graph of potential drop (vertical axis) *versus* current (horizontal axis), plot the results of the experiment. The independent variable is plotted along the vertical axis in this case, so that the slopes of the lines will correspond to other known quantities. There should be at least three distinct lines of best fit, one for each resistor used; each line should start from the origin. Calculate the slope of each line, and relate each slope to the resistance.
2. What is the relationship between current and resistance for a constant potential drop? Use experimental data to verify your answer.
3. What do you suppose is the relationship between the brightness of a lamp and the current through it? Explain your answer.

CONCLUSIONS . . .

18.7 Ohm's Law

It was the German physicist Georg Simon Ohm who discovered the relationship among electric current, potential difference, and resistance. His discovery, called **Ohm's law**, states:

> **The ratio of the potential drop across a resistor to the current through it is constant if the temperature remains constant.**

Because resistance is the ratio of potential drop to current, Ohm's law can be written in equation form:

$$\text{resistance} = \frac{\text{potential}}{\text{current}} \quad \text{or} \quad R = \frac{V}{I}$$

Here, V is measured in volts, I in amps, and R in ohms. Can you see that $1.0 \, \Omega = 1.0 \, \text{V/A}$?

SAMPLE PROBLEM 3

A heating coil on an electric stove draws 25 A of current from a 240 V circuit. What is the coil's resistance?

Solution

$R = \dfrac{V}{I}$

$= \dfrac{240 \text{ V}}{25 \text{ A}}$

$= 9.6 \, \Omega$

∴ The coil's resistance is $9.6 \, \Omega$.

Ohm's law is based on experimental observations and is true only for certain materials. Thus, it is not a fundamental law of nature such as the law of universal gravitation. Materials that obey Ohm's law are called **ohmic**; those which do not obey it are called **non-ohmic**.

The equation $R = V/I$ can be altered to find expressions for V and I. In fact, Ohm's law is often written in either of the following forms:

$$V = IR \quad \text{or} \quad I = \frac{V}{R}$$

PRACTICE

25. Calculate the value of the resistance in each case:
    (a) $V = 12$ V, $I = 0.25$ A
    (b) $V = 1.5$ V, $I = 30$ mA
    (c) $V = 2.4 \times 10^4$ V, $I = 6.0 \times 10^{-3}$ A
26. Find the unknown quantities.
    (a) $R = 30 \, \Omega$, $I = 0.45$ A, $V = ?$
    (b) $R = 2.2$ kΩ, $I = 1.5$ A, $V = ?$
    (c) $V = 6.0$ V, $R = 18 \, \Omega$, $I = ?$
    (d) $V = 52$ mV, $R = 26 \, \Omega$, $I = ?$
27. What current is drawn by a vacuum cleaner from a 115 V circuit having a resistance of 28 Ω?
28. Calculate the maximum rating (in volts) of a battery used to operate a toy electric motor which has a resistance of 2.4 Ω and runs at top speed with a current of 2.5 A.
29. A walkie-talkie receiver operates on a 9.0 V battery. If the receiver draws 300 mA of current, what is its resistance?
30. **Activity** Design and perform an experiment to determine whether the resistance element in an electric heater or lightbulb is ohmic or non-ohmic. (The light source in an optics ray box is convenient for this experiment.) Include a graph in your analysis.

Experiment 18B: Resistors in Series

INTRODUCTION

You have learned that ammeters must be connected in series, voltmeters must be connected in parallel, and cells may be connected either way. Like cells, resistors may be connected either in series or in parallel. In this experiment and the next one, you will learn about the characteristics of each type of connection.

To follow the instructions in this experiment you must understand the meanings of symbols with subscripts. The symbol I_a means the current through a conductor at point "a", so an ammeter is required at that point. The symbol V_{bc} means the potential drop (or rise) from point "b" to point "c", so a voltmeter is required from "b" to "c". These ideas are shown in Figure 18-26.

PURPOSE

To study the characteristics of an electric circuit with resistors connected in series.

Figure 18-26
Resistors in Series

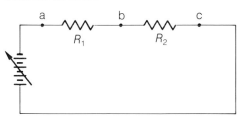

(a) Circuit with points a, b, and c

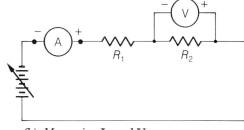

(b) Measuring I_a and V_{bc}

APPARATUS
2 fixed resistors of different sizes; variable power supply (or an alternative); voltmeter; ammeter; switch; connecting wires

PROCEDURE
1. Set up the series circuit shown in Figure 18-27 with an ammeter at point "a" and a voltmeter across "de". Adjust the supply potential to a moderate value, about 6.0 V or less, as advised by your teacher. Determine I_a in amps.

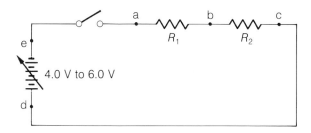

Figure 18-27
Circuit for Experiment 18B

2. Predict the value of the current at point "b", with V_{de} constant. Verify your prediction experimentally.
3. Predict the value of the current at "c", with V_{de} constant. Check your prediction.
4. With the supply potential constant, remove the voltmeter and use it to find the potential drops V_{ab} and V_{bc}. (If you leave the ammeter in the circuit, you can check that the current remains constant.)

ANALYSIS
1. Compare the currents measured at "a", "b", and "c".
2. Compare the potential rise (V_{de}) given by the supply with the sum of the potential drops ($V_{ab} + V_{bc}$) used by the resistors.
3. Use the results of this experiment in conjunction with Ohm's law to derive an equation for the total resistance of a series circuit in terms of the individual resistances. (*Hint*: Start by writing an equation for the potential rise in terms of the sums of the potential drops. Then substitute into the equation using Ohm's law.)
4. In a series circuit, will the current continue to flow if one resistor is suddenly removed? Explain.

CONCLUSIONS . . .

Experiment 18C: Resistors in Parallel

INTRODUCTION
When resistors are connected in parallel, the electrons in the circuit have more than one possible path to take. This gives the circuit characteristics that differ from those of series circuits. Can you predict what the differences are?

PURPOSE
To study the characteristics of an electric circuit with resistors connected in parallel.

APPARATUS
as in Experiment 18B

PROCEDURE
1. Set up the parallel circuit shown in Figure 18-28 with an ammeter at "a" and a voltmeter across "fa". Set the supply potential to an appropriate value, say 6.0 V. Find the total current (I_a) delivered by the power supply to the circuit.

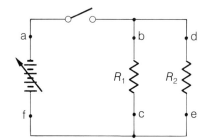

Figure 18-28
Circuit for Experiment 18C

2. Alter the circuit so that the ammeter is at "b". With the same supply potential (V_{fa}), find the current (I_b) through R_1.
3. Repeat Procedure Step 2 for the current (I_d) through R_2.
4. Leaving the supply potential constant, use the voltmeter to find V_{bc} and V_{de}.

ANALYSIS
1. Compare the total current (I_a) with the sum of the currents through R_1 and R_2.
2. Compare the potential drops V_{bc} and V_{de}.
3. Use the results of this experiment in conjunction with Ohm's law to derive an equation expressing the total resistance (R_T) of a parallel circuit in terms of the individual resistances. (*Hint*: Start by writing an equation for the total current in terms of the sum of the currents through the individual resistances. Then substitute into the equation using Ohm's law. The derived equation can be written in the form $\frac{1}{R_T} = \ldots$.)

4. Is the total resistance of resistors in parallel less than, greater than, or equal to any individual resistance? Explain why.
5. In a parallel circuit, will the current continue to flow if one resistor is suddenly removed? Should the loads in a household circuit be connected in series or in parallel? Explain.

CONCLUSIONS . . .

18.8 Analysing Electric Circuits

In this section, some of the important facts you have studied so far in this chapter will be applied to the analysis of electric circuits. Ohm's law indicates the manner in which current, potential, and resistance are related ($R = V/I$). Table 18-3 summarizes the characteristics of circuits with series and parallel resistors. These characteristics will be combined with Ohm's law in deriving equations for total resistance.

Table 18-3 Characteristics of Series and Parallel Circuits

	Series	Parallel
Sample		
Current	The current is constant throughout the entire circuit.	The total current equals the sum of the individual currents.
Potential	The total potential rise equals the sum of the individual potential drops.	The potential drop is constant across all loads in the circuit.
Resistance	The total resistance is greater than any individual resistance.	The total resistance is less than the smallest individual resistance.

First, consider a circuit having two resistors, R_1 and R_2, connected in series. The current (I) remains constant through both resistors. The total potential rise (V_T) equals the sum of the potential drops ($V_1 + V_2$). Using Ohm's law, we can calculate R_T, as shown in the following steps. (The explanations are in brackets.)

$$V_T = V_1 + V_2 \quad \text{(proved experimentally)}$$
$$IR_T = IR_1 + IR_2 \quad \text{(from Ohm's law, } V = IR\text{)}$$
$$\therefore R_T = R_1 + R_2 \quad (I = \text{constant, so it divides out)}$$

That is, the total resistance in a series circuit equals the sum of the individual resistances.

Next, consider a circuit with two resistors, R_1 and R_2, connected in parallel. In this case, the potential (V) remains constant across both resistors. The total current (I_T) equals the sum of the individual currents $(I_1 + I_2)$. Again using Ohm's law, we can find an expression for R_T, as shown below. (The explanations are shown in brackets.)

$$I_T = I_1 + I_2 \quad \text{(proved experimentally)}$$
$$\frac{V}{R_T} = \frac{V}{R_1} + \frac{V}{R_2} \quad \left(\text{from Ohm's law, } I = \frac{V}{R}\right)$$
$$\therefore \frac{1}{R_T} = \frac{1}{R_1} + \frac{1}{R_2} \quad (V = \text{constant, so it divides out})$$

That is, the reciprocal of the total resistance in a parallel circuit equals the sum of the reciprocals of the individual resistances.

SAMPLE PROBLEM 4

Find the total resistance when three resistors, having values of 5.0 Ω, 10 Ω, and 30 Ω, are connected (a) in series and (b) in parallel.

Solution

(a) In series: $R_T = R_1 + R_2 + R_3$
$$= 5.0 \, \Omega + 10 \, \Omega + 30 \, \Omega$$
$$= 45 \, \Omega$$

∴ The total resistance is 45 Ω.

(b) In parallel: $\dfrac{1}{R_T} = \dfrac{1}{R_1} + \dfrac{1}{R_2} + \dfrac{1}{R_3}$

$$\frac{1}{R_T} = \frac{1}{5.0 \, \Omega} + \frac{1}{10 \, \Omega} + \frac{1}{30 \, \Omega}$$

$$\frac{1}{R_T} = \frac{6}{30 \, \Omega} + \frac{3}{30 \, \Omega} + \frac{1}{30 \, \Omega}$$

$$\frac{1}{R_T} = \frac{10}{30 \, \Omega}$$

$$\therefore R_T = \frac{30 \, \Omega}{10} = 3.0 \, \Omega$$

Thus, the total resistance is 3.0 Ω.

Notice in Sample Problem 4(b) that the total resistance in parallel is less than even the smallest resistance. With added parallel resistors the electrons have more paths from which to choose, so the resistance to their flow is reduced. The situation is similar to the relationship between the resistance of a conductor and its cross-sectional area ($R \propto 1/A$).

The main goal of studying these concepts is to use them for analysing electric circuits. Information which is either given in a problem or found experimentally is used to find any related unknown in the circuit. The next three Sample Problems illustrate the basic process of analysing circuits. The techniques applied can be used to analyse much more complex electric circuits.

SAMPLE PROBLEM 5

In the series circuit shown in the diagram, $V_{ab} = 20$ V, $R_1 = 10\ \Omega$, and $I_a = 2.0$ A. Find values for the following:
(a) I_b (b) V_{bc} (c) R_T (d) R_2

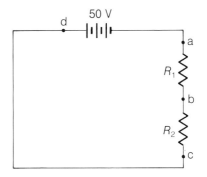

Solution

(a) $I_b = 2.0$ A, because the current in a series circuit is constant.

(b) $\quad V_{da} = V_{ab} + V_{bc}$
$\therefore V_{bc} = V_{da} - V_{ab}$
$\qquad = 50\text{ V} - 20\text{ V}$
$\qquad = 30\text{ V}$

Thus, V_{bc} is 30 V.

(c) $\quad R_T = \dfrac{V_T}{I}$
$\qquad = \dfrac{50\text{ V}}{2.0\text{ A}}$
$\qquad = 25\ \Omega$

Thus, R_T is 25 Ω.

(d) $\quad R_T = R_1 + R_2$
$\therefore R_2 = R_T - R_1$
$\qquad = 25\ \Omega - 10\ \Omega$
$\qquad = 15\ \Omega$

Thus, R_2 is 15 Ω.

SAMPLE PROBLEM 6

In the parallel circuit shown in the diagram, $V_{ab} = 20$ V, $I_a = 4.0$ A, and $I_c = 1.0$ A. Calculate values for the following:
(a) V_{cd} (b) I_e

Solution

(a) $V_{cd} = 20$ V, because the potential drop is equal in magnitude to the potential rise in a parallel circuit.

(b) $\quad I_a = I_c + I_e$
$\therefore I_e = I_a - I_c$
$\qquad = 4.0\text{ A} - 1.0\text{ A}$
$\qquad = 3.0\text{ A}$

Thus, I_e is 3.0 A.

ELECTRICITY AND ELECTROMAGNETISM

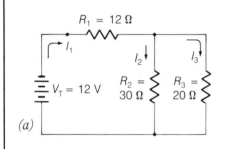

(a)

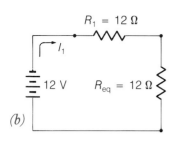

(b)

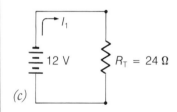

(c)

SAMPLE PROBLEM 7

For the circuit shown in diagram (a), find the current through, and the potential drop across, each resistor.

Solution

The first step is to find the total resistance of the circuit as "seen" by the 12 V supply. Resistances in parallel should be simplified and then added to the resistance(s) in series. In this case, let $R_{equivalent}$ or R_{eq} be the equivalent resistance of the R_2 and R_3 combination.

$$\frac{1}{R_{eq}} = \frac{1}{R_2} + \frac{1}{R_3} \qquad \therefore R_{eq} = \frac{60\ \Omega}{5}$$

$$= \frac{1}{30\ \Omega} + \frac{1}{20\ \Omega} \qquad = 12\ \Omega \text{ as shown in diagram (b).}$$

$$= \frac{2}{60\ \Omega} + \frac{3}{60\ \Omega}$$

$$= \frac{5}{60\ \Omega}$$

The total resistance (R_T) of the circuit is then $R_1 + R_{eq} = 24\ \Omega$, as shown in diagram (c). This value is used to find the total current (I_1) delivered by the source.

$$I_1 = \frac{V_T}{R_T}$$

$$= \frac{12\ V}{24\ \Omega}$$

$$= 0.50\ A$$

The total current, which passes through R_1 before it breaks into two parts, is used to find the potential drop (V_1) across R_1.

$$V_1 = I_1 R_1$$
$$= (0.50\ A)(12\ \Omega)$$
$$= 6.0\ V$$

Now the potential drop across R_2 equals that across R_3; it can be found using the fact that $V_T = V_1 + V_2$.

$$\therefore V_2 = V_T - V_1$$
$$= 12\ V - 6.0\ V$$
$$= 6.0\ V$$

The last step is to find I_2 and I_3 using Ohm's law.

$$I_2 = \frac{V_2}{R_2} \qquad I_3 = \frac{V_2}{R_3} \text{ or } I_3 = \frac{V_3}{R_3}$$

$$= \frac{6.0\ V}{30\ \Omega} \qquad \qquad\qquad\qquad = \frac{6.0\ V}{20\ \Omega}$$

$$= 0.20\ A \qquad\qquad\qquad\qquad = 0.30\ A$$

As a final check, notice that the total current (0.50 A) equals the sum of the currents in the parallel part of the circuit (0.30 A + 0.20 A).

CURRENT ELECTRICITY

In summary, electric circuits are analysed using a step-by-step process of applying the laws or rules of basic electric circuits. Much practice is needed to develop skill in analysing circuits.

There are numerous applications of current electricity and electric circuits, some of which have been described in this chapter. More are to be found in the next three chapters.

PRACTICE

31. Find the total resistance when the following resistors are connected in series:
    (a) 2.7 Ω, 9.8 Ω
    (b) 10 Ω, 10^2 Ω, 10^3 Ω
    (c) 1.0 Ω, 10^{-1} Ω, 10^{-2} Ω
32. Find the total resistance when the following resistors are connected in parallel:
    (a) 4.0 Ω, 4.0 Ω
    (b) 100 Ω, 100 Ω
    (c) 300 Ω, 300 Ω, 300 Ω
    (d) 150 Ω, 600 Ω, 600 Ω
33. Based on your answers to Practice Questions 32(a) and (b), state a rule that could be used to determine the total resistance when two resistors of equal size are connected in parallel.
34. Starting with the equation for resistors in parallel, prove that for two resistors, R_1 and R_2, the equivalent total resistance (R_T) is

$$R_T = \frac{R_1 R_2}{R_1 + R_2}.$$

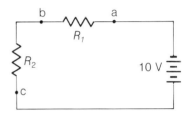

Figure 18-29

35. In the circuit shown in Figure 18-29, $V_{ab} = 4.0$ V, $I_a = 2.0$ A, and $R_1 = 2.0$ Ω. Find values for the following:
    (a) I_c (b) V_{bc} (c) R_T (d) R_2
36. In the circuit shown in Figure 18-30, $I_b = 2.0$ A and $I_c = 3.0$ A. Find values for the following:
    (a) V_{bd} (b) I_a (c) R_T
37. For each circuit shown in Figure 18-31, find the current through and the potential drop across each resistor.

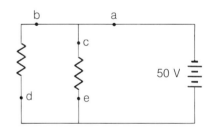

Figure 18-30

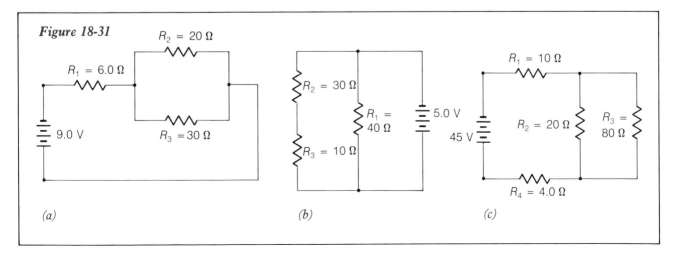

Figure 18-31

(a)   (b)   (c)

465

38. For the circuit shown in Figure 18-32, determine the value of the unknown resistor (R_3) and the potential rise (V_T) of the source.

Figure 18-32

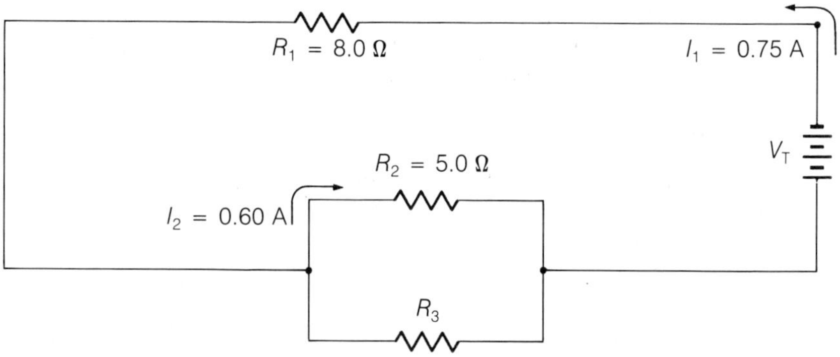

$R_1 = 8.0\ \Omega$ $I_1 = 0.75\ A$
$I_2 = 0.60\ A$ $R_2 = 5.0\ \Omega$
R_3
V_T

Review Assignment

1. What are the particles responsible for current electricity in conductors? What type of charge do the particles have?
2. Draw a labelled diagram of a voltaic cell used to light a small lamp. What are three possible electrolytes that could be used in the cell?
3. State a disadvantage of (a) a voltaic cell and (b) a primary cell.
4. State an advantage of (a) a primary cell and (b) a secondary cell.
5. The fact that a car engine is more difficult to start in cold weather results in a more rapid discharge of the battery. Often a battery from a second vehicle is connected to the discharged battery to recharge or "boost" it. Should the batteries be connected in series or in parallel? Explain your answer.
6. Calculate the current drawn by a microcomputer if 7.2 C of charge pass through it in 6.0 s.
7. A computer disk drive draws 0.48 A of current. How much charge is delivered to the drive each minute?
8. The current rating of a computer printer is 1.5 A. How long does it take to deliver 9.5 mC of charge to the printer?
9. A multi-scale ammeter reads 0.45 A on the 5.0 A scale. What would be the reading if the wire were moved to the terminal of the 500 mA scale?
10. A lightning stroke with a charge of 1.5×10^2 C releases 1.2×10^9 J of energy in 30 ms.
    (a) What is the potential difference across which the stroke moves?
    (b) What is the current delivered by the stroke?
11. Research physicists use high potential differences to accelerate particles they are studying. How much energy is given to a proton as it accelerates through a potential difference of 4.5×10^7 V? (The charge on a single proton is 1.6×10^{-19} C.)

Did You Know?
The Statue of Liberty in New York City was built over a century ago. Since then many of the iron bars that form a weblike structure supporting the skin of the statue have rusted away to about half their original thickness. For this reason, the statue was recently refurbished with hundreds of new stainless steel bars which were given extra strength by having an electric current sent through them. The current raised the temperature of the steel temporarily to over 800° C, a process that will prevent the weakening or deterioration of the metal.

12. An office manager estimates that the 10 A electric pencil sharpener is used for 50 s each working day. (The office is open 5 days per week all year.) The sharpener is connected to the 120 V outlet.
    (a) Calculate the charge delivered to the sharpener in one year.
    (b) Determine the amount of energy consumed by the sharpener in the year. Express your answer in megajoules.
13. What happens to the resistance of a conductor as the temperature increases? What name is given to materials that act in the opposite way?
14. Why is copper a good choice for electric wiring?
15. The length of a certain copper wire is L, its cross-sectional area is A, and its resistance is 50 mΩ. Assuming that the temperature remains constant, what changes could be made to the following in order to obtain a 25 mΩ resistor?
    (a) length   (b) cross-sectional area
16. State the function of each of these parts of an electric circuit:
    (a) source   (c) connecting wires
    (b) switch   (d) load
17. Calculate the total potential rise when three 9.0 V batteries are connected (a) in series and (b) in parallel.
18. An experiment is performed to find the relationship between current and potential for two resistors, A and B. The data collected are plotted in the graph shown in Figure 18-33. Find the resistance of each resistor for a current that does not exceed 500 mA. What do you suppose happens to B when the current exceeds 500 mA?

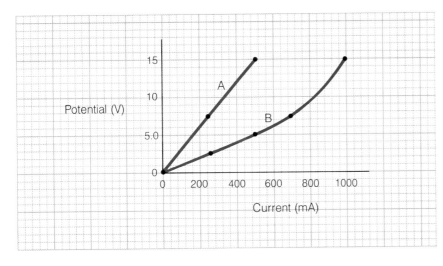

Figure 18-33

19. An electric can opener used in a 120 V circuit has a resistance of 110 Ω. How much current does it draw?
20. An electric razor has a resistance of 20 Ω and draws a current of 250 mA. What is the potential drop across the razor?
21. A 12 V battery, an ammeter, a 5.0 A fuse, and several 10 Ω lamps are used in an experiment to find the effect of connecting loads in parallel. See Figure 18-34 on page 468.
    (a) Determine the total resistance and current when the number of lamps connected in parallel is 1, 2, 3, 4, 5, and 6.

(b) What is the maximum number of lamps that can be connected before the fuse becomes overloaded and burns out?
(c) Write at least one conclusion for the experiment.

Figure 18-34

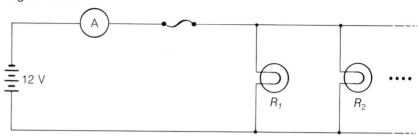

Figure 18-35

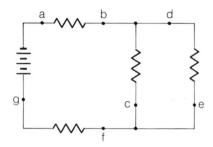

Figure 18-36

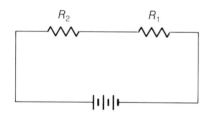

22. Redraw the circuit shown in Figure 18-35 in your notebook, but add the following:
    (a) ammeters to find I_b, I_c, and I_e
    (b) voltmeters to find the potential rise across the source and the potential drop across the two resistors in parallel
    (c) a fuse at "a"
    (d) arrows indicating the electron flow
    (e) all positive and negative terminals

23. In the circuit shown in Figure 18-36, $R_1 = 20\ \Omega$. The potential drop across R_1 is 10 V; across R_2, it is 20 V. Determine the following:
    (a) the total potential rise of the source
    (b) the current through the resistors
    (c) the resistance of R_2

24. For the circuit shown in Figure 18-37, find the following:
    (a) the current through R_1
    (b) the resistances of R_1 and R_2
    (c) the total resistance of the circuit

25. Find the current through and the potential drop across each resistor in the circuit in Figure 18-38.

Figure 18-37

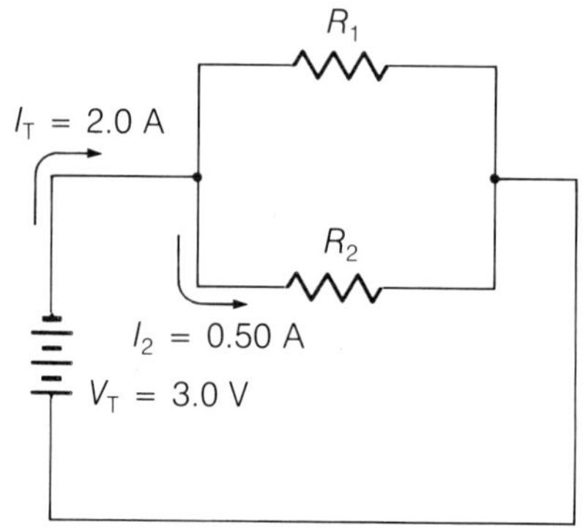

Figure 18-38

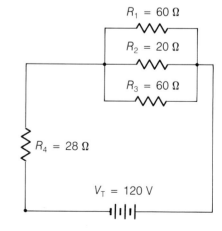

CURRENT ELECTRICITY

Key Objectives

Having completed this chapter, you should now be able to do the following:
1. Distinguish between static electricity and current electricity.
2. Name the components of a voltaic cell.
3. Describe sources of electrical energy.
4. Distinguish between primary and secondary cells.
5. Distinguish between a cell and a battery.
6. Define thermocouple, piezoelectricity, and photoelectric cell.
7. Define electric current in terms of charge and time, and measure it with an ammeter.
8. Given any two of current, charge, and time, determine the third quantity.
9. Define electric potential difference, and measure it using a voltmeter.
10. Given any two of electric potential energy, potential difference, and charge, find the third quantity.
11. Recognize the difference between potential rise and potential drop.
12. Distinguish between series and parallel connections for electric meters, cells, and resistors.
13. Define electric resistance, and describe what factors affect the resistance of an electric conductor.
14. Recognize the units used for electric charge, current, potential, and resistance.
15. Name the functions of the main parts of an electric circuit.
16. Use symbols to draw circuit diagrams.
17. Distinguish between open and closed circuits.
18. Set up electric circuits in the laboratory, and measure related quantities.
19. State Ohm's law in both word and equation form, and describe its limitation.
20. Given any two of resistance, potential difference, and current, determine the third quantity.
21. Describe the characteristics of circuits with resistors in series and in parallel.
22. Calculate the total resistance for resistors connected in series or in parallel.
23. Analyse electric circuits that have resistors connected in series, in parallel, or in a series-parallel combination.

ELECTRICITY AND ELECTROMAGNETISM

CHAPTER 19
Using Electrical Energy

This gondola lift, enjoyed by numerous sports enthusiasts and tourists, requires electricity for its operation.

At the beginning of each unit in this text is a list of careers that relate wholly or partially to the topics in the unit. The list for this unit on electricity and electromagnetism is lengthy, yet it is far from complete. The multiplicity of careers relating to electricity is just one indication of the benefits of electricity. There are many others, one of which can be seen in the photograph.

However, electricity can also be hazardous. Consider, for example, the true story of the boy who went directly from the family swimming pool into the kitchen, where he forced a misshapen piece of bread into an electric toaster. The toast became stuck and, without unplugging the toaster, the boy tried to pry the bread loose with a metal knife. That action ended his life.

Like so many tragedies related to the unsafe use of electricity, this death could easily have been prevented. In this chapter, you will learn how to avoid the common hazards of current electricity. You will also find answers to these questions: How is electrical energy distributed in household circuits? How is the cost of the energy calculated? What are some special scientific and technological uses of electrical energy?

19.1 Direct and Alternating Currents

You have learned that current electricity is the flow of electric charges; in metal conductors, it is the flow of electrons. If the electrons flow continuously in a path without reversing direction, the current is called **direct current** or **DC**. Battery-operated devices, such as electric watches, portable calculators, and electric systems in cars, use direct current, as did the experiments you performed in Chapter 18. See Figure 19-1.

If the electrons in a circuit are forced to reverse their direction periodically, the current is called **alternating current**, or **AC**. This type of current, which is used in most of our appliances and industries, is produced by large electric generators that force conduction electrons to move back and forth. See Figure 19-2. (Electric generators are discussed in greater detail in Chapter 21.)

Since the AC generators in North America force the electrons to repeat their back-and-forth motion 60 times per second, our AC is rated at a frequency of 60 Hz. Some countries, including those in Europe, use a frequency of 50 Hz. At lower frequencies, say around 20 Hz to 25 Hz, we can notice a flickering of lights, especially when the lights are first turned on and are still cool. (Some of the original AC in North America was produced at 25 Hz.)

The advantage of using AC rather than DC in electric systems is that, with present technology, AC is easier to transmit than DC. Electrical energy produced by generators can be transmitted over large distances with relative little energy loss by great networks of power lines and transformers. (Transformers are discussed in Chapter 21.)

The electricity available in most AC outlets is 120 V. However, there are some electric appliances that require DC rather than AC, and an electric potential other than 120 V. These needs can be met by using an adaptor, illustrated in Figure 19-3. The adaptor changes AC (120 V, 60 Hz) into DC (9.0 V) which can be used to operate a device such as a calculator or battery charger. In this case, the instrument can also be operated with a 9.0 V (DC) battery.

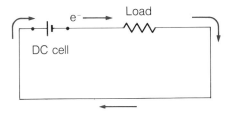

Figure 19-1
In a DC circuit, electrons flow continuously through the circuit without reversing direction.

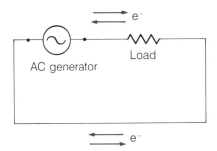

Figure 19-2
In an AC circuit, electrons reverse their direction periodically.

Figure 19-3
An adaptor changes AC into DC.

PRACTICE

1. Describe the motion of the electrons in the following:
   (a) a direct current circuit
   (b) an alternating current circuit

2. State whether AC or DC is used to operate each of the following devices:
   (a) a microwave oven
   (b) a portable radio
   (c) an automobile computer-control system
   (d) an incandescent lightbulb
3. **Activity** Observe the light from an electronic stroboscope reflected off a wall in a darkened room. Determine the minimum frequency at which the light appears to you to be steady.

19.2 Current Electricity in the Home

Electrical energy is delivered to homes either through underground cables or from utility poles. Three wires are fed into a meter and then into a fuse box or circuit breaker box, as shown in Figure 19-4. Two of the wires, covered with red insulation and black insulation (R and B respectively in Figure 19-5), are called "hot" or "live". The third wire, either bare or covered with white insulation (W), is called "neutral".

Notice in Figure 19-5 that neutral wires are connected to the neutral bar. They have a direct path to the ground because they are not connected to any fuses. The black and red cables pass through large cartridge fuses; then they are directed, through another set of cartridge fuses, to a circuit for large appliances. The potential difference between the black and red cables is 240 V, which is required for such major appliances as heat pumps, stoves, water heaters, clothes dryers, and some airconditioners.

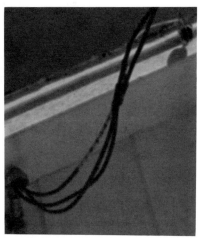

Figure 19-4
Three Wires Entering a Home Circuit

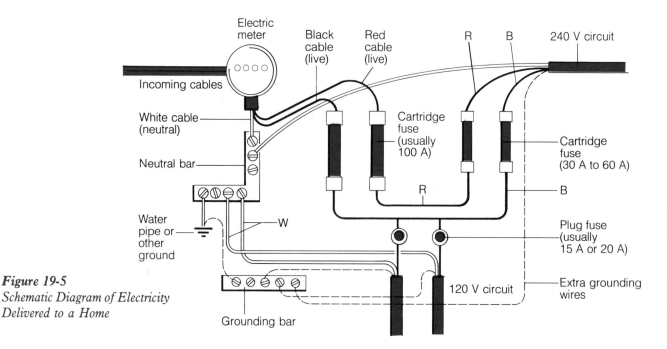

Figure 19-5
Schematic Diagram of Electricity Delivered to a Home

The black cable in Figure 19-5 is also connected through plug fuses to two 120 V circuits. The electric potential between B and W is 120 V, which is used for most lights, outlets, and small appliances. The 120 V circuits may also be connected between R and W, although this is not shown in the diagram.

The broken lines in Figure 19-5 represent grounding cables, which are included for extra safety in newer installations.

The fuses in electric circuits are used as protection against overheating and short circuits. Each fuse has a conductor that burns if the current exceeds a certain predetermined limit such as 15 A. The result is an open circuit; the electrons no longer flow, so the fuse must be replaced. (Section 19.5 deals with overheating and short circuits in greater detail.)

Circuit breakers have the same function as fuses, but are more convenient. One type of circuit breaker has a bimetallic bar, which is made of two adjacent strips of different metals that expand at different rates when heated. When a circuit overheats, the bimetallic bar bends far enough to trip a switch and open the circuit. The switch must then be reset to operate the circuit again.

Figure 19-6
A Simple 120 V Circuit

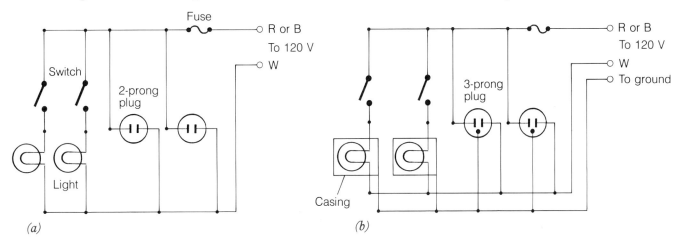

In each household there are several 120 V circuits connected either to B-W or R-W. A simple 120 V circuit is shown in Figure 19-6(a). It consists of two lights, two single outlets, and a plug fuse to the black or red wire. Figure 19-6(b) shows the same circuit with grounding added; (c) shows a three-prong plug (the third prong being the grounding wire) and its corresponding receptacle. The grounding wire is connected to the earth, often through water pipes. It provides protection against the hazard that would arise if a live wire contacted the metal body of an appliance.

In Figure 19-6 the loads are connected in parallel. As you learned in Experiments 18B and 18C, this type of connection has advantages over a series connection. In a series connection, if one load is burned out or removed, the other loads do not function. Furthermore, the potential drop across each load changes as more loads are added. In a parallel connection, each load can operate without the other loads, and the electric potential remains constant. Other applications of household circuits are found in the Practice Questions that follow.

(c) A three-prong plug and receptacle

ELECTRICITY AND ELECTROMAGNETISM

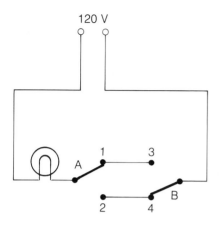

Figure 19-7

PRACTICE

4. What electric potentials are possible in household circuits in North America?
5. In a household electric circuit, what is the function of each of the following:
   (a) the electric meter
   (b) a fuse or circuit breaker
   (c) grounding wires
   (d) a switch
6. Figure 19-7 shows a pair of switches (A and B) used to control a single light in a household circuit. (This arrangement is often used at the bottom and top of a stairway.) The switches are in positions A1 and B4 respectively.
   (a) Name two sets of positions that cause the light to be off.
   (b) Name two sets of positions that cause the light to be on.
7. A dimmer switch is used in some household circuits to control the brightness of incandescent lightbulbs. Describe how this type of switch operates. (A diagram will aid your explanation.)
8. **Activity** Observe and describe a demonstration of the action of a fuse wire connected to a DC power supply.

19.3 Electric Power

Electric devices are labelled with the power in watts that they consume. For example, lightbulbs may be rated at 60 W, 100 W, and 200 W, stereo systems at 120 W, hair dryers at 1000 W, colour television sets at 200 W, and so on.

Recall from Chapter 6 that power is the rate at which energy is consumed or work is done. Thus,

$$\text{power} = \frac{\text{energy consumed}}{\text{time}} \quad \text{or} \quad P = \frac{\Delta E}{t}.$$

In Chapter 18, electric potential was defined as the energy given to or consumed by a unit charge. In equation form, $V = \Delta E/Q$. Therefore, $\Delta E = VQ$. Substituting this equation into the equation for power gives the following result:

$$P = \frac{\Delta E}{t} = \frac{VQ}{t} \quad \text{or} \quad P = V\frac{Q}{t}$$

Since $I = Q/t$ (also from Chapter 18), the equation for power can now be written as follows:

$P = VI$

Thus, the electric power of an appliance is simply the product of the potential drop across the appliance and the current through it. Power is measured in watts (W) when electric potential is measured in volts (V) and current in amps (A). A watt is a relatively small unit, so power is often stated in kilowatts (1.0 kW = 1.0×10^3 W), or megawatts (1.0 MW = 1.0×10^6 W). It is left to the Practice Questions to prove that the product volts × amps is equivalent to joules.

SAMPLE PROBLEM 1

A small colour television connected to a 120 V outlet draws 1400 mA of current. Calculate its power rating.

Solution

$P = VI$
$= (120 \text{ V})(1.4 \text{ A})$
$= 168 \text{ W or } 1.7 \times 10^2 \text{ W}$

∴ The power rating of the television is 1.7×10^2 W.

When combined with the Ohm's law equation ($R = V/I$), the equation for power ($P = VI$) can be written to include the electric resistance, as you will see in the Practice Questions.

PRACTICE

9. Calculate the power of each appliance:
   (a) A 120 V electric sander draws 2.9 A of current.
   (b) An electric can opener, used in a 120 V circuit, operates at 2.2 A.
   (c) A portable radio, using four 1.5 V cells in series, draws a current of 610 mA.
10. Calculate the electric potential drop across a 0.90 W calculator that draws a current of 100 mA.
11. (a) What is the current drawn by a 1.5 kW electric kettle in a 120 V household circuit?
    (b) How long will the kettle take to consume one megajoule of energy?
12. Prove that the product volts × amps is equivalent to joules. (*Hint*: Consider the defining equations for electric potential and current.)
13. Derive an equation for electric power in terms of the following:
    (a) electric potential and resistance
    (b) current and resistance
14. Rearrange the equations in Practice Questions 13(a) and (b) to express electric resistance in terms of the following:
    (a) electric potential and power
    (b) current and power
15. The heater in a waterbed has a resistance of 36 Ω. What is its power rating when operating in a 120 V circuit?
16. A 14 Ω electric handsaw draws 9.5 A of current. What is its power consumption?
17. Calculate the resistance of a 100 W lightbulb operating at 120 V.
18. The headlight of a car, rated at 180 W, draws 15 A of current. Calculate the light's resistance.

19.4 The Cost of Electrical Energy

Whether we use electrical energy or other forms of energy, the cost is important in our daily lives. To discover how electric power companies charge customers for energy, consider the defining equation for power ($P = E/t$), rearranged to express energy by itself. (The "Δ" is omitted for energy in these calculations.)

$$\text{energy} = \text{power} \times \text{time} \quad \text{or} \quad E = Pt$$

Thus, the energy (in joules) consumed by an appliance is the product of the power rating (in watts) and the time (in seconds) during which it is used.

After the energy has been calculated, it can be used in the following equation to determine the cost of the electrical energy:

cost = rate × energy

The rate is expressed in some convenient unit, as you will see in the Sample Problems that follow.

In Sample Problem 2, we assume that the power company charges 1.4¢ for each megajoule (1.0 MJ) of energy consumed. In other words, the rate charged is 1.4¢/MJ. (Eventually, charging for electrical energy by the megajoule will be common in Canada. You can now appreciate why so many "megajoule questions" are asked in this text.)

SAMPLE PROBLEM 2
A 1.0 kW hair dryer is used for 5.0 min. Calculate the following:
(a) the energy consumed (in joules and megajoules)
(b) the cost of the electricity

Solution

(a) $E = Pt$
$ = 1.0 \times 10^3 \text{ W} \times 300 \text{ s}$ (Recall that W = J/s.)
$ = 3.0 \times 10^5 \text{ J} \quad \text{or} \quad 0.30 \text{ MJ}$

$\therefore$ The energy consumed is 0.30 MJ.

(b) cost = rate × energy
$\phantom{\text{cost}} = 1.4¢/\text{MJ} \times 0.30 \text{ MJ}$
$\phantom{\text{cost}} = 0.42¢$ (Note the cancellation of units.)

$\therefore$ Thus, the cost of the electricity is 0.42¢.

In many regions, power companies still calculate energy in units called kilowatt hours (kW·h). Thus, in the equation $E = Pt$, power is stated in kilowatts (kW) and time in hours (h). In Sample Problem 3, we assume that the rate charged by a power company is 5.0¢/(kW·h).

USING ELECTRICAL ENERGY

SAMPLE PROBLEM 3
A 100 W lightbulb is turned on for 4.4 h. Calculate the following:
(a) the energy consumed (in kilowatt hours)
(b) the cost of the energy

Solution

(a) $E = Pt$
$= 0.10 \text{ kW} \times 4.4 \text{ h}$
$= 0.44 \text{ kW·h}$

∴ The energy consumed is 0.44 kW·h.

(b) cost = rate × energy
$= 5.0¢/(\text{kW·h}) \times 0.44 \text{ kW·h}$
$= 2.2¢$ (Note the cancellation of units.)

∴ Thus, the cost of the energy is 2.2¢.

An electric meter is used to determine the quantity of energy (in megajoules or kilowatt hours) consumed monthly by a household. Although some new meters are digital, many have dials. If the electric meter has five dials, the number is read directly. If, however, the meter has four dials, the number read is multiplied by 10. Figure 19-8(a) shows a typical four-dial electric meter, in which the dials are read from left to right. Notice that the dial on the left moves counterclockwise, the next one moves clockwise, and so on. In Figure 19-8(b) the reading is 2418 × 10 = 24 180 kW·h.

Figure 19-8
Reading a Dial Electric Meter

SAMPLE PROBLEM 4
(a) What is the meter reading in Figure 19-8(c)?
(b) Assume that one month elapsed between the readings in (b) and (c) in Figure 19-8. How much energy was consumed during the month?
(c) At a rate of 4.8¢/(kW·h), how much did the electrical energy cost?

Solution

(a) The meter reading is 24 530 kW·h.
(b) The energy consumed was
24 530 kW·h − 24 180 kW·h = 350 kW·h.
(c) cost = rate × energy
$= 4.8¢/(\text{kW·h}) \times 350 \text{ kW·h}$
$= 1680¢$ or $16.80.

∴ The energy cost $16.80.

(a) A typical dial electric meter

(b) March 1 reading

(c) April 1 reading

PRACTICE
19. For each appliance listed, calculate the energy consumed (in megajoules) and the cost of the energy. Assume a cost of 1.6¢/MJ.
   (a) A 120 W stereo system is operated for 15 h.

May 1

July 1

Figure 19-9

(b) A 200 W room airconditioner is run steadily for a day.
(c) A 1.5 kW coffee percolator is left on for 2.5 h.

20. For each appliance listed, calculate the energy consumed (in kilowatt hours) and the cost of the energy. Assume that the cost is 5.5¢/(kW·h).
    (a) A 300 W drill is used for 15 min.
    (b) A 1500 W oven is operated for 2.2 h.
    (c) An 800 W block heater is used for 8.0 h.

21. The diagrams in Figure 19-9 show meter readings taken two months apart. The electrical energy costs 5.2¢/(kW·h).
    (a) How much energy was used during the two months?
    (b) Calculate the cost of the energy.

22. Electric power companies charge a rate that decreases when the amount of energy consumed by a household exceeds a certain minimum quantity. Comment on the wisdom of this method.

23. **Activity** Read the electric meter in your home daily at approximately the same time for at least two weeks. Determine (by subtraction) the daily, weekly, and perhaps monthly consumption. Compare your consumption to the average values of about 20 (kW·h)/d to 30 (kW·h)/d or about 80 MJ/d to 100 MJ/d. Tabulate your data and discuss them in a report that includes the factors affecting the use of electrical energy as well as ways of conserving electrical energy in your home.

19.5 Electrical Safety

Each year many lives are lost and much property is damaged because of the careless use of electricity. Electricity can be hazardous if it is not treated with caution.

The human body reacts to even small electric currents, especially alternating currents. A current of about 9 mA (AC) received externally causes shock, and twice that amount causes great difficulty in breathing. A current of about 20 mA causes the muscles to become paralyzed, so if a person grasps a frayed electrical cord and receives that amount of current, the instinctive reaction to let go will not function. A current of about 100 mA causes *fibrillation*, a reaction in which the heart beats so rapidly and uncontrollably that death soon follows. Table 19-1 summarizes the reactions of the human body to both alternating and direct currents. Values in the table are approximate.

Table 19-1 Human Reactions to Electric Currents

Current (mA, AC)	Current (mA, DC)	Reaction
0.4	1	slight sensation
9	50	shock
20	70	muscles paralyzed
100	500	fibrillation and death

Moisture greatly affects the amount of current that passes through a human body. A person who is dry has a resistance on the order of 10^6 Ω, so the current resulting from contact with a 120 V circuit is about 12 mA ($I = V/R$), large enough to cause shock, but not necessarily death. However, a person with wet hands has a resistance of only about 1500 Ω, and a person sitting in bathwater has a resistance as low as 500 Ω. Under these circumstances, the current from a 120 V circuit is large enough to be fatal. (This explains the tragedy described at the start of the chapter.) Thus, wherever there is water or moisture, electricity should be avoided or at least treated with great caution. For instance, electric radios and electric shavers should not be used near a bathtub.

Like static shocks, electric currents are even more dangerous within the human body than externally. In hospitals, certain patients have apparatus connected to the heart. Even a tiny electric current of 30 μA directly to the heart can be fatal, so medical practitioners must exercise extreme caution in handling such apparatus.

Safety must also be kept in mind in using electric circuits in homes, schools, and offices. A common cause of electricity-related problems is an **overloaded circuit**. Overloading occurs when too many appliances are connected to a single circuit, resulting in increased current and overheating. Ideally, when the current exceeds the safe limit, the fuse should burn out or the circuit breaker should trip. When this occurs, the fault should be corrected, and the burned-out fuse should be replaced with a fuse of the correct size. The following Sample Problem illustrates how overloading causes a large current.

SAMPLE PROBLEM 5

A 20 A fuse is used to protect a 120 V kitchen circuit with several plug outlets.
(a) Calculate the amount of current required to operate the circuit as each appliance is added: (i) a 1000 W toaster, (ii) a 1400 W kettle, and (iii) a 60 W fruit juicer.
(b) Will the fuse burn out?

Solution

(a) (i) $I_1 = \dfrac{P}{V}$ (ii) $I_2 = \dfrac{P}{V}$ (iii) $I_3 = \dfrac{P}{V}$

$= \dfrac{1000 \text{ W}}{120 \text{ V}}$ $= \dfrac{1400 \text{ W}}{120 \text{ V}}$ $= \dfrac{60 \text{ W}}{120 \text{ V}}$

$= 8.3$ A $= 11.7$ A $= 0.5$ A

Because the appliances are connected in parallel, the currents add. Thus, with the toaster and kettle connected, the current is 8.3 A + 11.7 A = 20.0 A and with all three appliances connected, the required current is 20.5 A.

(b) The fuse will burn out when the third appliance is connected, because the current exceeds the 20 A limit.

ELECTRICITY AND ELECTROMAGNETISM

Another common cause of overheating is a short circuit. A **short circuit** occurs when a frayed electrical cord or a faulty electrical appliance allows the current to *arc* from one conductor to another with little or no resistance. In such a situation, the current increases rapidly, so the fuse or circuit breaker should function in its usual capacity. The problem should be corrected as soon as it arises.

Extension cords can also be a source of overheating and fire. Often the less expensive ones can carry no more than 7.0 A of current safely. Thus, only low-power appliances should be connected to ordinary extension cords.

Safety is an important consideration in the design of modern appliances. Appliances with metal chassis, such as clothes washers, have their electric circuits completely insulated from the metal. Some appliances provide added safety with a fuse-like device called a *ground-fault interrupter* in the ground wire, to prevent danger when the ground is not at zero potential.

Despite the technological advances in electrical safety, any electric device is only as safe as its user makes it. It is obvious that rules of safety and common sense should be followed when using this versatile form of energy.

PRACTICE

24. Determine the current received directly from contact with a 120 V (AC) circuit in each case, and state what the person experiences. Refer to Table 19-1.
    (a) A dry person has a resistance of 6.0×10^5 Ω.
    (b) A person with wet hands has a resistance of 1500 Ω.
    (c) A person in bathwater has a resistance of 500 Ω.
25. Compare the fatal internal current with the fatal external current for a human (Table 19-1). Explain why there is a difference.
26. A 20 A fuse is used to protect a 120 V household lighting circuit. How many 100 W bulbs can be connected to the circuit before the fuse blows?
27. Why is it dangerous and extremely foolish to replace a burned-out fuse with a copper penny?
28. The instructions accompanying a new extension cord state that the current is not to exceed 7.0 A. If the cord is used in a 120 V circuit, what is the maximum power of an appliance that can be used safely with this cord?
29. Assume that a 1200 W kettle is used with the extension cord described in Practice Question 28. Would a 15 A fuse provide adequate protection? Explain.

19.6 Electrical Energy for Optimists

Electrical energy is used in so many ways that it is difficult to imagine life without it. Most people tend to take for granted the common uses in which electrical energy is converted into some other form of energy.

Light energy is produced in lightbulbs; thermal energy in ovens, toasters, and hair dryers; mechanical energy (or energy of motion) in electric mixers, typewriters, and tools; and sound energy in telephones and radios. Chemical energy, although less common, is produced in such processes as the electroplating of silver onto less expensive nickel to make jewellery and silverware. There are hundreds of other common uses of electricity for domestic, commercial, and industrial applications, as well as for transportation and leisure activities.

Besides the common uses, however, there are numerous unique and exciting uses that scientists and technologists have discovered or designed. A small number of these specialized applications will be described here.

Consider first some of the applications of electricity that relate to the human body. Because the body has its own electrical system, many medical uses of electrical energy exist. Tiny currents of 1 nA to 3 nA (nano = 10^{-9}) promote the healing of broken bones and burned areas. Pacemakers stimulate the heart electrically and allow people with certain heart problems to lead normal lives. Devices called *defibrillators* send a large current through the body for a small fraction of a second to restore a heart that is beating uncontrollably. (Fibrillation was discussed in the previous section.) Radiation from heat lamps helps relieve certain ailments near the surface of the body. Other medical applications of electricity include the operation of heart and lung machines; the use of thermographs (temperature-sensitive photographs) to locate cancerous tumours near the skin surface; the use of electrodes attached to the body to locate tumours beneath the surface; and the tracing of the secrets of life with modern research apparatus. See Figure 19-10.

Did You Know?
Since the body's nervous system operates on electric impulses, electricity can be used to relieve pain as well as promote healing. The ancient Egyptians applied this principle when they used the electric charges generated by torpedo fish to ease pain. The Romans evidently used electric eels to treat headaches and arthritis. Today, physicians use electrical nerve stimulation to treat certain types of pain. A 9 V battery supplies a weak electric current across the patient's skin into the nerve cells beneath the skin's surface. The current stimulates the body's natural ability to fight pain.

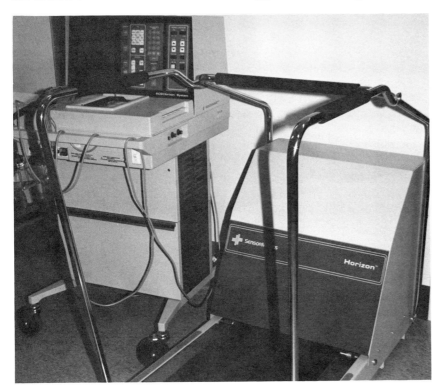

Figure 19.10
This apparatus is an example of the numerous application of electricity related to the human body. The machine on the left monitors the heart and lungs of a person exercising on the electrically-controlled treadmill. The information is used for both research and the recovery process of patients in a hospital.

PROFILE

Carla J. Miner, Ph.D. Experimental Solid State Physics
Member of Scientific Staff
Advanced Technology Laboratory
Bell Northern Research

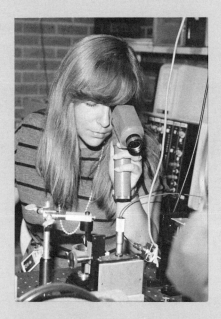

Without a doubt, the 1973 Canada Wide Science Fair was the principal pivot point in my whole career. Try to imagine a quiet, introspective, bookworm of a sixteen year old. There I was on my first trip away from home. For years, I'd put up with the teasing of classmates who thought my marks were too high, or my interests were weird. But among the other science fair entrants, marks didn't matter, and many interests were shared. They, as I, delighted in finding an underlying pattern in everyday phenomena, or an elegant solution to an engineering problem.

I was lucky enough to be able to participate in three more Canada Wide Science Fairs as an entrant, and later was a founding executive of the national Students Science Council. The approval of those science fairs peers put the seal on my plan for a career in science.

Of course, fleshing out the details of the plan took some time. At first, I wanted to be an astrophysicist, but a quick look at the limited opportunities for Canadian astrophysicists convinced me to choose something more down to earth. My second choice was meteorology. However, by the time I finished a B.Sc. at the University of Winnipeg, I had acquired a husband and baby boy. Since jobs for one's spouse can be rather scarce at some of Canada's high arctic meteorological stations, once again, I reset my sights. I applied for a job at Bell Northern Research, the place I had visited as a science fair prize winner five years earlier.

I was surprised to find that my application was accepted. It has been an exciting eight years so far. I have worked on a number of new electronic devices designed to be used in advanced telecommunications products: thin film selenium diodes, amorphous silicon photoconductors, metal-insulator-metal devices for liquid crystal displays, indium gallium arsenide phosphide laser diodes, two dimensional electron gas transistors, and quantum well devices. Most are made from semiconductors – those materials that are intermediate between insulators like glass and conductors like

Next consider the versatility of computers which, of course, require electrical energy to function. Computers are found in control systems in cars, airplanes, and spacecraft. They are used to help design automobiles and architectural structures, to operate mechanical robots that help handicapped people function independently, and to operate devices that let blind people read. Without computers, the instant banking systems that offer 24-hour access to your money as well as daily-interest accounts would not be possible. Computers are also used in education and in the metal. With the addition of trace impurities, or the application of an external stimulus like light, heat, or an electric field, semiconductors exhibit smooth changes from semi-insulating to semi-conducting states, and back again.

By studying the changes in conductivity when a semiconductor device is exposed to the stimuli mentioned, a physicist such as myself can put together a model of what he or she thinks the electrons inside the device are doing. The models are tested against further experiments, and are revised or rejected outright. The process continues until a consistent picture emerges that allows newer, higher performance devices to be designed and made. It isn't always easy. I will always remember an incident a few years ago that occurred at the point in this cycle when the models weren't standing up very well, and I was getting pretty frustrated. Then Dr. Zemel, an old hand at this type of work, gave me a reassuring smile and said, "Don't let it get to you; who has ever seen an electron?" My tension instantly dissolved, and eventually I was able to discern the underlying pattern in the data I was examining at the time.

The common goal in all of my research has been to understand what makes devices work, and how to make them work even better. I am pleased that my work in this area has led to a number of publications and conference presentations, and to eight patents.

In the last two years, the emphasis of my research has extended beyond the analysis of devices to the analysis of the material from which they are made. Extremely sensitive techniques are required since impurities at the level of one part in a thousand million are sufficient to ruin some types of semiconductor devices. One technique with the required sensitivity is low temperature photoluminescence. In my photoluminescence studies, a small laser is used to optically excite the semiconductor sample, which then gives off a characteristic light of its own. The light emitted is collected with a lens and analyzed with a spectrometer. If the experiment is done at room temperature, the observed spectrum is broad and featureless, but it *does* contain valuable clues as to the composition and purity of the sample. As the temperature is lowered to liquid helium temperatures (4 K above absolute zero), the spectrum becomes brighter and sharper, and eventually splits into distinct lines associated with the material itself and lines associated with impurities. In many cases, low temperature photoluminescence will identify a potential problem before it becomes serious. The trouble-shooting role that low temperature photoluminescence allows me to assume one that I find exciting and very rewarding.

It is great to have a job that keeps stretching your knowledge and technical abilities. In industrial research, ideas germinate, grow, and mature quickly in response to what we think will be needed three to five years from now. We have to keep learning all the time and be prepared to become familiar with a new topic on short notice. We are encouraged to pursue continuing education to help us keep current with a broad range of topics that may become important in the future. With Bell Northern Research's support and financial assistance, I have just completed a doctorate in physics at the University of Ottawa. Nearly all of the work was done on a part-time, one-course-a-term basis. Upon reflection, however, I must say that my nerves were probably stretched far more than my knowledge base during this period of being full-time scientist, part-time student, part-time wife, and mother.

There was one more aspect of the 1973 Canada Wide Science Fair that proved to be pivotal. It introduced me to my future career and future employer, and to a circle of good friends, which includes one of my best friends, my husband.

fields of art and music. Archaeologists are able to put together thousands of broken clay tablets in the proper order and decipher their ancient inscriptions with the help of computers. And astronomers study photographs of outer space that consist of millions of faint images combined by computers. Refer to Figure 19-11 on page 484.

Of course, there are countless other applications of electricity. Furthermore, there is reason for optimism that scientists and technologists will continue to discover and apply new uses for this amazing form of

ELECTRICITY AND ELECTROMAGNETISM

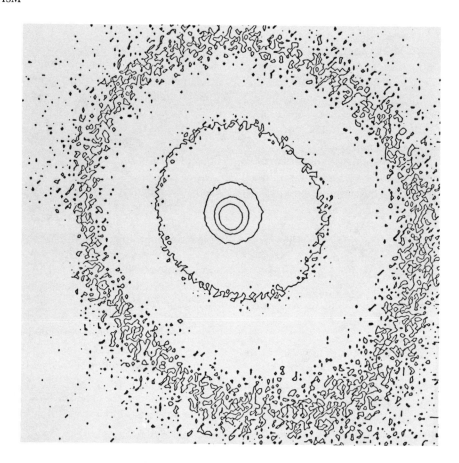

Figure 19-11
This computer print-out shows light extending 20 000 km from the nucleus of Comet Halley. As light from the comet fell on the 3.6 m Canada-France-Hawaii telescope in Hawaii for 5 min, approximately 150 000 numbers were received by electronic chips and interpreted by computer. In this interpretation, the light intensity at the inner circle is 20 times as great as the intensity at the outer-most regions. The comet nucleus, with a diameter of about 10 km, could be represented by a dot at the centre of the inner circle.

energy. Our optimism, however, must be combined with an awareness of the need to use electrical energy wisely, remaining mindful of energy conservation.

PRACTICE

30. List two devices other than those named in this section that convert electrical energy into each of the following forms of energy:
    (a) thermal energy
    (b) mechanical energy
    (c) sound energy
31. **Activity** Research and report on some advanced scientific uses of electrical energy.

Review Assignment

1. Describe the difference between alternating current and direct current.
2. What is the frequency of AC in North America?
3. What is the function of an adaptor?
4. State the electric potential between these pairs of wires in a household circuit:
    (a) black and white   (b) black and red   (c) red and white

5. What are the advantages of connecting appliances in parallel in a household circuit?
6. Name three household appliances that function at an electric potential of (a) 120 V and (b) 240 V.
7. Calculate the power rating in each case:
   (a) A 240 V water heater draws a current of 21 A.
   (b) An electronic toy using a 9.0 V battery draws a current of 150 mA.
8. Determine the current drawn by each device:
   (a) An electric clock uses 2.4 W of power in a 120 V circuit.
   (b) An electric typewriter used in a 120 V circuit is rated at 3.4 W.
9. What is the electric potential in each case?
   (a) A 30 W stereo tape-deck draws a current of 2.5 A.
   (b) A 300 W windshield-wiper motor draws a maximum current of 25 A.
10. A 20 Ω razor requires an electric potential of 5.0 V. What is the power of the razor?
11. A 24 Ω curling iron draws a current of 5.5 A.
    (a) Find the power of the curling iron.
    (b) Determine the time required to consume one megajoule of energy using the curling iron.
12. Calculate the effective resistance of a colour television rated at 210 W in a 120 V circuit.
13. A 1.1 kW electric lawn mower is used in a 120 V circuit. What is the mower's resistance?
14. A 2.2 kW stove burner is used for 24 min. The cost of the electrical energy is 1.6¢/MJ. Calculate the following:
    (a) the energy consumed in megajoules
    (b) the cost of the energy
15. Many lightbulbs are planned to burn out after about 3000 h of operation. Assume that a 100 W bulb is left on for 3000 h (about 4 months) and that the cost of electricity is 5.2¢/(kW·h). Find the following:
    (a) the energy consumed (in kilowatt hours)
    (b) the cost of the energy consumed
16. Read the meters in the diagrams in Figure 19-12 and find the cost of the electrical energy, assuming that the rate charged is 5.8¢/(kW·h).
17. Determine the number of megajoules in one kilowatt hour. (*Hint*: Apply the equation for energy in terms of power and time.) If the cost of electricity is 1.5¢/MJ, what is the cost in cents per kilowatt hour?
18. Estimate each of the following, showing your reasoning or calculations in each case:
    (a) the number of colour television sets in Canada
    (b) the average power rating of the colour television sets
    (c) the cost of the electrical energy consumed by all these colour television sets in one year
19. Calculate the maximum power available to each circuit:
    (a) A 15 A fuse is used for the signal light circuit of a car with a 12 V battery.

Sept. 15

Oct. 15

Figure 19-12

(b) A 12 A fuse is used in a 240 V circuit common in Europe.
20. What is the purpose of a fuse or circuit breaker?
21. Explain what precautions you would recommend (relative to the safe use of electricity) to a person operating an electric lawn mower.
22. Determine the maximum number of 200 W lightbulbs that can be connected (in parallel) into a 120 V household circuit protected by a 15 A fuse.
23. Explain how you would design any one of the following electric devices:
    (a) a hand dryer for use in a public washroom
    (b) a hair dryer with settings of 600 W, 900 W, and 1200 W
    (c) a curling iron that can be used in North America (120 V) or when you are travelling in most other countries (240 V)
24. Discuss the advantages and disadvantages of developing more uses of current electricity in our society.

Key Objectives

Having completed this chapter, you should now be able to do the following:
1. Compare direct current with alternating current.
2. Describe how a three-wire system can produce 240 V and 120 V household circuits.
3. State uses of 240 V and 120 V household circuits.
4. Describe ways of preventing electric circuits from causing fires.
5. State advantages of using parallel rather than series connections for electric circuits in homes and commercial buildings.
6. Derive and apply equations for electric power in terms of the following:
    (a) electric potential and current
    (b) electric potential and resistance
    (c) current and resistance
7. Given the power of an electric device and the amount of time for which it is used, calculate the energy consumed in joules, megajoules, or kilowatt hours.
8. Given the cost of electricity and the energy consumed, calculate the total cost of the electrical energy.
9. Determine the energy consumed in kilowatt hours, given two successive sets of readings on a dial electric meter.
10. Explain the need for safety when using current electricity.
11. Describe the function and proper use of fuses and circuit breakers.
12. Explain how overloading and short circuits cause overheating.
13. List common devices in which electrical energy is converted into other forms of energy.
14. Describe scientific and technological applications of electricity that illustrate its usefulness to humans.
15. Appreciate the need for conservation of electrical energy.

CHAPTER 20
Magnetism and Electromagnetism

Electromagnetic lifts like this one can lift up to 20 000 kg of metal.

The word "magnetism" conjures up images of little coloured magnets that attract certain metals, of experiments with iron filings in earlier science classes, or of using an orienteering compass when hiking in the woods. But magnetism is more common than these examples would indicate. Voltmeters and ammeters, electric motors and generators, tape recorders and computers, loudspeakers in stereos and televisions – all use magnetism. These devices, and numerous others, function because electric currents exert magnetic forces. This knowledge leads us to the study of electromagnetism.

In this chapter you will review the properties of magnets and learn how the magnetism of electric currents relates to the magnetism of permanent magnets. Then you will find out how the interaction of electricity and magnetism results in useful applications such as those named above, and others like the electromagnetic lift shown in the photograph on the previous page.

20.1 The Force of Magnetism

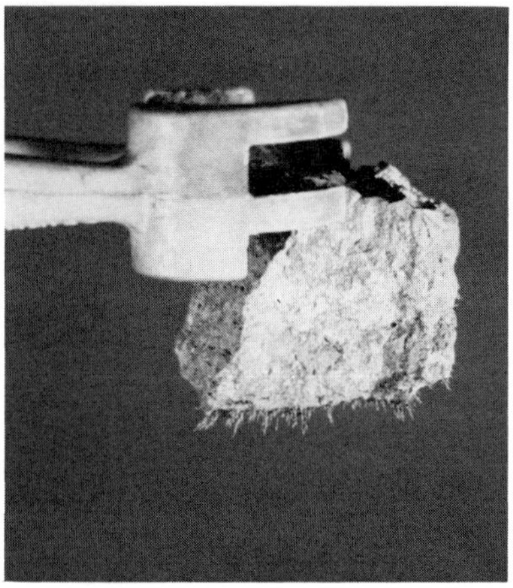

The force that attracts iron nails to a magnet or causes a compass needle to point in one direction is called **magnetic force** or the **force of magnetism**. (Magnetic force is related closely to electric force. Together the two forces constitute one of the four fundamental forces in nature, electromagnetic force.)

The ancient Greeks were perhaps the first to experiment with the force of magnetism. By about 600 B.C., in a district called Magnesia, they had discovered a mysterious mineral, later called "lodestone", which attracted iron and some other substances. The force it exerted, as illustrated in Figure 20-1, was called "magnetism" after Magnesia. The Greeks came to realize that one end of a suspended bar of lodestone always faced toward the North Star. In English, that star became known as the leading or "lode" star; this explains the origin of the word "lodestone".

By the second century A.D., the Chinese also knew about the force of magnetism, and in the eleventh century they used magnetic compasses to navigate at sea. During the twelfth century, the Chinese encountered European sailors, with whom they shared their knowledge of navigation. The compass made possible safer worldwide travel and exploration. See Figure 20-2.

Figure 20-1
This chunk of lodestone attracts small pieces of iron filings.

Figure 20-2
The navigation of ships similar to the one shown became more accurate with the use of a magnetic compass.

The first major account of magnetism was written by the Englishman William Gilbert. In his book *De Magnete*, published in 1600, Gilbert described experiments with lodestone and offered explanations of the earth's magnetic force. Many of his investigations are easily repeated using bar magnets.

If a bar magnet (or a compass needle, or even a chunk of lodestone) is suspended and allowed to swing freely, it will generally line up in the north-south direction, although the exact direction depends on the location. The end of the magnet that faces north is called the **north-seeking pole** or simply the **north pole**, and has the symbol **N**. The opposite end is called the **south-seeking pole** or **south pole**, and has the symbol **S**. Thus, every simple magnet has two poles, or locations of greatest concentration of magnetic force.

When the N pole of a magnet is brought close to the N pole of a second, suspended magnet, repulsion occurs. Two S poles also repel one another. However, N and S poles always attract each other. These facts are summarized both in Figure 20-3 and by the **laws of magnetic poles**:

Like poles repel.   Unlike poles attract.

A bar magnet can be used to determine whether a material is magnetic or non-magnetic. A **magnetic material** is one that is influenced by the force of magnetism, as evidenced by an attraction to a magnet. The most common magnetic materials are iron, nickel, cobalt, and their alloys. Most other materials, such as paper and glass, are **non-magnetic**. A **magnet** is an object composed of a magnetic material that is permanently magnetized.

Figure 20-3
The Laws of Magnetic Poles

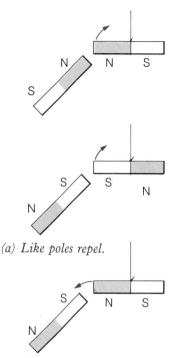

(a) Like poles repel.

(b) Unlike poles attract.

PRACTICE
1. Compare the laws of magnetic poles with the laws of electric charges (Experiment 17A).
2. Assume that you are given a bar magnet and an object, O. Describe how you would determine whether or not
   (a) O is magnetic
   (b) O is a magnet.

20.2 The Domain Theory of Magnetism

The theory explaining magnetism has undergone several changes since William Gilbert's first attempts in 1600. Certain experiments have helped develop our basic understanding of magnetism. Consider, for example, breaking a bar magnet first into two pieces and then into several pieces, as in Figure 20-4. Originally, the magnet has one set of N and S poles. Then each piece has its own new set of N and S poles, with every set still facing the same direction. If this breaking continues indefinitely, the results remain the same – each piece acts like a tiny magnet with N and S poles. This leads to the hypothesis that a magnet consists of many sets of particles aligned in the same direction. These sets of particles are

Figure 20-4
Breaking a Bar Magnet

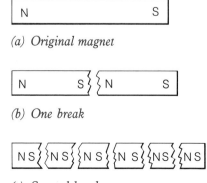

(a) Original magnet

(b) One break

(c) Several breaks

489

Figure 20-5
This set of microphotographs, with a magnification of 1000 times, reveals the domain arrangements of nickel at increasing temperatures. As the temperature of the nickel approaches 350°C, the domain area decreases dramatically, as shown at the far right.

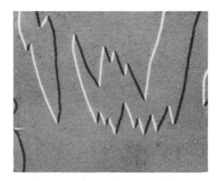
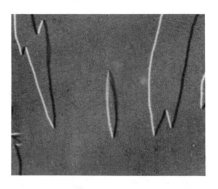

Figure 20-6
A magnetic material can become a magnet.

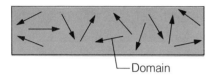

(a) *Random domains of an unmagnetized magnetic substance*

(b) *Several domains face the same direction, producing a weak magnet.*

(c) *All domains face the same direction, resulting in a strong magnet.*

called **domains**, and the theory of the structure of magnetic materials is called the **domain theory of magnetism**.

More recent research using sophisticated apparatus has helped verify the domain theory. Figure 20-5 shows photomicrographs of typical magnetic domains.

When the domains of a magnetic material are arranged randomly, the material does not act like a magnet, although it can be influenced by an external magnetic field. This situation is shown in Figure 20-6(a), in which each arrow represents a domain. (The diagram is drawn in two dimensions for ease of comprehension.) If several of the domains, but not all, are aligned in a single direction, as in Figure 20-6(b), the magnetic material is a weak magnet. If all the domains are aligned in one direction, as in Figure 20-6(c), a strong magnet results. If the latter arrangements lasts indefinitely, the magnet becomes permanent.

The domain theory can be used to explain observations related to magnets. For instance, one way to make an object composed of a magnetic material into a temporary magnet is to hold it close to a permanent magnet. The magnetic force of the permanent magnet influences many of the domains in the object. This process of magnetizing an object from a distance is called **magnetic induction**. (It is similar to electrostatic induction, studied in Chapter 17.) Figure 20-7 illustrates the magnetic induction of an iron nail, which is shown with its own N and S poles. When the nail is moved away from the magnet it loses its magnetism, because iron maintains its magnetism only if it is alloyed with other materials. Iron that loses its magnetism is called *soft iron* though it is not soft to the touch. A magnetic material that retains its magnetism is called *hard*.

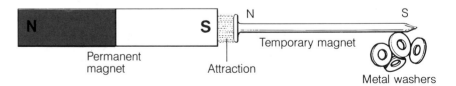

Figure 20-7
*Magnetic Induction of an Iron Nail
The domains of the nail become aligned in the same direction as the domains in the permanent magnet.*

A way to make an object composed of a magnetic material into a permanent magnet is to stroke it with a permanent magnet. The magnet is drawn over the object several times with a sweeping motion, as shown for the steel needle in Figure 20-8. The magnetic force of the permanent magnet causes many domains in the steel to become aligned in one

Figure 20-8
Making a Permanent Magnet by Stroking

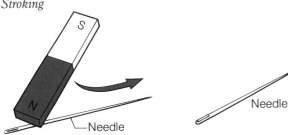

(a) Stroking the needle   (b) Testing the magnetic needle

Figure 20-9

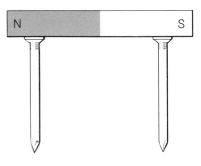

Figure 20-10

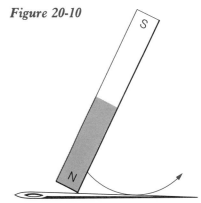

(a) Stroking with a single magnet

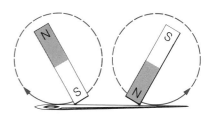

(b) Double-stroking, with two magnets

direction. Later, when the needle is tested for its magnetic strength, it is seen to have an N pole and an S pole. (Steel is an alloy consisting mainly of iron and a small amount of carbon.)

The domain theory can also be used to explain why heating a magnet causes it to lose its magnetism. Heating causes the motion of the particles to increase; thus, the domains become more randomly arranged and the total magnetic force is reduced.

PRACTICE

3. Two iron nails are held to a magnet, as shown in Figure 20-9. Predict what will happen when the nails are released, and explain your prediction. If possible, verify your prediction experimentally.
4. **Activity** Obtain a steel needle (or a large steel paper clip straightened out) and magnetize it by stroking it with the N pole of a magnet, using the sweeping motion shown in Figure 20-10(a). Which end of the needle becomes the N pole? Test your answer using a compass. As an optional exercise, try to determine whether the double-stroking technique shown in Figure 20-10(b) produces a stronger magnet.
5. Use the domain theory of magnetism to explain each of the followings statements:
   (a) When a magnet is being magnetized it reaches a point, called **saturation**, where it cannot become any stronger.
   (b) A magnet can be demagnetized by being hammered repeatedly.
6. Suggest a possible technique for making a magnet in the form of a flexible rubber strip.

20.3 Magnetic Fields

In Chapter 17 you learned how to map the electric field surrounding an electrically charged object. To determine the direction of the electric field lines, a small positive test charge was used. In a similar way, a mapping can be used to illustrate a **magnetic field**, the space in which a magnetic force exists. To determine the direction of the magnetic field lines, a small magnetic compass can be used. Magnetic fields always

ELECTRICITY AND ELECTROMAGNETISM

include at least two poles because a magnetic pole does not exist alone. (In this respect magnetic poles differ from electric charges, which *can* exist alone.) Figure 20-11 shows several test compasses around a bar magnet.

Figure 20-11
Test compasses indicate the presence of a magnetic field.

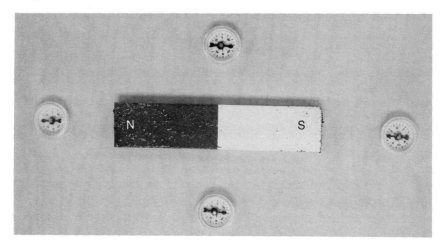

Figure 20-12
The Magnetic Field around a Bar Magnet

(a) Observed using iron filings

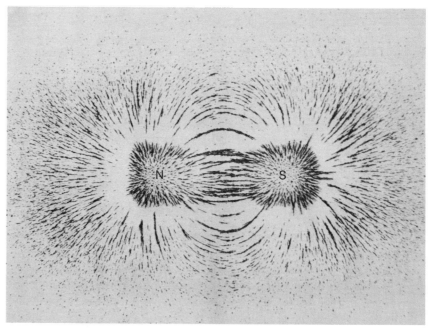

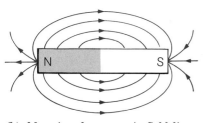

(b) Mapping the magnetic field lines

Now imagine the compasses in Figure 20-11 becoming smaller and smaller, until they reach the size of tiny iron filings. Each filing would align itself in the direction of the field lines around the magnet. Such is the case in Figure 20-12(a), which shows numerous iron filings sprinkled onto paper placed over a bar magnet. Figure 20-12(b) shows a mapping of the magnetic field lines around the bar magnet.

In Figure 20-12, the following characteristics of magnetic field lines can be observed:

(a) The spacing of the lines indicates the relative force. The closer the lines, the greater is the force.

MAGNETISM AND ELECTROMAGNETISM

(b) Outside a magnet, the lines are concentrated at the poles.
(c) By convention, the lines proceed from S to N inside a magnet and from N to S outside a magnet. (These directions are indicated by a plotting compass.)
(d) The lines do not cross one another.

The magnetic field around a U-shaped magnet is illustrated in Figure 20-13(a), using iron filings. Figure 20-13(b) shows what happens to that field when an iron bar called a **keeper** joins the poles. The keeper sustains the magnetism when the magnet is not in use.

Figure 20-13
The Magnetic Field near a U-shaped Magnet

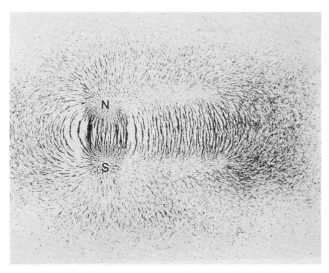

(a) Without a keeper

(b) With a keeper

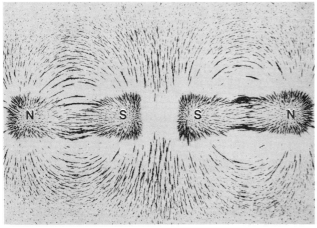

(a) Like poles repel.

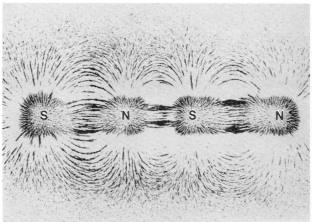

(b) Opposite poles attract.

Figure 20-14
Magnetic Fields near Pairs of Magnets

Iron filings are also useful for demonstrating the magnetic fields near pairs of magnets. In Figure 20-14(a) two like poles are repelling each other, while in (b) two unlike poles are attracting each other.

ELECTRICITY AND ELECTROMAGNETISM

The Earth's Magnetic Field

The earth has a magnetic field surrounding it. This field is concentrated at the earth's two magnetic poles. The north magnetic pole (which is actually south-seeking) is not located at the geographic North Pole of the earth. Therefore, for most longitudes, there is a difference between the direction the compass faces and true (geographic) north. The angle at a given location between the magnetic north and geographic north is called the **angle of declination**. The angles of declination for locations in Canada are given in Figure 20-15. Notice that the zero degree declination line passes through Lake Superior.

Figure 20-15
Canada's Magnetic Field

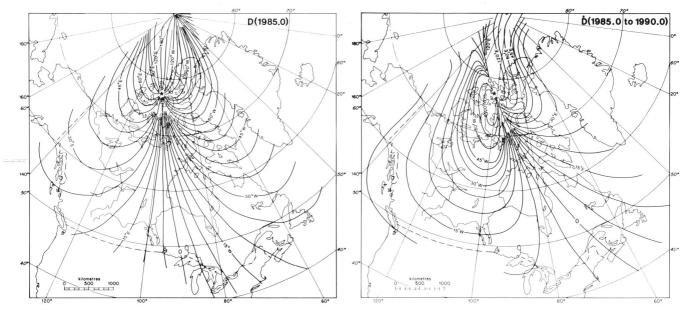

(a) Angles of declination in degrees as of 1985 (The geographic North Pole is at the top of the map.)

(b) Average expected rate of change of angles, in minutes per year, between 1985 and 1990

Did You Know?
Biomagnetism is the study of the relationship between magnetism and living organisms. Research has revealed that pigeons' skulls contain multiple-domain magnets which are connected via nerves to the brain. Thus, pigeons have a magnetic sense which can detect both the angle of declination and the angle of inclination. This ability gives them an excellent sense of direction.

The earth's magnetic field also affects the action of a compass differently at different latitudes. In regions near the equator, a compass will come to rest horizontally, but at northern and southern latitudes, the needle will dip away from the horizontal plane. The angle between the horizontal plane and the earth's magnetic field lines is called the **angle of inclination**, which varies with location. The "dipping needle" is a type of compass used to determine this angle. The needle rotates in the vertical plane rather than in the horizontal plane, as in ordinary compasses.

The information in Figure 20-15(a) varies slightly from year to year, as seen in Figure 20-15(b). This occurs because the earth's magnetic poles shift slowly but continuously. Much research is being carried out to discover the cause of these changes in the magnetic field. Data gathered during earthquakes provide an indirect means of determining the composition of the earth's core. The energy released in an earthquake is transmitted by different types of waves that can be measured and interpreted all around the world. The structure of the earth's core is

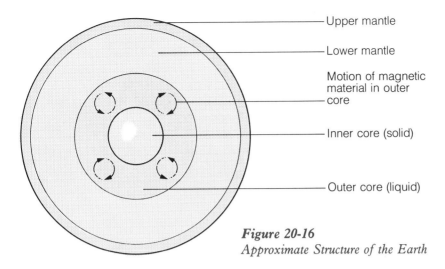

Figure 20-16
Approximate Structure of the Earth

thus relatively well known, and is the basis of the theory of the earth's magnetic field. It is now thought that the earth's magnetic field results from the continuous motion of magnetic material (mainly molten iron) in the liquid outer core of the earth. Figure 20-16 shows that a large part of the earth consists of a molten outer core surrounding a solid inner core. The motion of the liquid, which in effect creates a large electric current, is caused by various factors, including the earth's rotation on its own axis.

The earth's changing magnetic field has had many interesting effects. Evidence exists of hundreds or perhaps thousands of occasions when the change was so great that the poles became reversed, with the north magnetic pole becoming the south and *vice versa*. Such radical changes took place in a relatively short amount of time (a few thousand years), during which the strength of the field was reduced to about 10% of its average value. It is possible that a weak magnetic field could have affected the evolution of certain species, such as animals that depend on the magnetic field in their migration. A direct consequence of the north-south reversals is that scientists are able to use magnetic compasses to analyse layers of lava rock containing magnetic iron oxide that solidified in line with the earth's magnetic field at the time of a given eruption. Such analysis yields valuable information about the geological history of the earth.

The earth's magnetic field is also responsible for the display of auroras, the dancing lights that spread across the sky above the north and south magnetic poles. Energetic charged particles (such as electrons and protons) that stream from the sun are forced to spiral in along the magnetic field lines of the earth, and thus are concentrated in regions of the atmosphere above the north and south magnetic poles. These ions collide with atoms present in the upper atmosphere and give much energy to the atoms. In turn, the atoms give off their excess energy in the form of visible light. Colour Plate H shows a photograph of the *aurora borealis*, or Northern Lights, named after the north wind, Boreas, in Greek mythology. The Southern Lights, called *aurora australis*, are named after the south wind, Auster.

Did You Know?
Magnetic storms on the earth occur shortly after there are giant eruptions on the sun, which result in much larger than normal streams of charged particles from the sun. This so-called "solar wind" causes the aurora to become more brilliant and disturbs the upper atmosphere to such a degree that our communication systems can be disrupted for days.

ELECTRICITY AND ELECTROMAGNETISM

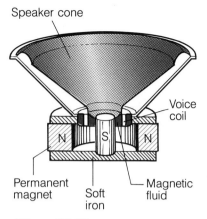

Figure 20-17
Magnetic Fluid in a Loudspeaker.
(See also Figure 20-41.)

Figure 20-18

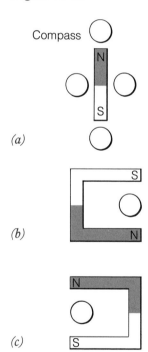

Magnetic Fluids

There are many modern applications of magnetism, several of which use fluids that have been magnetized. One interesting example is found in the attempt to develop better loudspeakers using a magnetic liquid placed between the voice coil of the loudspeaker and the permanent magnet surrounding it. (Refer to Figure 20-17.) Because the fluid is magnetic, it is held in place by the magnet. The advantages of using the fluid are that it helps to distribute heat efficiently, so the power output can be increased, and that it improves the frequency response of the speaker.

PRACTICE

7. In the diagrams in Figure 20-18, each circle represents a compass. Copy the diagrams in your notebook and show the direction of the needle in each compass.
8. Draw a diagram showing the magnetic field surrounding a U-shaped magnet. Include the directions of the field lines.
9. Are the north magnetic pole of the earth and the N pole of a magnet of the same polarity? Explain.
10. (a) What is the approximate angle of declination on Vancouver Island and in Newfoundland?
    (b) What is the angle of inclination near the equator, and at the north magnetic pole?
11. The planet Jupiter has a magnetic field about 13 times as strong as that of the earth, while the planet Venus has no detectable magnetic field. What can be predicted about the speed of rotation of these planets on their own axes? What can be predicted about their composition? Do research to verify your predictions. (One possible resource is the August 1983 issue of the periodical *Scientific American*.)

20.4 The Magnetic Effects of Electricity

In 1819, a Danish scientist named Hans Christian Oersted (1777–1851) (Figure 20-19) set up a demonstration to verify that magnetism has nothing to do with electricity. Much to the surprise of his audience and himself, he proved just the opposite – that electricity and magnetism cannot be separated. The study of the relationship between electricity and magnetism is called **electromagnetism**. Oersted's important discovery led to the development of the subject of electromagnetism and its countless applications. It also led to the understanding of how visible light and other parts of the electromagnetic spectrum are produced and transmitted.

In the remainder of this chapter you will learn about the characteristics and applications of the magnetic forces produced by an electric current. This knowledge will help you understand more about both the domain theory of magnetism and the earth's magnetic field. Then, in the next chapter, you will study how magnetism is used to produce electricity.

Experiment 20A: The Magnetic Field around a Straight Conductor

INTRODUCTION
In this experiment you will observe the characteristics of the magnetic field produced by an electric current. To help explain your observations, it will be useful to draw diagrams using the following symbols:

- ⊗ This represents a conductor with electrons travelling away from the observer, into the page. (Imagine the tail of an arrow moving away from you.)

- ⊙ This represents a conductor with electrons travelling toward the observer, out of the page. (Imagine the point of an arrow moving toward you.)

A plotting compass will be useful for finding the direction of the magnetic field lines around the conductor. Before you use the compass, always check that the N pole faces north, not south. Even if the compass is functioning properly, inaccuracies may still arise as a result of the interaction of the earth's magnetic field with the rather weak field around the conductor.

Since conductors can be held horizontally, vertically, or in any position in between, some of the steps in the experiment will be performed for both horizontal and vertical conductors.

PURPOSE
To study the characteristics of the magnetic field surrounding a straight current-carrying conductor.

APPARATUS
variable DC power supply; plotting compass (or compasses); switch (preferably a button switch); connecting wires

PROCEDURE
1. Place a plotting compass flat on a lab bench and place a conducting wire parallel to and just above the needle. Connect the wire to the switch and power supply, so that the electrons will travel the same direction in which the needle is facing (in other words, south to north). See Figure 20-20 on page 498. Close the switch momentarily and determine the effect.
2. Predict the effect of reversing the connections to the power supply, so that the electrons flow in the opposite direction (from north to south). Verify your prediction.

Figure 20-19
Hans Christian Oersted (1777–1851)

CAUTION:
(a) To prevent overheating, do not leave the switch closed for too long at any given time.
(b) If a power supply without a circuit breaker is used, connect a small resistor (5 Ω to 10 Ω) in series into the circuit.

Figure 20-20
Set-up for Experiment 20A,
Procedure Step 1

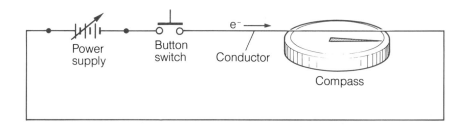

3. Predict the effect when the compass is held above the conducting wire and the electrons travel first south to north, then north to south. Check your predictions.
4. Predict what happens to the strength of the magnetic field around the conductor when the current increases. Verify your prediction by increasing the electric potential. How can you determine whether the field strength has changed?
5. Predict what happens to the strength of the magnetic field as the distance from the conductor increases. Check your prediction.
6. Hold the conductor vertically, keeping it straight and away from other parts of the electric circuit. Connect the circuit in such a way that the electrons are travelling up the conductor. Use the compass to determine whether the magnetic field lines are facing clockwise or counterclockwise when viewed from above. Draw a diagram of your observations, using the symbols shown in the Introduction to the experiment.
7. Predict the effect on the set-up in Procedure Step 6 if the electron flow is reversed. Verify your prediction.

ANALYSIS
1. A useful technique for checking the direction of magnetic field lines is to apply the **left-hand rule (LHR) for straight conductors**, which states:

   If you point the thumb of your left hand in the direction of the electron flow, then the fingers wrapped around the conductor indicate the direction of the magnetic field lines.

Figure 20-21
The LHR for Straight Conductors

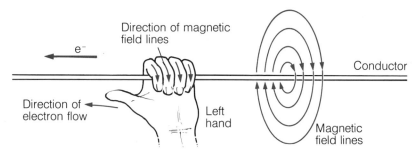

See Figure 20-21. Use this rule to check your observations in Procedure Steps 6 and 7.
2. Repeat Analysis Question 1 for Procedure Steps 1, 2, and 3.

CONCLUSIONS . . .

QUESTIONS

1. Electrons are travelling south in a straight, horizontal conductor. What is the direction of the magnetic field
   (a) below the conductor
   (b) above the conductor
   (c) east of the conductor
   (d) west of the conductor?
2. Each empty circle in Figure 20-22 represents a compass. Redraw the diagrams, indicating the direction of the needle of each compass.
3. Each empty circle in Figure 20-23 represents a conductor surrounded by a magnetic field. State whether a dot or an "x" should be placed in each circle.

Figure 20-22

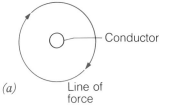
Compass — Conductor
(a)

(b)

Figure 20-23

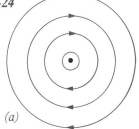
(a) Conductor / Line of force

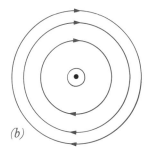
(b)

4. Choose the diagram in Figure 20-24 which best illustrates the strength of the magnetic field surrounding a conductor. Explain your answer.

Figure 20-24

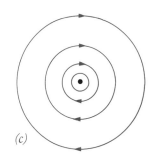
(a)   (b)   (c)

Experiment 20B: The Magnetic Field in a Coiled Conductor

INTRODUCTION

The magnetic field surrounding a straight conductor is quite weak. To increase the field strength, the conductor can be looped to form a coil. (A coil is also called a *helix*, and a long helix with many loops is called a *solenoid*.)

A **galvanoscope**, suggested for this experiment, is a device consisting of two or three sets of coils. Each set has a different number of loops. This device, named after Luigi Galvani, is shown in Figure 20-25(a). An apparatus which can be used as an alternative to a galvanoscope is shown in Figure 20-25(b).

As in the previous experiment, symbols will help make diagrams easier to draw. Figure 20-26 on page 500 shows two common ways of illustrating a coiled conductor with electrons travelling through it.

Figure 20-25
Galvanoscopes

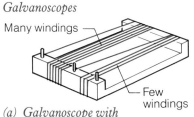
(a) *Galvanoscope with two sets of coils*

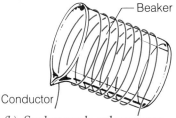
(b) *Student-made galvanoscope*

ELECTRICITY AND ELECTROMAGNETISM

Figure 20-26
Symbols for Coiled Conductors

(a)
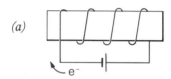

(b)
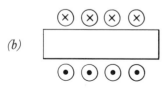

CAUTION:
(a) Do not leave the switch closed for too long at any given time.
(b) If the power supply does not have a circuit breaker, connect a small resistor (5 Ω to 10 Ω) in series into the circuit.

PURPOSE
To study the characteristics of the magnetic field in a coiled conductor.

APPARATUS
variable DC power supply; plotting compass(es); galvanoscope or set of coiled conductors; switch; connecting wires

PROCEDURE
1. Inspect the galvanoscope to learn how the wires are connected to the terminals. Place the compass inside the coil with the fewest number of loops in such a way that the needle and the wires are parallel. Connect the wires to the power supply so that the electrons flow toward the north above the compass. Refer to Figure 20-27(a). (If the compass does not fit inside the coil, you can use the alternate technique illustrated in Figure 20-27(b). In this case, two or more compasses should be used simultaneously.)
2. Based on your experience in the previous experiment, predict the direction of the magnetic field lines inside the coil when the switch is closed. Verify your prediction by closing the switch for a *very short time*.
3. Reverse the polarity of the connections to the power supply so that the electrons flow in the opposite direction, then repeat Procedure Step 2.

Figure 20-27
Set-up for Experiment 20B, Procedure Steps 1 and 2

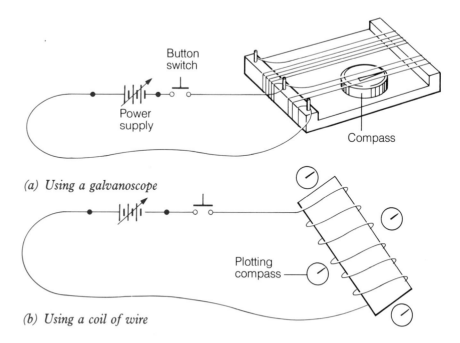

(a) Using a galvanoscope

(b) Using a coil of wire

CAUTION:
Do not exceed the maximum current indicated by your teacher, and do not leave the switch closed for more than a few seconds when doing Procedure Step 4(a).

4. Predict what happens to the strength of the magnetic field inside the coiled conductor when you do the following:
   (a) increase the current
   (b) increase the number of loops
   Verify your predictions.

ANALYSIS

1. The LHR for straight conductors can be applied to short segments of a coiled conductor to determine the direction of the magnetic field lines when the current is flowing. This can be extended to several segments, as shown in Figures 20-28(a) and (b). However, a more convenient form of the rule, called the **left-hand rule (LHR) for coiled conductors**, states:

   If you wrap the fingers of your left hand around the coil in the direction of the electron flow, then the thumb indicates the direction of the magnetic field lines inside the coil. In other words, the thumb points to the N pole of the coil.

   This is illustrated in Figure 20-28(c). Use this LHR to check your observations in Procedure Steps 2 and 3.

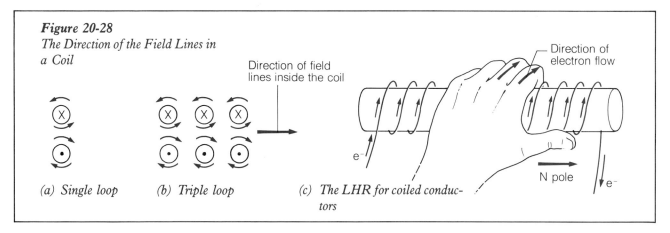

Figure 20-28
The Direction of the Field Lines in a Coil

(a) Single loop   (b) Triple loop   (c) The LHR for coiled conductors

CONCLUSIONS . . .

QUESTIONS

1. Each empty circle in Figure 20-29 represents a compass near one end of a coiled conductor. Redraw the diagrams, label the ends of the coils N or S, and show the direction of the compass needle in each case.
2. Two soft iron rods are placed inside a coiled conductor, as shown in Figure 20-30. State what happens to the rods when the switch is closed.
3. Is insulation necessary for the conducting wires of a coil? Explain.

Figure 20-29

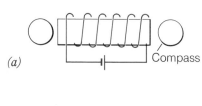
(a)

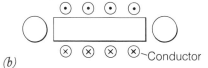
(b)

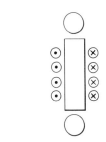
(c)

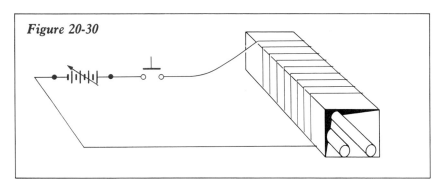

Figure 20-30

Figure 20-31
Comparing Magnetic Fields

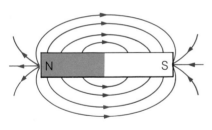

(a) Permanent magnet

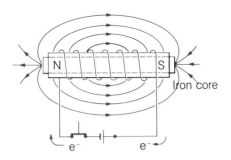

(b) Electromagnet

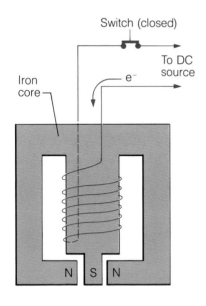

Figure 20-32
Design of a Powerful Electromagnetic Lift

20.5 Electromagnets

Both the straight and the coiled conductors used in the previous two experiments had relatively weak magnetic fields. To increase the magnetic field strength, a solid core made of a magnetic material is placed inside a coil. The magnetic field of the coil influences the domains of the core, and the core itself becomes a magnet. A common material used for cores is soft iron, which tends to lose its magnetism after the current in the coil stops flowing. An even more effective core material is *permalloy*, an alloy of iron and nickel.

A coiled conductor with a solid core is called an **electromagnet**. When the current is flowing, the magnetic field of the electromagnet resembles that of a permanent magnet, as Figure 20-31 demonstrates. One great advantage of an electromagnet is that its field can be turned on and off at will.

The results observed in the coiled conductor experiment also apply to electromagnets. The strength of an electromagnet increases under the following circumstances:

(a) if the number of loops of the coil increases
(b) if the current in the coil increases
(c) when the poles are closer together (occurs when the core is shaped like a C or a U)

Several devices use the force created by an electromagnet. The electromagnetic lift, the telephone receiver, and the computer storage system are described below. Other applications include the electromagnetic circuit breaker, the doorbell, the tape recorder, and the recording timer used in motion experiments.

Electromagnetic Lift

Huge electromagnets are used to lift or move iron sheets, old cars, or scrap iron, as the photograph at the start of the chapter indicates. Some electromagnets are able to hoist more than 20 000 kg of metal!

One way of making an electromagnetic lift is shown in Figure 20-32. The core passes through the coil and almost completely surrounds it. When the switch is closed, the core becomes magnetized, with the polarities as shown in the diagram. When the switch is opened, the core loses its magnetism, and the load drops.

Telephone Receiver

At the speaking end of a telephone, a microphone changes sound energy into electrical energy. At the listening end, the receiver uses an electromagnet to change the electrical energy back into sound energy. As the electric current passes through the coils of the electromagnet (Figure 20-33), the diaphragm is pulled toward the electromagnet over a distance that varies with the current. The diaphragm vibrates back and forth, reproducing the original sound.

Computer Storage System

A common and inexpensive method of storing information in computer systems is to use magnetic tapes and disks. A tape or disk is made by coating a non-magnetic substance with a very thin layer of a material containing millions of magnetic particles, such as iron oxide. The process of placing information onto the tape or disk begins when the computer changes the input data into electric currents. The currents are sent to the "write head", which is basically an electromagnet with a soft iron core and a small air gap, as shown in Figure 20-34. When current flows in the coil, a magnetic field is set up in the core. The field lines pass through the magnetic layer on the tape or disk at the location of the air gap. As the tape or disk passes rapidly by the air gap, the domains of the magnetic substance become aligned according to the signals received in the coil. This domain arrangement is actually the stored information.

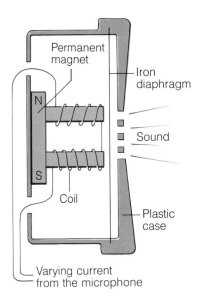

Figure 20-33
Telephone Receiver

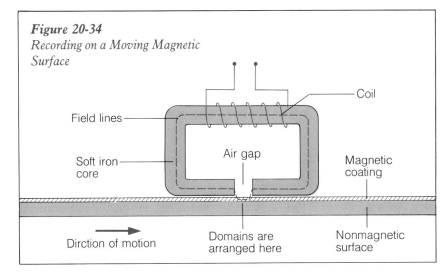

Figure 20-34
Recording on a Moving Magnetic Surface

To retrieve the stored information, the tape or disk is passed by the "read head", which resembles the write head. The domains on the tape or disk produce a variable magnetic field in the core, which in turn controls the output current of the coil. The output current is then interpreted by the computer.

Magnetic tapes used to record music operate in a similar fashion.

PRACTICE

12. **Activity** Design and perform an experiment using an electromagnet to lift iron objects such as washers or nails. Remember not to leave the current flowing for more than a few seconds at a time. Try the electromagnet both with and without the core, and with various currents and numbers of loops.
13. Figure 20-35 shows an electromagnet used in a doorbell. Explain how the bell operates.
14. Why is it wise to use plastic scissors rather than steel scissors when repairing a broken magnetic tape from a tape recorder?

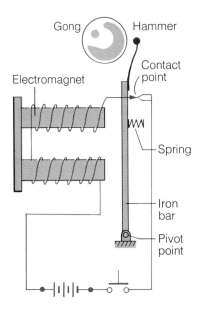

Figure 20-35

ELECTRICITY AND ELECTROMAGNETISM

Figure 20-36
Michael Faraday (1791–1867)

20.6 The Magnetic Force on Moving Charges

After Hans Christian Oersted discovered in 1819 that an electric current moving in a conductor produces a magnetic field around the conductor, many other scientists experimented with electricity and magnetism. One of the most interesting and important of these scientists was the Englishman Michael Faraday (1791–1867) (Figure 20-36). Although Faraday had no formal education, he had a great desire to learn. From age 13 to 21, he apprenticed as a bookbinder and read numerous books available through his work. Later he acquired a position as a laboratory assistant at the Royal Institution in London, where he began a series of important experiments. The story of Faraday's brilliant and successful life is fascinating.

Among other discoveries, Faraday found in 1821 that the magnetic field of a permanent magnet can exert a force on the charges in a current-carrying conductor. Faraday's discovery is called the "motor principle", because it is the basis of operation of all electric motors – from the tiny ones used in toys to the massive ones used to propel electric subway trains.

To learn how the motor principle works, consider Figure 20-37. Figure 20-37(a) shows the magnetic field of a U-shaped permanent magnet. Figure 20-37(b) shows the magnetic field around a conductor with electrons flowing away from the observer. (Remember the LHR for a straight conductor.) Figure 20-37(c) shows the two sets of fields when the conductor is between the poles of the magnet; the field of the magnet is here called the "external field". To the right of the conductor, the field lines are opposite in direction, so they tend to cancel. To the left of the conductor, the field lines are in the same direction, so they reinforce one another. These reinforced field lines force the conductor to the right, as shown in Figure 20-37(d).

The action of the force on the conductor can be summarized in the following statement, called the **motor principle**:

Figure 20-37
The Motor Principle

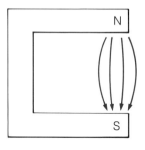

(a) *Magnetic field of the permanent magnet*

(b) *Magnetic field of the electron-carrying conductor*

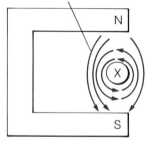

(c) *Shape of the magnetic field when the fields in (a) and (b) are superimposed*

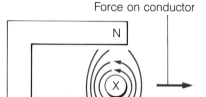

(d) *The direction of the force on the conductor is away from the region of concentrated field lines.*

The force exerted on a current-carrying conductor in the presence of an external magnetic field is perpendicular to both the direction of the current and the direction of the magnetic field lines.

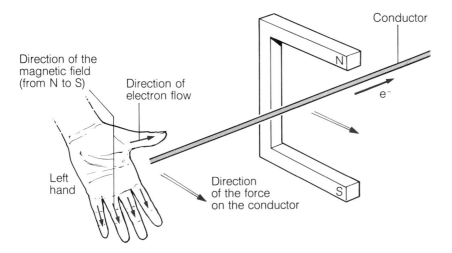

Figure 20-38
The LHR for the Motor Principle

Notice that this principle is verified in Figure 20-37, in which the three directions – the magnetic field of the magnet, the electron flow, and the force on the conductor – are all mutually perpendicular. This makes possible yet another LHR, the **left-hand rule for the motor principle**, which states:

If you point the thumb of the left hand in the direction of the electron flow and the outstretched fingers in the direction of the magnetic field of the magnet, then the palm will face in the direction of the push on the conductor.

This convenient rule is illustrated in Figure 20-38.

A set-up used to demonstrate the motor principle is shown in Figure 20-39. You should be able to verify theoretically that when the switch is closed, the conductor will move inward, toward the magnet. If either the electron flow or the magnet's polarity is reversed, the reaction will be different.

The motor principle is applied in the operation of several devices, some of which are described here.

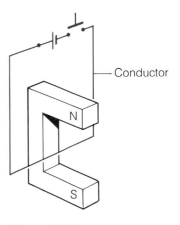

Figure 20-39
Demonstrating the Motor Principle

The Galvanometer

The galvanometer is a device that indicates the size and direction of a small direct current. To see how a galvanometer operates, refer to Figure 20-40(a). A movable coil surrounds a fixed iron core. When a current flows through the coil, it will experience a force caused by the field of the permanent magnet. In the situation shown, the N pole of the coil is repelled by the N pole of the magnet and attracted by the S pole. This causes the coil, with its attached needle, to rotate clockwise about the core and indicate the current on the scale. The greater the current, the greater is the rotation. When the current drops to zero, the coil is returned

ELECTRICITY AND ELECTROMAGNETISM

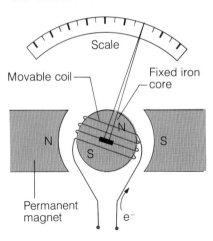

Figure 20-40
The Galvanometer

(a) Basic operation

(b) Actual view

to the middle position by a spring (not shown in the figure). If the current is reversed, the coil rotates counterclockwise. Figure 20-40(b) shows in detail the main components of a commercial galvanometer.

The operation of a galvanometer is applied in the operation of other electric meters as well. A voltmeter is a galvanometer with a high resistance, connected in series. An ammeter is a galvanometer with a low resistance, connected in parallel.

The Loudspeaker

The purpose of a loudspeaker is to change electric signals into sound vibrations quickly and accurately. The essential components of a loudspeaker are shown in Figure 20-41. A movable coil attached to a paper cone is placed over the central shaft of a permanent magnet. Variable currents from a radio or record player experience forces caused by the field of the permanent magnet. The coil is forced to vibrate according to the frequency and amplitude of the current variations. The paper cone vibrates with the coil, sending sound waves into the surrounding air.

Figure 20-41
The Loudspeaker

(a) Side view

(b) End view, showing that the field lines of the permanent magnet are always perpendicular to the current in the coil.

MAGNETISM AND ELECTROMAGNETISM

Defining the Ampere

The effect of the motor principle can be used to define the unit of electric current. The ampere, as a base SI unit, is measured in terms of force and distance. Thus, 1.0 A is the current in each of two long, parallel conductors, separated by a distance of 1.0 m, which produces a force of one conductor on the other of 2.0×10^{-7} N for each metre of length.

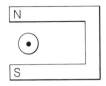

Figure 20-42

PRACTICE

15. Copy Figure 20-42 into your notebook, then use it to answer the following questions.
    (a) Draw the magnetic fields of the permanent magnet and the conductor.
    (b) Determine the direction of motion of the conductor.
16. Determine the direction of the force on each conductor shown in Figure 20-43.
17. A student sets up a successful demonstration of the motor principle, but notices that the force in the conductor is very weak. What could the student use to increase the force?
18. Go back to Figure 20-41(b), which shows a loudspeaker, and determine the direction of the force in the coil at the instant shown.
19. (a) Why is it necessary for a voltmeter to have a high resistance?
    (b) Why is it necessary for an ammeter to have a low resistance?

Figure 20-43

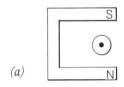

(a)

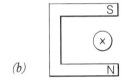

(b)

20.7 The Design of Electric Motors

An **electric motor** is a device that changes electrical energy into the mechanical energy of rotation. Small motors are used in toys and ripple tanks; larger motors are used in tools and many appliances around the home; and huge motors are used in trolley buses, subway trains, and electric locomotives. All such motors operate on the motor principle, Faraday's important discovery.

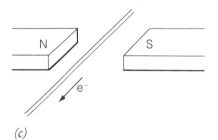

(c)

The two main parts of an electric motor are the current-carrying conductor and the magnet (either a permanent magnet or an electromagnet). In a demonstration of the motor principle, the conductor simply swings one way or another. In a motor, however, the conductor is wrapped into a loop which rotates around and around. To study how the loop in a DC motor is forced to rotate, refer to Figure 20-44 on page 508. Figure 20-44(a) shows the part of the loop with the electrons travelling toward the observer. According to the motor principle, the conductor experiences a downward force. Figure 20-44(b) shows that, at the same instant, the other part of the loop experiences an upward force. Figures 20-44(a) and (b) are combined in (c), and (d) shows a corresponding three-dimensional view of the conducting loop. You should prove to yourself that (c) and (d) are equivalent.

In Figure 20-44(d), the conducting loop is experiencing a clockwise force. After making half a turn (180°), the loop would be forced counter-

507

clockwise. Thus, a constant rotation of the motor would not be allowed to occur. This problem is solved by an important invention, the commutator.

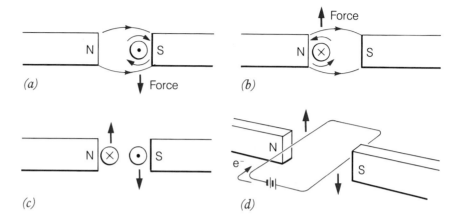

Figure 20-44
Applying the Motor Principle to a Loop of Wire

A **commutator** is a metal ring split into two parts which resembles a cylinder open at both ends and split lengthwise. Its purpose is to allow the loop in a DC electric motor to continue rotating in one direction. Figure 20-45 illustrates its operation. Figure 20-45(a) shows a commutator made of two parts, R_1 and R_2. The components touching the split ring, called brushes B_1 and B_2, are connected to a DC battery.

Figure 20-45(a) shows the same situation as Figure 20-44(d). The force on the loop is clockwise. In Figure 20-45(b), the loop has swung 90° to the vertical position. Here the current stops flowing for an instant because the brushes do not contact the split ring. However, because the loop is already in motion, it tends to continue rotating. In Figure 20-45(c), R_2 is contacting B_1. This situation resembles that in Figure 20-45(a), where R_1 is contacting B_1. Thus, the force on the loop is again clockwise. This process continues as the current reverses direction in the loop every half turn of the loop. Be sure you understand Figure 20-45 before you proceed to the next section, which describes more about electric motors.

Figure 20-45
Producing a Clockwise Force in a Single Loop of a DC Motor

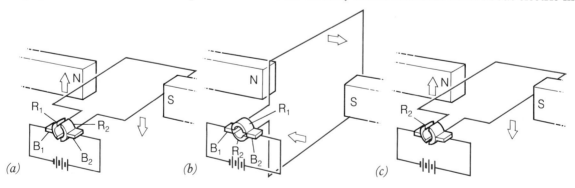

Figure 20-46

PRACTICE
20. The conductors shown in Figure 20-46 represent a loop in a magnetic field. Determine whether the force on the loop is clockwise or counterclockwise.

21. For the instant shown in Figure 20-47, is the force on the loop clockwise or counterclockwise? Explain your reasoning.
22. Describe two possible ways of forcing the loop in Figure 20-45 to rotate counterclockwise.

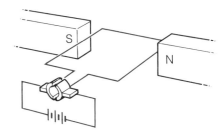

Figure 20-47

20.8 Constructing and Using Electric Motors

The description in the previous section of the operation of an electric motor did not indicate its construction. In an actual motor, the single loop is replaced by an **armature**, which is a coil of many loops wrapped around an iron core. Furthermore, the permanent magnets, which are called **field magnets** in a real motor, are often replaced by electromagnets. The numerous loops, the iron core, and the electromagnets combine to produce a motor with sufficient power to operate a device.

Special motors are available in science classrooms to help students understand the action of a motor. One such motor, called a **St. Louis motor**, is shown in Figure 20-48(a). The armature, commutator, field magnets, and one of the two brushes are visible. In Figures 20-48(b) and (c), it is possible to follow the electron flow from the source and through the coil. Using the LHR for coiled conductors, we find that the top of the armature in Figure 20-48(b) is N and the other end is S at the instant shown. Because N repels N and S repels S, the force on the armature is clockwise. Thus, the armature spins in that direction.

In operating a device, the armature is connected mechanically to a shaft which rotates with the armature. The rate of rotation is controlled by varying the current in the coils.

Figure 20-48
The St. Louis Motor

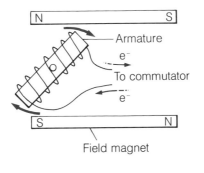

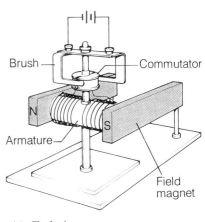

(a) General construction of the motor   (b) Top view   (c) End view

Up to this point, the description of how a motor operates has been limited to DC motors. AC motors, which are actually more common, operate in much the same manner as DC motors. The major difference is that AC motors use slip rings rather than commutators. Slip rings are shown in Figure 20-49 on page 510.

ELECTRICITY AND ELECTROMAGNETISM

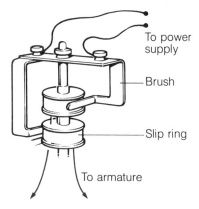

Figure 20-49
The Slip Rings of an AC Motor

The photograph in figure 20-50 shows a large electric motor used to operate a trolley in an urban transit system.

You have learned in this chapter about two great discoveries: Oersted's discovery that an electric current has a magnetic field around it, and Faraday's discovery that magnetic fields can interact. Both have had immense influence on our present technology and science. The question that naturally arises and leads to the next chapter is this: If electricity causes magnetism, can magnetism be used to produce electricity?

Figure 20-50
This photograph shows the internal structure of a large electric motor used to operate a trolley in a transit system.

PRACTICE

23. State the function of each of the following parts of a DC motor:
    (a) commutator   (c) brushes
    (b) armature   (d) field magnets
24. **Activity** Use a demonstration electric motor to learn first-hand how it operates. View the armature windings carefully, so you can determine the direction of the electron flow when the motor is connected to the energy source. Predict the direction of the armature rotation, then check your prediction. If possible, use an AC motor as well as a DC motor.
25. **Activity** Design and build your own DC electric motor.

Review Assignment

1. State the laws of magnetic poles.
2. What do you suppose caused the lodestones discovered originally in Magnesia to become magnetized?
3. Assume that you are given two apparently identical steel rods. One is magnetized and the other is not. Explain how you would discover which rod is magnetized, using the following:
   (a) a third object of your choice   (b) only the two steel rods

4. Describe the difference between the structures of magnetic and non-magnetic substances.
5. Compare the domains of the two steel rods described in Review Question 3 above.
6. Assume that a cassette tape of your favourite music dropped to the floor, but its case did not break. Use the domain theory of magnetism to explain why the music might not have the quality it had before the drop.
7. The pointed end of an iron nail is held close to the S pole of a magnet.
   (a) Which end of the nail becomes N?
   (b) Name the process that makes the nail a temporary magnet.
8. The needle of a plotting compass is found to be facing south rather than north. Explain how you would use the N pole of a magnet to magnetize the compass correctly.
9. Draw a bar magnet and the magnetic field around it. Use labelling to show that you understand the characteristics of magnetic field lines.
10. Figure 20-51 shows three similar magnets and some of the field lines of the resulting magnetic field.
    (a) Copy and complete the diagram to show the remaining field lines.
    (b) Label the polarities of the magnets.
11. A person walking in Winnipeg, Manitoba, where the angle of declination is 7° E, has a compass and heads due north according to the compass. In what direction does the person actually go?
12. **Activity** Design and carry out an experiment to discover whether or not magnetic field lines can pass through paper, aluminum, iron, or any other materials you wish to try. Discuss how you would design magnetic shielding to protect a sensitive electric instrument.
13. What is the direction of the magnetic field lines around a conductor when the electrons are travelling (a) toward you and (b) away from you?
14. A straight conductor is held near a plotting compass that is facing north. What direction of the electron flow in the conductor allows the compass needle to remain stationary if the compass is (a) beneath the conductor and (b) above the conductor?
15. Each diagram in Figure 20-52 represents two parallel electron-carrying conductors. In each case, determine whether the conductors attract or repel one another. Explain your reasoning.
16. Each empty circle in Figure 20-53 represents a plotting compass near a coiled conductor. Copy the diagrams, label the N and S poles of each coil, and indicate the direction of the needle of each compass.
17. The circles in Figure 20-54 represent a conductor of a coil. Determine which circles should have a dot and which a cross (×).
18. The domains of magnetic substances are composed of atoms of those substances. Which particles within the atoms (electrons or protons) are likely responsible for the magnetic properties of the domain? What experimental evidence supports your answers?
19. The diagrams in Figure 20-55 on page 512 show electromagnets. Determine which poles are S and which are N.

Figure 20-51

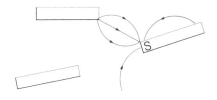

Figure 20-52

Figure 20-53

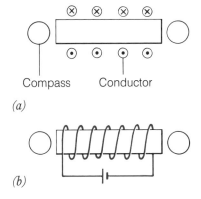

Figure 20-54

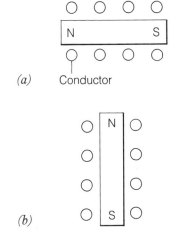

ELECTRICITY AND ELECTROMAGNETISM

Figure 20-55

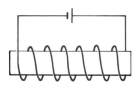

(a)

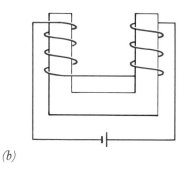

(b)

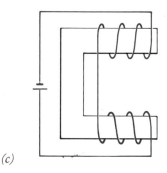

(c)

20. Is it logical to assume that a given electromagnet has a maximum magnetic field strength which cannot be increased by increasing the number of loops or the current? Explain.
21. How would you design a device that would efficiently magnetize worn-out magnets?
22. In the diagrams in Figure 20-56, the sizes of the components, the current through coils, and the number of windings are the same in all cases. The material of each component is labelled. State what happens to A, B, C, and D when the current is turned off.

Figure 20-56 →

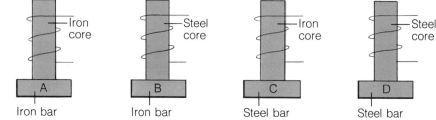

23. How would you make a compass without using any magnetic substance? (*Hint*: Is copper magnetic?)
24. In Figure 20-57, an electron-carrying conductor is in the magnetic field of a U-shaped magnet. With the aid of a diagram, determine the direction in which the conductor is forced.
25. An electric **relay** is a device that uses a small current in one circuit to control a large current in another circuit. For example, a large current is needed to start a car, but a small current is provided at the switch operated by the driver. Use the diagram of the relay shown in Figure 20-58 to explain its operation.

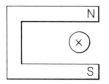

Figure 20-57

Figure 20-58
A Relay

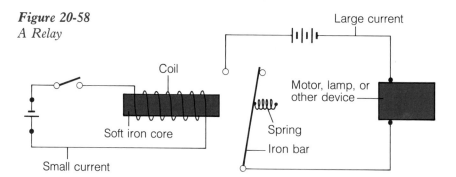

26. Describe how a galvanoscope (Experiment 20B) and a compass could be used together as a current-measuring device. How would its sensitivity compare to that of a galvanometer?
27. The diagram in Figure 20-59 represents a single loop in a DC electric motor. Determine the direction of the force on the loop.
28. Refer to Figure 20-60, which is a diagram of the St. Louis motor in Figure 20-48.
    (a) Name the parts of the motor labelled A, B, C, and D.
    (b) Determine which end of the coil is N.
    (c) In which direction will the armature spin?

Figure 20-59

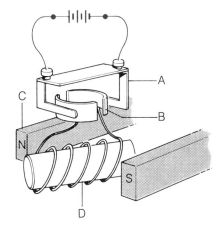

Figure 20-60

Key Objectives

Having completed this chapter, you should now be able to do the following:
1. Define magnetic N pole and magnetic S pole.
2. State the laws of magnetic poles.
3. Compare magnetic and non-magnetic substances.
4. Describe evidence that supports the domain theory of magnetism.
5. Apply the domain theory of magnetism to explain observations made in experiments using permanent magnets.
6. Recognize the difference between the magnetic properties of iron and steel.
7. Describe how to make an object composed of a magnetic substance into a magnet by induction or stroking.
8. Define magnetic field.
9. Describe how iron filings and a compass are used to find the shape of the magnetic field around a magnet.
10. List four characteristics of magnetic field lines in a magnetic field.
11. Draw the magnetic field lines around magnets of various shapes.
12. Describe the earth's magnetic field and the effects of its force.
13. Define electromagnetism.
14. Recognize the significance of the contributions of Oersted and Faraday to the development of electromagnetism.
15. Determine the direction of the magnetic field lines around a straight conductor and a coiled conductor.
16. Describe the factors affecting the strength of the magnetic field of either a straight or a coiled current-carrying conductor.
17. Describe the construction of an electromagnet and the factors affecting its strength.
18. Describe applications of the electromagnet.
19. Apply the motor principle to determine the direction of the force on a current-carrying conductor in a magnetic field.
20. Explain the operation of devices that use the motor principle, expecially those with motors.
21. In an electric motor, locate and state the function of the field magnets, armature, brushes, and commutator (DC motor) or slip rings (AC motor).

ELECTRICITY AND ELECTROMAGNETISM

CHAPTER 21
Electromagnetic Induction

This electric generating station is part of the extensive James Bay Project in Northern Québec.

After the discovery in 1819 that electricity produces magnetism, scientists sought to find out whether magnetism could produce electricity. A conductor was connected to a current-detecting device and placed near a magnet – no current flowed. Larger magnets and more sensitive detection devices were tried – still no current flowed. Then in 1831, the Englishman Michael Faraday made the breakthrough. He discovered that, when a conductor is in motion relative to a magnet, a current is produced. His momentous discovery eventually led to the development of the means of generating electrical energy in the vast quantities now used by our society. An example of one important contemporary method of generating electricity is illustrated in the photograph, which shows a hydroelectric generating station beneath a dam in Northern Québec. The material excavated from the station site was used to construct the dam. The photo shows only one part of the extensive endeavour called the James Bay Project.

In this chapter you will learn how electric currents are generated both in the laboratory and at electric generating stations. You will also read about how the electrical energy is distributed through vast networks of transformers and transmission lines.

21.1 Using Magnetism to Produce an Electric Current

The scientists who first tried to produce an electric current using magnetism probably tried experiments similar to the one shown in Figure 21-1. In Figure 21-1(a), the conductor held between the poles of a U-shaped magnet is connected to a galvanometer which detects small currents. The three arrows, B, C, and D, show three mutually perpendicular ways in which the conductor can be moved in the magnetic field. Figures 21-1(b), (c), and (d) indicate what happens for each motion.

Notice in Figures 21-1(b) and (c) that the conductor does not cross any magnetic field lines. In other words, relative to the conductor, the magnetic field is not changing. No current flows in the conductor. In Figure 21-1(d), however, the conductor cuts across the lines of force. Thus, relative to the conductor, the magnetic field is changing, and current flows in the conductor. The current is called an **induced current**. Therefore, **electromagnetic induction** is the creation of an electric current by means of a changing magnetic field.

In the example shown in Figure 21-1, the conductor is moved to cause the magnetic field around it to change. A second way to produce a changing magnetic field is to hold the conductor stationary and move the magnet in the appropriate direction. A third way is to change the size of the magnetic field around a stationary conductor. This is accomplished by using an electromagnet rather than a permanent magnet.

Because Michael Faraday discovered that a magnetic field can cause electricity, the **law of electromagnetic induction** is named after him.* **Faraday's law** states:

> Whenever the magnetic field in the region of a conductor changes, electric current is induced in the conductor.

---

* Joseph Henry, an American scientist, actually made the same discovery before Faraday did, but he neglected to publish his findings in time. Traditionally, credit for a scientific discovery is given to the person who first reports it in a scientific journal.

Figure 21-1
One Way to Create an Electric Current Using a Magnetic Field

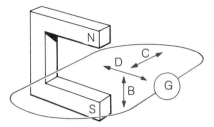

(a) *Experimental set-up to show how magnetism can create electricity*

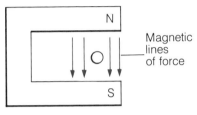

(b) *Motion in the direction of B: The conductor is moved vertically, parallel to the lines of force of the magnet. No current is produced in the conductor.*

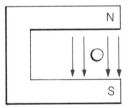

(c) *Motion in the direction of C: The conductor is moved parallel to itself. Again, no current is produced in the conductor.*

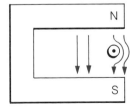

(d) *Motion in the direction of D: The conductor is moved horizontally, cutting across the magnetic lines of force. In this case, current is produced in the conductor.*

ELECTRICITY AND ELECTROMAGNETISM

Figure 21-2

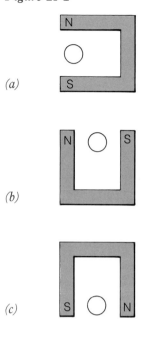

The application of Faraday's law helped change the industrial world. Most of the electricity we use today is created and transmitted using electromagnetic induction.

PRACTICE
1. What condition is necessary for a magnetic field to induce a current in a conductor?
2. Each circle in Figure 21-2 represents a conductor. In each case, state two ways of inducing a current in the conductor.
3. Based on the concepts studied in the previous chapter, predict ways of increasing the current induced in a conductor by a magnetic field.

Experiment 21A: Inducing Current in a Coiled Conductor

INTRODUCTION
The examples of electromagnetic induction given in Section 21.1 used a single conductor in which the induced current was very small. To obtain a larger current, a coiled conductor can be used, as in this experiment.

PURPOSE
(a) To study ways of inducing a current in a coiled conductor.
(b) To study factors affecting the size of the induced current.

APPARATUS
galvanometer; two coils having different numbers of windings but the same cross-sectional area; iron core that fits into one of the coils; two bar magnets (one stronger than the other); connecting wires

PROCEDURE
1. Connect the coil with the smaller number of windings to the galvanometer, as shown in Figure 21-3. Plunge the N pole of one magnet into the coil, and describe what happens.
2. Determine what happens to the current when the N pole of the magnet
    (a) is held stationary inside the coil
    (b) is moved in a circular fashion inside the coil.
3. Quickly withdraw the N pole from the coil and describe what happens.
4. Hold the magnet stationary, and move the coil in various directions. Describe the effects.
5. Place the iron core inside the coil, then touch the N pole of the magnet to the core and describe what you observe. Remove the magnet and again describe the effect.
6. Predict how the factors listed below affect the size of the current induced in a conductor by a changing magnetic field. Then use the set-up described in Procedure Step 1 to verify your predictions.

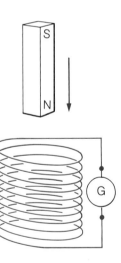

Figure 21-3
Set-up for Experiment 21A

(a) speed of motion of the magnet (Use fast, medium, and very slow speeds.)
(b) strength of the magnetic field (Use a strong magnet and a weak one.)
(c) number of turns of the coil (Use two different coils.)
7. Use any of the apparatus available to try to discover other facts about the current induced in the coil. Report on your findings, and record any questions you may have.

ANALYSIS
1. Relate your observations in this experiment to Faraday's law of electromagnetic induction.
2. Assume that you are asked to make a "generator" of electric current. Based on this experiment, what conditions would you use in order to obtain a relatively large current?

CONCLUSIONS . . .

QUESTIONS
1. Figure 21-4 shows one coil partially inserted into another coil and connected in series with a resistance, a switch, and a DC source. Predict what will be observed on the galvanometer when the switch
   (a) is suddenly closed
   (b) remains closed
   (c) is suddenly opened.
   If you want to verify your predictions experimentally, check with your teacher, and take note of the Caution in the margin.

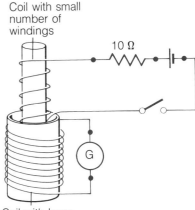

Figure 21-4

CAUTION:
The galvanometer is extremely sensitive and should *not* be connected in a circuit with a battery or power supply.

Experiment 21B: The Direction of Induced Current

INTRODUCTION
In the previous experiment, you learned about ways of inducing an electric current in a coiled conductor, as well as about factors affecting the size of the current. You may have wondered why the current flowed first in one direction and then in the other. In this experiment you will discover the answer.

The galvanometer used to determine current direction has a needle zeroed at the centre of the dial. It is important in this experiment to know the direction of the electron flow that causes the needle to swing either left or right. In many galvanometers, the needle swings to the right when the electrons enter the negative, or left-hand, terminal.

PURPOSE
To learn how to predict the direction of the induced current in a coiled conductor.

ELECTRICITY AND ELECTROMAGNETISM

APPARATUS

galvanometer; coiled conductor; magnet; connecting wires

PROCEDURE

1. Examine the coil carefully to determine the direction in which the wire is wound. Connect the coil to the galvanometer.
2. Plunge the N pole of the magnet downward into the coil, and answer the following questions.
   (a) In which direction does the galvanometer needle swing?
   (b) When viewed from above, what is the direction (clockwise or counterclockwise) of the electron flow in the coil?
   (c) Does the top of the coil become N or S when the N pole of the magnet is pushed toward it? (Apply the left-hand rule for electron flow in coiled conductors.)
   (d) Does the magnetic field of the coil help or hinder the motion of the magnet?

   Check your answers to these questions with your teacher before you proceed.
3. Predict each of the following, given that the N pole of the magnet is being withdrawn from the coil:
   (a) the direction in which the galvanometer needle swings
   (b) the direction of the electron flow in the coil
   (c) the polarity of the top of the coil
   (d) whether the motion of the magnet is helped or hindered
   Verify your predictions experimentally.
4. Repeat Procedure Step 3, forcing the S pole of the magnet first into and then out of the coil.

Figure 21-5

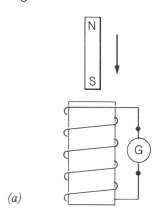

(a)

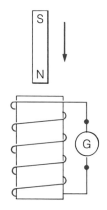

(b)

ANALYSIS

1. When a magnetic field is in motion near a conductor, an electric current is induced in the conductor. The current moves in a specific direction, causing the conductor to set up or induce its own magnetic field. How does the induced magnetic field interact with the magnetic field that caused it? (*Hint*: Does the induced field assist or oppose the action of the other field?)
2. Relate your observations to the expression, "Nature does not give something for nothing."

CONCLUSIONS . . .

QUESTIONS

1. Copy and complete the diagrams in Figure 21-5, indicating the polarity of each coil and the direction of the induced current.

21.2 Lenz's Law

In 1834 Heinrich Lenz (1804–1864) (Figure 21-6), a German physicist working in Russia, first explained the direction of the current induced by a changing magnetic field. Stated simply, Lenz proved that when a a force is exerted, nature opposes that force. To learn how this applies to a magnetic field moving near a coil, refer to Figure 21-7.

Figure 21-7(a) shows the N pole of a magnet being pushed into a coil. As the N pole approaches the coil, the magnetic field near the conductor increases. According to Faraday's law, this changing magnetic field induces a current in the conductor.

The induced current can flow in either of only two possible directions – clockwise or counterclockwise when viewed from above. If the electron current were to flow counterclockwise, the top of the coil would become an S pole, according to the LHR for coils. This S pole would attract the N pole of the magnet, increasing its speed. The result would be an increased current, and so on. Electrical energy would be produced with no effort. This is impossible, so we conclude that the current must flow clockwise. In this case, the top of the coil becomes an N pole, which opposes the N pole of the magnet. In other words, the induced magnetic field opposes the field that produced it. See Figure 21-7(b).

Figure 21-6
Heinrich Lenz (1804–1864)

Figure 21-7
Pushing the N Pole into a Coil

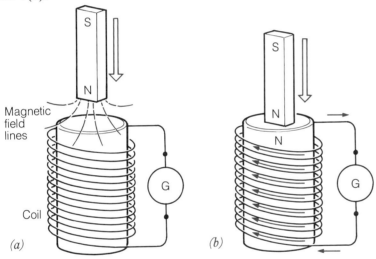

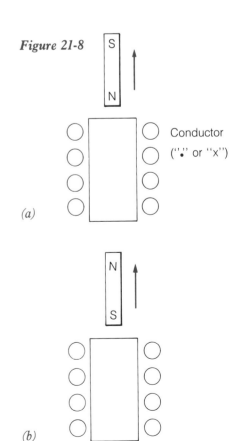

Figure 21-8

Heinrich Lenz summarized these concepts in a statement now called **Lenz's law**. It states:

> **For a current induced in a conductor by a changing magnetic field, the current is in such a direction that its own magnetic field opposes the change that produced it.**

Lenz's law is applied in the operation of electric generators and transformers, which are studied later in the chapter.

PRACTICE
4. The N pole of a magnet is being withdrawn from a coil. Using diagrams, explain step by step how to apply Lenz's law to determine the direction of the induced current.
5. Copy and complete each diagram in Figure 21-8, indicating the N and S poles of the coil and the direction of the induced current.
6. Comment on the statement, "Lenz's law is a direct consequence of the law of conservation of energy."

ELECTRICITY AND ELECTROMAGNETISM

Figure 21-9

7. A conductor (the circle in Figure 21-9) is being pushed to the left in the diagram. Use the following instructions and questions to determine the direction of the current induced in the conductor.
   (a) Copy the diagram and draw the magnetic field lines of the magnet.
   (b) According to Lenz's law, is the induced magnetic field (resulting from the induced current) clockwise or counterclockwise around the conductor?
   (c) Apply the LHR for straight conductors to determine the direction of the induced electron flow.

21.3 Electric Generators

In Experiments 21A and 21B, the mechanical energy of a magnet's motion was used to produce the electrical energy of electrons in motion. The magnet and coil acted as an **electric generator**, a device which converts mechanical energy into electrical energy.

Moving a magnet back and forth to induce a current is not as convenient as using rotation. Thus, practical generators are designed in such a way that either the coil or the magnetic field near it is rotating. The rotation is caused by a continuous source of energy such as wind, tides, falling water, or expanding steam.

The construction of a generator is identical to that of an electric motor, although its function is just the opposite. The main components of the former are field magnets, an armature, brushes, and either slip rings (for an AC generator) or a split-ring commutator (for a DC generator). These components are shown for demonstration AC and DC generators in Figure 21-10.

Figure 21-10
Demonstration Generators

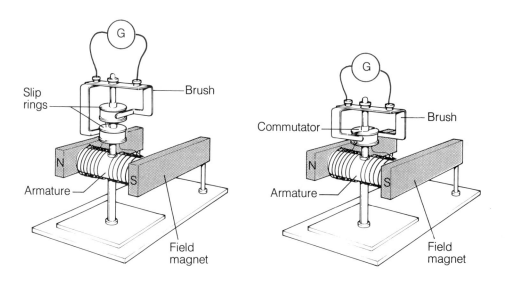

(a) AC generator   (b) DC generator

To explain the operation of AC and DC generators, a single-loop coil will be used. Therefore, it is necessary to understand how to determine the direction of the induced current in a single conductor. Figure 21-11(a) shows an upward force being exerted on a conductor in a magnetic field. The conductor is cutting across the field lines, so the magnetic field is changing. Thus, according to Faraday's law, current is induced in the conductor. The direction of the current is such that the induced magnetic field repels or opposes the magnetic field that produced it. Therefore, the induced magnetic field must be clockwise, and the electrons must be flowing out of the page. In Figure 21-11(b) the conductor is moving downward, and the electrons are moving into the page. Figures 21-11(a) and (b) are combined in (c), and (d) shows the entire loop, with directions indicated.

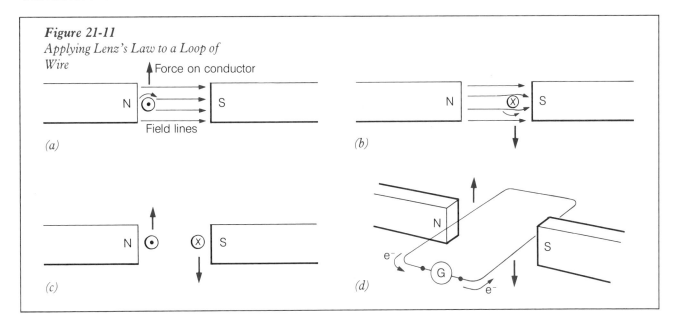

Figure 21-11
Applying Lenz's Law to a Loop of Wire

The AC Electric Generator

To study the operation of an AC generator, refer to Figure 21-12, on page 522. Figure 21-12(a) shows a simple AC generator with a coil of only one loop connected to the slip rings (R_1 and R_2), which rotate with the loop. The brushes (B_1 and B_2) touching the rings are stationary. The loop is being forced to spin clockwise by an external source of energy. In the position shown, the conductor is cutting across the field lines at a maximum rate, so the induced current is at a maximum. The direction of the current is found by applying Lenz's law; it is labelled in the figure.

As the loop rotates clockwise, the number of field lines crossed by the conductors drops gradually until it becomes zero when the loop is vertical. At this point, no current flows, as shown in Figure 21-12(b). When the loop rotates further, the current begins to flow again, this time in the opposite direction. The current reaches its second maximum value in the horizontal position, Figure 21-12(c). The current becomes zero in Figure 21-12(d), and the process begins again.

ELECTRICITY AND ELECTROMAGNETISM

Figure 21-12
A Single-loop AC Generator

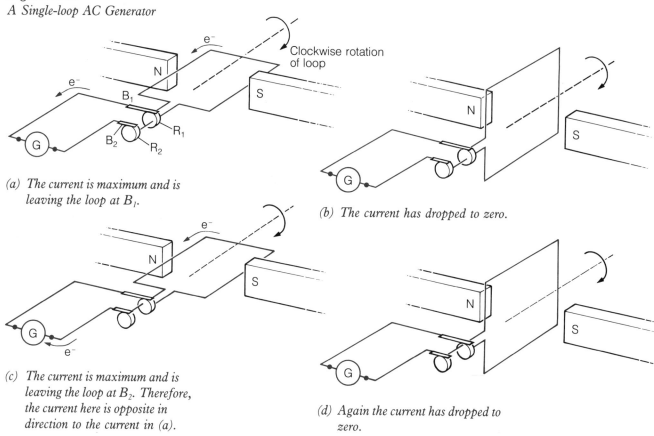

(a) The current is maximum and is leaving the loop at B_1.

(b) The current has dropped to zero.

(c) The current is maximum and is leaving the loop at B_2. Therefore, the current here is opposite in direction to the current in (a).

(d) Again the current has dropped to zero.

Because the current changes gradually, the output to the galvanometer is not constant. Figure 21-13 shows the output current that corresponds to the AC generator described. The points labelled A, B, C, and D in Figure 21-13 correspond to diagrams (a), (b), (c), and (d) in Figure 21.12.

Figure 21-13
Output Current of a Simple AC Generator

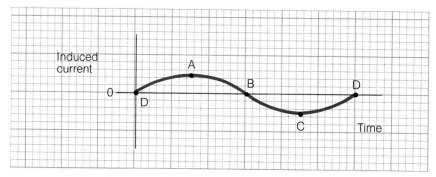

If the single-loop generator rotates at a certain frequency, the electrons in the conductor move back and forth at the same frequency. This produces an alternating current (AC). Commercial generators have several sets of coils, each with a great number of windings, and they often use electromagnets rather than permanent magnets for the field. Such generators produce a current much larger than the current produced by a demonstration generator. In North America, the frequency of the AC generation is 60 Hz. Figure 21-14 shows the AC generators at a coal-fired electrical generating plant.

Figure 21-14
A Commercial AC Generator

The DC Electric Generator

If the slip rings of an AC generator are replaced by a split-ring commutator, the generator is able to produce direct current (DC). Figure 21-15 shows the operation of a simple DC generator in which the external force on the loop is clockwise. In Figure 21-15(a), the electrons are leaving the loop through R_1 and B_1. In (b), the electrons stop flowing at the same instant as the splits in the commutator reach the brushes, and the current becomes zero. In (c), the electrons are leaving the loop through R_2 and B_1, so they travel in the same direction in the external circuit. In (d), the current once again drops to zero.

Figure 21-15
A Single-loop DC Generator

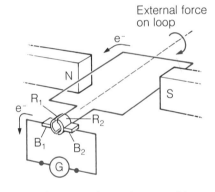

(a) The current is maximum and is leaving the loop through R_1.

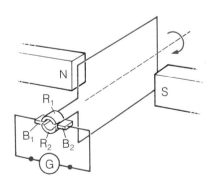

(b) The current is zero, R_1 is about to touch B_2, and R_2 is about to touch B_1.

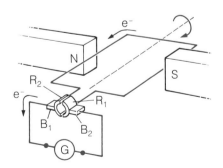

(c) The current is maximum again and is leaving the loop through R_2 and B_1. Therefore, the current is in the same direction as the current in (a).

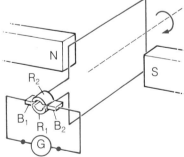

(d) Again the current is zero, and the cycle is about to start over.

The current produced by this simple DC generator is not the same as the constant DC current from a chemical source such as a battery. The electric generator produces a pulsating current in one direction, as illustrated in the graph in Figure 21-16. The points labelled A, B, C, and D on the graph correspond to (a), (b), (c), and (d) in Figure 21-15.

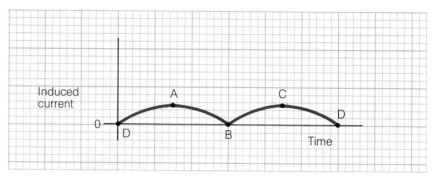

Figure 21-16
Output Current of a Simple DC Generator

Commercial DC generators have coils with numerous windings and iron cores, and they also have more than two commutator segments. In this way, the output current is kept very nearly constant, and its value is much greater than in a demonstration model.

Electric Generating Stations

Huge AC generating stations supply much of the demand for electrical energy in our society. At the same time, the generators at these stations require vast amounts of energy to sustain their operation. One source of energy is the kinetic energy of falling water (Figure 21-17). Another source is thermal energy which comes from chemical potential energy, as described back in Figure 7-4. The thermal energy for boiling water in this type of generating system can be obtained from coal, natural gas, oil, or nuclear reactions. A thermal generating plant, located at Nanticoke on Lake Erie in Ontario, is shown in Figure 21-18.

Figure 21-17
Hydroelectric Generating Plants

(a) Basic operation: At the top of the dam, the water has gravitational potential energy. The falling water gains kinetic energy which causes huge turbines to spin. The generator changes the mechanical energy of spinning into electrical energy.

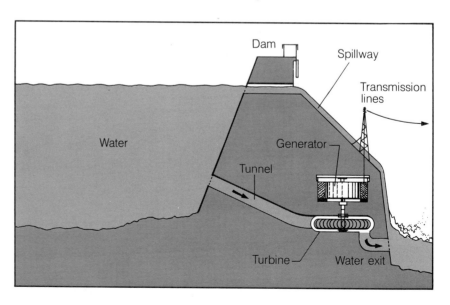

ELECTROMAGNETIC INDUCTION

(b) Construction of a water tunnel leading to the turbines

(c) Interior view showing the construction of the generators

(d) Aerial view of the "LG 3" generating station, which is part of the James Bay Project in Québec. The spillway is in the foreground, and the generating station is near the upper right-hand corner.

Figure 21-18
A Thermal-electric Generating Plant

PRACTICE

8. Figure 21-19 on page 526 shows a single-loop generator.
   (a) Copy the diagram and label the parts of the generator.
   (b) What type of generator is this?
   (c) Determine the direction of the induced current. Explain your reasoning.

ELECTRICITY AND ELECTROMAGNETISM

Figure 21-19

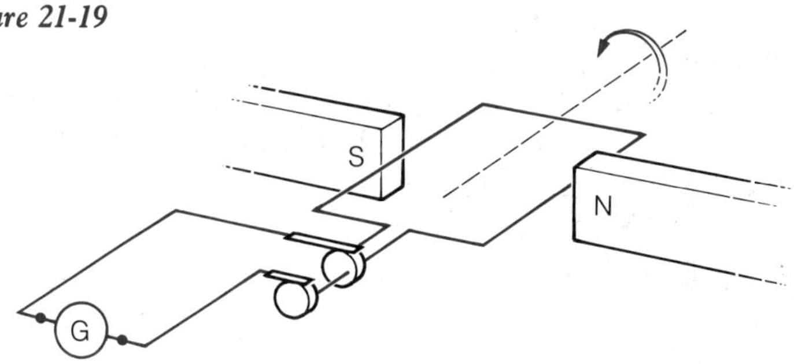

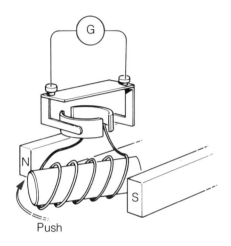

Figure 21-20

9. **Activity** Use a St. Louis motor (or a similar device) connected to a galvanometer to observe the action of a DC generator. (Refer to Figure 21-20.) Try to combine the concepts in this chapter in predicting the direction of the induced current when you spin the armature first in one direction and then in the other. If possible, replace the commutator with slip rings, and observe the action of an AC generator.
10. State the effect on the size of the current induced by a generator, given the following:
    (a) The number of loops of the coil is increased.
    (b) The rate at which the loops cut across the field lines increases.
    (c) The field strength increases.
    (d) An iron core is used in the coil rather than an air coil.
11. Natural gas can be used in a generating station to produce electricity. Describe the steps by which the chemical energy stored in the natural gas is converted into electrical energy.

21.4 Using Electricity to Generate Electricity

It has been shown several times in this chapter that a changing magnetic field near a conductor will induce a current in the conductor. Another device that illustrates this principle is a solid iron ring with a conductor looped around part of the ring, shown in Figure 21-21. When the magnet is suddenly touched to point A, the magnetic field of the ring changes. This induces a current in the coil for an instant. The current is detected by a galvanometer.

When the magnet is held stationary at A, the magnetic field does not change, so no current flows in the coil. However, when the magnet is

Figure 21-21
Another example of Electromagnetic Induction

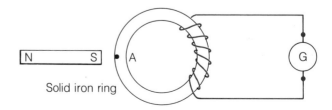

suddenly removed from A, the magnetic field in the ring changes, and again current is induced in the coil for an instant.

Instead of using a magnet touched to point A, it is possible to use current electricity to produce the same effect. This was first discovered by Michael Faraday when he designed a device now called *Faraday's ring*, illustrated in Figure 21-22. The two parts attached to the ring are the **primary circuit**, which includes the source of energy, and the **secondary circuit**, which uses the energy. When the switch in the primary circuit is suddenly closed, the magnetic field in the ring changes. This induces a current in the secondary circuit for an instant. As long as the switch remains closed, the magnetic field does not change, so no current flows in the secondary circuit. Then, when the switch is suddenly opened, the magnetic field in the ring changes. Again a current is induced in the secondary circuit for an instant.

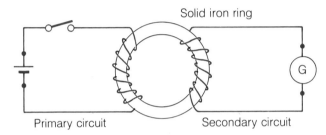

Figure 21-22
Faraday's Ring

In Faraday's ring, it is not convenient to open and close the switch every time current is needed in the secondary circuit. An easier method is to use AC electricity. This way, every time the electrons in the primary circuit change direction, the magnetic field in the ring changes. Thus, 60 Hz (AC) electricity in the primary circuit induces 60 Hz (AC) electricity in the secondary circuit. This process of inducing a current in one circuit by changing the current in another circuit is called **mutual induction**.

A very important application of mutual induction is the use of transformers. A **transformer** is a device that changes electricity at one electric potential (voltage) into electricity at a different potential. To achieve this, the primary coil of the transformer must have a different number of windings than the secondary coil.

An example of a transformer is shown in Figure 21-23. The primary coil has three windings and an electric potential of 10 V (AC). It is found experimentally that the secondary coil, with nine windings, has a potential of 30 V (AC). If the number of windings increases by a factor of three, the potential also increases, or steps up, by a factor of three.

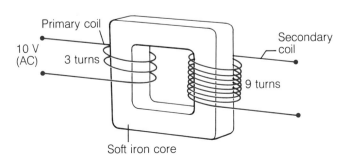

Figure 21-23
A Step-up Transformer

ELECTRICITY AND ELECTROMAGNETISM

This type of transformer is called a **step-up transformer**.

If the primary coil has more windings than the secondary coil, then the transformer is a **step-down transformer**. In the example shown in Figure 21-24, the primary potential is 20 V, and there are four times as many windings in the primary as there are in the secondary. Thus, the secondary potential is four times lower than the primary potential, or 5.0 V.

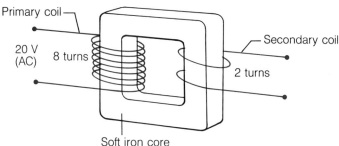

Figure 21-24
A Step-down Transformer

In these examples, the ratio of the secondary potential to the primary potential equals the ratio of the secondary windings to the primary windings. Experiments show that this relationship is true for all transformers. In equation form, the relationship is

$$\frac{\text{secondary potential}}{\text{primary potential}} = \frac{\text{secondary windings}}{\text{primary windings}} \quad \text{or} \quad \frac{V_s}{V_p} = \frac{N_s}{N_p}.$$

Here, V is electric potential, N is the number of windings, s means secondary, and p means primary. This equation can be used to find any one of the four quantities if the other three are known.

SAMPLE PROBLEM 1
A transformer used to operate a neon sign has 85 windings in the primary coil and 6800 windings in the secondary coil. If the transformer is used in a 120 V circuit, what is the potential required to light up the sign?

Solution

$$V_s = \frac{V_p N_s}{N_p}$$

$$= \frac{120 \text{ V} \times 6800 \text{ windings}}{85 \text{ windings}}$$

$$= 9.6 \times 10^3 \text{ V}$$

∴ The electric potential is 9.6×10^3 V or 9.6 kV.

In Sample Problem 1, the electric potential has increased by a factor of 80. It may appear that nature has been generous and provided something for nothing. Such is not the case! If the potential increases by a factor of 80, the current is reduced by the same factor. What remains constant is the product of the potential and the current, VI, which is the power, P. Constant power or zero power loss is an ideal situation (100% efficiency) and applies only in theoretical calculations. In practice, there are always some power losses associated with the use of transformers. Such losses are minimized in modern transformers by using a core material which quickly becomes magnetized and by constructing the core in thin, tightly bound layers separated by even thinner insulating material. See Figure 21-25. Efficiencies in the range of 95%–99% are now common for transformers. You will learn in the next section why transformers are so important in today's society.

Figure 21-25
This internal view of a polemount transformer shows the coil and layered core. At the top of the transformer are the leads that will be connected to conductors outside the transformer tank.

PRACTICE

12. **Activity** Figure 21-26 shows a transformer set-up used for student experimentation. Inspect and describe the apparatus, and perform measurements using various combinations of primary and secondary coils to determine whether the ratio $V_s : V_p$ is close to the ratio $N_s : N_p$.

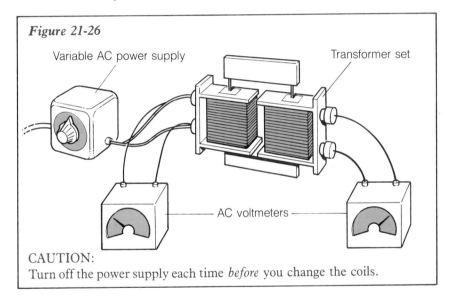

13. For each transformer shown in Figure 21-27, state whether it is a step-up or step-down transformer, and calculate the unknown quantity.

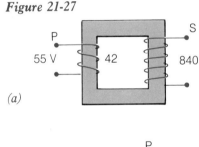

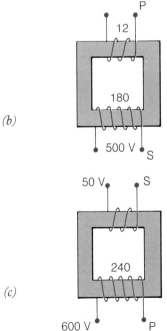

Figure 21-27

21.5 Using Transformers to Distribute Electrical Energy

Electric generating stations are often located far from the region where most of the electrical energy they produce is used. The energy must therefore be transmitted over long distances. The total amount of energy distributed is large, which means that the power ($P = E/t$) from the generating station is also large. Because $P = VI$, it is evident that there are two possible choices for distributing this large amount of power.

ELECTRICITY AND ELECTROMAGNETISM

One choice is a high current (I) and a low electric potential (V); the other is a low current and a high electric potential. The Sample Problems below reveal which choice is preferable.

SAMPLE PROBLEM 2

The transmission lines used to distribute 200 kW of power have a resistance of 0.50 Ω. If the power is transmitted at a potential of 1000 V (1.0 kV), calculate the following:
(a) the current in the lines
(b) the power loss (P_L) in the lines
(c) the fraction of the original power lost

Solution

(a) $I = \dfrac{P}{V} = \dfrac{2.0 \times 10^5 \text{ W}}{1.0 \times 10^3 \text{ V}}$

$= 200 \text{ A}$

Thus, the current is 2.0×10^2 A.

(b) $P_L = I^2 R$ (from Section 19.3, Practice Questions 13 and 16)
$= (200 \text{ A})^2 (0.50 \text{ Ω})$
$= 2.0 \times 10^4 \text{ W}$

Thus, the power loss is 2.0×10^4 W.

(c) $\dfrac{P_L}{P} = \dfrac{20 \text{ kW}}{200 \text{ kW}} = 0.10$

Thus, 10% of the original power is lost.

SAMPLE PROBLEM 3

Repeat Sample Problem 2, using the higher electric potential of 100 kV.

Solution

(a) $I = \dfrac{P}{V} = \dfrac{2.0 \times 10^5 \text{ W}}{1.0 \times 10^5 \text{ V}}$

$= 2.0 \text{ A}$

Thus, the current is 2.0 A.

(b) $P_L = I^2 R$
$= (2.0 \text{ A})^2 (0.50 \text{ Ω})$
$= 2.0 \text{ W}$

Thus, the power loss is 2.0 W.

(c) $\dfrac{P_L}{P} = \dfrac{2.0 \text{ W}}{2.0 \times 10^5 \text{ W}} = 1.0 \times 10^{-5}$

Thus, only 0.0010% of the original power is lost.

From the previous calculations, it is obvious that the best way to transmit electricity with little loss of power is to use a low current and a high electric potential. The device that increases the potential and simultaneously decreases the current is the transformer. Vast networks of transformers and transmission lines are utilized today to deliver electrical energy from generating stations to the consumer.

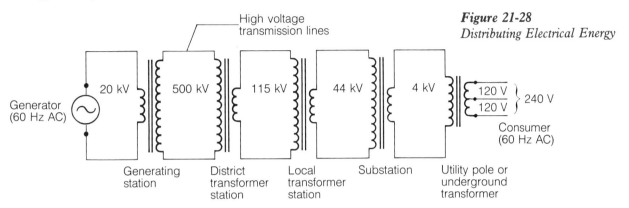

Figure 21-28
Distributing Electrical Energy

An example of an electricity distribution network can be seen in Figure 21-28. In this case, the potential is stepped up to 500 kV for transmission along the main lines. Then the potential is stepped down in stages until it enters the home at 120 V. In some circuits in the home, a potential of 240 V is required for major appliances. This potential is obtained by the use of an extra cable that enters the home at 120 V, but with its oscillations exactly out of phase with the other 120 V cable. When one cable is at a potential of +120 V, the other is at a potential of −120 V; thus, the potential difference is 240 V. Notice in the figure that the electrons in the household circuit do not originate at the generating station; rather, they simply vibrate back and forth in the very last circuit. In other words, it is the energy that is transmitted, not the particles. See also Figures 21-29, 21-30, and 21-31.

Figure 21-30
A Local Transformer Station

Figure 21-31
A Utility Pole Transformer. The energy enters the transformer at a high potential and leaves at a low potential.

Figure 21-29
Installing High-potential Transmission Lines

PRACTICE

14. The power distributed at 500 V on a set of transmission lines is 500 kW, and the resistance of the lines is 1.0 Ω. Determine the following:
    (a) the current in the lines
    (b) the power loss
    (c) the energy loss per hour
15. Repeat Practice Question 14, given that the potential is 500 kV.
16. Figure 21-32 shows the final transformer in a network which delivers electrical energy to a home. Assume that there are 150 windings from C to D. How many windings are there from A to B?

Figure 21-32

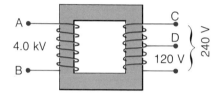

21.6 Electromagnetic Waves

You have studied how charged particles, namely electrons in the case of metal conductors, move back and forth in an alternating current. The electrons gain energy and accelerate, then they slow down, stop, and change direction. Each time they slow down they lose energy, some of which is emitted by the electrons and travels as a wave. Such a wave, which originates with the acceleration of a charged particle, is called an **electromagnetic wave**.

The electromagnetic waves produced by an alternating current of 60 Hz have a frequency of the same value, 60 Hz. The interference caused by such waves can be noticed on a car radio as the car passes under high-potential transmission lines.

The waves resulting from a 60 Hz alternating current are only one example of the entire set of electromagnetic waves. The entire set, the electromagnetic spectrum, was described in detail in Chapter 16. The discovery and subsequent study of this spectrum is part of the story of the development of physics, which continues in the final unit of this book.

Review Assignment

1. Assume that you are given a straight conductor and a U-shaped magnet. Describe how you would induce the maximum possible current in the conductor in these situations:
   (a) with the magnet held stationary
   (b) with the conductor held stationary

2. A bar magnet is used to induce a current in a coiled conductor. State the effect on the current in the following circumstances:
   (a) The number of coil windings is increased.
   (b) The speed of motion of the magnet relative to the coil is increased.
   (c) The strength of the magnet is increased.
   (d) The magnet is stopped relative to the coil.
3. In each case in Figure 21-33, determine which end of the coil is N, as well as the direction in which the electrons flow.
4. In the diagrams in Figure 21-34, the arrows indicate the direction of the mechanical force on a straight conductor. Determine the direction of the electron flow in each conductor. Show your reasoning.
5. State the function of (a) an electric generator and (b) an electric motor.
6. Describe the differences between AC and DC generators.
7. Describe how commercial generators produce an alternating current.
8. Suggest a design for an electric car which works in such a way that each time the brakes are applied, some of the car's kinetic energy is used by a generator to recharge the batteries.
9. Assume that you live in a cabin far from any source of current electricity.
   (a) What are some possible ways in which you could generate electricity?
   (b) Describe the energy conversions that occur in one of the ways mentioned in (a).
10. In a Faraday's ring apparatus, a DC source and a switch are connected in the primary circuit. Under what conditions is a current induced in the secondary circuit?
11. Why does the basic operation of a transformer depend on alternating current rather than on a constant direct current?
12. Describe the construction of (a) a step-up transformer and (b) a step-down transformer.
13. A transformer is required for operating a fluorescent light fixture. The primary coil has 260 windings and the secondary coil, 65 windings. If the primary coil is connected to a 120 V household circuit, what is the electric potential at which the fixture operates?
14. A model electric train operates at 8.0 V. A transformer with a primary coil having 930 windings is plugged into a 120 V circuit. What is the number of windings in the secondary coil?
15. The induction coil in an automobile is a solenoid and acts much like a transformer. It changes 12 V in the primary into perhaps 24 kV in the secondary. If the number of windings in the secondary is 2.4×10^5, find the number of windings in the primary. (The 12 V circuit in a car operates on DC. In order to make the induction coil work, the DC must be turned on and off quickly. This is controlled by the car's distributor.)
16. An electric doorbell uses a transformer to obtain 6.0 V. If the primary coil has 840 turns and the secondary coil has 42 turns, what is the primary electric potential?
17. What conditions minimize power loss in electric transmission lines?

Figure 21-33

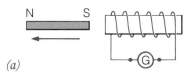
(a)

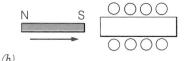
(b)

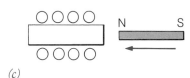
(c)

Figure 21-34

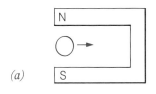
(a)

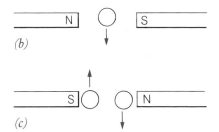
(b)

(c)

18. What is the source of the electrons that vibrate back and forth in the electric circuits of your home? What is their source of energy?
19. The power delivered by a set of transformer lines is 250 kW, and the line resistance is 1.5 Ω. If the lines are at 100 kV, find the following:
    (a) the current in the lines
    (b) the power loss
    (c) the fraction of the original power lost
20. Describe the origin of 60 Hz electromagnetic waves.
21. **Activity** One of the largest engineering projects in the world is the James Bay Project undertaken by Hydro Québec. Find information on this project and organize a debate or a class discussion; alternatively, write a report on it. Consider not only the physics aspects of the project, but also its environmental and societal implications.

Key Objectives

Having completed this chapter, you should now be able to do the following:
1. Define electromagnetic induction.
2. State Faraday's law of electromagnetic induction.
3. Demonstrate how to induce a current in a straight or coiled conductor.
4. List the factors affecting the size of the current induced in a conductor by a changing magnetic field.
5. Apply Lenz's law to find the direction of an induced current.
6. Describe the construction and operation of AC and DC generators.
7. Given the direction of the mechanical force on the armature of a generator, determine the direction of the induced current.
8. Distinguish between electric generators and motors.
9. Describe the energy conversions associated with various types of generating stations.
10. Describe the construction and operation of an electric transformer.
11. Distinguish between step-up and step-down transformers.
12. Given any three of the number of windings in the primary and secondary coils of a transformer and the electric potential of the primary and secondary coils, determine the fourth quantity.
13. Explain the advantages of using transformers to transmit electrical energy over long distances.
14. Given the electric potential, power, and resistance of a set of transmission lines, determine the current through the lines and the power loss.
15. Describe the origin of 60 Hz electromagnetic waves.

UNIT VII
Atomic Physics

The information in this Unit will be especially useful if you plan a career in diagnostic radiology, radiation therapy, nuclear medicine, nursing, teaching, family medicine, dentistry, X ray technology, nuclear research, nuclear power generation, chemistry, archaeology, forensic science, astrophysics, biophysics, elementary particle physics, or plasma physics.

ATOMIC PHYSICS

CHAPTER 22
Atoms and Radioactivity

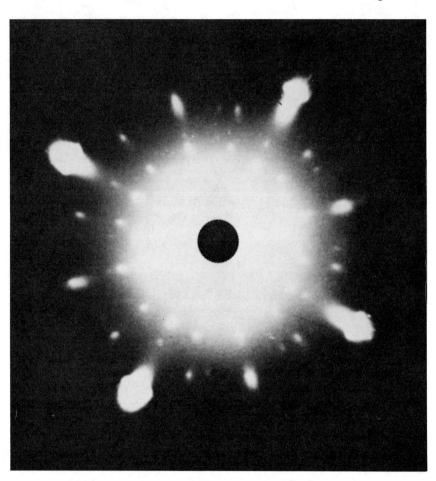

This diffraction photograph of a crystal of the metal palladium was taken with X rays. The interaction of the X rays with the crystal reveals the symmetric structure of the crystal.

Take a deep breath of air, keeping in mind that 10^{24} atoms make up that breath. Or consider that it would take five million hydrogen atoms, lined up beside each other, to stretch across the period at the end of this sentence. It is little wonder that many people have difficulty visualizing the size and appearance of an atom! Since no one has ever seen an atom, its appearance can only be deduced, from indirect evidence. Much of that evidence was discovered in a thirty-year period, starting in 1895 with the discovery of X rays. Recently, more indirect evidence has been gained by analysing photographs taken with special instruments that use high-energy particles and rays. One such photograph, shown here, reveals the regular structure of a metal. The bright spots are produced on a screen when charged particles spread out as they are repelled from a fine point of the metal. Thus, the atoms have an arrangement which corresponds to the bright spots.

In this chapter you will learn about the characteristics of high-energy particles and rays. Then you will be able to answer such questions as these: What evidence have scientists used in developing a successful model of the atom? How has the understanding of X rays and atomic radioactivity been applied to benefit humans? And how can the hazards of X rays and radioactivity be minimized?

22.1 From Classical to Modern Physics

In the early part of the nineteenth century, the discovery of electricity had a profound effect on the study of science. By the 1850s much was known about which solids and liquids were electric conductors or insulators. It was thought that all gases were electric insulators. Then, when vacuum pumps were improved, scientists soon discovered that gases under the low pressure produced by a vacuum pump did, in fact, conduct electricity. The set-up for testing the electric conductivity of gases in a vacuum tube is shown in Figure 22-1.

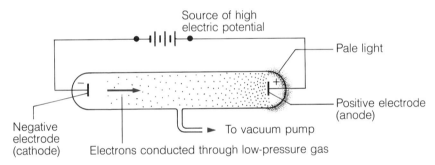

Figure 22-1
Conducting Electricity in a Gas under Low Pressure in a Vacuum Tube

The observation which puzzled scientists was that the glass at the positive end of the vacuum tube glowed with a pale green light. What type of invisible "ray" caused the glass to glow? Whatever it was must have originated at the negative electrode, or **cathode**, of the vacuum tube. Thus, the rays that caused the glow came to be known as "cathode rays", and the tube was named a **cathode-ray tube**.

The cathode-ray tube became an important device in scientific laboratories. Experiments showed that cathode rays could be bent using a magnet. Since light rays are not affected by a magnet, it was concluded that cathode rays are not light rays. Other experiments helped provide more conclusions about cathode rays in particular, and matter in general.

Toward the end of the nineteenth century, scientists began to think they understood nature completely, and all the important scientific laws had been discovered. Their key theories and laws make up what we now call *classical physics*.

This spell of confidence was shattered in 1895, when X rays were discovered through the use of a cathode-ray tube. Scientists now had to ask themselves why and how something was able to pass through substances previously thought to be solid. The year 1895 marked the beginning of what we now call *modern physics*. Its development is a fascinating story. Below is a chronology of the main events of modern physics.

- 1895 – discovery of X rays
- 1896 – discovery of radioactivity
- 1905 – special theory of relativity
- 1911 – discovery of the nucleus
- 1915 – general theory of relativity
- 1919 – observation of the first artificial nuclear reaction
- 1920 – naming of the proton
- 1932 – discovery of the neutron
- 1939 – discovery of nuclear fission
- 1942 – first sustained reaction in a nuclear reactor
- 1948 – invention of the transistor
- 1960 – invention of the laser

Modern physics has helped us expand our knowledge of nature and the universe. For example, today it is well known that there are many tiny particles making up the atom besides electrons, protons, and neutrons. Physicists no longer assume that they know all there is to know; on the contrary, they admit there is still much to learn.

PRACTICE

1. What are some characteristics of cathode rays?
2. What event in 1895 helped the transition from classical to modern physics?
3. Cathode-ray tubes were important in the development of modern physics, and they still have numerous applications today. Research and report on the design and operation of a cathode-ray tube in one of the following devices:
   (a) television   (c) computer monitor
   (b) oscilloscope   (d) radar screen
4. Research and report on the life and main contributions to physics of Sir William Crookes, an English physicist who invented a high-vacuum cathode-ray tube called a **Crookes tube**.

Figure 22-2
Wilhelm Roentgen (1845–1923)

22.2 The Discovery and Use of X Rays

It was in 1895, as you have read, that a German physicist named Wilhelm Roentgen (1845–1923) (Figure 22-2) was experimenting in his darkened laboratory with a cathode-ray tube which he had wrapped with black paper. Suddenly his attention was drawn to a dish of crystals nearby: they were glowing mysteriously in the dark. Roentgen immediately began a systematic search for the cause of the glow. See Figure 22-3.

After some experimenting, Roentgen discovered that the cathode rays (now known to be high-speed electrons) travelling in the glass tube struck the positive end of the tube, giving up much of their energy. Evidently that energy was then given off in the form of different rays that could

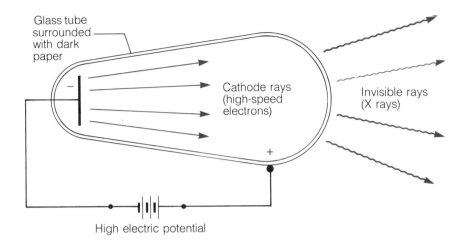

Figure 22-3
Roentgen's Experimental Set-up

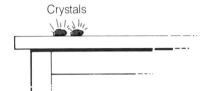

travel through black paper, as well as cause certain crystals to *fluoresce* (glow). The rays he had accidentally discovered had no name, so he decided to call them **X rays**, after the variable x in mathematics.

After further experimentation, Roentgen found that the X rays could expose photographic film and pass through certain low-density substances such as paper, wood, and flesh, and yet were absorbed by bone and metals such as lead. He even took the world's first X ray photograph by placing his wife's hand between the source of the X rays and a photographic plate. Figure 22-4 shows another famous X ray photograph taken by Roentgen.

Characteristics of X Rays

X rays are now known to have the following characteristics:

- They can pass through many substances.
- They can be emitted when high-speed electrons collide with matter.
- They can expose photographic film.
- They can cause certain substances to fluoresce.
- They travel at the speed of light in a vacuum.
- They travel in straight lines within a single substance.
- They can be reflected and refracted.
- They are not affected by electric or magnetic fields.

Many of these characteristics are identical to the characteristics of visible light. In fact, X rays belong to the high-energy portion of the electromagnetic (e.m.) spectrum. The e.m. spectrum, which was described in Chapter 16, includes radio and television waves, microwaves, infrared light, visible light, ultraviolet light, and gamma rays.

The Production of X Rays

The main components needed to produce X rays are as follows:
- a source of electrons (the cathode)
- a high electric potential (voltage) to accelerate the electrons

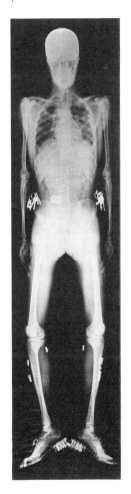

Figure 22-4
This interesting photograph, taken by Roentgen using X rays, shows the skeletal structure of a clothed man. Notice the keys in his jacket pockets, the metal clasps of the garters below his knees, and the nails in his shoes.

- a region where the air has been evacuated so the acceleration can occur more easily
- a target which the electrons strike, whereupon they lose their energy and produce X rays

The basic design of a modern X ray tube is shown in Figure 22-5. In this case the target is forced to rotate at a high frequency, to prevent overheating. The electric potential of the high-voltage source, which ranges from about 25 kV to 350 kV, depends on the required function of the X rays.

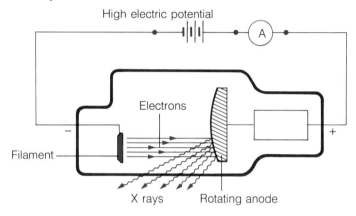

Figure 22-5
The Design of a Modern X Ray Tube

Medical Uses of X Rays

X rays are extremely useful for medical purposes, both diagnostic and therapeutic. Each year in North America hundreds of millions of X ray photographs are taken to let medical practitioners view people's teeth, bones, internal organs, and blood streams. Such photographs help diagnose problems as diverse as tooth cavities, broken bones, tumours, and enlarged arteries. When an X ray photograph of an internal organ or the blood stream is to be taken, a fluid that absorbs X rays is first injected into the appropriate system. A typical X ray photograph can be seen in Figure 22-6.

Multiple X ray images of a part of the body can be obtained by having an X ray machine revolve about the part. Numerous X ray "slices" are taken and then analysed by computer in what is called a *computerized tomography (CT)* scan.

X rays also have applications in treating certain types of cancers. The basic principle of this therapy is to try to destroy the unwanted cancerous tumour, without damaging the surrounding healthy tissue.

Because X rays are capable of damaging or killing living cells, great caution must be exercised in their use. The specialists who operate X ray machines usually leave the room when a photograph is being taken. In addition, they place lead shielding on the patient to prevent X rays from penetrating areas where they are not wanted. Patients should keep a record of the number of X ray photographs they have had in order to avoid excessive doses of radiation. Women who are pregnant should be particularly aware that X rays can cause permanent damage to an unborn child. (It is not practical to list the amount of radiation received per

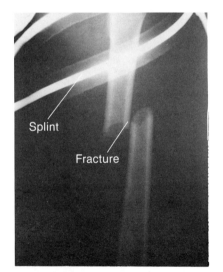

Figure 22-6
This X ray photograph shows a distinct fracture of a child's leg. The curved device near the top is a metal splint which will support the leg.

X ray because different machines emit different amounts of radiation. Machine operators should know what amount of radiation is "excessive" in a given situation.)

Other Uses of X Rays

Shortly after X rays were first discovered by Roentgen, people around the world had great fun seeing things that previously were invisible. For example, oyster farmers could "see" pearls in oysters without having to open them, customs inspectors could check baggage contents at borders, and jewellers could distinguish between artificial and genuine gems quickly. Such applications are still valuable, but numerous others have been developed.

In industry, X rays can be used to control the thickness of precision-made parts. They can also be used to detect flaws inside welds or metal parts, flaws that otherwise would remain undetected until a machine broke down during operation. Another important modern application is the production of the ever-smaller silicon chips used in microcomputer circuits. X rays, which have extremely short wavelengths, are used to etch these miniature electronic circuits, which contain up to ten times the number of components as corresponding circuits etched with ultraviolet light.

Scientific research also benefits from X rays. For example, to learn more about matter, scientists study crystal structure using the diffraction patterns created when X rays are beamed through a thin layer of a crystal. Also, X rays that pulse with a known frequency are used to take "movies" of substances undergoing chemical change. Scientists are constantly searching for new industrial applications of X rays. As you read through books or periodicals, try to find references to applications that were developed recently.

PRACTICE

5. By referring to Figure 16-17, compare X rays to visible light with reference to these criteria:
   (a) wavelength (c) speed in a vacuum
   (b) frequency   (d) energy

6. A colour television has the same components as an X ray tube. Although the television produces only low energy X rays at the front of the picture tube, it emits some harmful rays at the rear of the set.
   (a) What are the main components of an X ray tube?
   (b) Is any TV set you have at home oriented to avoid the effects of harmful radiation to the viewers?
   (c) **Activity** Visit the TV department of an appliance or department store and observe whether the sets on display are arranged in a way that minimizes the dangers of X rays reaching either the salespeople or the customers.

7. When was the most recent X ray photograph of your own teeth taken? (If you cannot recall having had it done, ask someone else

ATOMIC PHYSICS

this question.) Describe the precautions that were exercised when the photograph was taken.
8. In past centuries, one of the substances in artists' paints was lead. Today little or none of this poisonous substance is used in such paints. Could X rays be used to distinguish a genuine old masterpiece from a modern copy of that masterpiece? Explain.
9. **Activity** Research and report on the use of X rays in industry, scientific research, or medicine.

22.3 The Discovery of Radioactivity

Figure 22-7
Henri Becquerel (1852–1908)

Just a few months after Wilhelm Roentgen discovered X rays, a French scientist named Henri Becquerel (Figure 22-7) made another accidental discovery. Becquerel, who lived from 1852 to 1908, was experimenting with minerals which he thought might give off X rays when struck by sunlight. To test his hypothesis, he intended to use photographic film which, he knew, was sensitive to X rays.

One cloudy day in 1896 he stored some uranium samples on an unexposed, covered film in a drawer. Four days later, still a victim of cloudy weather, he developed the film just for interest, expecting to see nothing. To his amazement, the film had been exposed. The uranium had emitted something invisible, at first called "Becquerel rays", which had passed through the opaque cover of the film and had exposed it. Becquerel had discovered what later came to be called "radioactivity".

Unlike X rays, radioactivity does not depend on external factors such as temperature, electric or magnetic forces, light, or chemical activity to produce it. It simply occurs when the nucleus of a given atom does not remain together; that is, when the nucleus breaks apart or "decays". The decay is accompanied by the release of some types of radiation, which explains the prefix "radio". Thus, **radioactivity** is defined as the spontaneous decay of an unstable nucleus with the corresponding release of particles and/or energy.

Becquerel's experiment can be repeated in the laboratory. If a radioactive substance is placed on an unexposed photograph film for three or four days, it will cause the film to be exposed. An example of this is shown in Figure 22-8.

Figure 22-8
The photograph shows the faint outline of a radioactive watch built in the 1950s. The watch was left on an unexposed film for three days before the film was developed.

Shortly after Becquerel's discovery, Marie Curie (1867–1934) and her husband Pierre Curie (1859–1906) heard of Becquerel's rays and began chemically analysing the uranium compound which had caused them. See Figure 22-9. After much effort, they were successful in separating two previously unknown elements that were highly radioactive. They named these elements polonium (after Poland, Marie's home country) and radium.

Several other radioactive elements are now known. They include all the naturally occurring elements whose atomic numbers (number of protons) range from 83 (bismuth) to 92 (uranium). (Refer to the periodic table of the elements in Appendix B.) They also include some elements with fewer than 83 protons, as well as all the elements with an atomic number greater than 92. This latter group comprises the elements known as the *transuranium elements*.

Other radioactivity experiments revealed that there are several methods or modes of radioactive decay. The three most common modes are called alpha, beta, and gamma decays after the first three letters of the Greek alphabet. Their characteristics are described below. (The other decay modes are rare and are not significant for this discussion.)

Alpha (α) decay results in the release of an α particle. This type of particle has a double positive charge, can penetrate matter for only a short distance, and is relatively slow-moving. It is actually the nucleus of a helium atom.

Beta (β) decay results in the release of a β particle. This type of particle has a single negative charge, is much lighter than an α particle, and has a greater distance of penetration in matter. β particles are actually high-speed electrons, just like cathode rays are, though they originate from a different source.

Gamma (γ) decay results in the release of a γ ray. This type of ray has no electric charge. It is high-energy radiation belonging to the electromagnetic spectrum. Of the three radioactive emissions, γ rays can penetrate farthest into matter.

Figure 22-9
Marie and Pierre Curie

Each of these radioactive emissions acts differently in the presence of a magnetic field. The α particle, with its double positive charge, is deflected one way. The β particle, with its single negative charge, is deflected the other way. And the γ ray, with no charge, is not affected by a magnetic field. These concepts are illustrated in Figure 22-10.

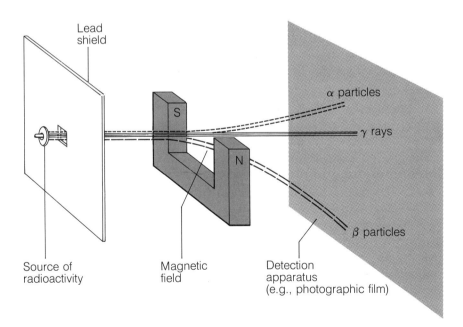

Figure 22-10
The Three Modes of Radioactive Decay in the Presence of a Magnetic Field

Methods of detecting and using these radioactive emissions, as well as writing decay equations, are discussed later in the chapter. As you study radioactivity, remember that the emissions (α, β, and γ) come spontaneously from the nucleus of an atom.

ATOMIC PHYSICS

CAUTION:
Use tongs or washable gloves to handle the source, and wash your hands after handling it. Never aim the source toward yourself or anyone else, and do not allow it to contact your mouth.

Figure 22-11
Ionization Caused by Radioactive Emissions

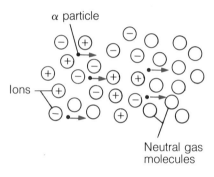

(a) Alpha particles cause intense ionization in a gas.

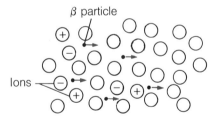

(b) Beta particles ionize a gas much less than alpha particles do.

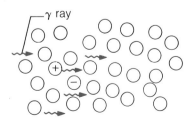

(c) Gamma rays ionize a gas even less than beta particles.

PRACTICE

10. **Activity** Design a way of exposing photographic paper, roll film, or instant film using a radioactive source. If you use photographic paper or roll film you will have to keep it and the source in the dark until the developing process is complete. If you use an instant film (such as Polaroid), you should leave the source on the film for a relatively long time because of the film's protective covering.

11. Set up a table summarizing the properties of X rays, α particles, and β particles. Include such criteria as speed, electric charge, ability to penetrate matter, and composition, as well as an example of the origin or source of each.

12. Research and report on the life of Marie Curie and/or her husband, Pierre. In your research, try to determine whether or not the time the Curies spent working with radioactive materials had any effect on their health.

13. Research and report on the way in which the following terms relate to radioactivity:
    (a) positron
    (b) neutrino, antineutrino
    (c) particle, antiparticle
    (d) carbon-14 decay

22.4 Detecting Radioactive Emissions

Since radioactive emissions (α, β, and γ) cannot be detected by any of our senses, what evidence proves their existence? One proof is that they expose photographic film, the characteristic of radioactive emissions which helped Becquerel discover them. A second proof is that when radioactive emissions strike certain crystals, they cause the crystals to glow or fluoresce. A third, important proof is that radioactive emissions cause particles in their path to become ionized, or electrically charged. (Ions were described in Section 17.2.) An ion pair, consisting of one negative ion (often one electron) and one positive ion, is created when a high-energy emission collides with a particle in its path and gives some of its energy to that particle (Figure 22-11). The ways in which ionized particles help detect radiation are described below for devices that measure radiation either qualitatively or quantitatively.

Radioactivity occurs naturally and spontaneously, not only in specific mineral samples in a science laboratory, but also in numerous other sources such as the sun, all plants and animals, and objects made of organic material. One process whereby living organisms become radioactive begins when high-energy cosmic rays from the sun strike our atmosphere. One of the results is the production of radioactive carbon, which joins with oxygen to become carbon dioxide. Plants absorb this radioactive carbon dioxide just as they absorb ordinary carbon dioxide. Thus, plants and the animals that eat plants, as well as materials made from plants, contain small amounts of natural radioactivity.

Radiation originating from natural sources is called **background radia-**

tion. When scientists want accurate readings of radioactive emissions from a source in the laboratory, they first measure the average background radiation, then subtract that amount from the recorded activity of the source.

Qualitative Methods of Detecting Radioactive Emissions

Since radioactive emissions cause nearby particles to become ionized, one device used to detect emissions is a charged electroscope. The radioactive source ionizes the air molecules near the electroscope, and the ions that have a charge opposite to the charge on the electroscope are attracted to the electroscope, causing it to become neutralized more rapidly. This process is shown in Figure 22-12.

Another common device used in schools and research laboratories to detect radioactivity is called a **cloud chamber**. One type of cloud chamber consists of a sealed container with a radioactive source and a layer of alcohol resting on a cold surface. The cloud chamber in Figure 22-13 uses solid carbon dioxide (dry ice) to maintain a cold temperature. As the alcohol evaporates, alcohol vapour (the "cloud") forms in the chamber. The cold temperature causes this vapour to be saturated and ready to condense when disturbed by radioactive emissions.

The radioactive emissions from the source cannot be seen directly in the alcohol vapour. However, as they travel through the vapour, they cause some of the vapour particles to become ionized. The ions attract other vapour molecules, which soon condense and form tracks of visible droplets. The effect is somewhat comparable to the vapour trail left by a jet aircraft high in a clear sky. Figure 22-14 shows a photograph of some typical tracks in a cloud chamber.

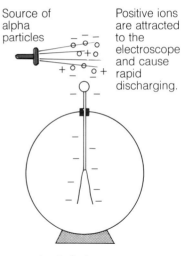

Figure 22-12
Discharging an Electroscope with a Radioactive Source

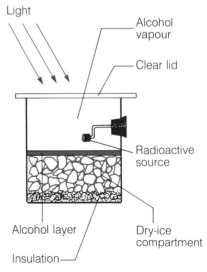

Figure 22-13
The Design of a Simple Cloud Chamber

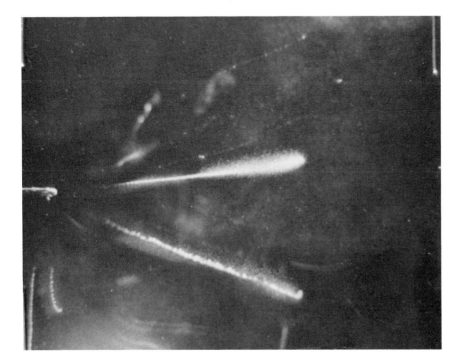

Figure 22-14
Vapour Trail Tracks Seen in a Cloud Chamber

The vapour tracks in a cloud chamber have various shapes, depending on what caused them. α particles are heavy and relatively slow-moving, so their tracks are relatively short and wide. β particles are light and very fast-moving, so their tracks are longer, thinner, and not necessarily straight. γ rays are high-energy electromagnetic radiations which do not directly produce tracks in a cloud chamber.

Other, more complex types of chambers are used in scientific laboratories. Physicists use the tracks observed in these chambers to study how particles interact. Characteristics of newly-discovered particles can be determined by applying known conservation laws, such as the law of conservation of energy. For example, if particle A decays into particles B and C, the sum of the energies of B and C equals the original energy of A. Figure 22-15 shows a typical cloud chamber photograph taken in a research laboratory.

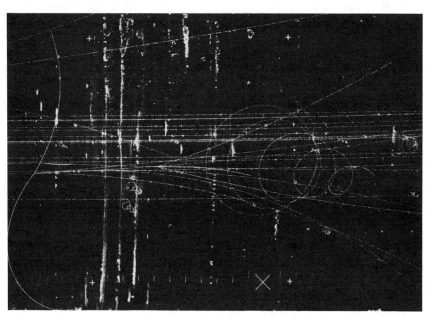

Figure 22-15
This photograph is one of millions taken in the hydrogen bubble chamber at the Stanford Linear Accelerator in California. Positively charged particles spiral in one direction, and negatively charged particles spiral in the opposite direction. Collisions are analysed to determine the characteristics of the particles.

Quantitative Methods of Detecting Radioactive Emissions

When quantitative measurements of radioactivity are made, the unit used is the becquerel (Bq), named after Henri Becquerel.

1 Bq = 1 emission/s

Thus, an activity of 2.5 Bq is the same as 2.5 emissions per second or 150 emissions per minute.

One becquerel is a very low rate of activity, so the kilobecquerel (kBq or 10^3 Bq) and megabecquerel (MBq or 10^6 Bq) are often used. For example, a patient undergoing diagnostic tests for a kidney disorder may have diluted radioactive iodine with an activity of 7 MBq injected into the bloodstream. This value is low compared to the extremely high activity of radium, one of the substances isolated from uranium by the

Did You Know?
The unit of radioactive emissions, the becquerel, is not appropriate when measuring the effect of radiation on the human body. Thus, a different unit of measurement, called the gray (Gy) was introduced in 1975. When a radiation beam gives 1.0 J of energy to 1.0 kg of live tissue, the *absorbed dose* of radiation is 1.0 Gy. Thus, 1.0 Gy = 1.0 J/kg. Radiation treatment of cancer typically involves absorbed doses in excess of 40 Gy. The gray is named after a British medical physicist, Harold Gray.

Curies. A 1.0 g sample of radium has an activity of 3.7×10^4 MBq, roughly three million times more active than the same amount of uranium.

You may have seen movies in which someone uses a clicking device to check for radioactivity levels in a hospital or nuclear generating station. This device is called a **Geiger-Müller tube**, after its inventors. (It is also called a *Geiger counter*. Don't confuse it with a metal detector, which emits similar clicking sounds.) The design of the tube is shown in Figure 22-16. A fine conducting wire is stretched down the middle of the partially evacuated tube and connected to a positive electric potential. The tube wall, which consists of copper, is connected to the negative electric potential. When a high-energy radioactive emission passes through the tube, it ionizes some of the gas molecules in the tube, freeing at least one electron, which is attracted toward the positive wire. Along its way, the electron causes more ionization by colliding with the gas molecules, and soon the resulting negative ions strike the wire. A current results which is detected and sent to a counter or loudspeaker.

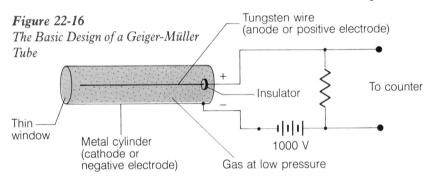

Figure 22-16
The Basic Design of a Geiger-Müller Tube

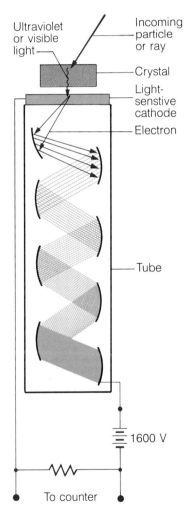

Figure 22-17
Scintillation Detector

Gamma rays are difficult to detect because they do not cause particles to ionize nearly as readily as α and β particles. This problem is overcome with a **scintillation detector**, which consists of a specially designed crystal (sodium iodide) and a series of metal electrodes, as shown in Figure 22-17. When a ray (or other radioactive emission) strikes the detector, it loses its energy to the crystal, which in turn emits light at a much lower energy than the original ray or particle. (This is another example of fluorescence.) This light then falls on the first electrode, which emits a few electrons toward the second electrode. The second electrode emits several electrons that strike the third electrode, and so on. Thus, the electron current is multiplied until a count can be made. A scintillation detector is more sensitive than a Geiger-Müller tube, because the crystal has a much higher density than the gas in the tube.

A recently-developed radiation detection device developed is the **semiconductor detector**. It consists of an electronic device called a junction diode, with the junction separating two different elements. An ionizing particle passing through the junction causes electric charges to be released. These charges produce short electric pulses that can be counted.

PRACTICE
14. Describe the basic principle upon which many of the detectors of radioactive emissions, among them the cloud chamber, operate.

ATOMIC PHYSICS

CAUTION:
Use extreme care when handling a radioactive source. Suggestions are given beside Practice Question 10 in the previous section.

CAUTION:
When handling dry ice, which is solid carbon dioxide at a very low temperature, use gloves or tongs. When handling a radioactive source, exercise care as suggested beside Practice Question 10.

CAUTION:
Only the teacher should handle the sources suggested for this activity.

15. **Activity** Design and perform an experiment to determine whether or not a known radioactive source can be used to neutralize a charged electroscope. Does the result depend on the type of charge on the electroscope or the type of radiaoctive source?

16. **Activity** Design and carry out an experiment to view the tracks created by a radioactive source in a cloud chamber. The chamber should have dry ice in the lower compartment, a level layer of alcohol in the upper compartment, a clear lid, and a radioactive source. Your report should include diagrams and descriptions of the tracks observed. (If very few tracks can be seen, try rubbing the container lid with a dry cloth.)

17. To prepare for a thyroid scan, a patient swallows a solution containing 4.0 MBq of radioactive iodine. Determine the number of emissions that come from the source in (a) 1.0 min and (b) 1.0 h.

18. Calculate the activity (in becquerels and kilobecquerels) of sources with the following:
    (a) 4300 emissions in 2.6 s
    (b) 3.8×10^6 emissions in 40 s
    (c) 4.2×10^3 emissions in 1.0 min

19. A 20 g sample of a substance thought to be radium is found to have an activity of 5.4×10^{11} Bq. Is the sample composed of pure radium? Explain your reasoning.

20. **Activity** Observe and describe a teacher demonstration to determine how certain factors affect the amount of radiation detected by a Geiger-Müller tube. The radiation is provided by a variety of sources such as strontium-90 (which undergoes β decay) and cesium-157 (which undergoes γ decay). The Geiger-Müller tube is connected to an electronic counter. Possible factors include background radiation, type of source, distance from the source, and type and thickness of absorbing material placed in front of the source. (Suggested absorption materials are paper, aluminum, and lead.) Can you think of other factors?

21. Research and report on one or more of the following radiation detectors:
    (a) Wilson cloud chamber (also called an expansion cloud chamber, originally designed by a British physicist, C.T.R. Wilson, 1869–1959)
    (b) bubble chamber (used to obtain Figure 22-15)
    (c) spark counter

22.5 Models of Atoms

With the discovery of X rays and radioactivity near the end of the nineteenth century, the concepts of classical physics had to be altered. There was no longer any doubt that atoms could undergo changes, and scientists plunged quickly into doing more research into the properties of atoms.

ATOMS AND RADIOACTIVITY

Thomson's Atomic Model

In 1897, an Englishman named Sir Joseph (J.J.) Thomson (1856–1940) (Figure 22-18) was performing experiments with cathode-ray tubes. It was he who discovered that cathode rays consist of extremely small particles all having the same mass and negative charge. Thus, cathode rays are fast-moving electrons.

Thomson reasoned that, since the electrons had come from atoms at the cathode which were originally neutral, the atoms must be divisible and must have a positive charge to neutralize the negative charge. He developed these concepts into a model of the atom in which negatively charged electrons were imbedded in a positively charged cloud, as shown in Figure 22-19. The model showed the whole atom in a state of equilibrium, with the negative electrons repelling each other but attracted to the positive regions.

Rutherford's Atomic Model

At the same time as Thomson was developing his model of the atom, one of the greatest experimental physicists of that era was designing investigations to delve further into atomic structure. Ernest Rutherford (1871–1937) (Figure 22-18) was born and educated in New Zealand and did most of his research in Canada and England. Between 1898 and 1907 he was at McGill University in Montréal, where he proved the existence of the three different modes of radioactive decay (α, β, and γ). It was α particles that played the major part in Rutherford's most famous experiment, the so-called "gold foil experiment", which he designed when he went to Manchester University in England.

In this important experiment, performed in 1911, Rutherford aimed positively charged α particles toward a thin sheet of gold foil. The paths of the α particles were determined by using a screen that could detect them, as shown in Figure 22-20(a). If the Thomson model of the atom were correct, the α particles would go straight through the foil or be deflected by very small amounts. Rutherford discovered that almost all the α particles did go straight through the foil, and some were deflected slightly to the sides. However, there were some which were deflected at very large angles, and a few which were actually reflected toward the source, as in Figure 22-20(b). Rutherford concluded that the major por-

Figure 22-18
J. J. Thomson, on the left, and Ernest Rutherford, two pioneers in the understanding of the atom

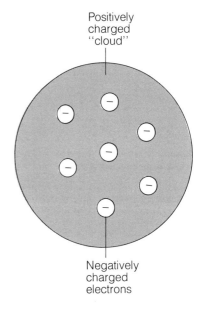

Figure 22-19
Thomson's Model of the Atom

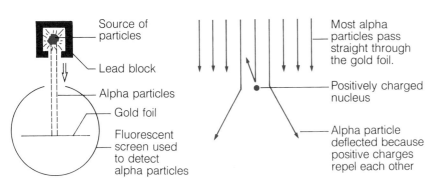

(a) *Experimental set-up*   (b) *Observed results*

Figure 22-20
Rutherford's Gold Foil Experiment

549

ATOMIC PHYSICS

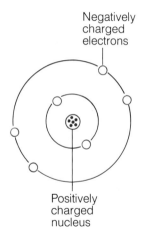

Figure 22-21
Rutherford's Atomic Model (not drawn to scale)

Figure 22-22
Niels Bohr (1885–1962)

Did You Know?
The Royal Society of Canada awards scientists the prestigious Rutherford Memorial Medal in Physics for outstanding research in any branch of physics. In 1985, Professor John Simpson of the University of Guelph won the award for his research to determine the mass of the particles called *neutrinos*.

tion of an atom is empty space, and the positively charged part of each atom must be concentrated in a small area, which he called the nucleus of the atom. Thomson's earlier model of the atom therefore needed to be modified to fit Rutherford's findings. The newer model resembled a tiny planetary system, as Figure 22-21 demonstrates.

The results of Rutherford's experiments also allowed him to compare the size of a nucleus to the size of an entire atom. The diameter of an atom was known to be approximately 10^{-10} m, which means it would take 10 million atoms lined up side by side to equal 1 mm. Rutherford estimated that the diameter of the nucleus was only about 1/10 000 of that of the atom, or about 10^{-14} m.

The Bohr Model of the Atom

The excitement generated by the discoveries in the early part of this century continued to grow as each new discovery created more problems to be solved. One of the important problems with the planetary model of the atom was that it did not work according to classical mechanics. The classical idea was that, as the electrons moved around the central nucleus, they would continually emit energy in the form of light. As their energy was being emitted, the electrons would spiral in toward the nucleus, and the atom would collapse. But atoms did not give up their energy or collapse, so a new theory had to be developed.

The scientist responsible for the next stage of solving the atomic puzzle was Niels Bohr (1885–1962) (Figure 22-22), a brilliant physicist from Denmark who began working with Rutherford in 1912. By 1913 he had developed a model of the atom which took into consideration the fact that atoms do not collapse when they emit energy. Bohr's atomic model, shown in Figure 22-23, made the following assumptions:

- Electrons travel in circular orbits around the nucleus of an atom.
- Each electron orbit contains a certain maximum number of electrons.
- Each orbit has a specific energy level associated with it.
- Atoms emit certain quantities of radiant energy when electrons move from one energy level to a lower one. These quantities of energy are called "quanta", which is the plural of "quantum". They are the basis of the theory within modern physics called *quantum mechanics*.

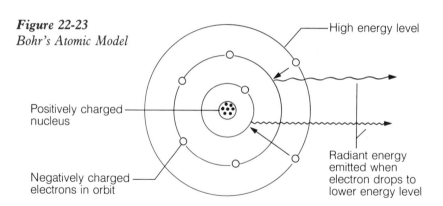

Figure 22-23
Bohr's Atomic Model

More Recent Atomic Models

By 1920, scientists were aware that an atom contains electrons travelling around a nucleus and that the nucleus contains positively charged protons and some other particles. Those particles were finally isolated in 1932 by James Chadwick (1891–1974), an English physicist (Figure 22-24). The particles are electrically neutral, so they are called "neutrons". The atomic model of oxygen based on the theories of 1932 is shown in Figure 22-25(a).

Since the 1930s, scientists have believed that electrons travel in regions, not in rigid orbits. Thus, they now often use a simplified model of the atom like the one shown in Figure 22-25(b). The electron arrangement in the regions closest to the nucleus was discussed in Section 17.2.

Figure 22-24
James Chadwick (1891–1974)

PRACTICE

22. Upon what evidence did J. J. Thomson base his hypothesis that an atom contains a positive charge?
23. **Activity** Figure 22-26 shows an apparatus called a "potential hill" which is used to simulate Rutherford's gold foil experiment. Design and perform an experiment in which a marble (or a steel ball) acts like an α particle and the potential hill acts like an atom of gold. Describe how your experiment relates to Rutherford's.

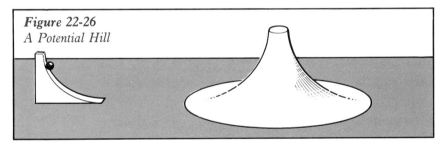

Figure 22-26
A Potential Hill

24. Four sponge cakes are baked, each with a solid metal object of unknown size and shape hidden inside. An experimenter shoots 8 round bullets at each cake and observes the paths shown in Figure 22-27 on page 552.
    (a) Describe what the experimenter is able to conclude about each cake.
    (b) Relate this experiment to Rutherford's gold foil experiment.
25. According to Rutherford's estimations, what is the ratio of the diameter of an atom to the diameter of its nucleus? Using this information, calculate the ratio of the volume of an atom to the volume of a nucleus, assuming that both are spherical. What percentage of an atom is thus empty space?
26. Research and report on the life and contribution to physics of one or more of the following:
    (a) Sir Joseph Thomson
    (b) Ernest Rutherford
    (c) Hans Gieger
    (d) Niels Bohr
    (e) James Chadwick

Figure 22-25
Models of an Oxygen Atom

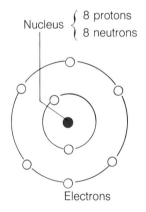

(a) A model used after the discovery of neutrons

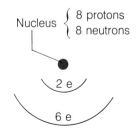

(b) A present-day model

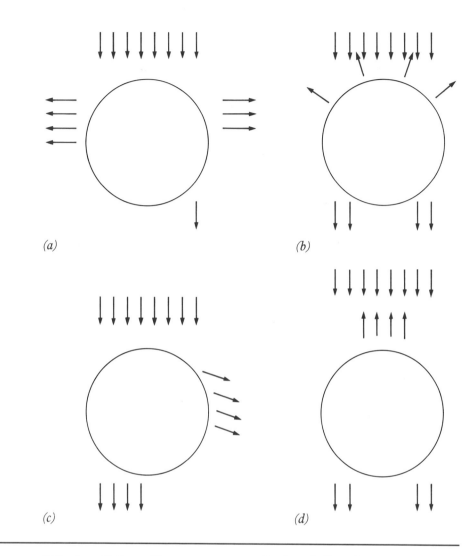

Figure 22-27

22.6 The Structure of the Nucleus

You can see that the model of the atom has changed drastically over the past century. Even today, scientists are not completely sure about all aspects of the structure of an atom. For example, they know that the nucleus is a tiny, relatively dense entity, but they admit there is much to be learned about the forces binding the nuclear particles together. Nevertheless, there are some important scientific terms that recur in discussions of the nucleus.

Because the structure of the nucleus is closely related to the concept of elements, the facts known about elements should now be reviewed and extended. An element is a substance that cannot be changed into other substances by ordinary chemical means. The periodic table of the elements, shown in Appendix B, lists the 92 naturally occurring elements, hydrogen to uranium, as well as the artificial elements, neptunium to hahnium. The elements are listed by atomic number. The **atomic number** (symbol Z) is, as we stated earlier, the number of protons in

ATOMS AND RADIOACTIVITY

the nucleus of each atom. It is also equal to the number of electrons in the atom when the atom is neutral. The atomic number is what determines an element; for example, every atom with atomic number 7 is an atom of nitrogen.

Another important feature of an element is its **atomic mass**, which is the average mass of the atoms of the element. This mass is so small in terms of kilograms that it is usually more convenient to express it in terms of a standard atomic mass. This standard mass, called an **atomic mass unit** (u), is based on a common carbon atom having 6 protons, 6 neutrons, and 6 electrons, and a mass of 1.992×10^{-26} kg. One atomic mass unit, or $1.000\ u$, is exactly 1/12 of the mass of this carbon atom; in other words, $1.000\ u = 1.660 \times 10^{-27}$ kg. A hydrogen atom with 1 proton and 1 electron has an atomic mass of $1.008\ u$, or 1.673×10^{-27} kg. The electron's mass is virtually insignificant, being only about 1/1800 of that of the proton or neutron.

An element's **atomic mass number** (symbol A) is the sum of the number of neutrons (symbol N) and protons in each nucleus. For instance, an atom of oxygen with 8 protons and 8 neutrons has an atomic mass number of 16. Table 22-1 lists several common elements, their chemical symbol, atomic number, atomic mass, and atomic mass number.

Table 22-1 Atomic Data for Some Common Elements

Element	Symbol	Atomic Number	Atomic Mass (u)	Atomic Mass Number (for the most common isotope)
Hydrogen	H	1	1.008	1
Helium	He	2	4.003	4
Lithium	Li	3	6.939	7
Carbon	C	6	12.011	12
Nitrogen	N	7	14.007	14
Oxygen	O	8	15.999	16
Fluorine	F	9	18.998	19
Sodium	Na	11	22.989	23
Magnesium	Mg	12	24.312	24
Aluminum	Al	13	26.982	27
Sulphur	S	16	32.064	32
Chlorine	Cl	17	35.453	35
Iron	Fe	26	55.847	56
Copper	Cu	29	63.546	64
Silver	Ag	47	107.868	108
Uranium	U	92	238.03	238

Since $A = Z + N$, it is possible to find the number of neutrons in the nucleus of an atom of known mass number and atomic number by rearranging the equation (*i.e.*, $N = A - Z$).

SAMPLE PROBLEM 1
What is the atomic mass number of an atom of carbon that has 7 neutrons?

Solution

$A = Z + N$, where $Z = 6$ for carbon
$ = 6 + 7$
$ = 13$

Thus, the atomic mass number is 13.

SAMPLE PROBLEM 2
Determine the number of neutrons in the nucleus of an atom of aluminum with an atomic mass number of 27.

Solution

$N = A - Z$, where $Z = 13$ for aluminum
$ = 27 - 13$
$ = 14$

∴ There are 14 neutrons in the nucleus of this atom.

Notice in Table 22-1 that for some elements the atomic mass number is very close to the atomic mass (see fluorine), but for other elements the numbers are quite different (see copper). This discrepancy results from taking an average of the atomic masses of the various isotopes of the elements. Two substances are **isotopes** if they have the same atomic number (Z) but a different atomic mass number (A). In other words, they have the same number of protons but a different number of neutrons in each atom. (The word "isotope" is derived from the Greek words *iso*, "equal" and *topos*, "place".) For example, most sulphur atoms have 16 protons and 16 neutrons, so their mass number is 32. However, some sulphur atoms have 16 protons and 19 neutrons, so their mass number is 35. The average mass number, weighted in accordance with their relative abundance in nature, is 32.064 – a number closer in value to the more common isotope. What does this tell you about the abundance of each of the isotopes of sulphur?

Two distinct symbols are used to describe isotopes of elements. In one symbol, the atomic number (Z) and the mass number (A) are shown in the arrangement $^{A}_{Z}X$, where X is the chemical symbol of the element (*e.g.*, $^{32}_{16}S$ and $^{35}_{16}S$). In the other symbol, the name or symbol of the element is followed by the mass number (*e.g.*, sulphur-32 or S-32, and sulphur-35 or S-35).

Table 22-2 lists isotopes of the elements hydrogen, carbon, and uranium. The isotopes which are radioactive are called **radioisotopes**. The next section deals with them in more detail.

SAMPLE PROBLEM 3

Write the symbol for each of the following isotopes in the form $^A_Z X$, and determine the number of neutrons in each atom.

(a) uranium-238 (b) uranium-235

Solution

(a) The atomic number (Z) of uranium is 92, so the symbol for U-238 is $^{238}_{92}U$. The number of neutrons is

$N = A - Z$
$ = 238 - 92$
$ = 146.$

∴ There are 146 neutrons.

(b) The symbol is $^{235}_{92}U$, and the number of neutrons is

$N = A - Z$
$ = 235 - 92$
$ = 143.$

∴ There are 143 neutrons.

Table 22-2 Isotopes of Three Common Elements

Element	Name	Symbol	Comment
Hydrogen	hydrogen	^1_1H	most abundant form of H
	deuterium	^2_1H	
	tritium	^3_1H	radioactive
Carbon	carbon-11	$^{11}_6C$	radioactive
	carbon-12	$^{12}_6C$	most abundant form of C
	carbon-13	$^{13}_6C$	
	carbon-14	$^{14}_6C$	radioactive
Uranium	uranium-232	$^{232}_{92}U$	radioactive
	uranium-233	$^{233}_{92}U$	radioactive
	uranium-235	$^{235}_{92}U$	radioactive
	uranium-236	$^{236}_{92}U$	radioactive
	uranium-238	$^{238}_{92}U$	radioactive; most abundant form of U
	uranium-239	$^{239}_{92}U$	radioactive

PRACTICE

27. For each of the elements listed below, state the number of protons and the number of neutrons in the nucleus of one atom. Refer to Table 22-1.

    (a) hydrogen-1 (b) aluminum-27 (c) silver-108

28. Carbon-14 and nitrogen-14 have the same atomic mass numbers, yet they are not the same chemical substance. Explain.
29. Calculate the number of neutrons in each isotope of carbon listed in Table 22-2.

22.7 Transmutations

For centuries, early researchers called *alchemists* searched for ways of chemically changing various substances into other substances, especially into gold. Little did they know that some elements change naturally, through radioactive decay or nuclear reactions rather than chemical reactions.

The changing of one element into another is called a **transmutation**. You studied two types of natural transmutations, α decay and β decay, earlier in this chapter. Now that you know about the symbols and mathematical relationships involving the nucleus, you can learn how to analyse these transmutations in more detail. (Other changes in elements, involving artificial transmutations and nuclear reactions, are described in the next chapter.) When Henri Becquerel discovered radioactivity, he actually observed a naturally occurring transmutation. All radioactivity transmutations result in the emission from the nucleus of some particle, such as an α particle or a β particle. The transmutation is called an α decay if an α particle is emitted and a β decay if a β particle is emitted. Some transmutations are also accompanied by the emission of a γ ray.

Alpha Decay

An α decay transmutation occurs for large nuclei which are unstable. The original nucleus, the **parent nucleus**, decays to a nucleus with a smaller atomic number and mass number called a **daughter nucleus**, with the release of an α particle. Since an α particle is the nucleus of a helium atom, its symbol in an nuclear decay equation is $^{4}_{2}\text{He}$.

An example of α decay occurs when the substance radium-226 decays into radon-222 and an α particle, helium-4, as shown in Figure 22-28. The nuclear decay equation for this decay is

$$^{226}_{88}\text{Ra} \rightarrow {}^{222}_{86}\text{Rn} + {}^{4}_{2}\text{He}.$$

Notice in this example that the total mass number is conserved (226 = 222 + 4), as is the total atomic number (88 = 86 + 2). This is true of all nuclear decays.

Figure 22-28
The Alpha Decay of Radium-226

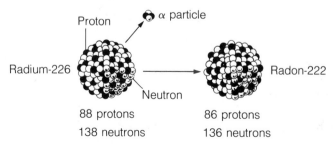

SAMPLE PROBLEM 4
Write the nuclear decay equation that represents the α decay of uranium-238 into thorium-234.

Solution
The atomic numbers of uranium and thorium are, respectively, 92 and 90. Thus, the equation is

$^{238}_{92}\text{U} \rightarrow \,^{234}_{90}\text{Th} + \,^{4}_{2}\text{He}.$

Beta Decay

Beta decay transformations also occur in large nuclei. In this case a high-speed electron (the β particle) is emitted from the nucleus, leaving behind an extra positive charge. The reaction can be thought of as a neutron emitting an electron and becoming a proton, although the modern explanation is not that simple. Since a β particle is an electron with a negligible mass and a charge of -1, its symbol in a nuclear decay equation is $_{-1}^{0}\text{e}$. The daughter nucleus has the same mass number as the parent nucleus, but a higher atomic number.

An example of a β decay occurs when thorium-234 decays into protactinium-234, with the release of a β particle, as shown in Figure 22-29. The nuclear decay equation for this decay is

$^{234}_{90}\text{Th} \rightarrow \,^{234}_{91}\text{Pa} + \,_{-1}^{0}\text{e}.$

Notice again in this example that both the total mass number ($234 = 234 + 0$) and the total atomic number ($90 = 91 + -1$) are conserved. (In order for energy and momentum to be conserved in a beta decay as well, another particle, a neutrino, must be emitted. However, the neutrino is of no importance to this discussion.)

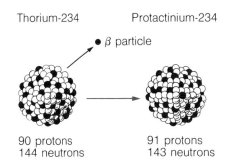

Figure 22-29
The Beta Decay of Thorium-234

SAMPLE PROBLEM 5
Write the nuclear decay equation that represents the β decay of protactinium-234 into uranium-234.

Solution
The atomic numbers of protactinium and uranium are 91 and 92 respectively. Thus, the equation is

$^{234}_{91}\text{Pa} \rightarrow \,^{234}_{92}\text{U} + \,_{-1}^{0}\text{e}.$

Radioactive Decay Series

Sometimes the daughter nucleus of a transmutation decays in its turn, creating yet another isotope. When this occurs, the result is a radioactive **decay series** which can involve α decays, β decays, or both. The series ends when a stable isotope is reached. One example of a decay

ATOMIC PHYSICS

series is shown in Figure 22-30. Here, the parent nucleus is uranium-238 and the final, stable nucleus is lead-206. Notice that the four decay examples (two α and two β) described previously in this section are included in the uranium decay series in Figure 22-30.

Figure 22-30
A Radioactive Decay Series Starting with Uranium-238

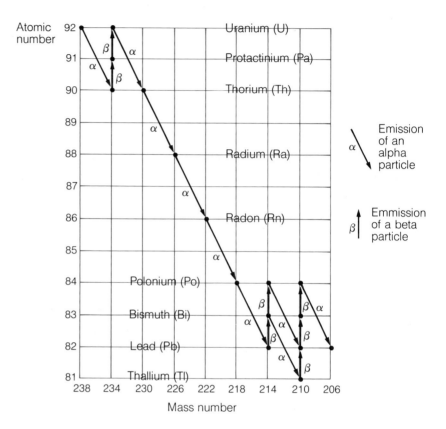

SAMPLE PROBLEM 6
Use the information in Figure 22-30 to write the nuclear decay equations for the following decays:
(a) from bismuth-214 to thallium-210
(b) from lead-210 to bismuth-210

Solution

(a) The transmutation of Bi-214 to Tl-210 involves an α decay.

$$^{214}_{83}\text{Bi} \rightarrow {}^{210}_{81}\text{Tl} + {}^{4}_{2}\text{He}$$

(b) The transmutation of Pb-210 to Bi-210 involves a β decay.

$$^{210}_{82}\text{Pb} \rightarrow {}^{210}_{83}\text{Bi} + {}^{0}_{-1}\text{e}$$

The particles, as well as any γ rays emitted during α and β decays, have relatively large amounts of energy. That energy can be both useful and harmful, as you will learn later in this chapter.

PRACTICE

30. State the mode of decay (α or β) that occurs in each transmutation listed below. See Figure 22-30.
    (a) from thallium-210 to lead-210
    (b) from uranium-234 to thorium-230
31. Use the information in Figure 22-30 to write the radioactivity decay equations for these transmutations:
    (a) from thorium-230 to radium-226
    (b) from bismuth-210 to polonium-210
32. Copy the following equations into your notebook and complete the missing information. See Figure 22-30.
    (a) $^{?}_{84}\text{Po} \rightarrow {}^{214}_{?}? + {}^{?}_{2}\text{He}$
    (b) $^{222}_{86}? \rightarrow {}^{?}_{84}\text{Po} + {}^{?}_{2}?$
    (c) $^{214}_{?}\text{Pb} \rightarrow {}^{?}_{83}? + {}^{?}_{-1}e$

22.8 Half-Life

As the parent nuclei of a radioactive isotope decay, they become the nuclei of a different element, the daughter element. The average length of time for half of the parent nuclei in a given sample to decay is called the **half-life** of the radioactive isotope. After one half-life of a certain sample of the isotope has elapsed, the activity or rate of emissions is half the original rate, because only half of the original number of particles remain. (Notice that the half-life does not indicate which particular nucleus will decay; rather, it indicates the statistical chances of decay.) Half-lives can vary from a small fraction of a second to billions of years.

Consider, for example, an isotope with a half-life of eight days. Assume that a certain sample gives off radioactive emissions at a rate of 1000 Bq. Eight days later, the activity will be half the original, 500 Bq. After another eight days, the activity will be 250 Bq, and so on, as shown in Figure 22-31(a). When plotted on a graph of activity as a function of time, this half-life example yields a smooth curve, as Figure 22-31(b) demonstrates.

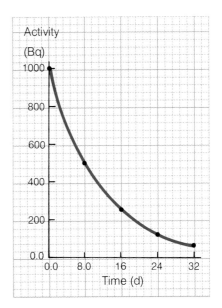

Figure 22-31
Graphing Half-life Activity

Time (d)	0.0	8.0	16	24	32
Activity (Bq)	1000	500	250	125	62.5

(a) Data of the activity of a radioactive isotope having an average half-life of 8.0 days

(b) Activity graph (half-life = 8.0 days)

ATOMIC PHYSICS

Experiment 22A: Simulation of Half-Life

INTRODUCTION

In this experiment, radioactive decay is simulated by means of cubes representing either the parent or daughter nuclei. A parent nucleus is represented by a cube having no dot on the side facing upward, and a daughter nucleus is represented by a cube with a dot on the side facing upward.

Because this experiment involves chance, statistics, and a relatively low number of trials, you should not expect the results to be highly accurate. However, there are ways of improving the accuracy; try to think of them as you perform and later analyse the experiment.

PURPOSE

To determine the average half-life of events that simulate nuclear decay reactions.

APPARATUS

large, open box made of cardboard or other suitable material; sets of 100 cubes made of sugar, wood, or other appropriate substance, and categorized as follows:

Set A – Each cube has a coloured dot on three sides.
Set B – Each cube has a coloured dot on two sides.
Set C – Each cube has a coloured dot on one side.

PROCEDURE

1. Set up a table of observations based on the one shown below. Predict the half-life (in shakes of the box) of each of the three sets of cubes.

Set	Predicted Half-life (in shakes)	Shake Number	Number of Parent Nuclei Remaining	Number of Daughter Nuclei Removed
A		0	100	0
		1		
		2		

2. Place the cubes of one set into the box and shake the box thoroughly. Count the number of daughter cubes (those with a dot facing upward). Repeat this five times using the 100 cubes and find an *average* of the five trials. Remove this average number of daughter nuclei from the box. Record the number of parent nuclei remaining and the number of daughter nuclei removed in your table of observations.
3. Repeat Procedure Step 2 with the remaining cubes. Continue doing this until only one or two cubes remain. Complete your table of observations for the first set of cubes.
4. Repeat Steps 2 and 3, using the other two sets of cubes available.

ANALYSIS

1. Beginning with set C, plot a graph of the parent nuclei remaining as a function of the shake number. Draw a smooth curve of best fit on the graph. Relate this curve to the half-life curve shown on the graph in Figure 22-31(b).
2. From your graph, determine the half-life in shakes of the parent nuclei in your experiment.
3. Repeat the Analysis Questions 1 and 2 for sets B and A.
4. To try to improve the validity of your results, borrow the data from other groups in the class and plot the data on your graph. What can you conclude?

CONCLUSIONS . . .

QUESTIONS

1. How could you improve the accuracy of the statistical results in this experiment?
2. Assume that objects with ten sides (rather than six) were used to perform a half-life simulation experiment. Predict the half-life (in shakes of the parent nuclei) if dots were placed on
   (a) five sides   (b) two sides   (c) one side.
3. A certain sample of a radioactive substance has an activity of 440 kBq at the start of an experiment. If the half-life of the substance is 2.0 h, calculate the activity at the following times:
   (a) 2.0 h   (b) 4.0 h   (c) 6.0 h

22.9 Uses of Radioactivity

Radioactive substances have many uses, some of which are described below. The choice of substance for a particular use is determined by the mode of decay and the half-life of the substance. Table 22-3 lists the half-lives of several radioactive isotopes.

Table 22-3 Half-Lives of Common Radioactive Isotopes

Isotope	Symbol	Half-Life	Radiation Produced
hydrogen-3 (tritium)	$^{3}_{1}H$	12.3 a	β
carbon-14	$^{14}_{6}C$	5730 a	β
sodium-24	$^{24}_{11}Na$	15.0 h	β, γ
cobalt-60	$^{60}_{27}Co$	5.27 a	β, γ
iodine-131	$^{131}_{53}I$	8.04 d	β, γ
lead-212	$^{212}_{82}Pb$	10.64 h	β, γ
polonium-210	$^{210}_{84}Po$	138 d	α, γ
uranium-235	$^{235}_{92}U$	7.04×10^8 a	α, γ
uranium-238	$^{238}_{92}U$	4.45×10^9 a	α, γ
americium-243	$^{243}_{95}Am$	7.37×10^3 a	α, γ

ATOMIC PHYSICS

Radioactive Dating

The process of using the half-life of a radioactive substance to find the age of an organism or object that was once living is called **radioactive dating**. It is a valuable aid to archaeologists, geologists, and historians.

Radioactive dating is possible because all living plants absorb carbon from the carbon dioxide they take in from the atmosphere. The isotopes carbon-12 ($^{12}_{6}C$) and radioactive carbon-14 ($^{14}_{6}C$) act the same way chemically. Plants therefore cannot distinguish between carbon-12 and carbon-14 and will absorb both isotopes.

Consider, for example, the problem of determining the age of a wooden bowl found at a burial site of an ancient civilization. During its lifetime, the tree from which the bowl was carved absorbed both carbon-12 and radioactive carbon-14. After the tree died, its carbon-12 content remained the same, because the tree was no longer taking in carbon dioxide. However, its carbon-14 content gradually decreased as a result of radioactive emissions. After about 5730 years the carbon-14 emissions would be reduced to half. (The half-life of carbon-14 is 5730 years.) Thus, scientists can determine the bowl's age by measuring the ratio of ^{14}C to ^{12}C in a sample of the bowl. See Figure 22-32.

Carbon-14 dating can be used on any object made of matter that was once living. However, it has certain limitations. The ratio of ^{14}C to ^{12}C in the atmosphere has not been constant. Thus, tree rings of very old

Figure 22-32
Radioactive Dating Using Carbon-14

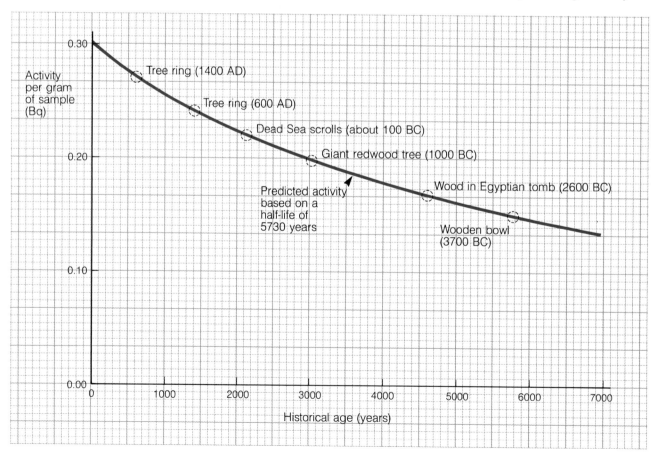

trees have been necessary for calibrating the radioactive dating method. Furthermore, the process is inaccurate for ages greater than about 30 000 years because the carbon-14 content becomes extremely small.

Radioactive dating can also be used to determine the approximate ages of rocks and mineral deposits. Uranium-238 ($^{238}_{92}U$), with a half-life of 4.5×10^9 years, has been used to determine that some rocks on the earth are about 4×10^9 years old.

Radioactive Tracers

Very small amounts of a radioactive substance injected into the liquid of a system have certain applications. As the isotope travels in the liquid through the system, it continually gives off emissions. These emissions allow a detector to trace the path of the isotope and thereby analyse the system. For example, to study kidney function, a radioactive sample is injected into the bloodstream. As the kidneys remove the radioactive substance from the blood, they begin to emit radiation which can be measured. The resulting data indicate whether or not the kidneys are functioning normally. Sodium-24 ($^{24}_{11}Na$), with a half-life of 15 h, is often used in biological systems such as the human body or plants to trace the flow of blood, food, or water. Iron-59 ($^{59}_{26}Fe$) can be used as a tracer in mechanical systems such as an engine to trace the flow of lubricating oil. See Figure 22-33.

Figure 22-33
This radioautograph of a fern frond is an example of the use of a radioactive tracer. Radioactive sulphur (S-35) is taken up by the plant and travels through the plant along with ordinary sulphur foods. When the plant is placed on a photographic film, the radiation emitted from the S-35 exposes the film with various intensities. In this case, the darkest part of the frond contains the highest concentration of S-35.

ATOMIC PHYSICS

Medical Sterilization and Therapy

Medical supplies must be sterilized to ensure that they are clean and germ-free. One way of sterilizing an object is to use γ radiation, a process which is now used on a large percentage of disposable medical supplies such as bandages and surgical equipment.

Radiation therapy or treatment of cancerous tumours is carried out using a radioactive source of γ rays, such as cobalt-60 ($^{60}_{27}$Co). Like X rays, γ rays travel through some cells but are absorbed by others. In radiation therapy, γ rays are aimed from several directions at a cancerous growth. If the radiation is successful, the cancer cells absorb the γ rays and are destroyed by them. See Figure 22-34.

Did You Know?
The first cobalt-60 therapy unit was made in Canada in 1951 by Harold Johns.

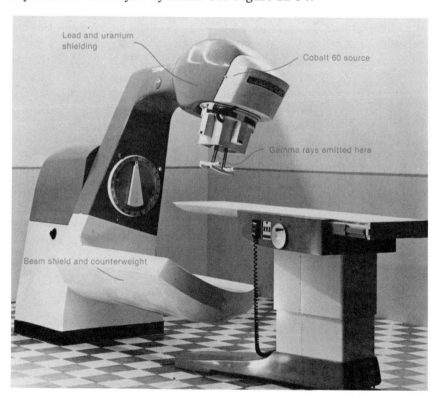

Figure 22-34
The cobalt-60 source in this machine emits γ rays that kill cancerous cells.

Industrial Applications

Radioactive emissions, especially γ rays, can be used to control the thickness of manufactured products such as paper, steel, and aluminum foil. If the product becomes thicker than desired, it absorbs more radiation, signalling the need for adjustments. (You will recall that X rays are used for the same purpose. However, now they are increasingly replaced by γ rays, which can penetrate more readily.) In a similar fashion, γ rays are used to detect flaws inside metal parts or welded joints. See Figure 22-35.

Ionization in Smoke Alarms

Many small smoke alarms in homes and apartments operate on the principle of ionization. The basic design of this type of alarm is shown in

Figure 22-35
Typical Industrial Applications of Radioactivity

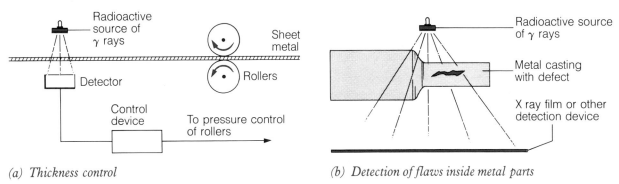

(a) Thickness control

(b) Detection of flaws inside metal parts

Figure 22-36. A radioactive source, americium-241, with a half-life of over 7000 years, emits α particles that ionize the air molecules within the chamber. These ions maintain a current in the chamber that is large enough to operate an electronic device which prevents the alarm from sounding. If smoke particles enter the chamber, the ionized air molecules attach themselves to the smoke particles, thereby reducing the current. The electric circuitry is designed in such a way that the alarm sounds when the current is reduced.

A smoke alarm requires a 9.0 V battery. When the battery becomes weak, the alarm starts to chirp at regular intervals.

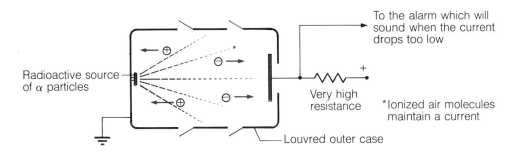

Figure 22-36
Ionization in a Smoke Alarm

Food Preservation

Much attention has been devoted recently to using γ radiation to help preserve certain food products. The process is used for providing familiar meals to astronauts aboard space flights. It is still being researched in laboratories throughout the world.

Food preservation using γ rays from a cobalt-60 or a cesium-137 source can be used for a variety of types of foods, including spices, grains, many vegetables, meat, poultry and fish. The γ rays ionize molecules, such as water molecules, within the food. The result is newly formed chemicals that react with proteins and other compounds in mould, bacteria, and insect larvae. This reaction either reduces the rate of growth of the harmful organism or, if the radiation is strong enough, kills it completely. See Figure 23-37 on page 566.

ATOMIC PHYSICS

Figure 22-37
This set of potatoes illustrates the effect of γ rays on food preservation. The photograph was taken 8.5 months after exposure to the γ rays. The potato at the upper left was not exposed to radiation, while the others were exposed to increasing dosages, to a maximum at the lower right. The best-preserved potato, at the lower left, was exposed to a medium dosage of radiation.

Did You Know?
One of the controversial aspects of using radiation to preserve food is finding means of telling the consumer which foods have been irradiated. Although irradiated food does not itself become radioactive, some people react negatively to buying, for instance, an apple labelled "Preserved with γ Radiation".

The advantages of food preservation using γ radiation are many. The energy required to refrigerate or freeze food is saved because irradiated food lasts for up to several years. Fewer chemical preservatives, some of which have harmful effects on humans, are required. World food shortages may be lessened through the prevention of spoilage, helping to solve the great hunger problem. Also, exotic foods from foreign sources can be made more readily available to North American consumers.

The disadvantages of using γ rays to preserve food are still uncertain. Research is being carried out to determine what chemical changes are produced by the γ rays. The nutritional value of the food may be reduced and, far more seriously, new chemicals that are harmful to humans may be created by the process.

PRACTICE

33. A certain sample of an ancient wooden log is compared with a sample of equal mass of wood from a newly-cut tree. Assume that the carbon-14 activity of the new sample is 600 Bq and the carbon-14 activity of the old sample is 150 Bq. What is the age of the old sample?
34. Assume that a 2000 Bq source of sodium-34, with a half-life of 15 h, is injected into a body. How long will it take the activity to decrease to 250 Bq? (Assume that none of the sodium is removed via urine, *etc.*)
35. Research and report on the modern use of γ radiation in one or more of these fields:
    (a) medicine   (b) industry   (c) agriculture.

22.10 Hazards of Radiation

The scientists who discovered and researched radioactivity around the beginning of the twentieth century did not at first realize the dangers of high-energy emissions (α, β, and γ) during radioactive decays. For instance, Marie Curie, who worked closely for several years with highly radioactive substances, contracted leukemia, a cancerous disease of the blood-forming organs. Today, much more is known about the health hazards of radiation, and everyone should be aware of their seriousness.

When a radioactive emission strikes a living cell, it causes certain molecules within the cell to become ionized. The ionized molecules may then prevent the cell from either growing or functioning normally, and may even prevent it from reproducing. Of course, bodily functions may deteriorate when normal cells are affected in this way. It is also possible for the ionized molecules to cause chemical changes in cells, so that they begin to multiply rapidly. This is one cause of the rapid reproduction of unwanted or damaged cells called **cancer**. Cancer which is not detected early or isolated in a single location is difficult to treat.

Long-term effects can also result from the ionization of molecules in living cells. If the cells are located in the reproductive organs, the damage can be passed on to the offspring; the effects on future generations are unpredictable.

It is obvious that radiation should be avoided unless exposure to it is absolutely necessary. People who work with or near radioactive sources take precautions to be sure they do not receive doses of radiation beyond the recommended safe limit. Patients who require radiation diagnosis or therapy should also be fully aware of the hazards involved.

Review Assignment

1. Compare and contrast cathode rays, electrons, and β particles.
2. What is one method used to produce X rays?
3. The X rays used to view luggage at an airport can damage high-speed film. How can the film be protected as it passes through the security check?
4. During the 1940s and 1950s, shoe stores all across North America had devices called fluoroscopes which used powerful X rays that allowed customers to see how their bones fit their new shoes. Offer a hypothesis describing why you think these devices were (a) allowed to be used and (b) later condemned as being unsafe.
5. Name the part of an atom from which radioactivity originates.
6. Describe the three modes of radioactive decay.
7. State the type of charge on each of the following:
   (a) α particle
   (b) β particle
   (c) X ray
   (d) γ ray
   (e) nucleus
   (f) neutron

8. What are some sources of background radiation? How is this type of radiation taken into account when radiation activity is being measured with a Geiger-Müller tube?
9. People who work near radioactive sources wear a type of photographic badge called a *radiation dosimeter* that is checked regularly. Describe how this badge would help prevent a worker from receiving more than the recommended amount of radiation.
10. Describe devices that measure radioactivity (a) qualitatively and (b) quantitatively.
11. Describe how to store a radioactive source safely.
12. At a rate of 85 Bq, how many emissions are sent out in (a) 5.0 s and (b) 1.0 min?
13. Describe a major weakness of the atomic theory developed by (a) Sir Joseph Thomson and (b) Ernest Rutherford.
14. Draw a model resembling the one in Figure 22-25(b) of one atom of each of the elements listed below. (Section 17.2 describes the electron arrangements.)
    (a) carbon-14   (b) carbon-12   (c) fluorine-19
15. State what happens to an element's atomic number and mass number when the element undergoes (a) α decay and (b) β decay.
16. Draw a graph similar to the one in Figure 22-30 showing the radioactive decay series of thorium-232. This radioactive substance undergoes the following ten decays before it reaches stability: α, β, β, α, α, α, α, β, β, α.
17. Write the nuclear equations for these decays:
    (a) from polonium-214 to lead-210
    (b) from thallium-210 to lead-210
18. It is possible, and in fact quite likely, that the so-called transuranium elements (those above atomic number 92) did exist when the earth was very young. Speculate on why they cannot be found in nature now, whereas a large amount of uranium can be found in the earth's crust.
19. The half-life of a certain radioactive isotope is 1.0 min. What fraction of a sample of this isotope remains after 10 min?
20. An experiment was performed to determine the half-life of iodine-131 ($^{131}_{53}\text{I}$). The following activities were recorded at noon on each observation day.

Observation day	0	4	8	12	16
Activity (Bq)	9900	7050	4950	3500	2510

(a) Plot a graph of the activity.
(b) According to the graph, what is the half-life of iodine-131?
(c) What would be the activity on day 24?

Figure 22-38
Alfred Nobel (1833–1896)

21. Alfred Bernhard Nobel (Figure 22-38), a rich Swedish chemist and inventor who lived from 1833-1896, set up an enormous fund of money from which interest could be generated and used to promote peace, culture, and research. The first Nobel prizes were awarded in 1901. Seven scientists named in this chapter won a Nobel

prize in physics or chemistry, and one scientist was awarded two Nobel prizes.
  (a) Make a list of all the scientists named in this chapter. Beside each name state the person's major contribution(s) and the year of each contribution.
  (b) Predict which scientists were likely winners of the Nobel prize and which one was the double winner.
  (c) Do research to determine how accurate your predictions were.

Key Objectives

Having completed this chapter, you should now be able to do the following:
1. Define cathode rays and state their origin.
2. Recognize some main reasons for the transition from classical physics to modern physics.
3. Recognize at least one important contribution made to physics by each of Wilhelm Roentgen, Henri Becquerel, and Marie and Pierre Curie.
4. Compare X rays and γ rays with each other and with other parts of the electromagnetic spectrum.
5. Describe the production, use, and dangers of X rays.
6. Define radioactivity and describe the three modes of radioactive decay.
7. Define background radiation.
8. Name devices used to detect radioactive emissions and describe their method of detection.
9. Define and apply the unit of radiation activity.
10. Name and compare materials that absorb radioactive emissions.
11. Describe reasons why the model of the atom changed as a result of contributions made by Thomson, Rutherford, and Bohr.
12. Describe the set-up, observations, and conclusions of Rutherford's gold foil experiment.
13. Define the atomic number and mass number of an isotope.
14. Given any two of the atomic number, mass number, and neutron number of an atom, determine the third quantity.
15. Define isotope and radioisotope.
16. State the meaning of transmutation, parent nucleus, and daughter nucleus.
17. Write nuclear equations for both α and β decay processes.
18. Understand how a radioactive series can be depicted in a diagram.
19. Define the half-life of a radioactive substance, and plot a half-life curve on a graph.
20. Given the activity of a radioactive sample, find the activity after any small number of half-lives.
21. Recognize how to handle and store a radioactive sample safely.
22. Appreciate the limitations of statistically analysing random events.
23. Describe uses and hazards of radioactivity.

ATOMIC PHYSICS

CHAPTER 23
Using Nuclear Energy

Damage at a nuclear generating station is repaired with great caution.

Every second, the mass of the sun is reduced by over four billion kilograms! What happens to the disappearing mass? It changes into energy as a result of nuclear reactions. These nuclear reactions create a spectacular amount of activity at the sun's surface.

Here on earth, scientists have learned how to produce and control some nuclear reactions. This knowledge has resulted in one of the most controversial issues the human race has ever faced: How can we benefit from using nuclear energy yet prevent our own destruction by that same energy? After studying how and why nuclear reactions produce vast amounts of energy, you will be able to answer such questions as these: How did scientists develop the ability to harness nuclear energy? What are the immediate and long-term benefits and dangers of using nuclear energy? What is Canada's role in the use of nuclear energy? What safety precautions are taken when damaged components at a nuclear generating station undergo repair, as in the photograph? Finally, what are some of the most recent developments in the understanding of matter?

23.1 Nuclear Reactions

When one element changes into another, a transmutation takes place. In the previous chapter, the type of transmutation studied was the decay of a nucleus. In a nuclear decay, no external infuence is required; the process occurs spontaneously and naturally. However, a transmutation can also occur because of external influences; for instance, a nucleus can be bombarded with a particle from outside itself, resulting in an interaction. An interaction that causes a change in a nucleus is called a **nuclear reaction**.

Ernest Rutherford, whose work was described in Chapter 22, was the first person to observe a nuclear reaction. In 1919 he aimed α particles ($^{4}_{2}$He) from a radioactive source at a sample of pure nitrogen gas. A diagram of the apparatus he used is shown in Figure 23-1. When Rutherford analysed the products of the reaction, he found that protons ($^{1}_{1}$H) had passed through the silver foil and were detected at the screen. Within the tube he found that some of the nitrogen gas had changed into oxygen gas. Rutherford had performed the first artificial transmutation.

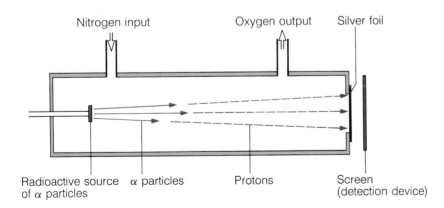

Figure 23-1
Experimental Set-up for the World's First Artificial Transmutation

The nuclear reaction equation representing the initial and final products observed in Rutherford's transmutation experiment is

$$^{4}_{2}\text{He} + ^{14}_{7}\text{N} \rightarrow ^{17}_{8}\text{O} + ^{1}_{1}\text{H}.$$

Notice that, just as in nuclear decays, both the atomic number (2 + 7 = 8 + 1) and the mass number (4 + 14 = 17 + 1) are conserved. The conservation of atomic and mass numbers holds true for all nuclear reactions. See Figure 23-2 on page 572.

ATOMIC PHYSICS

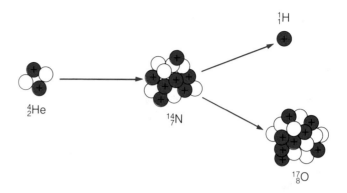

Figure 23-2
The Nuclear Reaction in Rutherford's Transmutation Experiment

Rutherford's success at creating different substances by bombarding nuclei marked the beginning of numerous experiments involving artificial transmutations. Scientists could now do in the laboratory what the alchemists had hoped to do in previous centuries and what nature has been doing since the beginning of time.

An example of a nuclear reaction in nature was mentioned in the previous chapter, but the reaction equation was not stated there. Cosmic rays from outer space produce neutrons in the atmosphere which react with nitrogen-14 nuclei, causing a transmutation to radioactive carbon-14 nuclei, with an ejection of a proton. See Figure 23-3. The equation for this nuclear reaction is

$${}_0^1 n + {}_7^{14}N \rightarrow {}_6^{14}C + {}_1^1 H.$$

The resulting carbon-14 reacts chemically in the same way as carbon-12, as stated earlier. It produces carbon dioxide (CO_2), a gas which is absorbed by all living plants. The plants use the carbon and expel the oxygen. This process explains why plants, as well as the animals that eat plants, have small amounts of carbon-14 within them. (We discussed in Chapter 22 how this fact makes radioactive dating possible.)

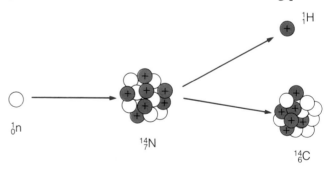

Figure 23-3
The Production of Carbon-14

One example of an important artificial transmutation occurred in 1932, when James Chadwick demonstrated the existence of the neutron. Chadwick, a co-worker of Rutherford in England, used a radioactive source of particles in an experiment illustrated in Figure 23-4. He used a detector to observe certain charged particles (α particles) in region A and different charged particles (protons) in region C. But there were no charged particles in region B. Further investigation proved that there *were* particles in B, and that they had a mass almost identical to that of a proton, but no charge. These neutral particles were neutrons.

USING NUCLEAR ENERGY

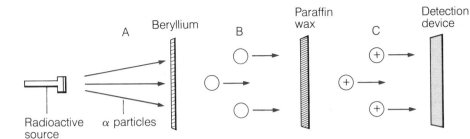

Figure 23-4
The Experimental Set-up That Led to the Discovery of the Neutron

In Chadwick's experiment, evidently the α particles had caused neutrons in the beryllium target (see Figure 23-4) to be driven out. Those neutrons in turn had struck the wax and dislodged protons there. The nuclear reaction in the beryllium is given by this equation:

$$_{2}^{4}\text{He} + {_{4}^{9}\text{Be}} \rightarrow {_{6}^{12}\text{C}} + {_{0}^{1}\text{n}}$$

With Chadwick's discovery, scientists had a way of creating and controlling the production of neutrons. This led to a whole new series of investigations. The advantage of using neutrons as "bullets" to bombard nuclei is that, because they have no charge, they are not repelled. (The disadvantage of using α particles or protons is that they are repelled by the positive charges in a nucleus.)

One scientist who applied this principle and contributed greatly to nuclear research was Enrico Fermi (1901–1954), an Italian physicist who moved to the United States in 1939 (Figure 23-5). Fermi hypothesized that a neutron, having no charge, would penetrate a nucleus more easily than a proton or α particle and thus initiate a nuclear reaction more readily. Between 1934 and 1936, Fermi produced several previously unknown isotopes by bombarding elements with neutrons. He also made the discovery that water placed between the neutron source and the target slowed down the neutrons by absorbing some of their energy. The slower speed gave the neutrons a greater chance to interact with the nuclei of the target element. This discovery later proved to be important in the development of nuclear reactors, described later in this chapter.

In the Practice Questions that follow, you are asked to write equations involving various nuclear reactions. Table 23-1 summarizes the common symbols for the particles that either are used to bombard nuclei or are emitted as a reaction product.

Figure 23-5
Enrico Fermi (1901–1954)

Did You Know?
Neutron activation analysis is a peaceful application of nuclear energy. Slow neutrons are absorbed by a sample and cause it to decay. The resulting γ ray spectrum allows identification of the elements present, using very small amounts. This is a valuable tool for forensic scientists (who do work connected with law enforcement) and archaeologists.

Table 23-1 Symbols for Particles Commonly Involved in Nuclear Reactions

Particle	Chemical Symbol	Nuclear Symbol
proton (hydrogen nucleus)	H	$_{1}^{1}\text{H}$
neutron	n	$_{0}^{1}\text{n}$
β particle (electron)	e⁻	$_{-1}^{0}\text{e}$
α particle (helium nucleus)	α	$_{2}^{4}\text{He}$
heavy hydrogen nucleus (deuteron or deuterium nucleus)	D	$_{1}^{2}\text{H}$

ATOMIC PHYSICS

PRACTICE

1. Distinguish between a nuclear decay transmutation and a nuclear reaction transmutation.
2. Explain why neutrons are more effective than protons or α particles as "bullets" in nuclear reactions.
3. In the nuclear reaction equations shown, each question mark represents a single particle or nucleus. Complete each equation.
   (a) $^{2}_{1}H + ^{200}_{80}Hg \rightarrow ^{198}_{79}Au + ?$
   (b) $? + ^{10}_{5}B \rightarrow ^{7}_{3}Li + ^{4}_{2}He$
   (c) $^{4}_{2}He + ? \rightarrow ^{17}_{8}O + ^{1}_{1}H$
4. Write the nuclear reaction equations for the following:
   (a) A proton strikes a lithium-7 atom, resulting in the production of two helium-4 nuclei.
   (b) Two deuterium nuclei collide, forming a helium-3 nucleus and a neutron.
   (c) An α particle bombards an aluminum-25 nucleus, creating a phosphorus-28 nucleus and a neutron.
5. Research and report on the life of Enrico Fermi.

23.2 Mass and Energy in Nuclear Reactions

In 1905, long before the discovery of nuclear reactions, Albert Einstein (1879–1955) (Figure 23-6), a German scientist who moved to the United States in 1933, proposed his now-famous special theory of relativity. A consequence of the theory is that matter and energy are not considered separate entities; rather, they are known to be different forms of one another. The equation that indicates how much energy is produced when a certain amount of mass changes to energy is

$$E = mc^2.$$

Here, $c = 3.0 \times 10^8$ m/s is the speed of light in a vacuum; m is the mass in kilograms that has disappeared; and E is the energy in joules that results.

Figure 23-6
Albert Einstein (1879–1955)

SAMPLE PROBLEM 1

If 1.0 g of a substance changes entirely into energy, how much energy is produced?

Solution

$E = mc^2$
$= (1.0 \times 10^{-3}$ kg$)(3.0 \times 10^8$ m/s$)^2$
$= 9.0 \times 10^{13}$ J

Thus, the amount of energy produced is 9.0×10^{13} J, which is enough to light 30 100 W lightbulbs for about 10 000 years!

Until 1905, scientists believed that matter could not be created or destroyed (law of conservation of mass) and that energy could not be created or destroyed (law of conservation of energy). When Einstein proposed the relation $E = mc^2$, the conservation laws had to be combined into one law, the **law of conservation of mass-energy**. It states:

For an isolated system, the total amount of mass-energy remains constant.

How does this law relate to nuclear reactions? The answer is found in careful measurement of the mass of particles involved in such reactions. For example, a helium-4 nucleus (or α particle), which is 2 protons and 2 neutrons bound together, has a mass that is less than the mass of 2 free protons and 2 free neutrons. In other words, when the 2 protons and 2 neutrons join together to form a helium nucleus, their total mass decreases. According to the law of conservation of mass-energy, the remaining mass appears as energy. See Figure 23-7.

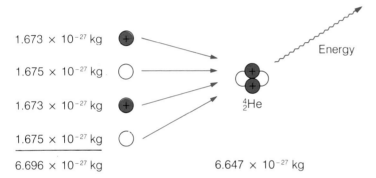

Figure 23-7
Conserving Mass-Energy. When two protons and two neutrons join to form a helium atom, some mass is lost and appears as energy.

In general, when a nuclear reaction occurs, the mass of the product is less than the mass of the original nuclei by the amount that has changed into energy ($E = mc^2$). The origin of this energy is the force that binds the nucleons together. This force, called the **strong nuclear force**, is an attractive force that exists between all nucleons (protons and neutrons) when they are very close together. At close range, this nuclear force is much stronger than the electric force of repulsion between the protons, which have positive electric charges. The relationship between the strong nuclear force and the energy released in a nuclear reaction is described later in more detail.

PRACTICE

6. Determine the amount of energy released in each case:
   (a) 600 g of matter changes entirely into energy.
   (b) 0.50% of a 20 kg sample of uranium changes into energy.
   (c) Two protons, each with a mass of 1.673×10^{-27} kg, combine with two neutrons, each with a mass of 1.675×10^{-27} kg, to produce an α particle with a mass of 6.647×10^{-27} kg.
7. If one megajoule of energy was created during a nuclear reaction, how much mass was converted into energy?
8. For a stable nucleus containing several protons and neutrons, compare the electric forces with the strong nuclear forces.

9. Scientists believe there are four fundamental forces in nature: electromagnetic, gravitational, weak nuclear, and strong nuclear.
   (a) List these forces in order of strongest to weakest.
   (b) Do research to determine whether your hypothesis is correct.

23.3 Nuclear Fission

Enrico Fermi's work in using slow-moving neutrons to create new isotopes led to an amazing discovery in 1939 which, because it led to the development of nuclear reactors and bombs, changed the course of world history. In that year, two German scientists, Otto Hahn (Figure 23-8) and Fritz Strassmann, following up on Fermi's work, found they could cause uranium nuclei to split into much smaller fragments. The result was the release of large amounts of energy. Two other scientists, Lise Meitner (Figure 23-8), with whom Otto Hahn had earlier worked, and her nephew, Otto Frisch, working in Sweden, heard of the discovery and soon explained it.

The splitting of a large nucleus into two nuclei of much smaller size is called **nuclear fission**. It is accompanied by the release of neutrons, as well as energy. This is illustrated in Figure 23-9(a).

The reason that energy is released is that the mass decreases as the nucleus splits up. The smaller nuclei are more tightly bound than the large nucleus, so the extra energy must be given up during their creation.

An element commonly used for nuclear fission is uranium. Most uranium nuclei (U-238) do not undergo fission easily. However, one isotope, U-235, can be made to undergo fission by bombarding it with a neutron. The extra neutron is absorbed by the U-235, creating an unstable isotope which almost immediately splits into two smaller nuclei, releasing neutrons and energy.

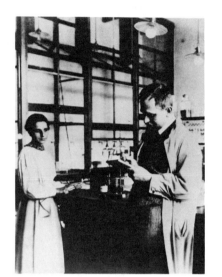

Figure 23-8
Lise Meitner (1878–1968) and Otto Hahn (1879–1968)

Figure 23-9
The Products of Nuclear Fission

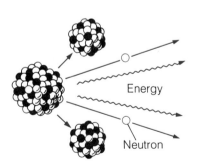

(a) *The fission of a large nucleus into two smaller nuclei is accompanied by the release of energy and at least one neutron.*

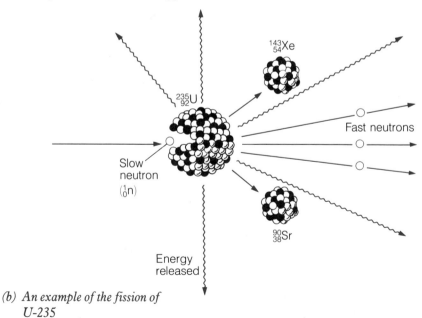

(b) *An example of the fission of U-235*

In one example of a uranium fission reaction, shown in Figure 23-9(b), the smaller nuclei created are strontium-90 and xenon-143. Using symbols, the equation for this reaction is

$$^{235}_{92}U + ^{1}_{0}n \rightarrow ^{236}_{92}U \rightarrow ^{90}_{38}Sr + ^{143}_{54}Xe + 3^{1}_{0}n + \text{energy}.$$

The amount of energy produced can be calculated by determining the loss of mass during this reaction and applying the relation $E = mc^2$.

In the above fission reaction, the three neutrons are capable of joining with other U-235 nuclei. Every time such a nucleus absorbs one neutron, it splits up and emits three more neutrons, each capable of initiating another fission reaction. Soon thousands and then millions of nuclei are splitting up, resulting in a **chain reaction** (illustrated in Figure 23-10). You can imagine that such a chain reaction can release tremendous amounts of energy.

Did You Know?
The fission of 1.0 g of U-235 yields 2.5×10^6 times as much energy as the burning of the same mass of carbon.

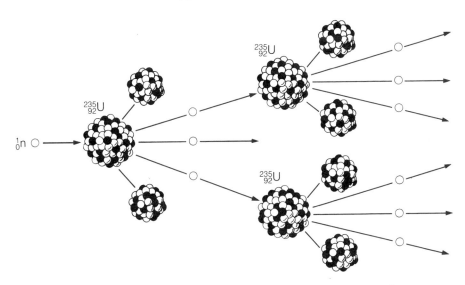

Figure 23-10
The Beginning of a Chain Reaction

From the time of its discovery, the use of nuclear fission has been controversial. As you study the applications of this powerful source of energy in the next section, try to maintain an analytical attitude which is unbiased, yet informed of the many viewpoints on the topic.

PRACTICE
10. What product results from the fission of a nucleus?
11. Use the data given below to determine the energy released during the following fission reaction:

$$^{1}_{0}n + ^{235}_{92}U \rightarrow ^{90}_{38}Sr + ^{136}_{54}Xe + 10^{1}_{0}n + \text{energy}$$

Particle	Mass ($\times 10^{-27}$ kg)
neutron	1.675
U-235	390.989
Sr-90	149.301
Xe-136	225.687

23.4 The First Nuclear Fission Reactor

It is important to view the 1939 discovery of nuclear fission in the context of Europe's political situation at the time. Under Adolf Hitler, Germany was already beginning its aggressive invasion of other nations, and many European scientists chose to emigrate to the United States. Soon after World War II began, research into nuclear fission became veiled in secrecy. In the United States, physicists who understood the potentially destructive power of nuclear energy in the hands of the Nazis urged Albert Einstein to inform President Roosevelt of the situation. As a result, the President authorized a secret program known as the Manhattan Project, whose purpose was to develop a nuclear bomb before the Nazis did.

One of the leading scientists in the Manhattan Project was Enrico Fermi, who first heard of the discovery of nuclear fission while he was still living in Europe. Fermi hypothesized that a controlled fission reaction might be possible. On moving to the United States at the outbreak of the war, he went to the University of Chicago (Illinois), where he was joined by other scientists from Europe, Canada, and the United States. Fermi led this team of scientists in designing and building a **nuclear fission reactor**, a device in which a controlled fission reaction could occur.

The scientists realized that several matters had to be considered before they could design a nuclear reactor. They had four main concerns.

- What substance should be used?
- How much of the substance should be used?
- What method should be employed to ensure that the neutrons would travel slowly enough to be absorbed by the target nuclei?
- How could the reaction be prevented from going out of control, causing excess heat and radiation?

The substance that Fermi decided to use in the reactor was the same as the substance used in the discovery of fission – natural uranium. It consists of about 99.3% U-238, an isotope which is not highly fissionable, and only about 0.7% of the highly fissionable U-235. It is possible to **enrich** natural uranium to obtain a higher concentration of U-235, but the process is difficult and expensive.

To ensure that a chain reaction would be self-sustaining, a certain minimum mass of the fissionable material, called the **critical mass**, was needed. If the mass was too small, the neutrons produced in each fission reaction likely could escape from the sample without causing additional fissions. As a result, the chain reaction would not sustain itself. The critical mass for a nuclear reactor could range from a few kilograms to several thousand kilograms, depending on the material used. In the first reactor, over 40 000 kg of uranium oxide and uranium metal were used.

To slow the neutrons down without stopping them, a substance of appropriately low density, called a **moderator**, had to be found. Three possible moderators were ordinary water (H₂O), also called *light water* in technological terms, *heavy water* or deuterium oxide (D₂O), and carbon. In the first reactor, the moderator chosen was carbon in the form of graphite.

Finally, to ensure that the fission reaction could be prevented from going wild once it had begun, **control rods** were used. The rods, made of cadmium, could be moved into and out of the reactor. Their function was to absorb neutrons, so if the reaction was proceeding too quickly, as indicated by the heat produced, the rods could be inserted further into the reactor.

With these design features, Fermi and the other members of the Manhattan Project built the world's first nuclear reactor under a squash court at the University of Chicago (Figure 23-11). On December 2, 1942, three years after construction had begun on the reactor, they were successful in producing the first controlled chain reaction.

Figure 23-11
This is an architect's drawing of the world's first nuclear reactor, located at the University of Chicago.

Most unfortunately, what Fermi's team had constructed quickly led to the building of the first nuclear bombs. Several of the scientists later regretted that their research had resulted in such immense destructive powers. However, their success also led to important peaceful applications, one of which you will read of in the next section. (A discussion of nuclear weapons is reserved for later in the chapter, after you have studied fusion, the other type of nuclear reaction.)

PRACTICE

12. What it the meaning of each of the following terms?
    (a) enriched uranium   (c) critical mass
    (b) moderator   (d) control rods
13. When in 1939 Enrico Fermi learned that uranium was fissionable, he realized that it was just a matter of luck that he had not discovered the phenomenon five years earlier. His comment was that, in light of the European situation, the world was fortunate that nuclear fission was not discovered until 1939. Explain some possible reasons for this opinion.

23.5 Using Nuclear Fission to Generate Electricity

Although the first nuclear reactors were built to obtain materials for making nuclear bombs, it soon became apparent that the energy released in a controlled fission process could be used for two major peaceful purposes. The purpose of **research reactors** is to study the structure of matter and produce isotopes for medical applications. The purpose of **generation reactors**, discussed in detail here, is to convert nuclear energy into electrical energy. Today, the use of nuclear power reactors to generate electricity is an important yet controversial technology in our society.

There are many designs for nuclear reactors, which differ in the type of process and the substances used as coolant and moderator. Some of the more common reactors are examined below.

The CANDU Reactor

The fission reactor designed and built in Canada is called the **CANDU reactor**. Its name indicates that the reactor is **CAN**adian in design, uses **D**euterium oxide (heavy water) as its moderator, and has **U**ranium as its fuel.

The history of Canadian reactors began in 1942, when the physicists Halban and Kawarski escaped from Paris, France, which was under German occupation. They went to England and smuggled with them a supply of heavy water, which they thought would make an excellent moderator for a nuclear reactor. They soon convinced several British scientists, including James Chadwick, the discoverer of the neutron. He in turn was able to persuade the Canadian government to support the building of a heavy water reactor in Canada as a secret, Allied war effort.

Scientists and engineers from Canada, Britain, and the United States joined together to design and build a reactor along the Ottawa River in Ontario. This reactor, built at the town of Chalk River and called ZEEP (Zero Energy Experimental Pile), was the first nuclear reactor in the world to operate outside the United States. It began to produce small amounts of energy in September, 1945, shortly after the end of World War II, and is still being used as a research reactor.

By the early 1950s, after having solved many scientific and technological problems, Canadian scientists began to plan the first CANDU reactor, which was to be capable of generating electricity on a commercial basis. This reactor, located at Rolphton, also on the Ottawa River, began operation in 1962. By the 1970s several more CANDU reactors, some of them very large, were in operation. Typical of these reactors are the four known as Pickering A, located at Pickering, Ontario. They have generated more electrical energy than any other nuclear generating station in the world since they began production in 1970.

Because they were built at different times and in different locations,

Did You Know?
Heavy water, which was discovered in the early 1930s, looks and tastes like ordinary water. However, its density is about 10% greater than that of ordinary water, and its freezing and boiling points also differ. In nature, about 1 part in every 7000 parts of ordinary water is heavy water.

the various CANDU reactors have design differences. However, they have several important common features. The **fuel** used in a CANDU reactor is natural uranium (in contrast with enriched uranium) in the form of uranium dioxide (UO_2). The fuel is pressed into pellets (Figure 23-12(a)) and placed in long metal tubes sealed at the ends. Several such tubes are assembled into bundles, each bundle having a mass of about 22 kg. See Figure 23-12(b).

Figure 23-12
Fuel for a CANDU Reactor

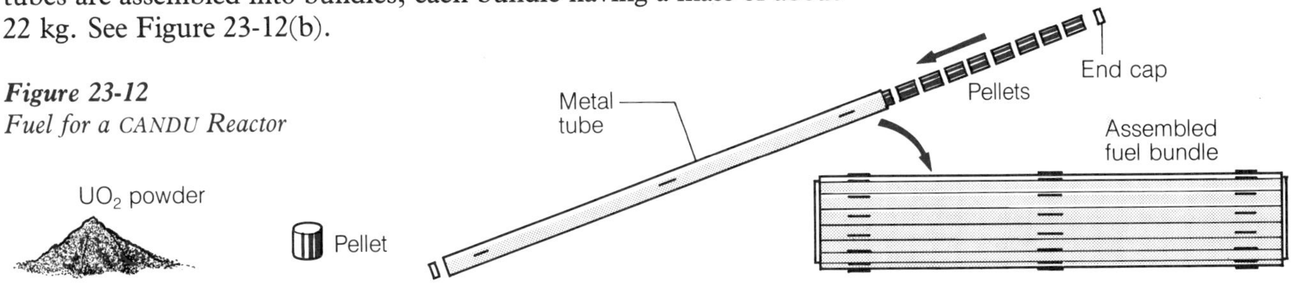

(a) *Uranium oxide powder is pressed into a pellet.*

(b) *Uranium oxide pellets are placed in tubes which are assembled in bundles.*

Each reactor contains approximately 5000 fuel bundles placed horizontally in an assembly called a **calandria**. The horizontal arrangement has the advantage that the bundles can be replaced one at a time without shutting down the entire reactor for several days to refuel. In fact, the CANDU reactor was the world's first reactor capable of being refueled at full power.

To regulate the number of neutrons striking the fissionable U-235 nuclei, CANDU reactors use control rods, usually made of cadmium like those in Fermi's reactor, to absorb neutrons. These rods can be moved into and out of the calandria to adjust the rate of neutron absorption.

Although the control rods determine the number of neutrons striking the U-235 nuclei, the speed of the neutrons is also important. Since the natural uranium used in a CANDU reactor contains only about 0.7% of the fissionable U-235, a good moderator must be used to slow down the neutrons by absorbing some of their energy. Slow neutrons have a much greater chance of being absorbed by the U-235 nuclei than fast neutrons. The moderator used in a CANDU reactor is heavy water, D_2O. Deuterium (D or ^2_1H) is an isotope of hydrogen containing one proton and one neutron. It is the extra neutron in each hydrogen atom that helps prevent heavy water from absorbing the fission neutrons, making it a much better moderator than ordinary or light water (H_2O). See Figure 23-13.

In a CANDU reactor, heavy water is also used as the **coolant**, a substance that absorbs much of the energy produced by the nuclear reaction. After absorbing the energy, the coolant, which is under pressure to prevent its boiling, is circulated to a **heat exchanger**, where it gives energy to ordinary water in a boiler. This ordinary water boils, producing steam. The steam is directed through huge **turbines**, forcing them to spin. Refer to Figure 23-14, as well as to the photo essay on the construction of a CANDU nuclear generating station in Figures 23-15 to 23-22 on pages 582–584.

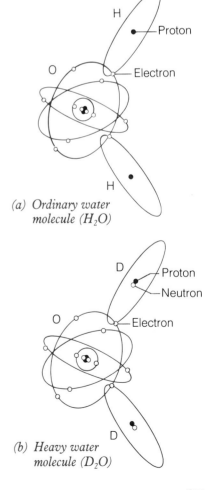

Figure 23-13
A Comparison of Heavy Water and Ordinary Water

(a) *Ordinary water molecule (H_2O)*

(b) *Heavy water molecule (D_2O)*

ATOMIC PHYSICS

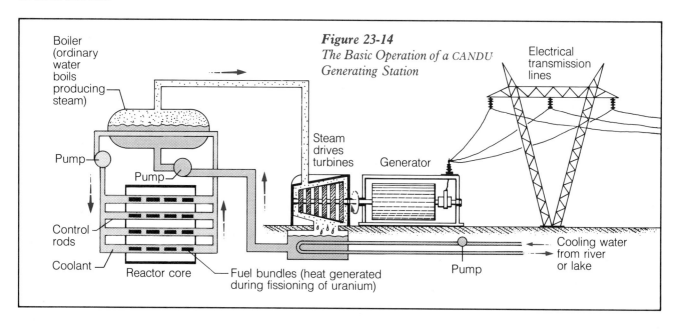

Figure 23-14
The Basic Operation of a CANDU Generating Station

Figure 23-15
Materials and apparatus, including the calandria shown, are delivered to the construction site. →

↑
Figure 23-16
Construction begins on the vacuum building, which is part of the safety containment system needed if an emergency, such as the bursting of a coolant pipe, were to occur.

Figure 23-17
The vacuum building, which will be linked to all the reactor buildings at the site, nears completion.

Figure 23-18
Feeder pipes are connected to the calandria. These pipes are used to carry the heavy water coolant which takes the heat away from the horizontal tubes in the calandria.

Figure 23-19
The finished calandria is shown, with its automated fueling machine.

Figure 23-20
A giant rotor is lowered into a turbine in the turbine hall. Eventually, pressurized steam will force the rotor to spin and thus drive the electrical generators.

ATOMIC PHYSICS

Figure 23-21
The turbine hall is completed.

Figure 23-22
This view shows the Pickering generator station (A and B) on Lake Ontario, with its eight reactor buildings and large vacuum building.

As you know, the ultimate purpose of a nuclear generating station is to produce electrical energy. As in other generating stations, the spinning turbines are connected to alternating current (AC) generators which produce electricity for delivery to consumers.

Pressurized Water Reactor (PWR)

The PWR is used widely throughout the world, except in Canada. It was developed from the system used to operate nuclear-powered submarines. The moderator, which also acts as the coolant, is ordinary water, H_2O, which tends to absorb neutrons far more easily than heavy water. This creates the need to enrich the uranium fuel so that it contains between 2% and 4% of the fissionable U-235. To achieve a high temperature without boiling, the coolant is under high pressure (hence the name of the PWR). Just as in the CANDU reactor, the coolant delivers its energy to a water-and-steam circulation system. The steam drives turbines, and so on. See Figure 23-23.

A disadvantage of this type of reactor is that it must be shut down for refueling every 12 to 18 months.

Did You Know?
The neutrons ejected during a fission reaction travel at speeds up to about 42 000 km/s. In order to be absorbed by U-235 nuclei, they must be slowed down to about 3 km/s.

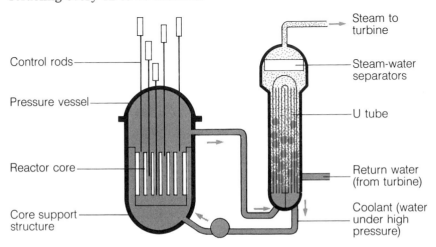

Figure 23-23
The Basic Design of a Pressurized Water Reactor (PWR)

Boiling Water Reactor (BWR)

The BWR is similar in design to the PWR, but the coolant is under about half as much pressure. The coolant, which is ordinary water, undergoes boiling and becomes steam. The steam is then forced directly to the turbines, skipping the heat-exchanger stage found in other types of reactors. See Figure 23-24.

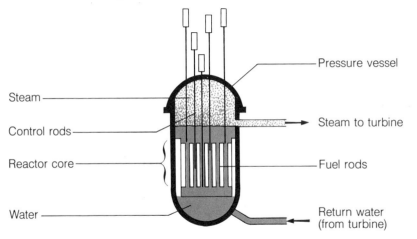

Figure 23-24
The Basic Design of a Boiling Water Reactor (BWR)

PROFILE

C. MacKay-Lassonde, M.Sc., M.B.A., P.Eng.
Manager, Load Forecasts Department
Ontario Hydro

Over most of my career I have done very little of what is conventionally thought of as engineering – the calculating, designing, and drawing of any structure or equipment. Neither have I overseen the construction of any building or the setting up of any equipment.

In spite of the fact that I may not now appear to be practising "engineering", my career so far is not at all atypical of most engineering graduates. It all started in high school where I was doing quite well in mathematics, physics, and chemistry. I was not too sure what I wanted to be, but I knew I should probably make use of these three subjects.

One day, I had just read about the fascinating work of Marie Curie when I heard the boy next door discussing a possible future career in engineering. The idea intrigued me, although, to be honest, I had no idea what engineering was all about. My father encouraged me to try it, despite the fact that everyone else said engineering was "too difficult for girls".

My years studying chemical engineering at University of Montreal were wonderful. After graduation, though, the situation changed. My first job consisted of advising manufacturers about the types of oils and greases they should use in their machinery. In the process I faced discrimination, open and hidden. Callers always asked to talk to my boss. When they realized that I was supposed to be the expert, they often tried to get the information they needed from other sources. Fortunately, whenever they tried, they were always referred to my office.

Soon my husband and I decided to leave Montreal for Utah where he was to go to graduate school and I was to have our first child, Julie. I began graduate studies, too, shortly after our child was born. I completed my Master's degree in nuclear engineering in one year, sometimes working on my take-home exams with Julie sitting on my lap. Fortunately, she was a quiet baby.

Then came the question of my getting another job. It was at that point that I became fully aware of my non-traditional status as a woman nuclear engineer. After eighteen unsuccessful applications, I managed to get an interview with an engineering construction firm in San Francisco. I received an offer right then and there – but I do not believe it was because I was so good, but rather because the firm needed women engineers in order to meet their Affirmative Action

High Temperature Gas-Cooled Reactor (HTGR)

The HTGR, first developed in Britain, uses enriched uranium as its fuel and helium gas as its coolant. The coolant is kept at a high temperature because gases are not as efficient as liquids in transferring heat energy, especially at low temperatures. The coolant circulates to a heat exchanger and gives its excess energy to water that turns to steam to drive the turbines. The moderator used is graphite, a form of carbon. Most reactors of this type must be shut down for refueling one or two weeks each year. An advantage of the HTGR is that, in the event of an accidental coolant loss, the reactor would cool down more efficiently program quota.

My job was to design cooling and purification systems for nuclear plants. As part of this work, I had to calculate how much heat would be generated as the irradiated fuel used in the nuclear reactors decayed. I also had to estimate how much radioactive material would escape into the atmosphere in different types of (hypothetical) accidents, as well as judge where in the atmosphere this radioactive material would be dispersed.

By the time I returned to Canada, I had become a specialist in radioactive material dispersion through atmospheric releases. I continued to work on highly technical nuclear engineering problems, and was responsible for much of the research that lay behind a CSA (Canadian Standards Association) standard for estimating, calculating, and determining the dispersion of radioactive material into the atmosphere.

I was responsible for directing the development work for an improved mechanical shut-off rod. Such rods terminate the fission reaction, and must be extremely reliable, *i.e.*, dropping when needed, not before. (Otherwise the power would be shut off.) We were looking into ways of increasing the initial acceleration, weight holding-capacity, reactivity effect (characteristic required to terminate the fission chain reaction), and reliability. At the same time we were considering design changes to reduce the cost, neutron absorption, and the size of the drive package. The work was being done primarily for the next generation of nuclear stations in Canada. Because of the significant slowdown in the annual increase of electricity demand beginning in the mid 70s, most of the work related to the next generation of nuclear reactors was put on hold, including the new mechanical shut-off rod. However, the knowledge acquired during the development and initial testing of the device remains and will most likely be used in future generations of CANDU reactors.

The second mechanism on which I worked at that time was just as interesting. Instead of using a solid rod to terminate the fission reaction, we were experimenting with the injection of liquids that were capable of absorbing neutrons. We tested the mechanism by means of a pilot project, but when the need and interest in nuclear electrical generating stations waned, the work did not progress any further. However, this technique is being used in experimental nuclear reactors in other countries.

At this stage of my career, I decided to study part-time for a degree in business administration because I felt I was not being promoted as rapidly as I would like. The extra skills I acquired in such areas as accounting, marketing, and organizational behaviour helped me win the promotions I was after.

Recently I have become President Elect of the Association of Professional Engineers of Ontario. I am very pleased about this recognition of my achievements.

Since the late 1970s, I have been very active in promoting women in science and engineering. In 1977, I called a meeting of the then eleven women engineers and scientists working in Ontario Hydro. That was the beginning of WISE – Women in Sciences and Engineering. Later, in 1980, I convinced a few of my friends to help me organize the first Canadian Convention of Women Engineers.

One of my major goals has been to promote and give support to other women in engineering. At the best of times, it is difficult for people to relate to engineers, let alone women engineers. Few members of this profession are seen in TV series or shows as is the case with doctors and lawyers; but my own experiences have shown that engineering can be a stimulating and worthwhile career.

than other reactors because it has a large surface area exposed to the air. (The HTGR is almost identical in appearance to the PWR, shown in Figure 23-23.)

Fast Breeder Reactor (FBR)

The FBR does not use a moderator, so the neutrons ejected from the nuclei undergoing fission are not slowed down. (This explains the word "fast".) Since there is no moderator, the fuel for a FBR is different from that used in other reactors. It consists of at least 90% uranium-238, with the remainder being highly fissionable isotopes such as plutonium-239. Most of the fissionable material is placed in the central **core** of the fuel bundle; then the core is surrounded by a U-238 **blanket**. As fission occurs in the core, fast neutrons from the fission reactions escape from the surface of the core. Almost immediately they strike U-238 nuclei in the blanket, causing a series of β decays that result in the production of plutonium (Pu-239). Thus, the FBR actually produces, or breeds, more fissionable fuel (Pu-239) than it consumes; hence the name "breeder". The coolant used is a liquid metal, sodium, which does not have to be under high pressure and also does not absorb or slow down neutrons. The main advantage of a FBR is that it utilizes more than 50% of the original fuel (uranium), compared with only about 1% or less for most other reactors. This could help solve the problem of the earth's limited supply of U-235. See Figure 23-25. The disadvantages of the FBR are mentioned in the next section.

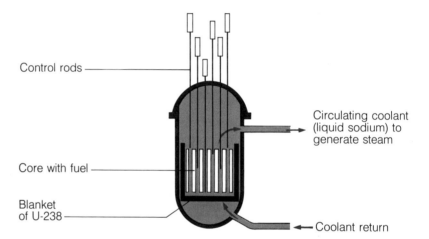

Figure 23-25
The Fuel Arrangement in a Fast Breeder Reactor (FBR)

PRACTICE

14. In what ways are nuclear generating stations and fossil-fuel generating stations (a) similar and (b) different?

15. The five nuclear generating reactors described in this section are named CANDU, PWR, BWR, HTGR, and FBR. Compare these reactors in a table using these headings: Name; Meaning of Name; Fuel; Coolant; Moderator; and Distinguishing Features(s).

16. The following word equations describe the process whereby a fast breeder reactor creates plutonium from uranium. Beside each product is its half-life, listed in brackets for interest only. Write the equation in symbols for each nuclear reaction or decay. (See the periodic table of the elements in Appendix B for the required information.)
uranium-238 + neutron → uranium-239 (23.5 min)
uranium-239 → β particle + neptunium-239 (2.35 days)
neptunium-239 → β particle + plutonium-239 (24 400 years)

23.6 The Nuclear Reactor Debate

One person brags that the nuclear reactor industry has great potential for alleviating Canada's energy problems. A second person argues that the peaceful use of nuclear energy is a curse as terrible as the nuclear bombs that could destroy the human race. Which person, if either, is correct? The nuclear reactor debate cannot be solved here, but we can analyse it objectively.

First consider that the purpose of a nuclear reactor is to convert nuclear energy into electrical energy. Since Canadians are unlikely to stop using electricity, the electrical energy must come from somewhere. In many ways, the best alternative is to develop localized sources of energy that require only renewable resources and cause no pollution or environmental damage. Scientists are researching such alternatives (see Chapter 7), but a long time will be needed to develop them. Another alternative is to use more fossil fuels to generate electricity at centralized generating stations. Fossil fuels are in limited supply, are costly to transport, and create much environmental pollution. In some cases, they can also provide hazardous working conditions, as evidenced by disasters in coal mines and offshore oil rigs.

A third alternative is to use nuclear energy to generate electricity. The fuel is clean-burning and relatively easy to transport, but this method too has several disadvantages. Some of them include the possibility of safety hazards for the workers (see Figure 23-26), environmental damage near the plant, the problem of storing highly radioactive wastes, and a limited supply of raw materials. Is anything being done to remedy these difficulties?

Figure 23-26
This international symbol is used to warn people of dangerous ionizing radiation.

Career Risks

People whose jobs involve working close to uranium and other radioactive substances are subject to certain dangers. For example, workers in uranium mines are exposed to ionizing radiation that occurs naturally as the uranium decays. As we discussed in Chapter 22, the radiation can ionize living cells and lead to cancerous growths. Thus, the miners have a higher-than-normal risk of developing lung cancer. To date, this problem has not been solved, although several recommendations have been made or implemented. Some of these are as follows:

- sealing off unused sections of the mine
- circulating fresh air throughout the mine
- keeping track of each worker's exposure to radiation and, if an established limit is reached, changing the person's job location

Miners are not the only workers at risk from possible exposure to radiation. Personnel at all stages of the nuclear industry are exposed to varying degrees of hazards. These stages include milling the uranium ore, refining the fuel, operating the reactor, and handling the radioactive wastes.

ATOMIC PHYSICS

Figure 23-27
The radiation monitor and shoe cleaning equipment shown help ensure protection against radiation.

At the reactor site, several precautions are taken to ensure that the personnel are subjected to the least possible risk. The fuel is contained in well-designed and -maintained equipment, and the air in the reactor building is constantly filtered and checked for radiation levels. Monitors are used to check radiation on personnel as they work or move from certain areas to others (Figure 23-27). Furthermore, the workers wear badges sensitive to radioactive emissions; the badges are examined regularly for type and amount of emissions received.

Accidental Risks

Nuclear reactors have built-in safety features intended to prevent the possibility of serious accidents. In most reactors including the CANDU reactor, a nuclear explosion is impossible, because there is not enough fissionable material gathered close together. However, other problems, such as the leakage of radioactive substances to the environment, could occur if the safety features malfunctioned because of human error, equipment failure, or an unforeseen event such as an earthquake.

One important consideration is to prevent the heat generated by the nuclear reaction from melting the container and controlling materials, a condition called **meltdown**. If these materials were to melt, the hot radioactive substances would come into contact with the coolant liquid. The result would be either a steam explosion if the coolant is water, or a chemical explosion if the coolant is liquid sodium. Dangerous radioactive materials could then be released into the atmosphere.

The CANDU reactors have among the world's best designs for the prevention of meltdown or radioactive emissions. First, if a loss of coolant occurs, the **emergency core cooling system** begins to act. Then, if further precautions are required, the **containment system** starts to operate. These devices provide both short-term and long-term control of any leaks. Finally, if a severe emergency arises for any reason, the reactor can be shut down completely, by means of as many as two or three independent systems. These shutdown systems act as described below. (In the more recent CANDU reactors, only the last two systems are available.)

Figure 23-28
Shutdown Systems

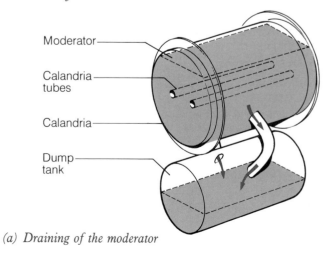

(a) *Draining of the moderator*

(b) *Using shutoff control rods and liquid poison*

First, the moderator (heavy water) can be quickly drained to a large dump tank beneath the calandria. Without the moderator, the chain reaction will cease.

Second, shutoff control rods can be dropped into the calandria, absorbing neutrons and thereby slowing the reaction. This is the preferred method of emergency shutdown.

Finally, liquid "poison" can be inserted directly into the moderator if other systems fail. This liquid absorbs the neutrons and slows the fission reaction. It is called "poison" because it dilutes the liquid moderator, which then has to be replaced. See Figure 23-28.

Thermal Effects

All fossil fuel and nuclear generating stations generate waste heat that is either recycled or discarded. Some of the discarded heat goes into the atmosphere near the station, but most of it is carried away by water dumped into a nearby river, lake, or ocean. The hot water emitted from generating stations alters the ecosystem and can cause difficulties for underwater life, expecially in small rivers and lakes. For example, aquatic life often increases in the warm waters near a station. Then, if for some reasons the station shuts down temporarily, the water cools rapidly and the life in it may not survive. One way of reducing this problem is to use a recirculating system of water in large cooling towers which give off the heat into the atmosphere.

Radioactive Wastes

Probably the most controversial and difficult problem facing the nuclear industry is the disposal of radioactive wastes. In general, there are two levels of wastes.

Reactor wastes, also called *low-level wastes*, are associated with various materials used in maintenance operation. These constitute only a small fraction of 1% of the radioactive wastes at a nuclear reactor. Currently, contaminated objects such as mops and filters are placed in containers and buried at disposal sites, with the hope that the radioactive byproducts will not seep into the local water system.

Spent-fuel wastes, or *high-level wastes*, result from the fission of the uranium fuel. About 99% of all the wastes at a reactor are found in the spent (used) fuel bundles. (About 1% is found in the reactor itself.) It is the spent fuel that constitutes a storage problem, because it generates a great amount of heat and emits ionizing radiation.

Currently, when used fuel bundles are removed from the calandria, they are stored under water in nearby pools (Figure 23-29). During the first several years, the bundles cool, and there is a reduction in the activity level of all radioactive isotopes, particularly iodine-131, a dangerous isotope with a half-life of 8.0 days. However, the spent fuel is still highly radioactive. (One of the products is plutonium, whose half-life is 24 400 years!) Therefore, a solution to the problem of waste disposal or reprocessing must be found.

Did You Know?
The disaster that occurred at the nuclear generating plant at Chernobyl in the U.S.S.R. in April, 1986, provides proof that people around the world should be aware of the consequences of using nuclear energy. The Chernobyl reactor, which used graphite as its moderator, overheated and burned, sending deadly radiation into the atmosphere. The first reports of the radiation were received from Sweden, hundreds of kilometres away from the disaster. There, a worker at a nuclear plant had received radiation from the fallout and set off an alarm in a monitoring device similar to the one shown in Figure 23-27. The environment near Chernobyl will take many years to recover from the disaster.

Did You Know?
Thermal effects are a necessary consequence of all types of electrical generating stations. The efficiency of a generating system can be calculated using the relation $Eff = 1 - (T_{out}/T_{in})$ where T_{out} is the lower output temperature and T_{in} is the higher input temperature. For all generating stations near large bodies of water, T_{out} is approximately the same. Thus, the operating temperature (T_{in}) can be used to compare efficiencies. In general, T_{in} is lower for nuclear generating stations than for fossil fuel generating stations. This means that nuclear generating stations operate at a lower efficiency.

Figure 23-29
Underwater Storage of Spent Fuel

Permanent storage is a problem to which there is no satisfactory solution as yet. Research is being carried out to discover the viability of placing spent fuel in special containers (Figure 23-30), or in remote places such as abandoned mines, ocean depths, and even outer space. In Canada, one of the more likely solutions is to drill shafts 3000 m down into the solid rock in the Canadian shield region and store the wastes in containers there. See Figure 23-31. Of course, most of these possible solutions have disadvantages related to transportation (Figure 23-32), cost, environmental damage, possible earthquakes, as well as political decision-making.

Reprocessing spent fuel concentrates the reusable fuel in reactors (or nuclear bombs). But the byproducts of reprocessing are even more difficult to store than the original spent fuel. Reprocessing is not carried out in Canada because we have a large supply of the natural uranium used in CANDU reactors.

The Use of Fast Breeder Reactors

One of the concerns of using nuclear fission to generate electricity is that the earth has a limited supply of uranium. Fast breeder reactors will help reduce this problem by creating plutonium-239, which can be used in nuclear reactors to generate electricity. However, although plutonium is relatively stable, it is highly toxic. Furthermore, it can be used to make nuclear bombs, so the fear exists that terrorists or unfriendly nations might use it for purposes that would endanger human life. Breeder reactors have other design problems. The fuel in the core needs to be highly enriched, and the moderator, liquid sodium, reacts violently with water or air. As a result, this type of reactor puts great demands on the design specifications and control systems used.

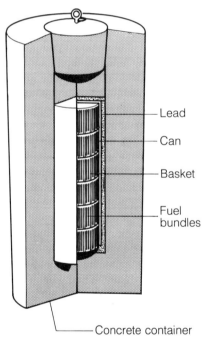

Figure 23-30
Spent Fuel Containment

USING NUCLEAR ENERGY

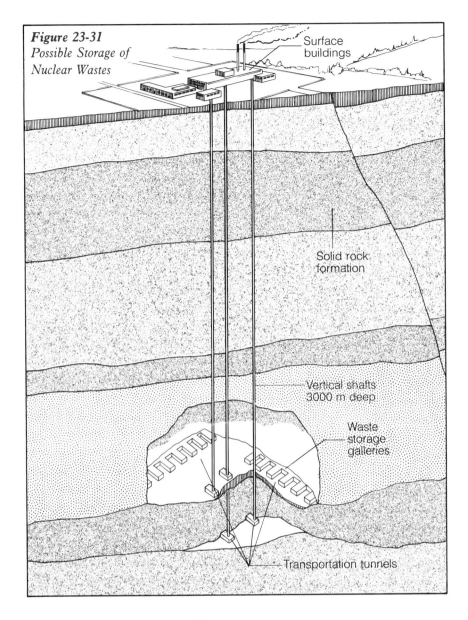

Figure 23-31
Possible Storage of Nuclear Wastes

Figure 23-32
Transporting Nuclear Wastes

(a) Current method of transporting wastes

(b) Tests are constantly being carried out to improve transportation methods. Here, a truck carrying a newly designed container and travelling at a speed of 130 km/h crashes into a concrete barrier.

(c) After the crash, the essential part of the container remains undamaged.

The Future of Nuclear Reactors

Clearly, there are both advantages and disadvantages to generating electricity using nuclear reactors. Opponents of reactors have been successful at forcing the number of proposed reactors to be reduced. In fact, in some regions such as California, the law dictates that no further reactors can be built until the problem of safe permanent storage of wastes has been solved.

Promoters of nuclear reactors point to their relatively good safety record in Canada and the problems caused by other sources of energy. Obviously, there is no consensus of opinion.

It is the responsibility of all scientists, politicians, and citizens to ensure that future generations do not have to solve the problems caused by our technology.

PRACTICE

17. Should moral issues about scientific discoveries or applications be discussed in a physics class? Explain your opinion.
18. In a discussion, compare the hazards of generating electricity using fission, fossil fuels, and hydroelectric power (from damming rivers).
19. Describe why spent-fuel storage is a large problem for the nuclear industry.
20. Research and report on methods used at CANDU reactors to ensure the safety of (a) the operating system and (b) the workers.
21. Research and report on the incidents that occurred at Three Mile Island (March, 1979) and/or Chernobyl (April, 1986). Discuss the causes and the short- and long-term effects of the incidents.

23.7 Nuclear Fusion

As the world's population increases, and as that population consumes more and more energy, the answer to our future energy needs may lie in the very process which provides much of the energy in the universe – nuclear fusion.

Nuclear fusion is the fusing (joining together) of two small nuclei to make a larger nucleus. The mass of the product nucleus is less than the sum of the masses of the original nuclei. In other words, some of the mass disappears during fusion. As in nuclear fission reactions, the lost mass changes into energy ($E = mc^2$). The energy from a fusion reaction is even greater per gram of reactants than the energy from a fission reaction, because a higher percentage of the original mass is converted into energy.

One example of a fusion reaction is the collision of two high-energy deuterium nuclei, which creates a helium nucleus, a neutron, and energy (Figure 23-33). The equation for this reaction is

$${}^2_1H + {}^2_1H \rightarrow {}^3_2He + {}^1_0n + \text{energy.}$$

Nuclear fusion is, as we have said, the basic source of energy in the universe. It goes on continuously in the sun and stars, where hydrogen and other light elements fuse into heavier elements. For instance, in every second on the sun, over 4×10^9 kg of mass disappears and changes into energy. Fortunately, the sun's mass is so vast (about 2×10^{30} kg) that we need not worry about the "death" of the sun for billions of years.

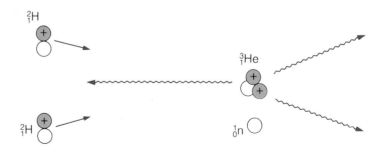

Figure 23-33
The Fusion of Two Deuterium Nuclei

Energy is given off during fusion reactions involving elements up to iron, atomic number 26. Elements of higher atomic number have nuclei with relatively strong bonds, so they need to absorb energy in order to undergo fusion. Thus, fusion results in energy emission only when the fusing nuclei are small. This may be contrasted with nuclear fission, in which energy is given off if the fissioning nuclei are very large. In other words, the middle elements are the stable, strongly-bound products of *both* fusion and fission. See Figure 23-34.

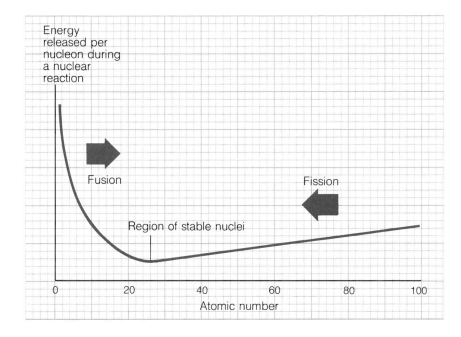

Figure 23-34
Energy Released during Fission and Fusion (Energy units are omitted.)

Since the 1950s, scientists have been searching for ways of controlling nuclear fusion in devices called **fusion reactors**. Sustained fusion in such reactors requires the reactants to be at temperatures in the millions of degrees. (Extremely high pressures, like those on the sun, would also help, but they are not possible here on earth.) Trying to maintain the reactants at such high temperatures is extremely difficult, because if they contact the solid parts of the reactor, they soon lose their energy. Developing a fusion reactor which can successfully confine the fuel and thus produce electricity commercially may be the greatest challenge facing technology in the latter part of this century.

Canadian scientists are playing an important role in researching and developing the fuel and the methods of fuel confinement in nuclear fusion reactors. The fuel for the first commercial reactors is expected to be deuterium (^2_1H) and tritium (^3_1H). Canada is a world leader in extracting deuterium from heavy water, the moderator used in CANDU reactors. Tritium, a radioactive byproduct of CANDU reactors, has useful applications in fusion research and in self-luminating industrial products.

One method of confining the fuel in a fusion reactor is called **inertial confinement**. The idea is to heat a fuel pellet to a high temperature by means of a focusing implosion. Some Canadian fusion research facilities use intense laser beams aimed from several directions toward the fusing materials (either deuterium or a combination of deuterium and tritium).

The electromagnetic energy of the laser beams increases the temperature of the fuel to such a high level that electrons are stripped from its atoms; the result is the creation of separate negative and positive charges. This ionized substance, called **plasma**, undergoes further temperature increases until fusion occurs. The energy released during the brief fusion reaction can initiate further reactions. This type of reactor is being used to investigate the interaction of laser beams and solid targets. See Figure 23-35.

Figure 23-35
The Basic Design of a Laser Fusion Reactor

(a) *The fuel, which consists of frozen fuel pellets, is injected into the fusion reactor.*

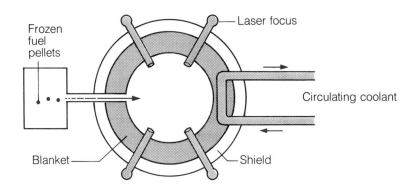

(b) *Intense laser beams cause the fuel pellets to become plasma and then to implode.*

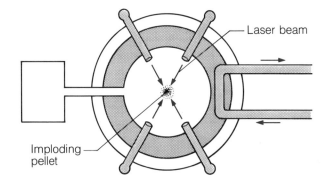

(c) *The fusion of nuclei causes a release of energy that is transferred to the coolant to be used to operate a steam turbine and generator.*

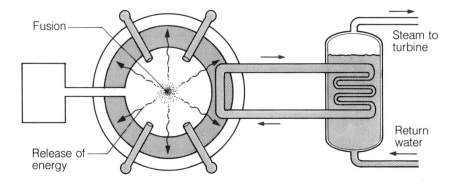

Another means of confining the fuel is **magnetic confinement**. In this method, large magnetic fields confine ionized plasma to a small space where fusion can occur. One type of reactor using magnetic confinement is a **tokamak**, which is a Russian name meaning "doughnut-shaped magnetic chamber". The design of the first Canadian tokamak, built for research purposes, is illustrated in Figure 23-36. Huge copper coils around the doughnut chamber carry electric current that produces a powerful magnetic field. The field forces the positively charged nuclei to accelerate to very high speeds as they travel around the loop, thus increasing their temperature. In the tokamak, temperatures as high as 10^8 degrees are required if the fusion reaction is to generate more energy than it consumes. Most of the energy released in the reaction is in the form of high-energy neutrons, which are used to heat the walls of the surrounding container. These walls are cooled by water that vaporizes into steam, which is then used to drive turbines to generate electricity.

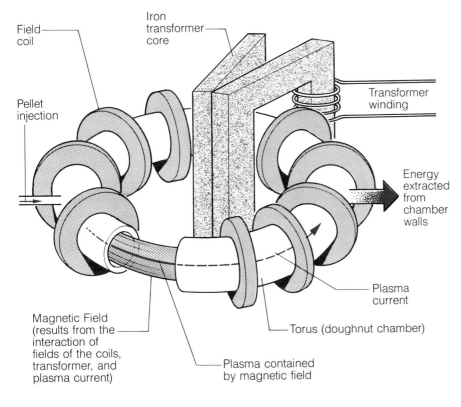

Figure 23-36
The Design of the Canadian Tokamak Fusion Facility Located at Varennes, near Montréal, Québec

The supply of the main fusion material (deuterium, which is found in ordinary water) is enormous here on earth. Thus, if nuclear fusion reactors are successful, they will provide a feasible solution for our energy supply problems in the future. Fusion reactors also have the advantage that they create far fewer radioactive wastes than fission reactors. However, the technical problems of building fusion reactors are, at least for now, difficult to overcome.

PRACTICE

22. The mass of the sun, about 2×10^{30} kg, is reduced by an estimated 4×10^9 kg each second.
    (a) Calculate the amount of energy created by the sun each second.
    (b) What is the power rating of the sun? (*Hint*: Recall the equation for power in terms of energy and time.)
    (c) Determine how long the sun will last, assuming that it continues to lose mass at the current rate.

23. Once the fusion reactions occurring in stars create enough helium nuclei, further fusion reactions take place. Some examples of these reactions are summarized below. For each example, write the full fusion reaction equation.
    (a) helium-4 + helium-4 → beryllium-8
    (b) beryllium-8 + helium-4 → carbon-12
    (c) carbon-12 + hydrogen-1 → nitrogen-13

24. For each fusion reaction below, determine the change in mass and the energy released. The masses of the isotopes are as follows:

deuterium (H-2)	3.3446×10^{-27} kg
tritium (H-3)	5.0085×10^{-27} kg
He-3	5.0084×10^{-27} kg
He-4	6.6467×10^{-27} kg
neutron	1.6750×10^{-27} kg

    (a) $^2_1H + ^2_1H \rightarrow ^3_2He + ^1_0n +$ energy
    (b) $^2_1H + ^3_1H \rightarrow ^4_2He + ^1_0n +$ energy

    (This reaction takes place in fusion reactors that use deuterium and tritium as their fuel.)

25. An alternate design to the tokamak type of fusion reactor uses a "magnetic bottle" in a process called *tandem magnetic mirror confinement*. Research and report on this method of controlling fusion.

26. Heating water to produce steam that forces turbines to spin is a relatively inefficient means of generating electricity. A more efficient method, called *magnetohydrodynamics* (MHD), creates the electrical energy directly. Research and report on how MHD can be applied to fusion reactors.

23.8 Nuclear Weapons

One of the results of the secret Manhattan Project was the invention of the nuclear bomb. The world's first nuclear weapon, a test device that used the fission of plutonium, exploded in the desert of New Mexico in the United States, on July 16, 1945. See Figure 23-37.

The world saw its first example of human destruction by a nuclear weapon on August 6, 1945, when the United States Air Force dropped a uranium fission bomb over Hiroshima, Japan. The destruction had both immediate and long-term effects. Seconds after the bomb was

Figure 23-37
Test of the First Fission Bomb

Figure 23-38
The Aftermath of the Fission Bomb at Hiroshima, Japan

dropped, the sky lit up as the chain-reaction brought temperatures within a small region to millions of degrees. Direct radiation, travelling at the speed of light in all directions after the explosion, seared human flesh. Shortly after, an expanding shock wave increased the number of people killed to about 100 000. Residents who survived the initial blast suffered long-term effects from the ionizing radiation from the bomb. See Figure 23-38.

Three days after the destruction of Hiroshima, the Americans dropped a plutonium fission bomb over Nagasaki, Japan. This bomb had much the same effect as the previous one. On August 15, 1945, Japan surrendered, ending the Second World War.

Since 1945, the race to develop nuclear weapons has been a controversial political reality involving mainly the superpowers – the United States and Russia – and, to a lesser extent, other countries such as France, England, India, China, and Israel.

Certain alarming facts about nuclear weapons indicate how seriously the subject should be treated.

- The annual world-wide military expenditure is about twice the entire annual earnings of the poorer half of the human race.
- The nuclear explosive power carried by a single aircraft bomber is greater than the total destructive power unleashed in all the wars in the history of the human race.
- One kilogram of fissionable uranium is equivalent to 7.3×10^7 kg (73 megatons) of the chemical explosive TNT (trinitrotoluene).

Did You Know?

The development of the hydrogen bomb in the U.S.A. began as a secret military operation in 1950. The first test device was set off in the Pacific Ocean in 1951, and the first fusion bomb was detonated in 1954. The U.S.S.R. exploded its first fusion test device in 1953.

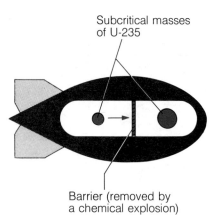

Figure 23-39
The Structure of a Uranium Fission Bomb

Did You Know?
Another method of isotope separation is gaseous diffusion. For this method to work, an absorbing cover material was needed. It is to fill this need that the material teflon was developed.

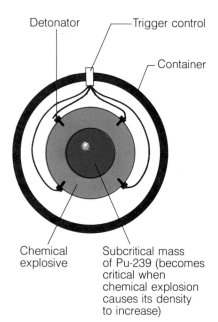

Figure 23-40
The Structure of a Plutonium Fission Bomb

Fission Bombs

The basic structure of a uranium fission bomb is shown in Figure 23-39. The uranium must be enriched to nearly pure U-235. This allows the critical mass required to initiate a chain reaction to be very low, approximately 15 kg (which has a volume about the size of a grapefruit). The U-235 is stored in two sections in a bomb, each section having a mass less than the critical mass. To explode the bomb, the two sections must be brought together by a chemical explosion.

The process of enriching natural uranium to become almost pure U-235 is extremely difficult and expensive because the two substances, U-238 and U-235, behave the same chemically. One process of doing so, **centrifuge separation**, involves spinning a gaseous compound of uranium in a high-rotation centrifuge. The heavier molecules (containing U-238) move away from the middle of the centrifuge, and the lighter molecules (containing U-235) remain near the middle, where they are drained off. Recent developments have shown that the two substances can be separated using laser light with a frequency that agitates one of the isotopes. This **laser separation** process may help military powers obtain U-235 much more cheaply than other processes.

Plutonium, also highly fissionable and used in bombs, is easily obtained because it is a byproduct in nuclear fission reactors. The design of a plutonium bomb is illustrated in Figure 23-40.

Fusion Bombs

A **fusion bomb**, also called a **hydrogen bomb**, is more technically advanced and even more devastating than a fission bomb. Its structure is shown in Figure 23-41. To obtain the high temperature required for fusion, this type of bomb begins with a fission explosion. The fission reaction releases neutrons that combine with lithium-6 to form a highly radioactive isotope of hydrogen, tritium or H-3. The tritium nuclei then fuse with the deuterium nuclei (H-2), producing α particles, neutrons, and vast amounts of energy. The equations for this process are

$$^{6}_{3}\text{Li} + ^{1}_{0}\text{n} \rightarrow ^{3}_{1}\text{H} + ^{4}_{2}\text{He}$$
$$\text{and} \quad ^{3}_{1}\text{H} + ^{2}_{1}\text{H} \rightarrow ^{4}_{2}\text{He} + ^{1}_{0}\text{n} + \text{energy}.$$

Another method of causing the fusion of hydrogen nuclei uses a laser beam to obtain the high temperatures required. This eliminates the need for a fission reaction and the subsequent production of tritium. Since much of the radioactive fallout from a fusion bomb results from the fission reaction, a laser fusion bomb can destroy a military target without showering the surrounding area with large amounts of radioactive fallout. (Apparently, this type of bomb has not yet been tested.)

Yet another type of bomb, called a **neutron bomb**, produces a relatively small explosion, but sends out an intense shower of neutrons produced in the fusion of hydrogen-2 and hydrogen-3. Although structures near the explosion are destroyed, structures beyond about 200 m remain unharmed. The high-energy neutrons, having no electric charge, can actually pass right through concrete, metal, and other materials without having any effect. However, they do great damage to biological cells, so living organisms in a range six times as far as the explosion – 1200 m – die of radiation sickness within a few hours or days. Little radiation fallout remains behind, so the area can be reinhabited after a short time. Unfortunately, this type of bomb increases the probability of nuclear warfare, because military powers are more likely to use it than the highly radioactive types of nuclear bombs.

The Arms Race

The development of nuclear arms is a result of both fear and aggression. The two superpowers (U.S.A. and U.S.S.R.) compete with one another in developing nuclear warheads, missiles to carry those warheads, and defence mechanisms to prevent destruction by the opponent's warheads. Other nations, such as those in western Europe adjacent to the Soviet Bloc, fear that a war could be staged in their own countries. There are many moral, political, economic, and philosophical questions related to the arms race. Only informed and caring citizens can make wise judgements about the factors that affect all of us.

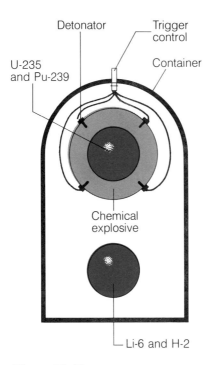

Figure 23-41
The Structure of a Fusion Bomb

PRACTICE

27. Before the United States dropped the fission bomb over Japan, some scientists, who perhaps realized the bomb's potential danger to innocent civilians, suggested that the Japanese might surrender if they were allowed to view a testing of the bomb. Discuss your opinion of the scientists' suggestion.

28. Compare fission and fusion bombs, using the following questions as a guide.
    (a) What materials are required to make the bomb?
    (b) What is the source of the materials?
    (c) How is the reaction initiated?
    (d) What is the end result of the explosion?

29. Calculate the loss of mass and the energy produced in the two reactions given for the fusion bomb. (The mass of lithium-6 is 9.99×10^{-27} kg and other required masses were given in Practice Question 24 in this chapter.)

30. Discuss or debate the following questions, taking into consideration both scientific and moral issues.
    (a) Two applications of the laser are mentioned in this section – one to enrich U-238 to obtain U-235, and the other to initiate a fusion bomb reaction. How valuable are these applications of the laser?
    (b) Should Canadian manufacturers sell heavy water to other countries? (Heavy water contains deuterium, H-2, which can be used to produce tritium, H-3.)
    (c) Should the testing of devices such as American cruise missiles over Canadian terrain be supported? (A cruise missile is a small jet airplane carrying a nuclear bomb but no pilot.)

31. Research and report on one or more of the following topics.
    (a) intercontinental ballistic missiles (ICBMs)
    (b) sea-launched ballistic missiles (SLBMs)
    (c) anti-ballistic missiles (ABMs)
    (d) strategic arms limitation talks (SALT)
    (e) tactical nuclear weapons (based on the use of neutron bombs)

23.9 The Ever-Changing Theories of Physics

Figure 23-42
This first stage of a particle accelerator gives the initial boost of energy to protons. The high-speed protons will later collide with a target and produce new particles to be studied.

Figure 23-43
The Interaction of a Particle and Its Antiparticle

(a) An electron (e^-) and a positron (e^+) approach each other.

(b) They collide and destroy one another, producing energy in the form of γ radiation.

The study and applications of modern physics produce more questions than answers – both those related to theory and practical applications, and those related to human values and the philosophy of life.

Today's physicists, like their counterparts in previous centuries, continue to search for explanations of their surroundings. In the quest to delve deeper into the nucleus, scientists have built large devices called **particle accelerators**, such as the one shown in Figure 23-42, at great expense. The purpose of an accelerator is to give as much energy as possible to particles and force them to collide so that an interaction can take place. The products of the interaction help physicists speculate on the structure of matter.

Besides studying matter in accelerators, physicists also study antimatter. **Antimatter** consists of **antiparticles**, which are produced in nuclear reactions and other processes when enough energy is available. Antiparticles are not ordinarily observed (in our part of the universe) because they cannot exist long in the presence of ordinary matter. Very shortly after an antiparticle is created, it is likely to meet with its corresponding particle, and the two will destroy each other, producing energy. Every known particle is assumed to have its own antiparticle. For example, the antiparticle of an electron is a positron, which has the same mass as an electron but the opposite charge. Although antimatter lasts for only a small fraction of a second in our part of the universe, it is possible that it leads a stable existence somewhere else. See Figure 23-43.

As a result of research using particle accelerators, physicists say there are apparently two general categories of particles. One category, called **leptons**, consists of particles which can be found alone, are influenced by the weak nuclear force, and usually have zero or little mass. (The word "lepton" is from the Greek and means "light in weight".) There may be up to six elementary particles which are leptons, the most common one being the electron. Two others are listed in Table 23-2.

The second general category of particles consists of more massive particles called **hadrons**. ("Hadron" means "bulky".) These are particles that are influenced by the strong nuclear force. Hadrons are further subdivided into two other categories – **mesons**, which are lighter than any nucleons, and **baryons** such as protons, neutrons, and other

USING NUCLEAR ENERGY

heavier particles. ("Meson" means "middle" and "baryon" means "heavy".) Hundreds of hadrons have been discovered. Some of the more common ones are listed in Table 23-2. See also Figure 23-44.

Table 23-2 Partial List of Elementary Particles

Category	Particle Name	Symbol	Electric Charge
leptons	electron	e	-1
	muon	μ	-1
	neutrino	ν	0
mesons	pi meson or pion	π	$+1, 0,$ or -1
	k meson or kaon	K	$+1, 0,$ or -1
baryons	proton	p	$+1$
	neutron	n	0
	lambda	Λ	0
	sigma	Σ	$+1, 0,$ or -1

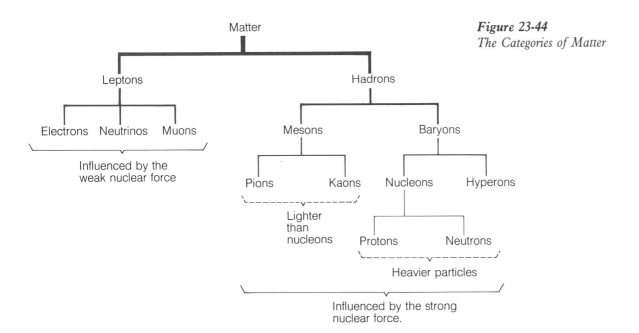

Figure 23-44
The Categories of Matter

Because so many hadrons were discovered, physicists began to call the collection of particles the "particle zoo". At first there appeared to be no sense or order within the particle zoo, so scientists began to search for a simple set of particles that would help explain it. Thus in 1963 began the theory of quarks, which combine in various ways to form protons, neutrons, and other particles. A **quark** is a point-like particle that is influenced by the strong nuclear force and, at least to date, has never been found alone. (The word "quark" was taken from the phrase "three quarks for Muster Mark" in the novel *Finnegans Wake* by James Joyce.) There appear to be six types (or *flavours*) of quarks: up, down, strange, charmed, beauty (also called bottom), and truth (also called top). One of the properties of quarks is that they have a fractional charge, such as +2/3 or −1/3. (Refer to Table 23-3.) Another unusual property is that the farther they are pulled apart, the greater is their force of attraction for one another. This *might* explain why they do not appear by themselves, but no one knows for certain.

Table 23-3   Quarks

Name	Symbol	Charge
up	u	+2/3
down	d	−1/3
strange	s	−1/3
charmed	c	+2/3
beauty (or bottom)	b	−1/3
truth (or top)	t	+2/3

Quark theory is used to try to explain the composition of the two types of hadrons, mesons and baryons. It is believed that a meson consists of a quark and its antiquark, while the heavier baryon consists of three quarks. Consider, for example, a proton, which is a stable baryon. It consists of three quarks, two "up" quarks, each with a charge of +2/3, and one "down" quark, with a charge of −1/3. Thus, the total charge of a proton (uud) is +4/3 − 1/3 or +1. See Figure 23-45.

As you read these statements, you may be thinking that physics and physicists have become rather strange. Students in the nineteenth century probably felt the same way when they learned that experiments proved that atoms could be divided. And students in the early part of the twentieth century might have thought that physicists were a confused lot when they described how a nucleus could be divided. But those events seem relatively commonplace to today's students, so who can predict what will be discovered in the future? As the twenty-first century approaches, and as devices like particle accelerators become more sophisticated, perhaps scientists will learn that leptons and quarks, too, can be divided.

Thus, physics provides a means of satisfying our basic need to try to understand ourselves and our universe. It also provides applications that, for the most part, better our lives. Physics is often exciting, sometimes frightening, and *always* open to new ideas and discoveries.

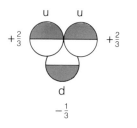

Figure 23-45
The Composition of a Proton, According to Quark Theory

PRACTICE
32. Name the general categories of particles influenced by (a) the weak nuclear force and (b) the strong nuclear force.
33. What is the purpose of a particle accelerator?
34. A neutron, which consists of three quarks, has zero charge. How can down and up quarks combine to make a neutron?
35. A lambda particle consists of three quarks, uds. What is the charge of this type of baryon?

Review Assignment

1. What are three main ways in which a nucleus can undergo a transmutation? (One way was described in the previous chapter.)
2. Write the nuclear symbols for the following particles:
   (a) α particle   (d) proton
   (b) β particle   (e) deuterium nucleus
   (c) neutron   (f) tritium nucleus
3. Why are neutrons more effective than protons in bombarding nuclei?
4. Complete the following nuclear reaction equations:
   (a) $^{133}_{55}Cs + ^{1}_{0}n \rightarrow ^{132}_{54}Xe + ?$
   (b) $^{6}_{3}Li + ? \rightarrow ^{9}_{5}B + ^{4}_{2}He$
5. An α particle (He-4) is used to bombard an oxygen nucleus (O-16). Write the nuclear reaction equation if one of the products is (a) neon (Ne-19) and (b) beryllium (Be-9).
6. Determine the amount of energy created when one microgram (1.0 μg) of a substance changes entirely into energy.
7. For the nuclear reaction given below, calculate the mass lost and the energy created. (Data for the particle masses are found in Practice Questions 24 and 29 of this chapter.)

   $^{6}_{3}Li + ^{1}_{0}n \rightarrow ^{4}_{2}He + ^{3}_{1}H$

8. Why is energy released when the nuclei of a heavy element undergo fission?
9. A nuclear fission reaction that uses uranium as its fuel requires the use of a moderator.
   (a) Why is the moderator necessary?
   (b) In general, why is heavy water a more effective moderator than ordinary water?
10. What conditions are required for a sustained chain reaction? Do you think it is possible for chain reactions to occur in natural deposits of uranium ore?
11. Describe the function of the following components of a CANDU reactor:
    (a) calandria   (c) coolant
    (b) control rods   (d) turbines
12. State an advantage and a disadvantage of a fast breeder reactor.
13. In what ways do a nuclear fission reactor and a nuclear fission bomb differ?
14. Which cause greater concern to the nuclear reactor industry, reactor wastes or spent-fuel wastes? Why?
15. If you were on a citizens' action committee, what arguments would you give for and against locating a nuclear waste disposal site in your area?
16. What is nuclear fusion? What is the source of energy in a fusion reaction?
17. Explain why the materials used for nuclear fusion cannot be confined in an ordinary container.

Figure 23-46
The Orion Nebula

18. What two main techniques are currently being researched to contain nuclear fusion?
19. Plasma is often called the fourth state of matter. How does it compare with the other states (solid, liquid, and gas)?
20. The evolution of stars is an interesting topic in the study of astronomy. As stars undergo changes, nuclear reactions occur. For example, the final explosion of some stars is called a **supernova**. A star to which this happens becomes extremely bright temporarily, then becomes a nebula (Figure 23-46). The gravitational forces in the middle of the star become so great that the star collapses. The resulting extremely high temperatures in its core initiate fusion reactions which create elements beyond element 26 (iron, Fe). First, several neutrons are liberated (equation A below); then they are captured by other Fe-56 nuclei (equation B). The resulting Fe-57 nuclei undergo β decay, creating Co-57 (equation C). This process of neutron capture and β decay can continue to create the elements heavier than iron.

   A. $^{56}_{26}\text{Fe} \rightarrow 26\,^{1}_{1}\text{H} + 30\,^{1}_{0}\text{n}$

   B. $^{56}_{26}\text{Fe} + ^{1}_{0}\text{n} \rightarrow ^{57}_{26}\text{Fe}$

   C. $^{57}_{26}\text{Fe} \rightarrow ^{57}_{27}\text{Co} + ^{0}_{-1}\text{e}$

   (a) In order to produce arsenic (As-75), how many such sets of reactions (neutron capture followed by beta decay), starting with Fe-56, must occur? (Refer to the periodic table of the elements, Appendix B.)
   (b) Repeat (a) for uranium (U-238).
   (c) Do research on supernovas and/or nebulas, and give a report on your findings.

21. When a supernova occurs, it is hypothesized that the electrons in the plasma are forced into the nucleus where they are captured by protons to form neutrons. This process results in the formation of a **neutron star**, which is a dense ball of neutrons. Astronomers think that neutron stars are the basis of **pulsars**.
    (a) What do you think holds a neutron star together?
    (b) How many neutrons result when each of the following nuclei absorbs electrons to become electrically neutral?
       (i) $^{14}_{7}\text{N}$ (ii) $^{19}_{9}\text{F}$ (iii) $^{56}_{26}\text{Fe}$
    (c) Do research on pulsars and/or neutron stars, and give a report on your findings.

22. Is there a limit to the amount of fuel that can be stored in one lump if the fuel is used for (a) a fission bomb and (b) a fusion bomb? Explain.
23. Any country using a nuclear fission reactor has the means of obtaining materials for making a bomb. What type of bomb can be made from materials obtained from (a) a CANDU reactor and (b) a breeder reactor?
24. Comment on the statement, "All our energy resources are nuclear in origin."
25. What happens when a particle and its antiparticle meet?

26. Determine the energy released when an electron and a positron (each of mass 9.11×10^{-31} kg) annihilate each other.
27. Repeat Review Question 26 for a proton and an antiproton, each having a mass of 1.67×10^{-27} kg.
28. What is the charge of a particle composed of these sets of quarks?
    (a) ddu (b) uud
29. The transuranic elements are those whose atomic numbers exceed 92 (uranium). By referring to the periodic table of the elements in Appendix B, list the transuranic elements that are named in honour of scientists mentioned in this chapter. State a major contribution to physics by each scientist named.

Key Objectives

Having completed this chapter, you should now be able to do the following:
1. Define a nuclear reaction, and give an example of both natural and artificial nuclear reactions.
2. Describe why neutrons are more apt to initiate a nuclear reaction than protons or α particles.
3. Write balanced equations for given nuclear reactions.
4. Apply the law of conservation of mass-energy to explain why energy is released in nuclear reactions.
5. Given the masses of particles before and after a nuclear reaction, determine the loss of mass and the amount of energy created.
6. Name and describe the force that binds nucleons together.
7. Define nuclear fission.
8. Describe how a chain reaction is set up.
9. Describe the functions of the main components of a nuclear fission reactor.
10. Trace the energy changes that allow nuclear energy to be converted into electrical energy.
11. Draw a labelled diagram illustrating the basic operation of a CANDU reactor.
12. Recognize the common characteristics of the various types of nuclear fission reactors.
13. Explain both the advantages and the disadvantages of generating electrical energy using nuclear fission reactors.
14. Define nuclear fusion, and compare it to nuclear fission.
15. Describe the fundamental design of a nuclear fusion reactor.
16. Compare the design and operation of a nuclear fission bomb with that of a nuclear fusion bomb.
17. Recognize the societal implications of the spread of nuclear devices, especially warheads.
18. Distinguish between matter and antimatter; leptons and hadrons; mesons and baryons.
19. Recognize how the quark theory tries to simplify the explanation of numerous particles.
20. Appreciate that research in modern physics carries both advantages and disadvantages.

Epilogue An Overview of Physics

What do the two photographs shown here have in common? First, although one appears much more modern than the other, they were both taken in the year this book was written. Second, they both involve physics principles. In the one, a gasoline pump operates on the mechanical principles of simple machines. This pump is located in a remote area of Africa, far from any source of electricity. In the other photograph, numerous physics principles are combined in attaining the stage at which a satellite is launched from the cargo bay of a space shuttle high above the earth. Third, both photographs epitomize the flavour of this book. Physics is more than just equations and problem-solving; it is also an awareness of examples and applications, whether simple or complex, which can be observed either close to home or around the world.

As you studied the principles of physics and their applications in this book, you were exposed to many related concepts. Historical developments were included to help you understand the process of science and appreciate the increased rate at which knowledge is accumulating. Societal implications were introduced, often in the form of questions, to help make you aware of the concerns and responsibilities we should adopt in order to ensure a healthy and happy world for future generations. The biographical summaries of Canadian scientists were added to provide glimpses of physics-related careers.

Having considered the past and present, what can we predict for the future? No one knows for certain what lies ahead, and it is difficult to predict how the negative effects of pollution and the depletion of our non-renewable energy resources will influence scientific progress. However, it is interesting to consider several of the exciting possibilities that some scientists predict may be a part of the future.

(a) *Space Exploration*   Space probes assisted by computers will gather detailed information about our solar system. Minerals will be mined on the moon and the asteroids, and living in space could become a reality.

(b) *Artificial Intelligence*   Machines that talk, make decisions, and even think may change the very pattern of human thought.

(c) *Astronomy and Astrophysics*   The most ancient of the sciences, astronomy, will continue to be important. Both radio telescopes which receive signals from space in the range of radio frequencies and new types of telescopes in space will gather information for analysis by computers. Astrophysicists will gain much new knowledge about phenomena such as pulsars, neutron stars, and black holes, and about the origin of the universe.

(d) *Particle Physics (or High-Energy Physics)*   This branch of physics will contribute to our understanding of the building blocks of nature.

(e) *Fundamental Forces*   There is evidence that two of the fundamental forces in nature, the electromagnetic force and the weak nuclear force, are manifestations of a single fundamental force. Physicists will continue to do research to determine whether all four forces, including the strong nuclear force and the gravitational force, may be unified into one grand theory.

(f) *Plasma Physics*   As the world's non-renewable resources become depleted, this branch of physics will become more important as physicists attempt to control nuclear fusion.

(g) *Lasers* Tunable lasers and holography will become more sophisticated and common.

(h) *Biophysics* As the understanding of biological systems improves, many more medical applications will be developed.

(i) *Solid State Physics* This branch of physics, which made possible portable electronic devices and space travel, will continue to develop as the technology is extended to liquids.

(j) *Cryogenics* Research in this branch of physics, which utilizes temperatures approaching absolute zero, will result in applications of the materials called superconductors.

These predictions certainly represent a vast increase in knowledge since the origins of science and technology among the ancient civilizations. Although the amount of knowledge has increased, the reason for its existence remains the same as in the past: it is part of human nature to strive to understand human life and the universe, and to apply that understanding for the benefit of all.

Review Assignment

1. State which scientist has been honoured in the name of the unit for each of the following quantities:
   (a) pressure
   (b) force
   (c) frequency
   (d) electric charge
   (e) electric potential
   (f) electric current
   (g) electric resistance
   (h) power
   (i) energy
   (j) radiation activity
   (k) sound intensity level
   (l) common temperature
   (m) thermodynamic temperature

2. Determine a logical estimate for the number of ballpoint pen refills it took to write the original manuscript of this book.

3. An interesting statistic to consider is that, historically, whenever people set out to explore unknown territories, the average maximum time away from home was about four years. (A few centuries ago four years was the amount of time it took to circumnavigate the earth.) Using the data listed below, determine the approximate point at which a space probe would have to begin its return flight to ensure that the humans aboard need not stay away longer than four years.

Maximum speed of rocket:	1.5×10^5 km/h
Circumference of the earth:	4×10^4 km
Distance to the moon:	4×10^5 km
Distance to Venus, the nearest planet:	4×10^7 km
Distance to the farthest planet:	4×10^9 km
Distance to Proxima Centauri, the nearest star beyond the solar system:	4.3 light years

4. Assume that you are part of a team researching the feasibility of sending a space probe to Proxima Centauri (distance given in the previous question).
   (a) At an acceleration equal in magnitude to the average acceleration due to gravity on earth, how long would it take a rocket to reach a speed equal to 98% of the speed of light in a vacuum?
   (b) How far will the rocket in (a) travel in the time calculated? How much farther must it travel to reach Proxima Centauri? What is the total time it will take to get there?
   (c) If a radio signal, which is an electromagnetic wave, is sent from the earth to the space probe near Proxima Centauri, what minimum time will elapse before a reply can be expected back on earth?
   (d) Comment on the feasibility of travelling to other stars.

5. **Activity** For this activity, you will need a known quantity of cold water, a thermometer, an electric kettle of known power, a stirring rod, and a timing device. Add the water to the kettle and determine the initial temperature. Plug the kettle in and determine the time needed to increase the water temperature to 80°C. Stir the water for best results. Answer the following questions, which involve concepts from Chapters 2, 6, 7, and 19.
   (a) Determine the output work done by the kettle in heating water of mass m and specific heat capacity c to change the temperature by ΔT.
   (b) Calculate the input work done by electricity at a known power rating P for a measured amount of time t.
   (c) Find the percent efficiency of the kettle in the situation described by calculating the ratio of the output work to the input work.
   (d) Compare the efficiency of the kettle with that of an electric motor (about 90%) and a gasoline engine (about 25%).
   (e) Suggest some ways in which people could improve the efficiency of the use of electric appliances in the home.

6. Make a list of questions you would ask a physicist whom you were interviewing.

7. Give arguments in favour of, and opposed to, the statement, "Physicists should become involved in political decision-making."

8. Choose one of the predictions made in this epilogue and discuss the current situation as well as implications for the future. (Research may be necessary to answer this question.)

APPENDICES

Appendix A The Metric System of Measurement

1. Metric Prefixes and Their Origins

Prefix	Abbreviation	Meaning	Origin
exa	E	10^{18}	Greek *exa* – out of
peta	P	10^{15}	Greek *peta* – spread out
tera	T	10^{12}	Greek *teratos* – monster
giga	G	10^{9}	Greek *gigas* – giant
mega	M	10^{6}	Greek *mega* – great
kilo	k	10^{3}	Greek *khilioi* – thousand
hecto	h	10^{2}	Greek *hekaton* – hundred
deca	da	10^{1}	Greek *deka* – ten
Standard Unit	—	10^{0}	
deci	d	10^{-1}	Latin *decimus* – tenth
centi	c	10^{-2}	Latin *centum* – hundred
milli	m	10^{-3}	Latin *mille* – thousand
micro	μ	10^{-6}	Greek *mikros* – very small
nano	n	10^{-9}	Greek *nanos* – dwarf
pico	p	10^{-12}	Italian *piccolo* – small
femto	f	10^{-15}	Greek *femten* – fifteen
atto	a	10^{-18}	Danish *atten* – eighteen

2. Système International (SI) Base Units of Measurement and Their Definitions

Quantity	Unit	Symbol
Length	metre	m
Mass	kilogram	kg
Time	second	s
Electric current	ampere	A
Thermodynamic temperature	kelvin	K
Amount of substance	mole	mol
Luminous intensity	candela	cd

DEFINITIONS

The *metre* is the distance that light travels in a vacuum in 1/299 792 458 of a second.

The *kilogram* is equal to the mass of the international prototype of the kilogram kept at the International Bureau of Weights and Measures.

The *second* is the time for 9 192 631 770 vibrations of a cesium-133 atom.

The *ampere* is that current which, if maintained in two straight parallel conductors of infinite length, of negligible circular cross-section, and placed 1 m apart in vacuum, would produce between these conductors a force equal to 0.20 μN/m of length.

The *kelvin*, the unit of thermodynamic temperature, is the fraction 1/273.16 of the thermodynamic temperature of the triple point of water. (*Note*: The triple point of water has a temperature of 0.01°C; hence, 0°C corresponds to 273.15 K.)

The *mole* is the amount of substance of a system which contains as many elementary entities as there are atoms in 0.012 kg of carbon-12.

The *candela* is the luminous intensity, in the perpendicular direction, of a surface of 1/600 000 m² of a black body (full radiator) at the temperature of freezing platinum under a pressure of 101.325 kPa.

3. Derived Units Often Used in Physics

Quantity	Unit	Symbol	Relation to Other Units
Force	newton	N	$1.0 \text{ N} = 1.0 \text{ kg} \cdot \text{m/s}^2$
Energy and work	joule	J	$1.0 \text{ J} = 1.0 \text{ N} \cdot \text{m}$
Power	watt	W	$1.0 \text{ W} = 1.0 \text{ J/s}$
Pressure	pascal	Pa	$1.0 \text{ Pa} = 1.0 \text{ N/m}^2$
Frequency	hertz	Hz	$1.0 \text{ Hz} = 1.0 \text{ s}^{-1}$
Electric charge	coulomb	C	$1.0 \text{ C} = 1.0 \text{ A} \cdot \text{s}$
Electric potential	volt	V	$1.0 \text{ V} = 1.0 \text{ J/C}$
Electric resistance	ohm	Ω	$1.0 \text{ }\Omega = 1.0 \text{ V/A}$
Radiation activity	becquerel	Bq	$1.0 \text{ Bq} = 1.0 \text{ s}^{-1}$

4. Using the Metric Memory Aid

E P T G M k h da — d c m μ n p f a

(a) *Linear Measure*: Each line in the memory aid represents a move of a single decimal place. (Refer to Sample Problems 4, 5, and 6 in Chapter 2.)
e.g., $2.4 \times 10^2 \text{ mm} = 2.4 \times 10^{-1} \text{ m}$

(b) *Area Measure*: Each line in the memory aid represents a move of a double decimal place.
e.g., $2.4 \times 10^2 \text{ mm}^2 = 2.4 \times 10^{-4} \text{ m}^2$

(c) *Volume Measure:* Each line in the memory aid represents a move of a triple decimal place.
e.g., $2.4 \times 10^2 \text{ mm}^3 = 2.4 \times 10^{-7} \text{ m}^3$

Appendix B Periodic Table of the Elements

1 **H** Hydrogen 1								
7 **Li** Lithium 3	9 **Be** Beryllium 4							
23 **Na** Sodium 11	24 **Mg** Magnesium 12							
39 **K** Potassium 19	40 **Ca** Calcium 20	45 **Sc** Scandium 21	48 **Ti** Titanium 22	51 **V** Vanadium 23	52 **Cr** Chromium 24	55 **Mn** Manganese 25	56 **Fe** Iron 26	59 **Co** Cobalt 27
85 **Rb** Rubidium 37	88 **Sr** Strontium 38	89 **Y** Yttrium 39	90 **Zr** Zirconium 40	93 **Nb** Niobium 41	98 **Mo** Molybdenum 42	99 **Tc** Technetium 43	102 **Ru** Ruthenium 44	103 **Rh** Rhodium 45
133 **Cs** Cesium 55	138 **Ba** Barium 56	See 'A' below 57–71	180 **Hf** Hafnium 72	181 **Ta** Tantalum 73	184 **W** Tungsten 74	187 **Re** Rhenium 75	192 **Os** Osmium 76	193 **Ir** Iridium 77
223 **Fr** Francium 87	226 **Ra** Radium 88	See 'B' below 89–103	261 **Rf** Rutherfordium 104	262 **Ha** Hahnium 105				

Legend:
- 28 — Mass number (protons + neutrons) of the most stable isotope
- **Si** — Symbol
- Silicon — Name
- 14 — Atomic number (protons)

A The Rare Earth Series

139 **La** Lanthanum 57	140 **Ce** Cerium 58	141 **Pr** Praseodymium 59	142 **Nd** Neodymium 60	147 **Pm** Promethium 61	152 **Sm** Samarium 62	153 **Eu** Europium 63

B The Actinide Series

227 **Ac** Actinium 89	232 **Th** Thorium 90	231 **Pa** Protactinium 91	238 **U** Uranium 92	237 **Np** Neptunium 93	244 **Pu** Plutonium 94	245 **Am** Americium 95

APPENDICES

									4 **He** Helium 2
			11 **B** Boron 5	12 **C** Carbon 6	14 **N** Nitrogen 7	16 **O** Oxygen 8	19 **F** Fluorine 9	20 **Ne** Neon 10
			27 **Al** Aluminum 13	28 **Si** Silicon 14	31 **P** Phosphorus 15	32 **S** Sulphur 16	35 **Cl** Chlorine 17	40 **Ar** Argon 18
58 **Ni** Nickel 28	64 **Cu** Copper 29	64 **Zn** Zinc 30	69 **Ga** Gallium 31	74 **Ge** Germanium 32	75 **As** Arsenic 33	80 **Se** Selenium 34	79 **Br** Bromine 35	84 **Kr** Krypton 36
106 **Pd** Palladium 46	108 **Ag** Silver 47	114 **Cd** Cadmium 48	115 **In** Indium 49	120 **Sn** Tin 50	121 **Sb** Antimony 51	130 **Te** Tellurium 52	127 **I** Iodine 53	132 **Xe** Xenon 54
195 **Pt** Platinum 78	197 **Au** Gold 79	202 **Hg** Mercury 80	205 **Tl** Thallium 81	206 **Pb** Lead 82	209 **Bi** Bismuth 83	209 **Po** Polonium 84	210 **At** Astatine 85	228 **Rn** Radon 86

158 **Gd** Gadolinium 64	159 **Tb** Terbium 65	164 **Dy** Dysprosium 66	165 **Ho** Holmium 67	168 **Er** Erbium 68	169 **Tm** Thulium 69	174 **Yb** Ytterbium 70	175 **Lu** Lutetium 71

248 **Cm** Curium 96	249 **Bk** Berkelium 97	249 **Cf** Californium 98	252 **Es** Einsteinium 99	253 **Fm** Fermium 100	256 **Md** Mendelevium 101	254 **No** Nobelium 102	257 **Lw** Lawrencium 103

Appendix C Physical Constants

Quantity	Symbol	Approximate Value
Speed of light in a vacuum	c	3.00×10^8 m/s
Gravitational constant	G	6.67×10^{-11} N·m²/kg²
Coulomb's constant	k	9.00×10^9 N·m²/C²
Charge on electron	$-e$	-1.60×10^{-19} C
Charge on proton	e	1.60×10^{-19} C
Electron mass	m_e	9.11×10^{-31} kg
Proton mass	m_p	1.673×10^{-27} kg
Neutron mass	m_n	1.675×10^{-27} kg
Atomic mass unit	u	1.660×10^{-27} kg

Appendix D Terrestrial Data

Quantity	Approximate Value
Mass of earth	5.98×10^{24} kg
Radius of earth at equator	6.38×10^6 m
Acceleration due to gravity	9.80 m/s²
Mean earth-moon distance	3.84×10^8 m
Mean earth-sun distance	1.50×10^{11} m
Standard atmospheric pressure	1 atm or 1.013×10^5 Pa

Appendix E Trigonometry

Trigonometry is the branch of mathematics that deals with the relations between the sides and angles of triangles. In the right-angled triangle, $\triangle ABC$, in the diagram, there are three angles, A, B, and C, and three sides a, b, and c. The six possible ratios of sides in this triangle are a:b, a:c, b:c, and their reciprocals b:a, c:a, and c:b. Each ratio is given a specific name, but for use in the topic of light in this text, only one ratio is needed. This ratio is called the *sine ratio* or *sine function*; it has the symbol *sin*.

In $\triangle ABC$, the sine of angle A is the ratio a:c or $\frac{a}{c}$. Since $\triangle AB'C'$ is similar to $\triangle ABC$, this ratio is the same as the ratio a':c'. Thus,

$$\sin A = \frac{a}{c} = \frac{a'}{c'}.$$

In general, for any angle θ in a right-angled triangle,

$$\sin \theta = \frac{\text{length of the side opposite } \theta}{\text{length of the hypotenuse}}.$$

Sine ratios for angles between 0° and 90° are tabulated in Appendix F. Two other common ratios, *cosine* and *tangent*, are also tabulated for interest.

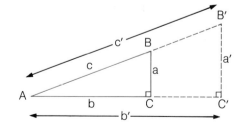

Figure E-1

Appendix F Table of Trigonometric Values

Degree	Sine	Cosine	Tangent	Degree	Sine	Cosine	Tangent
0°	0.000	1.000	0.000				
1°	.017	1.000	.017	46°	0.719	0.695	1.036
2°	.035	0.999	.035	47°	.731	.682	1.072
3°	.052	.999	.052	48°	.743	.669	1.111
4°	.070	.998	.070	49°	.755	.656	1.150
5°	.087	.996	.087	50°	.766	.643	1.192
6°	.105	.994	.105	51°	.777	.629	1.235
7°	.122	.992	.123	52°	.788	.616	1.280
8°	.139	.990	.140	53°	.799	.602	1.327
9°	.156	.988	.158	54°	.809	.588	1.376
10°	.174	.985	.176	55°	.819	.574	1.428
11°	.191	.982	.194	56°	.829	.559	1.483
12°	.208	.978	.212	57°	.839	.545	1.540
13°	.225	.974	.231	58°	.848	.530	1.600
14°	.242	.970	.249	59°	.857	.515	1.664
15°	.259	.966	.268	60°	.866	.500	1.732
16°	.276	.961	.287	61°	.875	.485	1.804
17°	.292	.956	.306	62°	.883	.470	1.881
18°	.309	.951	.325	63°	.891	.454	1.963
19°	.326	.946	.344	64°	.899	.438	2.050
20°	.342	.940	.364	65°	.906	.423	2.145
21°	.358	.934	.384	66°	.914	.407	2.246
22°	.375	.927	.404	67°	.921	.391	2.356
23°	.391	.920	.424	68°	.927	.375	2.475
24°	.407	.914	.445	69°	.934	.358	2.604
25°	.423	.906	.466	70°	.940	.342	2.747
26°	.438	.899	.488	71°	.946	.326	2.904
27°	.454	.891	.510	72°	.951	.309	3.078
28°	.469	.883	.532	73°	.956	.292	3.271
29°	.485	.875	.554	74°	.961	.276	3.487
30°	.500	.866	.577	75°	.966	.259	3.732
31°	.515	.857	.601	76°	.970	.242	4.011
32°	.530	.848	.625	77°	.974	.225	4.331
33°	.545	.839	.649	78°	.978	.208	4.705
34°	.559	.829	.674	79°	.982	.191	5.145
35°	.574	.819	.700	80°	.985	.174	5.671
36°	.588	.809	.726	81°	.988	.156	6.314
37°	.602	.799	.754	82°	.990	.139	7.115
38°	.616	.788	.781	83°	.993	.122	8.144
39°	.629	.777	.810	84°	.995	.104	9.514
40°	.643	.766	.839	85°	.996	.087	11.43
41°	.656	.755	.869	86°	.998	.070	14.30
42°	.669	.743	.900	87°	.999	.052	19.08
43°	.682	.731	.933	88°	.999	.035	28.64
44°	.695	.719	.966	89°	1.000	.017	57.29
45°	.707	.707	1.000	90°	1.000	.000	∞

Appendix G The Greek Alphabet

A	α	alpha	I	ι	iota	P	ρ	rho
B	β	beta	K	κ	kappa	Σ	σ	sigma
Γ	γ	gamma	Λ	λ	lambda	T	τ	tau
Δ	δ	delta	M	μ	mu	Υ	υ	upsilon
E	ϵ	epsilon	N	ν	nu	Φ	ϕ	phi
Z	ζ	zeta	Ξ	ξ	xi	X	χ	chi
H	η	eta	O	o	omicron	Ψ	ψ	psi
Θ	θ	theta	Π	π	pi	Ω	ω	omega

Answers to Numerical Problems

Chapter 2

PRACTICE
4. (a) $6.272\,640 \times 10^6$ in^2
6. (a) kg·m^2/s^2
   (b) kg·m^2/s^3
   (c) kg/(s^2·m)
7. (a) 3.2×10^7 s
   (b) 6.25×10^{18} electrons/s
   (c) 9.192×10^9 vibrations/s
   (d) 3.8×10^{-10} kg
8. (a) 420 000 kg
   (b) 0.000 000 31 L
   (c) 82 s
   (d) 956 400 000 m
9. (a) 42.8×10
   (b) 0.64×10^6
   (c) $93\,000 \times 10^6$
   (d) 0.0012×10^6
10. (a) 3.8272×10^{10} g
    (b) $4.451\,23 \times 10^5$ m
11. (a) 4.3387×10^6 cm
    (b) 3.1×10^{-7} s
12. (a) 6.0×10^{25}
    (b) 4.8×10^{11}
13. (a) 2.0×10^6 kg/m^3
    (b) 2.0 cycles/s
14. (a) 38 hm   (d) 0.57 dL
    (b) 49 hg   (e) 3.05 m
    (c) 400 dW (f) 0.125 dag
15. (a) 9.4×10^{10} m
    (b) 8.5×10^{-3} g
    (c) 7.1×10^{23} mm
    (d) 2.9×10^2 Ms
    (e) 6.5×10^{-15} mm
21. (a) 3
    (b) 4
    (c) 2
    (d) 3
22. 3.06% or 3.1%
23. (a) 6.4   (f) 3.5
    (b) 8.5   (g) 5.1
    (c) 2.8   (h) 7.2
    (d) 8.2   (i) 2.8
    (e) 4.6   (j) 3.2
24. (a) 11.3 cm
    (b) 60.9 kg
    (c) 4 mL
25. (a) 5.3×10^2 m
    (b) 2.5 cm/s
27. $d = a - bc$
    $c = (a - d)/b$
28. (a) $x = 3 - y$
    (b) $x = 2y/3$
    (c) $x = (m - n)/(4y)$
    (d) $x = aby$
29. (a) 30 m
    (b) 84 m
31. (a) 3.14 or 3.1;
    0.33 m/year;
    6.0 m/s

REVIEW ASSIGNMENT
2. (a) 6.2×10^5 m
   (b) 8.9×10^7 kg
   (c) 9.8×10^4 mm
3. (a) 3.7×10^{12} cm
   (b) 4.1×10^{-8} mm
   (c) 6.75×10^{-3} kW
   (d) 5.6×10^5 km/s
4. (a) 1.1 g/cm^3
   (b) 1.1×10^3 kg/m^3
   (c) 8.3%
5. (a) $t = E/P$
   (b) $t = (v_f - v_i)/a$
   (c) $t = \sqrt{2d/a}$
6. 2 678 400 s
7. (b) 21 m/s
8. (a) 2.0 mg per grain
   (b) (i) 1.2×10^5 grains
       (ii) 1.2×10^6 grains
       (iii) 1.2×10^9 grains

PRACTICE (SUPPLEMENTARY)
36. (c) 1.0×10^3 m
37. (c) 3.14

Chapter 3

PRACTICE
5. (a) 720 m
   (b) 280 m [N]
6. (a) 12 m, 17 m, 29 m, 34 m
   (b) 12 m [N],
       13 m [N 23° W],
       5.0 m [W], 0.0 m
7. (a) 628 m
   (b) 0.0 m
10. (a) 11 m/s
    (b) 2.5 m/s
    (c) 28 m/s
11. (a) 0.64 km/h
    (b) 36 km/h
    (c) 86 km/h
12. 6.0×10^2 m/s
13. $\times 10^3$ m/s
14. (a) 49 km/h
    (b) 31 km/h [N]
15. (a) 27 m/min
    (b) 19 m/min [W 39° N]
16. (a) $d = vt$
    (b) $t = d/v$
    (c) $\vec{d} = \vec{v}t$
    (d) $t = \vec{d}/\vec{v}$

619

ANSWERS TO NUMERICAL PROBLEMS

17. 4.0×10^7 m/s;
    1.2×10^3 m;
    7.5×10^1 s
18. 6.0 cm
19. 26 h
22. (c) 60 m/s [S]
    (e) 720 m [S]
23. (a) 200 m [E]
    (b) 200 m [E]
    (c) 0.0 m
    (d) -200 m [E] or
        200 m [W]
    (e) 0.0 m
24. (a) 100 m/s [E]
    (b) 0.0 m/s
    (c) -33 m/s [E] or
        33 m/s [W]
26. 0.0 m

REVIEW ASSIGNMENT
3. 38.7 m at an angle of 45° to the first base line
4. (b) 4.2 km/h
   (c) 2.0 km/h [E 40° S]
5. 12.5 m/s or 12 m/s;
   45 km/h
6. 17 m/s
7. 3.0×10^8 m/s
8. 4.0×10^3 m
9. 3.84×10^8 m
10. 1.12×10^2 s
13. (b) 250 cm/s [W]; 0.0 cm/s;
        750 cm/s [E]
    (d) 75 cm [W]
14. (b) 7.48 times
15. (b) 29 mm/year

PRACTICE (SUPPLEMENTARY)
30. (a) 8.0 m/s [W]
    (b) 8.0 m/s [E]
    (c) 4.0 m/s [W]
31. (a) 20 m/s [S]
    (b) 20 m/s [N]
    (c) 61 s
    (d) 100 s
33. (a) 6.3 h
    (b) 7.1 h
34. (a) 3.2 m/s relative to the shore
    (b) 1.2 m/s relative to the shore
35. 41 m/min [N 14° W]
36. (a) 255 km [E 11° S]
    (b) 102 km/h [E 11° S]

Chapter 4

PRACTICE
3. (a) 12 m/s² [W]
   (b) 168 m/s² [N]
   (c) -0.65 (km/h)/s [S]
4. (a) 47 (km/h)/s [fwd]
   (b) 13 m/s² [fwd]
5. 6.0 (km/h)/s [N]
6. -5.0×10^{-2} m/s² [S]
7. (a) $\vec{v}_f = \vec{v}_i + \vec{a}t$
   (b) $\vec{v}_i = \vec{v}_f - \vec{a}t$
   (c) $t = (\vec{v}_f - \vec{v}_i)/\vec{a}$
8. (a) 59 m/s [E]
   (b) 75 m/s [E]
   (c) 45 s
9. 300 s
11. 22 m/s [up]
14. (a) 2.0 m/s [E]; 1.0 m/s [E]
    (b) -10 m/s² [E]
    (c) 5.0 m/s² [E]
    (d) 5.0 m/s² [E]
16. 12 (km/h)/s [fwd];
    5.4 (km/h)/s [fwd];
    1.8 (km/h)/s [fwd];
    -27 (km/h)/s [fwd]
17. (d) 4.0 m/s² [N];
        6.0×10^2 mm/s² [W]
18. (a) 5.9 m/s [down]
    (b) 12 m/s [down]
    (c) 24 m/s [down]
19. (a) 9.9 m/s [down]
    (b) 16 m/s [down]
    (c) 28 m/s [down]
20. (a) 9.0 m/s
    (b) 45 m
21. 1.8 m/s²
22. -4.5×10^3 m/s²
23. (a) 64 m/s [E]
    (b) 2.8×10^2 m
24. 10 s
25. (a) 1.0 s
    (b) 2.0×10^1 m
27. -5.3 m/s²; 28 m/s

REVIEW ASSIGNMENT
4. 0.44 m/s²; 0.30 m/s²
5. (a) 7.5 m/s [W];
       15 m/s [W];
       7.5 m/s [W]
   (b) 7.5 m/s² [W];
       0.0 m/s²;
       -15 m/s² [W]
6. 6.5 s
7. 40 m/s² [S]
8. final velocity is 32 cm/s [forward]
   final displacement is 128 cm [forward]
9. (a) 0.81%
   (b) 1.3%
10. (a) 9.7 m/s²
    (b) 8.3×10^{-1} m/s²
    (c) 4.2×10^{-2} m/s²
    (d) 2.5 m/s²
12. (a) 9.8 m/s² [down]
    (b) 9.8 m/s² [down]
    (c) 9.8 m/s² [down]
13. (a) 14 m/s [up]
    (b) 10 m [up]
15. 4.0×10^3 m/s²
16. 11 m/s [up]
17. (a) 1.8 m/s
    (b) 0.93 s
18. 0.31 m
19. 7.3 m
20. (a) 45 s (b) 75 s
    (c) 9.0×10^2 m
21. 3.1×10^7 s
24. 0.10 m

Chapter 5

PRACTICE

4. (a) 4.4 N [S]
   (b) 101 N [up]
14. (a) 0.24 m/s² [W]
    (b) 2.4 m/s² [forward]
    (c) 3.2 × 10² m/s² [up]
15. (a) 2.5 × 10⁴ N [forward]
    (b) 1.2 × 10² N [E]
    (c) 1.9 × 10⁵ N [S]
16. $m = \vec{F}_{un}/\vec{a}$
17. 58 kg
18. $\vec{F}_{un} = m(\vec{v}_f - \vec{v}_i)/t$
19. 2.0 × 10⁻² N
20. (a) 40 m
    (b) 3.8 s
22. (a) 1.5 × 10² N [down]
    (b) 1.5 × 10² N [up]
24. (a) 4.4 × 10² N [down]
    (b) 1.3 × 10³ N [down]
26. (a) 1.8 × 10³ kg
    (b) 2.8 m/s² [E]
    (c) 5.0 × 10¹ m [E]
32. High tide 10:32 h
    Low tide 16:45 h
    High tide 22:58 h
    Low tide 5:10 h
35. 4 times; 2062
36. 2 × 10¹¹ suns
39. (a) 10 N
    (b) 39 N [N]
    (c) 88 N [W]
40. (a) 1.8 × 10² N [forward]
    (b) 3.0 m/s² [forward]
    (c) 6.0 m
41. (a) 7.2 × 10² N [down]
    (c) 73 kg

REVIEW ASSIGNMENT

5. (a) 5.2 N [forward]
   (b) 4.3 m/s² [forward]
8. 1.0 × 10² N [forward]
9. (a) 49 N; 98 N
11. −7.3 m/s² [down]
13. 6.0 × 10⁻⁴ N
14. (a) 8.8 × 10¹⁵ m/s²
    (b) 1.9 × 10⁷ m/s
15. (a) 8.0 × 10³ N [N]
16. (a) 3.3 m/s²
    (b) 2.9 m/s²
18. (a) 8.0 m/s² [forward]
    (b) 0.16 m
19. (a) 1.6 × 10³ kg
    (b) 2.0 × 10³ N [W]
    (c) 1.2 m/s² [W]
22. 21 m/s [N]

PRACTICE (SUPPLEMENTARY)

43. 1.7 × 10⁻⁴ N
44. 6.0 × 10²⁴ kg
45. 1.99 × 10²⁰ N
47. (a) 1.0 × 10¹ kg·m/s [E]
    (b) 2.9 kg·m/s [N]
    (c) 3.0 × 10⁴ kg·m/s [W]
48. 5.0 × 10¹ s
49. 1.0 × 10¹ kg
50. (a) 8.0 × 10⁴ N·s [N]
    (b) 6.7 × 10² N [N]
51. 23 m/s
53. 0.18 m/s [S]
54. 1.9 × 10² m/s [forward]
55. 1.0 × 10⁻²⁶ kg
56. 2.8 m/s [E]
57. 1.7 × 10⁻¹¹ m/s

Chapter 6

PRACTICE

3. (a) 64 J
4. (a) 1.5 N
   (b) 3.0 J
5. 0.0 J
6. 0.0 J
7. 0.0 J
10. (b) 2.5 × 10² J
11. kg·m²/s²
12. (a) $F = W/d$
    (b) $d = W/F$
13. (a) 2.5 × 10³ N
    (b) 2.6 × 10² kg
14. 9.1 m
15. 1.4 × 10³ J
16. (a) 0.0 J
    (b) 2.9 J
18. (a) 850 J
    (b) 820 J
19. (a) $m = E_p/gh$
    (b) $g = E_p/mh$
    (c) $h = E_p/mg$
20. 1650 m
21. 1.5 × 10⁴ kg
22. (a) 3.3 m/s² [down]
    (b) 1.3 s
    (c) 4.3 m/s [down]
23. (a) 5.2 × 10² J
    (b) 1.4 × 10⁴ J
25. (a) $m = 2E_K/v^2$
    (b) $v = \sqrt{2E_K/m}$
26. 0.17 kg
27. (a) 10 m/s
28. (a) 450 MJ
    (b) 3.3 × 10² m
30. 13 m/s
31. 3.1 m
34. 2.0 m/s
38. kg·m²/s³
39. 1.7 × 10² W
40. (a) 1.6 × 10² J
    (b) 9.6 × 10³ J
    (c) 2.4 × 10² W
43. (a) $E = Pt$
    (b) $t = E/P$
44. 3.3 × 10⁵ s
45. 16 s
46. 1.9 × 10⁸ MJ
47. 2.4 × 10² m

REVIEW ASSIGNMENT

2. 2.5 × 10⁴ J
4. (a) 2.6 × 10³ N
   (b) 5.4 × 10⁵ m/s²
   (c) 7.0 × 10¹ m/s

ANSWERS TO NUMERICAL PROBLEMS

5. (a) 4.5 km
6. (a) 2.8×10^4 J
   (b) 2.8×10^4 J
7. 2.1 kg
8. 3.0 m
9. (a) 8.8×10^2 J
10. Increases by a factor of 4; increases by a factor of 9.
11. 1.6 kg
12. (a) 26 m/s
    (b) 34 m
14. 250 W
15. (a) 9.4×10^5 J
    (b) 9.4×10^6 J
    (c) 6.5×10^2 W
16. 3.6×10^5 J
17. 2.4 MJ
18. (b) 1.7×10^3 s
    (c) 14 h
20. 9.5 m/s
21. (a) 9.5×10^5 MJ
    (b) $28 000

Chapter 7

PRACTICE
4. (a) 1.1×10^3 W
   (b) 1.2×10^4 J
6. 9.0 W/kg
10. 42 years
11. about 25%
14. 4.9×10^{18} J
15. 8.6×10^{15} J
17. 4.4×10^9 J

REVIEW ASSIGNMENT
2. 15 W/kg
4. 52 kg

5. (a) 2 (b) 4 (c) 32
6. 1.4%/a
12. (a) 3.7×10^{11} J

PRACTICE (SUPPLEMENTARY)
35. 373 K; 293 K; 310 K
36. (a) 120°C
    (b) 200°C
    (c) −109°C
39. (b) 9.0×10^{-7} °C^{-1}
42. (a) 2.1×10^5 J
    (b) 1.7×10^1 J
    (c) 2.2×10^4 J

43. (a) 1.1×10^5 J
    (b) 8.3×10^3 J
    (c) 3.0×10^4 J
44. (a) $c = E_H/(m\Delta T)$
    (b) $m = E_H/(c\Delta T)$
    (c) $\Delta T = E_H/(mc)$
45. (a) 5.0×10^4 J
    (b) 1.0×10^3 J/(kg·°C)
46. 2.2×10^2 kg
47. 21°C
49. (a) 450 J/(kg·°C)
53. (a) 3.4×10^7 J
    (b) 3.8×10^6 J
54. 2.5×10^5 J/kg
55. 2.1×10^6 J/kg

Chapter 8

PRACTICE
3. (a) 1.3×10^3 kg/m³
   (b) 7.9×10^2 kg/m³
   (c) 1.03×10^3 kg/m³
6. (a) 8.4×10^2 Pa
   (b) 3.1×10^4 Pa
   (c) 1.2×10^4 Pa
7. 2.5×10^6 Pa
8. 2.4×10^6 Pa
9. (a) $F = pA$
   (b) $A = F/p$
10. (a) 0.55 N
    (b) 8.0×10^{-3} m²
11. 8.0×10^4 N
12. 3.5 m²
15. 10.8 km
20. 306 kPa
21. 361 kPa
22. (a) 67 kPa
    (b) 34 kPa
25. (a) $F_s = (F_L A_s)/A_L$

(b) $F_L = (F_s A_L)/A_s$
(c) $A_s = (F_s A_L)/F_L$
(d) $A_L = (F_L A_s)/F_s$
26. 3.0×10^2 N; 8.0×10^{-2} m²; 4.3×10^3 N; 1.1 m²
27. (a) 4.4×10^4 N
    (b) 4.5×10^3 kg
28. 5.0×10^2 N
29. (a) 980 N
    (b) 580 N
31. (a) 53 N
    (b) 50 N
    (c) 47 N
    (d) 49 N
32. 0.90 N
33. 5.5 N
34. (a) 8.2×10^7 N
    (b) 8.4×10^6 kg
35. 5.0 cm
38. 0.75
42. (a) 66%
    (b) 3.5×10^3 N

REVIEW ASSIGNMENT
3. (a) 3.9×10^4 Pa
   (b) 3.9×10^3 Pa
4. 3.6×10^5 N
5. (a) 0.22 m²
10. (a) 8.0×10^6 N
    (b) 8.2×10^5 kg
11. 47 kPa
13. (a) 2.0 kPa
    (b) 104 kPa
14. (a) 14 kPa; 9.0 kPa
20. (a) 2.0×10^1 kPa
    (b) 2.7×10^2 kPa
21. 1.2×10^8 Pa
23. 1.2×10^2 N; 1.1×10^3 N
24. 750 N
27. 0.50 N [up]
28. (a) 14.7 N
    (b) 1.2 N
    (c) 13.5 N
    (d) 12
29. (a) 4.6 N (b) 4.6 N

# ANSWERS TO NUMERICAL PROBLEMS

Chapter 9

PRACTICE
12. to the left when viewed from above

REVIEW ASSIGNMENT
5. 85 cm/s

Chapter 10

PRACTICE
2. (b) 8.5 cm
   (c) 170 cm
3. (b) 3.5 cm
   (c) 49 cm
4. (a) 30 Hz; 3.3×10^{-2} s
   (b) 90 Hz; 1.1×10^{-2} s
   (c) 7.7 Hz to 10.7 Hz; 1.3×10^{-1} s to 9.4×10^{-2} s
5. (a) 5.9 s
   (b) 2.5×10^{-7} s
   (c) 8.3×10^{4} s
6. (a) 1.0×10^{2} Hz
   (b) 5.0×10^{7} Hz
   (c) 1.16×10^{-5} Hz
8. $A = 1.0$ cm; $\lambda = 4.0$ cm
9. 5.7 cm
12. (a) 49 m/s
    (b) 4.2×10^{7} m/s
    (c) 2.0×10^{8} m/s
    (d) 1.7×10^{6} m/s
13. (a) $f = v/\lambda$
    (b) $T = \lambda/v$
    (c) $\lambda = v/f$
    (d) $\lambda = vT$
14. (a) 2.0×10^{3} Hz
    (b) 0.20 s
    (c) 34 m
    (d) 84 m
15. (a) 2.0×10^{3} Hz
    (b) 5.0×10^{-4} s
16. 0.12 m
20. $\lambda = 4.0$ m
21. (a) 0.40 Hz
    (b) 0.80 Hz
    (c) 1.6 Hz

REVIEW ASSIGNMENT
2. 3.5 cm
3. (a) 0.60 s
   (b) 4.0 s
   (c) 1.3 s
4. (a) 440 Hz
   (b) 60 Hz
   (c) 0.56 Hz
5. (b) 0.46 s
   (c) 2.2 Hz
7. 5.0×10^{-11} s
8. 250 Hz
14. 1.5×10^{3} m/s
15. (a) 25 s; 4.0×10^{-2} Hz
    (b) 44 s; 2.3×10^{-2} Hz
16. 0.67 m
17. 6.0 m/s
18. 2.0×10^{-4} s
19. (a) 20 m/s
    (b) 5.0 m
    (c) 1 node; 2 antinodes
25. 7.0×10^{2} m/s

Chapter 11

PRACTICE
6. about 1.2 cm
8. 3.3×10^{2} m/s
9. 2.8 km
10. 2.9 s
11. (a) 336 m/s
    (b) 346 m/s
    (c) 330 m/s
13. 304 m
14. $T = (v - 332 \text{ m/s})/(0.6 \text{ m/s}/°C)$
15. (a) 27°C
    (b) −20°C
16. (a) 345 m/s
    (b) 0.91 m
18. The time is about 1/15 as long.
19. (a) 1500 m/s
    (b) 349 m/s
21. (a) 2 Hz (b) 6 Hz
    (c) 5 Hz
22. 2, 6, 7, 9, 13, and 15 Hz
23. 437 Hz
26. 1440 m/s
27. 150 m
29. (a) 210 m/s
    (b) 1260 m/s
30. 325 m/s
32. (a) 0.67 m
    (b) 365 m/s; 315 m/s
    (c) 545 Hz; 470 Hz

REVIEW ASSIGNMENT
5. (a) 340 m/s
   (b) 13°C
6. 20.5 s
7. 1.02×10^{3} m or about 1.0 km
8. 172 m
10. (a) 0.39 s
    (b) 1.6 s
11. 571 s
14. 3, 5, and 8 Hz
20. 1.4×10^{-4} m
21. 1.4×10^{3} m/s
22. 87 m
24. 3.1×10^{2} m/s
25. (a) A: 1.28×10^{3} km/h [W]; B: 2.13×10^{3} km/h [E]
    (b) A relative to B: 3.41×10^{3} km/h [W]
    B relative to A: 3.41×10^{3} km/h [E]
27. 5.7×10^{3} s or 1.6 h

ANSWERS TO NUMERICAL PROBLEMS

Chapter 12

PRACTICE
5. (a) 420 Hz (b) 636 Hz
   (c) 1180 Hz
6. (a) 155 Hz (b) 342 Hz
   (c) 6.0×10^3 Hz
7. (a) 1.2×10^3 Hz
   (b) 2048 Hz
8. (a) 3:2
   (b) 5:4
   (c) 2:1
   (d) 41:40
9. B_5: 987.8 Hz; C_6: 1046.6 Hz
10. 300 Hz; 337.5 Hz;
    375 Hz; . . . 600 Hz
11. 500 Hz; 574 Hz;
    660 Hz; . . . 2000 Hz
13. (a) 10^1 (b) 10^3 (c) 10^2
    (d) 10^4
14. 90 dB
15. (a) 52 dB
    (b) 1 or 2 dB
16. (a) 134 dB
    (b) 120 dB
18. (a) 880 Hz
    (b) 1320 Hz
    (c) 2200 Hz
22. (a) 440 Hz
    (b) 880 Hz
24. (a) 436 Hz
27. (a) 43.2 cm
    (b) 43.2 cm
29. (a) 344 m/s
    (b) 0.344 m
    (c) 0.172 m
43. 60 dB
44. (a) 16 dB
    (b) 7.0 dB
    (c) 0.0 dB
45. 17 m

REVIEW ASSIGNMENT
3. (a) 7200 Hz (b) 225 Hz
4. 13:12; 5:4; 3:2; 1:2; 1:1
5. (a) 682.6 Hz
   (b) 960.0 Hz
   (c) 65.4 Hz
   (d) 784.0 Hz
6. 200 Hz; 211.9 Hz;
   224.5 Hz; . . . 400 Hz
8. 10^4 times
9. 50 dB
10. 1200 Hz to 5000 Hz
12. 300 Hz; 600 Hz; 1200 Hz
14. $f \propto 1/\text{length}$;
    $f \propto 1/\text{diameter}$;
    $f \propto \sqrt{\text{tension}}$;
    $f \propto 1/\sqrt{\text{density}}$
15. (a) 880 Hz
    (b) 1056 Hz
    (c) 825 Hz
    (d) 1760 Hz
20. 17.5 cm; 52.5 cm; 87.5 cm
22. 54 cm
24. (a) 660 Hz; 990 Hz
    (b) 990 Hz; 1650 Hz
32. 12 dB

Chapter 13

PRACTICE
4. 9.5×10^5 MJ
9. 23%
10. 1.5×10^{11} m
11. 2.56 s
12. 2.99×10^8 m/s
15. (a) 9.47×10^{15} m
    (b) 2.0×10^6 light years
16. 7.48 times
25. (a) 50 cm (b) 100 cm

REVIEW ASSIGNMENT
5. (a) 3.00×10^8 m/s
   (b) 3.00×10^5 m
6. 2.0×10^{-2} s
7. 9.47×10^{20} m
12. 120 cm
16. (b) 9:25
19. (b) 7 images
20. (a) 1:2 (b) 2:1
25. 2.4 m

Chapter 14

PRACTICE
1. (a) 1.25
   (b) 1.67
2. (a) 0.800
   (b) 0.600
3. $n_{A \to M} = 1/n_{M \to A}$
4. 0.625
5. (a) 1.50×10^8 m/s
   (b) 1.20×10^8 m/s
   (c) 2.73×10^8 m/s
9. (a) 1.89; 1.29; 1.66
11. 2.26×10^8 m/s
12. 41.1°; 37.3°
20. 6.0 cm

REVIEW ASSIGNMENT
3. (a) 1.38
   (b) 0.723
   (c) 1.08
5. 0.649
6. (a) 1.65×10^8 m/s
   (b) 2.76×10^8 m/s
9. (a) 0° (b) 0°
12. (a) 50°
    (b) 1.53
    (c) 1.96×10^8 m/s
    (e) 40.8°
    (f) 12 mm
16. (a) 31.7°
    (b) 50.3°
17. (a) $n_{A \to S} = 1.74$
    (b) $n_{A \to S} = 1.31$
25. 18 cm

ANSWERS TO NUMERICAL PROBLEMS

Chapter 15

PRACTICE
9. (a) 4 times lower
   (b) 16 times higher
   (c) 64 times lower
10. (a) 1/30 s
    (b) 1/480 s or 1/500 s
11. (a) 7.1 mm
    (b) f/2.8
22. 2.1

REVIEW ASSIGNMENT
10. (a) increases by a factor of 32
    (b) decreases by a factor of 8
11. (b) f/22
12. (b) 1/960 s or 1/1000 s

Chapter 16

PRACTICE
5. (a) 25.1°
   (b) 24.9°
7. 41.3°
8. 1.98×10^8 m/s; 1.96×10^8 m/s
9. (a) red and green
   (b) red and blue
   (c) red, green, and blue
10. (a) red
    (b) blue and green
    (c) nothing
    (d) red
11. (a) red and blue
    (b) nothing
    (c) blue
    (d) green
12. (a) red
    (b) green
    (c) blue
    (d) black
13. (a) west
    (b) east
15. (a) red and blue
    (b) blue and green
17. (a) red and green
    (b) red, blue, and green
    (c) none
20. (a) 7.32×10^{14} Hz
    (b) violet
22. (a) 1.5×10^{-10} m
    (b) 6.0×10^4 m
    (c) 5.0×10^6 m

REVIEW ASSIGNMENT
6. (a) 41.1°
   (b) 1.97×10^8 m/s
8. 49.3°
9. (a) yellow
   (b) white
   (c) white
10. (a) red and green
    (b) blue
11. (a) black
    (b) red
    (c) yellow
13. Mix magenta and cyan pigments.
26. (a) 6.56×10^{-7} m
    (b) red
30. 6.06×10^{-7} m; 4.95×10^{14} Hz; orange

Chapter 17

PRACTICE
3. (a) positive
   (b) none
   (c) negative
   (d) positive
   (e) negative
   (f) positive
4. (a) excess
   (b) deficiency
   (c) deficiency
6. (a) 7 (b) 0 (c) 16
9. (a) lead (−); wool (+)
   (b) fur (+); acetate (−)
   (c) wax (−); glass (+)
   (d) sulphur (−); silk (+)
12. neutral
25. The types of charge are the same.
27. The types of charge are opposite.

REVIEW ASSIGNMENT
8. (a) A (+); B (−)
9. (a) 5 (b) 17
10. (a) 8 (b) 22
11. (a) 2 (b) 12
16. (a) rubber (−); wax (+)
    (b) silk (−); lead (+)
22. (a) positive

PRACTICE (SUPPLEMENTARY)
36. 2.0 N
37. (a) -2.9×10^2 N
    (b) -7.2×10^1 N
38. 2.6×10^{-19} N
39. (a) -8.0×10^{-19} C
    (b) 1.6×10^{-16} C
    (c) -10 C
40. 6.25×10^{18} elementary charges
41. 3.0×10^{12} electrons

ANSWERS TO NUMERICAL PROBLEMS

Chapter 18

PRACTICE
8. (a) 14 A
   (b) 7.5 A
   (c) 0.85 A
9. (a) 1.8×10^2 C
   (b) 2.7×10^2 C
   (c) 0.52 C
10. (a) 4.3 s
    (b) 5.0×10^{-2} s
    (c) 5.0×10^{-1} s
11. (a) 150 mA
    (b) 1.5 A
    (c) 7.5 A
12. (a) 12 V
    (b) 120 V
    (c) 1.5 J
    (d) 3.6 J
    (e) 0.50 C
    (f) 18 C
13. 120 V
14. 5.5×10^{-5} J
15. 6.5×10^{-5} C; 65 µC
16. (a) 3.5 V
    (b) 10.5 V
    (c) 210 V
19. 1.2×10^6 J; 1.2 MJ
21. (a) 0.11 Ω
    (b) 0.11 Ω
25. (a) 48 Ω
    (b) 50 Ω
    (c) 4.0×10^6 Ω
26. (a) 14 V
    (b) 3.3×10^3 V
    (c) 0.33 A
    (d) 2.0 mA
27. 4.1 A
28. 6.0 V
29. 30 Ω
31. (a) 12.5 Ω
    (b) 1110 Ω or 1.1×10^3 Ω
    (c) 1.11 Ω or 1.1 Ω
32. (a) 2.0 Ω
    (b) 50 Ω
    (c) 100 Ω
    (d) 100 Ω
35. (a) 2.0 A
    (b) 6.0 V
    (c) 5.0 Ω
    (d) 3.0 Ω
36. (a) 50 V
    (b) 5.0 A
    (c) 10 Ω
37. (a) For R_1: $I = 0.50$ A, $V = 3.0$ V
       For R_2: $I = 0.30$ A, $V = 6.0$ V
       For R_3: $I = 0.20$ A, $V = 6.0$ V
    (b) For R_1: $I = 0.12$ A, $V = 5.0$ V
       For R_2: $I = 0.12$ A, $V = 3.8$ V
       For R_3: $I = 0.12$ A, $V = 1.2$ V
    (c) For R_1: $I = 1.5$ A, $V = 15$ V
       For R_2: $I = 1.2$ A, $V = 24$ V
       For R_3: $I = 0.30$ A, $V = 24$ V
       For R_4: $I = 1.5$ A, $V = 6.0$ V
38. 20 Ω; 9.0 V

REVIEW ASSIGNMENT
6. 1.2 A
7. 29 C
8. 6.3 ms
9. 450 mA
10. (a) 8.0×10^6 V
    (b) 5.0×10^3 A
11. 7.2×10^{-12} J
12. (a) 1.3×10^5 C
    (b) 16 MJ
15. (a) take $\frac{1}{2}$ the length
    (b) double the area
17. (a) 27 V
    (b) 9.0 V
18. $R_A = 30$ Ω; $R_B = 10$ Ω
19. 1.1 A
20. 5.0 V
21. (a) 10 Ω; 5.0 Ω; 3.3 Ω; 2.5 Ω; 2.0 Ω; 1.7 Ω
    (b) 4 lamps
23. (a) 30 V
    (b) 0.50 A
    (c) 40 Ω
24. (a) 1.5 A
    (b) $R_1 = 2.0$ Ω; $R_2 = 6.0$ Ω
    (c) 1.5 Ω
25. For R_1: $I = 0.60$ A, $V = 36$ V
    For R_2: $I = 1.8$ A, $V = 36$ V
    For R_3: $I = 0.60$ A, $V = 36$ V
    For R_4: $I = 3.0$ A, $V = 84$ V

Chapter 19

PRACTICE
4. 120 V; 240 V
6. (a) A1 and B4; A2 and B3
   (b) A1 and B3; A2 and B4
9. (a) 3.5×10^2 W
   (b) 2.6×10^2 W
   (c) 3.7 W
10. 9.0 V
11. (a) 12.5 A or 12 A
    (b) 6.7×10^2 s or about 11 min
13. (a) $P = V^2/R$
    (b) $P = I^2R$
14. (a) $R = V^2/P$
    (b) $R = P/I^2$
15. 400 W
16. 1.3 kW
17. 144 Ω or 1.4×10^2 Ω
18. 0.80 Ω
19. (a) 6.5 MJ; 10¢
    (b) 17 MJ; 27¢
    (c) 14 MJ; 22¢
20. (a) 0.075 kW·h; 0.41¢
    (b) 3.3 kW·h; 18¢
    (c) 6.4 kW·h; 35¢

ANSWERS TO NUMERICAL PROBLEMS

21. (a) 580 kW·h
    (b) $30.16
24. (a) 0.20 mA
    (b) 80 mA
    (c) 240 mA
26. 24 bulbs
28. 840 W

REVIEW ASSIGNMENT
 4. (a) 120 V
    (b) 240 V

(c) 120 V
 7. (a) 5.0 kW
    (b) 1.4 W
 8. (a) 20 mA
    (b) 28 mA
 9. (a) 12 V
    (b) 12 V
10. 1.2 W
11. (a) 730 W
    (b) about 23 min
12. 69 Ω

13. 13 Ω
14. (a) 3.2 MJ
    (b) 5.1¢
15. (a) 300 kW·h
    (b) $15.60
16. $26.68
17. 3.6 MJ; 5.4¢/(kW·h)
19. (a) 180 W
    (b) 2.9 kW
22. 9 bulbs

Chapter 20

PRACTICE
10. (a) about 22° E;
        about 25° W
    (b) 0°; 90°
15. (b) to the left, or away from the magnet
16. (a) to the right
    (b) to the left
    (c) downward

18. into the speaker, or away from the observer
20. counterclockwise
21. clockwise

REVIEW ASSIGNMENT
 7. (a) the pointed end
11. about 7° E
13. (a) clockwise
    (b) counterclockwise
14. (a) west
    (b) east

15. (a) repel
    (b) attract
19. (a) left pole is N
    (b) left pole is S
    (c) both poles are S
24. outward, or to the right
27. clockwise
28. (b) left end is N
    (c) counterclockwise when viewed from above

Chapter 21

PRACTICE
 7. (b) clockwise
    (c) toward the observer
 8. (c) counterclockwise through the loop when viewed from above
10. In all cases the induced current increases.
13. (a) 1.1 kV
    (b) 33 V
    (c) 20 windings
14. (a) 1.0×10^3 A

    (b) 1.0×10^6 W
    (c) 3.6×10^9 J
15. (a) 1.0 A
    (b) 1.0 W
    (c) 3.6×10^3 J
16. 5000 windings

REVIEW ASSIGNMENT
 2. (a) (b) (c) Current increases.
    (d) Current stops.
 3. (a) left end is N
    (b) right end is N
    (c) right end is N
 4. (a) toward the observer

    (b) away from the observer
    (c) away from the observer in the left conductor, opposite in the right conductor
13. 30 V
14. 62 windings
15. 120 windings
16. 120 V
19. (a) 2.5 A
    (b) 9.4 W
    (c) 0.0038% of the original power is lost.

Chapter 22

PRACTICE
17. (a) 2.4×10^8 emissions
    (b) 1.4×10^{10} emissions
18. (a) 1.7 kBq

    (b) 95 kBq
    (c) 70 Bq
25. 10^4:1; 10^{12}:1; almost 100%
27. (a) 1; 0
    (b) 13; 14
    (c) 47; 61

29. 5; 6; 7; 8
30. (a) β
    (b) α
31. (a) $^{230}_{90}\text{Th} \rightarrow {}^{226}_{88}\text{Ra} + {}^{4}_{2}\text{He}$
    (b) $^{210}_{83}\text{Bi} \rightarrow {}^{210}_{84}\text{Po} + {}^{0}_{-1}\text{e}$

ANSWERS TO NUMERICAL PROBLEMS

32. (a) 218; 82; Pb; 4
    (b) Rn; 218; 4; He
    (c) 82; 214; Bi; 0
33. about 11 500 years
34. 45 h

REVIEW ASSIGNMENT
7. (a) + (b) − (c) 0
   (d) 0 (e) + (f) 0
12. (a) 4.2×10^2 emissions
    (b) 5.1×10^3 emissions

17. (a) $^{214}_{84}Po \rightarrow {}^{210}_{82}Pb + {}^{4}_{2}He$
    (b) $^{210}_{81}Tl \rightarrow {}^{210}_{82}Pb + {}^{0}_{-1}e$
19. 1/1024 or about 1/1000
20. (b) 8.0 days
    (c) about 1200 Bq

Chapter 23

PRACTICE

3. (a) $^{4}_{2}He$
   (b) $^{1}_{0}n$
   (c) $^{14}_{7}N$
4. (a) $^{1}_{1}H + {}^{7}_{3}Li \rightarrow {}^{4}_{2}He + {}^{4}_{2}He$
   (b) $^{2}_{1}H + {}^{2}_{1}H \rightarrow {}^{3}_{2}He + {}^{1}_{0}n$
   (c) $^{4}_{2}He + {}^{25}_{13}Al \rightarrow {}^{28}_{15}P + {}^{1}_{0}n$
6. (a) 5.4×10^{16} J
   (b) 9.0×10^{15} J
   (c) 4.4×10^{-12} J
7. 1.1×10^{11} kg
11. 8.3×10^{-11} J
16. $^{238}_{92}U + {}^{1}_{0}n \rightarrow {}^{239}_{92}U$
    $^{239}_{92}U \rightarrow {}^{0}_{-1}e + {}^{239}_{93}Np$
    $^{239}_{93}Np \rightarrow {}^{0}_{-1}e + {}^{239}_{94}Pu$
22. (a) 3.6×10^{26} J
    (b) 3.6×10^{26} W
    (c) 1.6×10^{13} years

23. (a) $^{4}_{2}He + {}^{4}_{2}He \rightarrow {}^{8}_{4}Be$
    (b) $^{8}_{4}Be + {}^{4}_{2}He \rightarrow {}^{12}_{6}C$
    (c) $^{12}_{6}C + {}^{1}_{1}H \rightarrow {}^{13}_{7}N$
24. (a) 5.2×10^{-13} J
    (b) 2.8×10^{-12} J
29. 9.8×10^{-30} kg;
    8.8×10^{-13} J;
    2.9×10^{-29} kg;
    2.6×10^{-12} J
32. (a) leptons
    (b) hadrons
34. udd
35. zero

REVIEW ASSIGNMENT
2. (a) $^{4}_{2}He$
   (b) $^{0}_{-1}e$
   (c) $^{1}_{0}n$
   (d) $^{1}_{1}H$
   (e) $^{2}_{1}H$

   (f) $^{3}_{1}H$
4. (a) $^{2}_{1}H$
   (b) $^{7}_{4}Be$
5. (a) $^{4}_{2}He + {}^{16}_{8}O \rightarrow {}^{19}_{10}Ne + {}^{1}_{0}n$
   (b) $^{4}_{2}He + {}^{16}_{8}O \rightarrow {}^{9}_{4}Be + {}^{11}_{6}C$
6. 9.0×10^7 J
7. 9.9×10^{-30} kg;
   8.9×10^{-13} J
20. (a) 7 sets
    (b) 66 sets
21. (b) (i) 7 new neutrons for a total of 14
        (ii) 9 new neutrons for a total of 19
        (iii) 26 new neutrons for a total of 56
26. 1.6×10^{-13} J
27. 3.0×10^{-10} J
28. (a) zero (b) +1

Epilogue

REVIEW ASSIGNMENT

3. about 2.6×10^9 km from the earth (just over half the distance to the farthest planet)
4. (a) 3.0×10^7 s or 347 d
   (b) 4.4×10^{12} km;
       3.7×10^{13} km;
       1.6×10^8 s or 4.9 years
   (c) 8.6 years

PHOTO CREDITS

All photographs not specifically credited to another source are included by courtesy of the author, Alan J. Hirsch. The photographs in the Profiles are provided by the contributors, with these three exceptions: H. W. Wevers, page 112, *courtesy* Lisa Lowry; Dr. Bruce Pennycook, page 295, *courtesy* Sparks & Associates; Dr. Alexander Szabo, page 400, *courtesy* National Research Council.

PREFACE	
Preface photo	*Courtesy* Mount Wilson and Palomar Observatories
CHAPTER 1	
Opening photo	*Courtesy* NASA
Figure 1-10	*Courtesy* New York Public Library Picture Collection
Figure 1-13	*Courtesy* The Bettmann Archive, Incorporated
CHAPTER 2	
Opening photo	*Courtesy* Ontario Hydro
Figure 2-2	*Courtesy* Optech Inc, Downsview, Ontario
Figure 2-7	*Courtesy* National Research Council
CHAPTER 3	
Opening photo	*Courtesy* Air Canada
Figure 3-8	*Courtesy* Dr. H. E. Edgerton, MIT, Cambridge, Massachusetts
Figure 3-13	*Courtesy* The Library of Congress, Washington, D.C.
CHAPTER 4	
Figure 4-15	*Courtesy* Dr. W. R. Franks
Figure 4-16	*Courtesy* NASA
Page 83	*Courtesy* General Motors Corporation
Figure 4-17	*Courtesy* General Motors Corporation
Figure 4-18	*Courtesy* Kitchener-Waterloo Record
CHAPTER 5	
Opening photo	*Courtesy* NASA
Figure 5-2	*Courtesy* Yerkes Observatory, University of Chicago
Figure 5-16(b)	*Courtesy* NASA
Figure 5-21	*Courtesy* D. Bannister
Figure 5-22(b)	*Courtesy* NASA
Figure 5-23	*Courtesy* Mount Wilson and Palomar Observatories
Figure 5-27(b)	*Courtesy* British Hovercraft Corporation
Figure 5-32	*Courtesy* Amray Incorporated
CHAPTER 6	
Figure 6-1(a)	*Courtesy* The Toronto Star
Figure 6-3	*Courtesy* AIP, Niels Bohr Library
Figure 6-11	*Courtesy* M. C. Escher Heirs, c/o Cordon Art, Baarn, Holland
Figure 6-12	*Courtesy* Fisher Scientific Collection

PHOTO CREDITS

CHAPTER 7	
Opening photo	*Courtesy* INDAL Industries Inc
Figure 7-5	*Courtesy* Syncrude Canada Ltd
Figure 7-8(a)	*Courtesy* Environment Canada
Figure 7-9	*Courtesy* Nova Scotia Power Corporation
Figure 7-11	*Courtesy* Pacific Gas and Energy, San Francisco
Figure 7-13	*Courtesy* Los Alamos National Laboratory
CHAPTER 8	
Figure 8-6	*Courtesy* The Toronto Star
CHAPTER 9	
Opening photo	*Courtesy* General Motors Corporation
Figure 9-3	*Courtesy* National Research Council
CHAPTER 10	
Opening photo	*Courtesy* University of Washington College of Engineering, Library Services
Figure 10-19(a), (b)	*Courtesy* University of Washington College of Engineering, Library Services
CHAPTER 11	
Opening photo	*Courtesy* Leo Burnett USA Advertising
Figure 11-13	*Courtesy* Brüel & Kjaer Canada Ltd, Bramalea, Ontario
Figure 11-16	*Courtesy* NASA
CHAPTER 12	
Opening photo	*Courtesy* The Toronto Symphony, Andrew Davis, Musical Director; photo by Larry Miller
Figure 12-6	*Courtesy* The Bettmann Archive, Incorporated
Figure 12-13	*Courtesy* Joanna Jordan, CLAZZ, Toronto
Figure 12-28	*Courtesy* Yamaha Canada Music Ltd
Figure 12-29	*Courtesy* Yamaha Canada Music Ltd
Figure 12-31	*Courtesy* Yamaha Canada Music Ltd
Figure 12-35(a)	*Courtesy* Bell Northern Research Ltd
CHAPTER 13	
Figure 13-2	*Courtesy* Goway Travel Ltd
CHAPTER 14	
Opening photo	*Courtesy* Diamond Information Centre
Figure 14-2	*Courtesy* The Bettmann Archive, Incorporated
Figure 14-10(d)	*Courtesy* Bell Northern Research Ltd
CHAPTER 15	
Opening photo	*Courtesy* Ontario Hydro
Figure 15-1	*Courtesy* Hale Observatories
Figure 15-5(b)	*Courtesy* Pentax Canada Incorporated, Credit David McQueen
Figure 15-17	*Courtesy* Imperial Optical Canada
Figure 15-18	*Courtesy* Imperial Optical Canada
CHAPTER 16	
Opening photo	*Courtesy* NASA
Figure 16-12	*Courtesy* Dr. David Williams
Figure 16-14	*Courtesy* The Bettmann Archive, Incorporated
Figure 16-15	*Courtesy* AIP, Niels Bohr Library

PHOTO CREDITS

Colour Plate 1	*Courtesy* John Wiley & Sons, New York
Colour Plate 2	*Courtesy* John Wiley & Sons, New York
Colour Plate 3	*Courtesy* John Wiley & Sons, New York
Colour Plate 4	*Courtesy* A. & J. Verkaik, Beamsville, Ontario
Colour Plate 5	*Courtesy* Jack Finch, Fairbanks, Alaska
Colour Plate 6	*Courtesy* Dr. G. A. Kenny-Wallace, Photo by Rudi Crystal
Colour Plate 7	*Courtesy* General Motors Corporation, Warren, Michigan
Colour Plate 8	*Courtesy* Optech Inc, Downsview, Ontario

CHAPTER 17
Figure 17-1	*Courtesy* Gordon R. Gore
Figure 17-2	*Courtesy* Fisher Scientific Collection
Figure 17-27	*Courtesy* Ontario Science Centre
Figure 17-28	*Courtesy* Wellcome Institute for the History of Medicine, London, England
Figure 17-34	*Courtesy* General Motors Corporation

CHAPTER 18
Opening photo	*Courtesy* IBM Canada Ltd
Figure 18-1	*Courtesy* The Bettmann Archive, Incorporated
Figure 18-2	*Courtesy* AIP, Niels Bohr Library
Figure 18-5	*Courtesy* Union Carbine "Eveready"
Figure 18-6	*Courtesy* Union Carbide "Eveready"
Figure 18-10	*Courtesy* NASA
Figure 18-13	*Courtesy* The Bettmann Archive, Incorporated
Figure 18-20	*Courtesy* Smith Collection Centre for History of Chemistry

CHAPTER 19
Figure 19-4	*Courtesy* Gordon R. Gore
Figure 19-11	*Courtesy* The National Research Council

CHAPTER 20
Opening photo	*Courtesy* Dofasco, Hamilton
Figure 20-5	*Courtesy* Tak Sato, Department of Physics, University of Toronto
Figure 20-15	*Courtesy* These maps are based on information from MCR.701 Isogonic Chart D © Her Majesty the Queen in right of Canada with permission of Energy Mines and Resources Canada.
Figure 20-19	*Courtesy* AIP, Niels Bohr Library
Figure 20-36	*Courtesy* The Bettmann Archive, Incorporated

CHAPTER 21
Opening photo	*Courtesy* Hydro Québec
Figure 21-6	*Courtesy* AIP, Niels Bohr Library, E. Scott Barr Collection
Figure 21-14	*Courtesy* Ontario Hydro
Figure 21-17(b), (c), (d)	*Courtesy* Ontario Hydro, Ontario Hydro, Hydro Québec
Figure 21-18	*Courtesy* Ontario Hydro
Figure 21-25	*Courtesy* Federal Pioneer Limited
Figure 21-29	*Courtesy* Ontario Hydro

CHAPTER 22
Opening photo	*Courtesy* Dr. F. D. Manchester, Department of Physics, University of Toronto
Figure 22-2	*Courtesy* Niels Bohr Library, Lande Collection
Figure 22-4	*Courtesy* The Deutches Museum, Munich

PHOTO CREDITS

Figure 22-7	*Courtesy* Niels Bohr Library, William G. Myers Collection
Figure 22-9	*Courtesy* Niels Bohr Library
Figure 22-15	*Courtesy* Research team of physicists from the University of Toronto in collaboration with the Brookhaven National Laboratory
Figure 22-18	*Courtesy* Niels Bohr Library, Bainbridge Collection. Photo by D. Schoenberg
Figure 22-22	*Courtesy* Niels Bohr Library
Figure 22-24	*Courtesy* The Bettmann Archive, Incorporated
Figure 22-33	*Courtesy* Brookhaven National Laboratory (photo now out of print)
Figure 22-34	*Courtesy* Atomic Energy of Canada Ltd
Figure 22-37	*Courtesy* Brookhaven National Laboratory (photo now out of print)
Figure 22-38	*Courtesy* The Bettmann Archive, Incorporated

CHAPTER 23

Opening photo	*Courtesy* Ontario Hydro
Figure 23-5	*Courtesy* AIP, Niels Bohr Library
Figure 23-6	*Courtesy* Niels Bohr Library
Figure 23-8	*Courtesy* AIP, Niels Bohr Library, from *Otto Hahn, A Scientific Autobiography*, New York: Charles Scribner's Sons, 1966
Figure 23-11	*Courtesy* Los Alamos National Laboratory
Figures 23-15 to 23-22, 23-27, 23-29, 23-32(a), (b), (c)	*Courtesy* Ontario Hydro
Figure 23-37	*Courtesy* Los Alamos National Laboratory
Figure 23-38	*Courtesy* Official US Navy photo
Figure 23-42	*Courtesy* Los Alamos National Laboratory
Figure 23-46	*Courtesy* NASA

EPILOGUE

Second opening photo	*Courtesy* NASA

INDEX

Note: Pages listed in boldface indicate a definition or, in the case of scientists' names, an illustration.

A

Aberration
  chromatic 346, 347
  in mirrors 323
Absolute zero **173**
Acceleration **66**
  and force and mass 97–100
  and Newton's Second law 101
  due to gravity 76, 77, 84–86
  uniform 65–89
Accelerator, particle 166, 602
Accelerometer 90
Accommodation (of the eye) 367
Acoustics 298–301
Action force 105, **106**, 107
Adaptor 471
Additive colour mixing **387**,
  Colour Plate 2
Adhesion **204**
Air columns 285–289
Air reed instruments 289
Airplane wings 219
Alchemists 556
Alhazen 335, 346, 347
Alpha decay **543**, 556
  tracks 546
Alternate energy sources – see
  Energy or Renewable energy
  resources
Alternating current (AC) **471**
  motors, 509, 510
Altimeter 191
Ammeter 449
Ampere **448**, 507, 613
Ampère, André-Marie **448**
Amplitude
  of a pendulum **225**
  of a wave **234**, 235
Anechoic chamber **300**
Aneroid barometer 191

Angle
  critical **340**
  of declination **494**
  of deviation **385**
  of emergence **336**, 337
  of incidence **315**, 336, 337
  of inclination **494**
  of reflection **315**
  of refraction 336, 337
Annular eclipse **310**
Antimatter 602
Antinode **245**, 286, 287
Antiparticles 602
Aperture **364**
Applied physics 14
Arabic language 9
Archimedes 8, **186**
Archimedes' principle **200**
Area of graphs 48–50
Aristotle **8**
Armature **509**
Astigmatism **370**
Astronomical telescope **374**
Athabaska Tar Sands 158
Atmosphere, as a source of heat 164
Atmospheric pressure **189**, 190,
  191
Atomic mass **553**
Atomic mass number **553**
Atomic mass unit **553**
Atomic number **415**, 552
Atomic theory 414–416
Atomists 8
Atoms 8, **414**
  and radioactivity 536–570
  models of **416**, 417, **548–551**
Attack of a sound 298
Attitude (of image) **317**
Attraction
  electric 413
  magnetic 489

Audible range **249**, 263
Aurora **495**
  Australis 495
  borealis **495**, Colour Plate 5

B

Balanced forces **44**
Barometer **191**
Barrier, sound **265**
Baryons 602, 603
Base unit **19**, 612
Baseball curve 220
Bathyscaphe **209**
Battery **443**
Bearing 113
Beats 256, 257
Becquerel **546**
Becquerel, Henri **542**, 546, 556
Bed of nails 188
Bel 276
Bell, Alexander Graham **276**, 277
Bell timer – see Recording timer
Bernoulli, Daniel 218
Bernoulli's principle **218–221**
Beta decay **543**
  tracks 546
Bimetallic strip **176**
Binoculars 342, 343
Bioluminescent materials 307
Biomagnetism 494
Biomass (energy) **163**
Bitumen **158**
Blood pressure 193, 194
Body waves **247**
Bohr, Niels **550**
Bohr's atomic model **550**
Boiling water reactor 585
Boldt, Arnold 136
Brass instruments 290
Breeder reactor, 587, 588, 592

INDEX

British System – see Imperial System
Bronze Age 5
Buoyancy **197**, 198, 200–203
Buoyant force – see Buoyancy

C

Calandria **581**, 582, 583
Calorie 137
Camera
  lens **361**–366
  obscura 361
  pinhole **358**, 361
  single-lens reflex 362–365
Cancer 567
CANDU reactor **580**–583, 590, 591, 595
Capacitor 431
Capillary
  tube **206**
  action **206**
Carburetor 220
Cathode 537
Cathode-ray tube 537
Cavendish, Henry 120
Cell
  electric 433–445
  in parallel 455
  in series 455
  photoelectric 446
  photovoltaic 159
  symbols 455
Celsius, Anders 173
Celsius scale 173
Centre of curvature **321**
Chadwick, James **551**, 572, 573, 580
Characteristics of images 317, 318
Charge
  elementary **438**
  in motion **441**
Chemiluminescent materials **307**
Chernobyl 591, 594
Chinese influence 10
Chromatic aberration 346, 347
Circuit
  closed **455**
  electric **454**, 461–465
  household 472, 473
  open **455**

  overloaded **479**
  parallel 461
  primary 527
  secondary 527
  series 461
Circular mirrors **321**
Classical physics 537
Cloud chamber **545**, 546
Coefficient
  drag 216
  of linear expansion **174**, 175
Cogeneration **168**
Cohesion **204**
Collisions 123, 124
Colour(s)
  blindness **371**, 394
  complementary light **386**
  primary light 385, 386, Colour Plate 2
  secondary light 386, Colour Plate 3
  spectral 383, Colour Plate 1
  television 394
  vision **393**, 394
Comet 109
  Halley's 109, 484
  Mbros 110
Complementary light colours **386**
Compound bar 176
Compression **234**, 251
Compression forces 93
Computer
  timing 52
  storage stystem 502
Computerized tomography 540
Conductor
  and magnetic fields 497–501
  electric **418**, 422–424
Conservation of energy (law of) **138**-140
Conservation of momentum (law of) **124**
Consonance **272**, 273
Constructive interference **238**
Control rods 579
Converging lens **345**, 346
  ray diagrams 350, 351
Converging mirror **321**
  applications 326, 327
  ray diagrams 325

Convex lens – see Converging lens
Convex mirror – see Diverging mirror
Coolant **581**
Cost of electrical energy 476, 477
Coulomb **437**
Coulomb, Charles 436
Coulomb's law **437**
Crest **234**
Critical angle 340
Critical mass **578**
Crookes' radiometer 313
Crookes, Sir William 313, 538
Crookes' tube 538
Crova's disk 253
Curie, Marie 542, **543**
Curie, Pierre 542, **543**
Current electricity 440, **441** – 469
  alternating 471
  conventional 442
  direct 471
  in the home 472, 473
  induced 515 – 518
Cylindrical mirrors **321**

D

Da Vinci, Leonardo 361
Decay (radioactive)
  alpha **543**, 556
  beta **543**, 557
  gamma **543**
  series **557**, 558
Decay of sound 298
Decibel 276
Declination, angle of **494**
Deductive reasoning 7
Democritus 8
Density 186
  see also Relative density
Dependent variable **31**
Depth of field 365
Derived units 20, 93, 613
Destructive interference **238**
Deviation, angle of **385**
Diamond 343
Diastolic pressure **193**, 194
Diffraction
  of light 314
  of waves 241
Diffuse reflection **314**

Dimensional analysis 230
Direct current (DC) **471**
   motors 509
Direct variation **31**
Discharge **411**
Dispersion **383**
Displacement **44**
   and velocity 46
   antinode 286, 287
   lateral 338
   node 286, 287
Dissonance 272
Distance 44
   and speed 46
Diverging lens **345**, 349
   ray diagram 351
Diverging mirror **321**
   applications 327, 3287
   ray diagrams 325
Domain **490**
Domain theory (of magnetism) 490
Doppler, Christian 264
Doppler effect **264**, 265
Dosimeter **526**, 568
Double mechanical reed instruments 290
Doubling time **154**
Drag coefficient **216**
Dry cell **443**
Dynamics **95**

E

Ear (human) 260–263
Earth's magnetic field 494, 495
Echo finding **263**
Eclipse **309**, 310
Eddies 212
Egyptians 5, 92, 111
Einstein, Albert 12, **574**
Elastic forces 93
Electric
   circuit **454**, 455
   conductor **418**, 419, 422–424
   current **448**
   fields **420**, 421
   forces 92, **411**
   generating stations 524, 525
   generators **520**–524
   insulator **418**, 419, 422–424

   meters 477
   motors 507–510
   potential difference **450**
   power 474
   relay **512**
   resistance **452**, 453
   safety 478–480
Electrical instruments 295
Electricity
   cost of **476**, 477
   current 440, **441**–469
   generation 157, 520–525
   static **411**–439
Electrodes 441
Electrolyte 441
Electromagnet 487, **502**, 503
Electromagnetic induction 514, **515**–535
Electromagnetic spectrum 397
Electromagnetic waves **396**, 532
Electromagnetism **496**–513
Electromotive series 447
Electron **415**
Electronic instruments 296
Electronic stroboscope – see Stroboscope
Electroscope **413**
   charging by conduction **425**
   charging by induction **426**
   discharging with a radioactive source 544
Element(s) **415**
   periodic table of 415, 416, 614, 615
Elementary charge **438**
Energy **128**
   and force 129, 130
   and power 143
   biomass 163
   conserving 138, 169
   consumption 151–155
   converters 157
   efficiency of use of 168
   electrical sources 443
      – using 470–486
   forms of 128, **129**
   geothermal 163
   gravitational potential **134**, 135
   hydraulic 160
   importance of 128, 129

   in a modern world 148
   in nuclear reactions 574
   law of conservation of **134**–140
   law of conservation of mass-energy **575**
   light 307
   solar 160, 161
   sound 250, 251
   thermal 173, 446
   tidal **160**–162
   transfer of 230
   wave 162
   wind 160, 161
Equation
   lens 352
   mirror 326
Equitempered scale 275
Error
   random 26
   systematic 26
Escher, E.M. 142
Estimating quantities 35, 36
Expansion 174
   coefficient of linear **174**, 175
Exponents, 21, 22
Extrapolation 32

F

Fairing 216
Far-sightedness 370
Faraday, Michael **504**, 514, 515
Faraday's law 515
Faraday's ring 527
Fast breeder reactor 587, 588, 592
Fermi, Enrico, **573**, 578, 579
Fermi question 36
Fibre optics **342**, 343, 401
Fibrillation 478
Field
   depth of **365**
   electric **420**, 421
   magnetic **491**–495, 497–501
   magnets 509
First law of motion (Newton's) 95, **96**, 97
Fission bomb 600
Fission, nuclear 157, **576**
Fixed-end reflection 232, 233
Flow meter (venturi) 222

INDEX

Fluid friction 110
Fluidic control devices 220, 221
Fluids
   at rest 184–210
   buoyancy in 197
   in motion 211–222
   synovial 113
Fluorescent sources **307**
Focal length **321**
Focal point **321**
   primary (lens) **346**
   secondary (lens) **346**
Force 91–126
   and energy 129, 130
   and pressure 187
   and work 130
   balanced **94**
   definition and types of **92**, 93
   electromagnetic 411
   frictional 110–114
   fundamental 92, 411, 609
   of electricity **411**, 437
   of gravity 103, 104, 108, 109
   of magnetism **488**
   strong nuclear **575**
   unbalanced **94**
   see also Buoyancy, Adhesion, Cohesion, Surface tension
Fossil fuels 156
Franklin, Benjamin **412**, 424
Franks, W.R. **84**
Free fall 76, **85**
Frequency **227** 236
   and musical scales 273–275
   and pitch 272
   fundamental **245**, 278
   of AC electricity 471
   of beats 257
   of singer's voices 294
   of vibrating strings 279
   recording timer 53
Frictional forces – see force
Frisch, Otto 576
F-stop number **364**
Fuel cell **168**
Fuels, fossil 156
Fundamental frequency **245**, 278
Fusion bomb 600, 601
Fusion (melting) 178, 179, 181
Fusion, nuclear 164, **594**

G

Galilean telescope 377
Galilei, Galileo **10**, 76, 84, 95, 191, 376
Galvani, Luigi 440, **441**, 499
Galvanometer **441**, 505, 506
Galvanoscope 499
Gamma decay **543**
Gasohol 163
Gauge pressure **191**, 192
Geiger-Müller tube **547**
Generator
   AC 521, 522
   DC 523
   electric **520**–525
   static electricity **428**
   Van de Graaff **429**
Geothermal energy 163
Gilbert, William 412, 489
Giraffe 193
Gold-foil experiment 549
Graphing 31–33
   and uniform acceleration 68–72
   and uniform motion 48–50
Gravitational potential energy **134**, 135
   see also Energy
Gravity 92, 103, 104, 108, 109
   acceleration due to 76, 77
   universal 119, 120
Gray **546**
Gray, Harold 546
Greek alphabet 618
Greek civilization 7, 8
Grounding **419**
Growth rate 153

H

Hadrons 602, 603
Hahn, Otto **576**
Halban 580
Half-life **559**, 560, 561
Halley, Edmund 109
Halley's comet 109, 484
Harmonics **278**
Harris, Gretchen L.H. 12, 13
Hearing 259, 260
   threshold of **276**
Heat **173**–180

capacity (specific) **176**, 177
   exchange, principle of **177**
   of fusion (specific latent) **179**, 181
   of vaporization (specific latent) **179**, 181
   transfer of 176–180, 230
Heat pump 164, 165
Heavy water 579–581
Helix 499
Henry, Joseph 515
Hertz 53
Hertz, Heinrich 53, **396**
High-temperature gas-cooled reactor 589
Hiroshima, Japan 598, 599
Holography **403**, 404
Horsehead nebula 357
Horsepower 144
Human ear 259–263
   and sensitivity 227
Human eye 367–369
Human voice 293–295
Hydraulic
   brakes 196
   energy 160
   jack 196
   press 195
Hydrocarbons 156
Hydrogen bomb 600
Hydrometer 202
Hyperopia (or hypermetropia) **370**
Hypothesis 11

I

Image characteristics 317, 318
Imaginary image – see Virtual image
Imperial System of Measurement 18
Impulse 121, **122**
Incandescent sources **307**
Incident ray 315
Inclination, angle of **494**
Independent variable **31**
Index of refraction **333**, 334, 383, 384
Induction
   electromagnetic **515**
   magnetic **490**
   mutual **527**
   of electric charges 427

Inductive reasoning 7
Inertial confinement 595
Infrared radiation 385
Infrasonic **263**
Instantaneous velocity 71
Instruments
   air reed 289
   brass 290
   double-mechanical reed 290
   electrical 295, 296
   electronic 296
   lip reed 290
   optical 356-381
   percussion 292
   single-membrane reed 290
   stringed 281-285
Insulator, electric **418**, 419
Intensity (sound) **276**
Intensity level (sound) **276**
Interference
   of pulses 237, 238
   of sound waves 255
   of waves 243, 244
Internal reflection 340
Interpolation 32
Invar 176
Inverse variation **38**, 39
Ion **416**, 544
Ion drive propulsion 146
Irregular reflection **314**
Islam 9, 10
Isotopes **554**, 555

J

Jet engine 107
Joule 130
Joule, James **130**

K

Kadeidoscope 320
Kawarski 580
Keeper (magnetic) 493
Kelvin 613
Kelvin scale 173
Kelvin, William Thompson 173
Kerogen **158**
Kinematics **43**
Kinetic energy **136**
   see also energy

Kinetic friction 110
Kunov, Hans 260, 261

L

Lambda 235
Laminar flow 212
Laser 398-404, 610
   characteristics 398
   holography **403**, 404
   operation of 399
   pulsed 401
   uses 400, Colour Plates 6, 7, and 8
Latent 179
Lateral displacement 338
Latin language 9
Law
   Coulomb's **437**
   Faraday's **515**
   Lenz's **519**
   of conservation of energy **138**-140
   of conservation of mass-energy **575**
   of conservation of momentum **124**
   of electric charges **413**, 414
   of exponents, 21, 22
   of magnetic poles **489**
   of motion (Newton's) **96**, **101**, **105**
   of universal gravitation (Newton's) **119**, 120
   Ohm's **457**
   Pascal's **195**
   Snell's **338**
Left-hand rules
   for coiled conductor **501**
   for motor principle **505**
   for straight conductor **498**
Legault, France 82, 83
Length 19
   of a pendulum 225
Lens **345**
   camera 361-366
   converging 345
   diverging 345
   equation 352
   erector 374
   eyepiece 373
   objective 373

Lenz, Heinrich **518**, 519
Lenz's law **519**
Leptons 602, 603
Level 114
Leyden jar 430, 431
Lidar 401, Colour Plate 8
Light **307**
   rays 309
   sources of 307
   spectrum 383
   speed of 310-312
   theories 395-398
   transmission of 307
Light year **313**
Lightning 411, 412
   rod **424**
Line of best fit 31
Linear expansion – see Expansion
Lip reed instruments 290
Local consumption 168
Longitudinal
   sound waves 251
   vibrations **225**, 226
Loudspeakers 295, 496, 506
Luminous sources 307
Lunar eclipse **309**

M

Mack, Ernst 265
Mack number **265**
MacKay-Lassonde, C. 586, 587
Magnet **489**
Magnetic
   field **491-495**, **497-501**
   field of the earth 494
   induction 490
   material 489
   poles 489
   saturation 491
Magnetism 487-496
   domain theory of 489, 490
   forces of **488**
Magnification **318**
Manhattan Project **578**, 498
Manometer 193
Mass 20
   and acceleration 100
   and Newton's second law 101
   and weight 103
   critical 578

INDEX

law of conservation of mass-energy 575
Mass number **415**
Mbros comet 110
Measurement
  accuracy of 26, 17
  and calculations 28, 29
Mechanical forces 93
Mechanical resonance **242**
Mechanics 41–147
Meitner, Lise **576**
Meldown 590
Meniscus **206**
Mesons 602, 603
Metabolic rate **152**, 153
Metre 19, **612**
Metric
  memory aid 24, 613
  prefixes 23, **24**, 612
Metric system 19, 612, 613
  see also Système International
Michelson, Albert 311, 312
Microscope 372, 374, 375
Middle Ages 9, 10
Middle ear 261, 262
Miner, Carla J. 482, 483
Mirage **344**
Mirror
  circular **321**
  converging **321**
  curved 321–327
  cylindrical **321**
  diverging **321**
  equation 326
  parabolic **321**
  plane 315–320
  spherical **321**
Moderator **578**
Modern physics 537
Momentum **121**, 122, 123, 137
  law of conservation of **124**
Moslems – see Islam
Motion
  accelerated 65
  laws of (Newton's) **96**, **101**, **105**
  non-uniform **43**
  relative 58–61
  uniform 43
  see also Speed, Velocity, Acceleration

Motor
  electric 507–510
  St. Louis 509
Motor principle 504, **505**
Musical instruments – see instruments
Musical scales 272–275
Musician's scale 273–275
Myopia **369**

N

Nagasaki, Japan 599
Near-sightedness **369**
Nebula
  Horsehead 357
  Orion 606
Negative acceleration **65**
Neolithic Era **3**, 4
Neuron 450
Neutron **415**
Neutron star 606
New Stone Age **3**, 4
Newton 93, 101
Newton, Sir Isaac 11, **93**, 383, 395
Newton's laws
  first law of motion **96**, 97
  second law of motion 101
  third law of motion 105–**107**
  universal gravitation **119**, 120
Nobel, Alfred **568**
Nobel prize 568, 569
Node **243**, 286, 287
Noise pollution 278
Non-ohmic materials 458
Non-renewable resources – see Resources
Non-uniform motion **43**
Normal **315**
North-seeking pole 489
Northern Lights – see Aurora
Nuclear energy 570–607
Nuclear force
  strong 92, 575
  weak 92
Nuclear fission 157, **576**
  reactor 578–593
Nuclear fusion 164, **594**
  reactor 596
Nuclear reactions **571**, 574, 575, 594, 595

Nuclear weapons 598–601
  fission bomb 600
  fusion bomb 600
  hydrogen bomb 600
Nucleus **415**, 552–555
  daughter 556
  parent 556

O

Octave **272**
Oersted, Hans Christian 496, **497**
Ohm **453**
Ohm, George Simon **453**
Ohmic materials 458
Ohm's law **457**
Old Stone Age 3
Opaque materials **313**
Open-end reflection 232, 233
Opera glasses 376
Ophthalmoscope 319
Optical instruments **356**–381
  functions of **357**
Oscilloscope **271**
Ostrich 138
Overloaded circuit **479**

P

Paint sprayer 219
Paleolithic Era **3**, 4
Parabolic mirrors **321**
Parallax **26**
Parallel connection
  cells 455
  resistors 460, 461
  voltmeter 451
Partial eclipse **310**
Particle accelerator 166, 602
Particle physics 609
Pascal 187
Pascal, Blaise 187
Pascal's law **195**
Passive solar heating 159
Pendulum 225, 229
Pennycook, Bruce W.J. 297
Penumbra **309**
Percussion instruments 292
Period
  of vibration **227**
  of waves 236

INDEX

Periodic table of the elements **415, 614, 615**
   partial 416
Periodic waves **234**
Periscope 342, 343
Phase 233
Phosphorescent materials **307**
Photocopier 377
Photoelectric cells 446
Photoelectric effect 396
Photographic enlarger 377
Photon theory of light 396
Photons 396
Photovoltaic cells 159
Physics
   applied 14
   classical 537
   modern 537
   particle 609
   plasma 609
   pure 13
Piezoelectricity **446**
Pile driver 135, 138, 139
Pinhole camera 358–361
Pitch 272
Planck, Max 396
Plane mirrors 315–320
Plasma 596
   physics 609
Plumb bob 108
Poles, magnetic **489**
Pollution
   control 169
   noise 278
   thermal 167
Polygraph 453
Potential difference — see Electric
Potential energy — see Energy
Power **143**
   electric **474**
Preferred SI units 93
Presbyopia **371**
Pressure **187**
   atmospheric **189**, 190, 191
   blood 193, 194
   diastolic **193**, 194
   gauge **191**, 192
   systolic **193**, 194
Pressurized water reactor 585
Primary cell **443**, 444

Primary circuit (transformer) **527**
Primary light colours **385**, 386, Colour Plate 2
Primary pigment colours 391, Colour Plate 3
Primary rainbow 392
Primary waves **247**
Principal axis **321**
Principle
   Archimedes' **200**
   Bernoulli's **218**–221
   motor 504, **505**
   of heat exchange **177**
   of superposition **238**, 279
Projector 377
Proof plane 422
Proton **415**
Ptolemy 335, 336
Pulfrich effect 407
Pulsar 606
Pulse **231**
   interference of 237, 238
Pure physics 13
Pyramids 5, 185, 186
Pythagoras 7

Q

Qualitative descriptions 17
Quality (of musical sounds) **278**
Quanta (quantum) of energy 396, 550
Quantitative descriptions 17
Quantity
   scalar **44**
   vector **44**
Quantum mechanics 550
Quark 603
Quintessence 8

R

Radiation
   background **544**
   hazard 567
   infrared 385
   monitor 590
   transfer of 231
   ultraviolet 385
Radioactive
   dating 562
   tracers 563
   wastes 591, 592, 593

Radioactivity **542**
   decay modes 543
   detection of 544, 545
   discovery of 542
   half-life **559**, 560, 561
   uses 561–566
Radioisotopes **554**
Radiometer **313**
Radiosonde 203
Radius of curvature **321**
Rainbow
   primary 392
   secondary 392, 393, Colour Plate 4
Random error **26**
Rarefaction **234**, 251
Rate
   growth 153, 154
   metabolic **152**, 153
Ray diagrams
   for curved mirrors 325
   for lenses 350, 351
   for plane mirrors 317
Reaction force 105, **106**, 107
Reaction time 86
Reactor
   boiling water 585
   CANDU **580**
   fast breeder **587**, 592
   fission 578
   fusion 595
   future of 593
   generation 580
   high temperature gas-cooled 586
   meltdown **590**
   pressurized water 585
   research 580
   wastes 591
Real image **318**
Reasoning
   deductive 7
   inductive 7
Recording timer **52**, 53
Reflected ray **315**
Reflection
   diffuse **314**
   fixed-end 232, 233
   irregular **314**
   laws of 315, 316, **317**
   open-end 232, 233

INDEX

reflex or retro **314**
regular or specular **314**
total internal **340**
Reflex reflection **314**
Refracting telescope **372**, 373
Refraction of light **333**
  index of **333**, 334, 383, 384
Refrigerant 164
Regular reflection **314**
Relative density 199, **202**
Relative velocity 58–61
Relay, electric 512
Renaissance 11
Renewable resources — see resources
Repulsion
  electric 413
  magnetic 489
Resistance, electric 453
Resistor 453
  parallel 460
  series 458
Resonance
  and standing waves 243, 244
  apparatus 285, 288
  in sound 158
  mechanical **242**
Resonant frequency 242
Resonator 281, 293
Resources
  renewable 159–164
  non-renewable 156–158
Retinal fatigue 394
Retro-reflection **314**
Reverberation time **299**
Rheostat 453
Ripple tank **240**
Roemer, Olaf 311
Roentgen, Wilhelm **538**, 539
Rolling friction 110
Roman Empire 9
Rounding off 29
Roy Thomson Hall 270
Rutherford, Ernest **549**, 571, 572
Rutherford's atomic model 550
Rutherford's gold-foil experiment 549

S

Sailboat 220

Satellite 106, **109**
Saturation, magnetic 491
Scalar Quantity **44**
Scientific method **11**
Scientific notation 21
Scientific Scale (musical) 273
Scintillation detector **547**
Second 20, 613
Second law of motion (Newton's) **101**
Secondary cell **443**, 445
Secondary circuit (transformer) **527**
Secondary light colours **386**, Colour Plate 2
Secondary pigment colours 391, Colour Plate 3
Secondary rainbow 392, 393, Colour Plate 4
Secondary waves **247**
Semiconductor **420**
  detector 547
Series connection
  ammeter 449
  cells 455
  resistors 458
Shadows 309
Shearing forces 93
Shock wave **260**
Shutdown systems 590, 591
Significant digits **26**, 27
Silo 208
Single-membrane reed instruments 290
Skiis 219
Sliding friction **110**
Slope **32**
  see also Graphing
Snell, Willebrord **335**, 336
Snell's law of refraction **338**
  and critical angle 340
Solar eclipse **309**
Solar energy **159**
Solar heating
  active **160**
  passive **159**
Solar system 86
Solenoid 499
Sonar 263
Sonic boom **266**
Sonometer 279

Sound
  barrier **256**
  intensity **276**
  intensity level **276**
  production 251
  resonance 258
  speed of 253, 254
  traces 271
  transmission 251
  wave interference 255
South-seeking pole **489**
Southern Lights — see Aurora
Specific heat capacity **176**, 177
Specific latent heat of fusion **179**, 181
Specific latent heat of vaporization **179**, 181
Spectrum
  electromagnetic 397
  visible 383, 396, **397**, Colour Plate 1
Specular reflection **314**
Speed **46**
  and refraction of light 333, 334
  of light 310–312
  of sound 253, 254
  supersonic 266
  subsonic 265
Spherical mirrors **321**
Spring scale 93
Squid 107
Standard form — see Scientific notation
Standing waves 243, 244, 286, 287
Starting friction **110**
Static electricity **411–439**
  generator 428
  series 417, 418
Static friction **110**
Step-down transformer 527, **528**
Step-up transformer **528**
Stopwatch 52
Storage cell 443
Strassmann, Fritz 576
Streamlining 215
Stringed instruments 281–285
  resonator 281
Strings (vibrating) 279
Stroboscope 52, 308
Subsonic speeds **265**

Subtractive colour mixing **389**–391, Colour Plate 3
Superconductors **174**
Supercrest **238**
Supernova 606
Superposition
    principle of **238**, 279
Supersonic speeds **266**
Supertrough **238**
Surface tension **204**
Sympathetic vibrations **259**
Synovial fluid 113
Système International d'Unités (SI) **19**, 93, **612**, 613
Systematic error **26**
Systolic pressure **193**, 194
Szabo, Alexander **402**, 403

T

Table of trigonometric values 617
Tacoma Narrows Bridge 242, 243
Tangent **71**
Tangent technique **71**
Tar Sands 158
Telescope
    astronomical 374
    Galilean 377
    reflecting 326, 327
    refracting 373
    terrestrial 374, 375
Temperature **173**
Temperature scales 173, 174
Tension forces 93
Terminal velocity 85
Tesla, Nicola **429**
Thales of Miletus 7, 411
Thermal energy 173, 446, 591
Thermal pollution 167, 591
Thermocouple **446**
Thermometer **173**
Thomson, J.J. **549**
Thomson's atomic model 549
Three-Mile Island 594
Threshold of hearing **276**
Threshold of pain 276
Tidal bore **235**
Tidal energy **160**, 161
Tidal node 243
Tides 108, 109

Time, 20
    doubling 154
    measuring 51–53
    reaction 86
    see also Period
Timer — see Recording timer
Tokamak **597**
Torricelli, Evangelista 191
Torsional vibration **225**, 226
Total eclipse 310
Total internal reflection **340**
    applications of 342, 343
Transformer **527**–531
    step-down **528**
    step-up 527, **528**
Translucent materials **313**
Transmission of light 309
Transmutation **556**, 571, 572
Transparent materials **313**
Transuranium elements 543
Transverse vibration **225**
Tremelo **294**
Triboluminescence **308**
Trigonometry **616**
Trough 234
Tsunami **247**
Tuning fork 252, 256
Turbulance **212**

U

Ultrasonic **263**
Ultraviolet radiation 385, 397
Umbra **309**
Unbalanced force **94**
Uniform acceleration 64, 65–89
Uniform motion **43**–58
Unison 272
Units 612, 613
    base 19, 613
    derived 20, 93, 612
    preferred SI 93
    SI 19, 20
Universal gravitation 119, 120
Universal wave equation **236**
Uranium, enriched 578

V

Van de Graaff generator 429
Van de Graaff, Robert 429
Vaporization 179, 181
Variable
    dependent **31**
    independent **31**
Variation
    direct **31**
    inverse **38**, 39
Vector Quantity **44**
Velocity **46**
    and acceleration 66–69
    instantaneous 71
    relative 58–61
    terminal 85
Venturi flowmeter 222
Vertex 321
Vibrations 225
    sympathetic 259
Vibrato **294**
Vibrator 281, 293
Virtual focal point 345
Virtual image 318
Viscosity **212**
Visible spectrum 383, 396, 397, Colour Plate 1
Vision 367, 368
    colour 368, **393**, 394
    defects 369–371
Voice (human) 293, **294**
Volt 450
Volta, Count Alessandro **441**, 450
Voltage 450
Voltaic cell **441**, 442, 443
Voltmeter 451, 452

W

Wastes
    radioactive **591**–593
Watt 143
Watt, James **143**
Wavelength **235**, 236
    and sound in air columns 286, 287
Waves **230**
    body 247
    diffraction of **241**
    electromagnetic **396**, 532
    on water 240
    periodic **234**
    primary 247
    secondary 247
    sound 251, 255
    standing 243, 244, 286, 287

INDEX

tsunami 247
universal equation 236
Weak nuclear force 92
Weapons, nuclear 598–601
Weight **103**
Wevers, H.W. 112

Wind energy **160**, 161
Wind generator, vertical axis 148, 161
Wind instruments 289–292
Work **130**, 132
and power 143

X

X ray tube 540
X rays **539**
characteristics 539
production 539
uses 540, 541